SATELLITE SYSTEMS FOR PERSONAL APPLICATIONS

Wiley Series on Wireless Communications and Mobile Computing

Series Editors: Dr Xuemin (Sherman) Shen, *University of Waterloo, Canada*
Dr Yi Pan, *Georgia State University, USA*

The 'Wiley Series on Wireless Communications and Mobile Computing' is a series of comprehensive, practical and timely books on wireless communication and network systems. The series focuses on topics ranging from wireless communication and coding theory to wireless applications and pervasive computing. The books provide engineers and other technical professionals, researchers, educators and advanced students in these fields with invaluable insight into the latest developments and cutting-edge research

Other titles in the series:

Misic and Misic: *Wireless Personal Area Networks: Performance, Interconnection, and Security with IEEE 802.15.4*, Janyary 2008, 978-0-470-51847-2

Takagi and Walke: *Spectrum Requirement Planning in Wireless Communications: Model and Methodology for IMT-Advanced*, April 2008, 978-0-470-98647-9

Pérez-Fontán and Espiñeira: *Modeling the Wireless Propagation Channel: A simulation approach with MATLAB®*, August 2008, 978-0-470-72785-0

Ippolito: *Satellite communications Systems Engineering: Atmospheric Effects, Satellite Link Design and System Performance*, August 2008, 978-0-470-72527-6

Lin and Sou: *Charging for Mobile All-IP Telecommunications*, September 2008, 978-0-470-77565-3

Myung and Goodman: *Single Carrier FDMA: A New Air Interface for Long Term Evalution*, October 2008, 978-0-470-72449-1

Wang, Kondi, Luthra and Ci: *4G Wireless Video Communications*, April 2009, 978-0-470-77307-9

Cai, Shen and Mark: *Multimedia Services in Wireless Internet: Modeling and Analysis*, June 2009, 978-0-470-77065-8

Stojmenovic: *Wireless Sensor and Actuator Networks: Algorithms and Protocols for Scalable Coordination and Data Communication*, February 2010, 978-0-470-17082-3

Liu and Weiss, *Wideband Beamforming: Concepts and Techniques*, March 2010, 978-0-470-71392-1

Hart, Tao and Zhou: *Mobile Multi-hop WiMAX: From Protocol to Performance*, October 2010, 978-0-470-99399-6

Qian, Muller and Chen: *Security in Wireless Networks and Systems*, January 2011, 978-0-470-512128

SATELLITE SYSTEMS FOR PERSONAL APPLICATIONS

CONCEPTS AND TECHNOLOGY

Madhavendra Richharia
Knowledge Space Ltd, UK

Leslie David Westbrook
QinetiQ Ltd, UK

A John Wiley and Sons, Ltd., Publication

This edition first published 2010

© 2010 John Wiley & Sons Ltd

Registered office

John Wiley & Sons, Ltd, The Atrium, Southern Gate, Chichester, West Sussex PO19 8SQ, United Kingdom

For details of our global editorial offices, for customer services and for information about how to apply for permission to reuse the copyright material in this book, please see our website at www.wiley.com.

Library of Congress Cataloging-in-Publication Data

Richharia, M. (Madhavendra)
  Satellite systems for personal applications : concepts and technology / Madhavendra Richharia, Leslie Westbrook.
    p. cm.
  Includes bibliographical references and index.
  ISBN 978-0-470-71428-7 (cloth)
  1. Artificial satellites in telecommunication. 2. Personal communication service systems. I. Westbrook, Leslie. II. Title.
  TK5104.R5323 2010
  621.382′5 – dc22

                                                                                                            2010003320

A catalogue record for this book is available from the British Library

ISBN 978-0-470-71428-7 (H/B)

Typeset in 9/11pt Times by Laserwords Private Limited, Chennai, India
Printed and Bound in Singapore by Markono Print Media Pte Ltd

Contents

About the Series Editors

Xuemin (Sherman) Shen (M'97–SM'02) received a BSc degree in Electrical Engineering from Dalian Maritime University, China, in 1982, and MSc and PhD degrees (both in Electrical Engineering) from Rutgers University, New Jersey, United States, in 1987 and 1990 respectively. He is a Professor and University Research Chair, and the Associate Chair for Graduate Studies, Department of Electrical and Computer Engineering, University of Waterloo, Canada. His research focuses on mobility and resource management in interconnected wireless/wired networks, UWB wireless communications systems, wireless security and ad hoc and sensor networks. He is a coauthor of three books, and has published more than 300 papers and book chapters in wireless communications and networks, control and filtering. Dr Shen serves as a Founding Area Editor for *IEEE Transactions on Wireless Communications*, as Editor-in-Chief for *Peer-to-Peer Networking and Application* and as Associate Editor for *IEEE Transactions on Vehicular Technology; KICS/IEEE Journal of Communications and Networks, Computer Networks, ACM/Wireless Networks, Wireless Communications and Mobile Computing* (John Wiley & Sons), etc. He has also served as Guest Editor for *IEEE JSAC, IEEE Wireless Communications* and *IEEE Communications Magazine*. Dr Shen received the Excellent Graduate Supervision Award in 2006, the Outstanding Performance Award in 2004 from the University of Waterloo, the Premier's Research Excellence Award (PREA) in 2003 from the Province of Ontario, Canada, and the Distinguished Performance Award in 2002 from the Faculty of Engineering, University of Waterloo. He is a registered Professional Engineer of Ontario, Canada.

Dr Yi Pan is the Chair and a Professor in the Department of Computer Science at Georgia State University, United States. He received his BEng and MEng degrees in Computer Engineering from Tsinghua University, China, in 1982 and 1984 respectively, and his PhD degree in Computer Science from the University of Pittsburgh, in 1991. His research interests include parallel and distributed computing, optical networks, wireless networks and bioinformatics. Dr Pan has published more than 100 journal papers, with over 30 papers published in various IEEE journals. In addition, he has published over 130 papers in refereed conferences (including IPDPS, ICPP, ICDCS, INFOCOM and GLOBECOM). He has also coedited over 30 books. Dr Pan has served as Editor-in-Chief or as an editorial board member for 15 journals, including five IEEE Transactions, and has organized many international conferences and workshops. He has delivered over 10 keynote speeches at many international conferences. Dr Pan is an IEEE Distinguished Speaker (2000–2002), a Yamacraw Distinguished Speaker (2002) and a Shell Oil Colloquium Speaker (2002). He is listed in Men of Achievement, Who's Who in America, Who's Who in American Education, Who's Who in Computational Science and Engineering and Who's Who of Asian Americans.

Preface

People in our modern society are profoundly dependent on technology for their work, well-being and quality of life. In recent years, satellite systems have introduced a universal dimension to this technological landscape – although the individual may not always be aware of the extent of the contribution of satellite systems. Satellite technology is today accessible and affordable by individuals, and this book has been created to lay a strong technical foundation towards understanding the role and functioning of existing and emerging satellite systems for personal (i.e. end-user) applications.

Whereas previous books have addressed satellite technology and the personal role of satellite systems in individual service areas – notably personal satellite communications – this book spans the entire breadth of satellite-enabled end-user applications. The aim has been to present the subject matter in a clear and concise manner with key illustrative examples.

After an introductory chapter, the book presents fundamental concepts applicable generally across all the systems. Subsequent chapters delve into techniques and examples of specific systems and services available directly from personal satellite terminals. Such applications encompass broadcasting, communications (narrowband and wideband, commercial, military and amateur), navigation and satellite-based distress services. The book additionally covers those services that are gradually permeating into the personal domain–in particular, satellite imaging and remote sensing.

Finally, the authors explore the trends and evolution of such satellite systems, taking into consideration the influences, user expectations, technology evolution, regulatory efforts and characteristics of satellite systems.

Readers wishing to glean further useful information about the book, to obtain a list of errata, and/or provide feedback to the authors may wish to visit the website at http://www.SatellitesAndYou.com.

Acknowledgements

The authors gratefully acknowledge those individuals and organizations who have kindly given permission for their material to be included in this book. Every effort has been made to obtain permission from the appropriate copyright holders but in the event that any have been inadvertently overlooked they should contact the authors in the first instance (via the publisher), who will endeavor to make appropriate arrangements at the earliest practical opportunity.

Grateful thanks are extended to Tiina Ruonamaa, Sarah Tilley and the team at John Wiley & Sons for their invaluable support, patience and timely guidance throughout the project. The authors express their gratitude to the series editors for their support and thank the anonymous reviewers for their constructive assessment and critique which helped the authors to present the book in its present form.

The authors would also like to take the opportunity to acknowledge those colleagues and fellow researchers who have so enthusiastically shared their extensive knowledge of satellite technology over the years, and who have thus contributed indirectly to the content of this book.

Lastly, and most importantly, some personal thanks:

I (MR) would like to thank my wife Kalpana wholeheartedly for bearing my long absences patiently during the family's prime time.

I (LDW) thank my wife Eva for her enduring patience and continued support, without which I could not have completed this project.

1

Introduction

1.1 Scope

The past two decades have seen a quiet revolution in satellite-based services. Once the preserve of governments, international bodies, public utilities and large corporations, today the majority of satellite service users are *individuals*, who can now access, *directly*, a wide range of satellite services – typically using personal, mass-market and even handheld devices. These satellite systems now fulfil a variety of personal necessities and aspirations spanning telecommunications, broadcast services, navigation, distress and safety services and (indirectly) remote sensing, in the commercial, military and amateur sectors. It therefore seems an appropriate time for a book that addresses these services from the perspective of their support for, and functionality delivered to, individual users.

This book therefore aims to:

- enhance awareness regarding the expanding role of satellite systems in individuals' daily lives;
- lay a strong technical foundation of the basic principles and functioning of these satellite systems for personal communications, navigation, broadcasting and sensing applications;
- illustrate current practice using selected example systems in each field;
- review current trends in relevant satellite and related technology.

The book aims to address an audience that is inquisitive and keen to understand the role of satellites in our daily lives and the underpinning concepts, and, in contrast to alternative offerings, the focus in this book is on the *individual* and the *end-user application*. It aims to provide all of the relevant concepts, in a clear and concise manner, together with descriptions of key systems as illustrations of their implementation in practice.

Satellite services are formally categorized by the International Telecommunications Union (ITU) according to their broad service types. For example, the Broadcast Satellite Service (BSS) addresses recommendations and specifications related to satellite-enabled broadcasts. This book, instead, attempts to address all the services with respect to a user's application perspective – be it telecommunications, broadcast, navigation, amateur, military or safety-related systems.

Space technology comprises a number of branches – satellite communications, satellite aids to the amateur, space exploration, radio astronomy, remote sensing/earth observation, military reconnaissance/surveillance, deep-space communication, launch technology, interplanetary exploration, radio astronomy, space tourism, etc. This book focuses on those technologies where individuals benefit, in a direct or tangible way, from a satellite system. A user interacts directly with a personal satellite broadband terminal when communicating via satellite or interacts with a direct-to-home television

Satellite Systems for Personal Applications: Concepts and Technology   Madhavendra Richharia and Leslie David Westbrook
© 2010 John Wiley & Sons, Ltd

receiver when viewing a programme directly from a broadcast satellite. Similarly, an individual using satellite navigation interacts directly with a Global Positioning System (GPS) receiver.

In some cases the user may not interact directly but nevertheless benefits from information obtained (only) through the use of a satellite system, with some aspects of user hardware or software typically tailored to exploit that system's capabilities, and such applications are also included in the scope of this book. An application in this category would be viewing images of the Earth's weather system appearing daily on our television and computer screens. Here, the pictures transmitted from the satellite are processed elsewhere for the intended audience. Nevertheless, in such instances the individual is conscious that a satellite system is involved.

Those applications and systems where satellites remain in the background are not addressed here, although the same technical concepts apply in the majority of the cases. Examples of this category are interconnection between telecommunication traffic nodes or terrestrial base stations, remote sensing for government (e.g. monitoring vegetation), military surveillance and communications dealing with weapons delivery, television programme distribution between broadcasters, etc. Space tourism (personal spaceflight) is not included in this edition of the book.

1.2   Perspective

Modern society leans heavily on technology for its personal needs – be it entertainment, communications, travel, safety services or domestic appliances. This book deals with the role of satellites in the consumer (or individual) technology paradigm. Consequently, generic user terminal technologies such as terrestrial mobile systems, personal digital assistants, personal computers, etc., are discussed where relevant to personal satellite systems use.

The dependency on satellites in the developed world is quite remarkable. Furthermore, it continues to increase in both the developing and the underdeveloped world owing to falling technology costs together with a growing awareness of the accruing benefits. It must be remarked here, though, that there is a significant difference in priorities in each sector. In an affluent modern society, a majority of people expect a ubiquitous voice service with broadband Internet access, whether they are at home, away or travelling. Many individuals also now aspire to owning a converged handset encompassing some or all of the complementary features such as computing and database functionalities, a hi-fi digital music player, a camera, including video, a radio receiver and mobile television.

In the less developed world, individual requirements and aspirations are curtailed by lower affordability, infrastructure limitations and social conditions. It has been observed that the Gross Domestic Product (GDP) of an economy increases in direct proportion to the improvements to the communications infrastructure. Therefore, there is a great interest in the developing world for deploying wired and wireless technologies such as mobile telephony, the wireless local area network (WLAN) and satellite communications. In the developing world, there is typically minimal fixed infrastructure, with the result that satellites offer an attractive means to build up services, before it becomes economic to introduce fixed assets. One also expects some modifications to mainstream technologies for them to be cost effective and relevant in this environment. The notion that a personal handset is unaffordable, or that the average daily use of such terminals is miniscule, is offset by the fact that such resources are often shared by groups or communities. An example of technical adaptation in a developing region is the extended WLAN trials reported by Raman and Chebrolu (2007) where WLAN coverage was extended to a much wider area than in developed countries, to support scattered rural communities.

Computation, television, broadcast and navigation solutions continue to converge rapidly, enabled by digitization, the vast strides in large-scale integration and mass production techniques resulting in attractively priced converged handsets and accompanying infrastructure enhancements, as the operators reposition themselves in this new paradigm. A number of enabling technologies are instrumental in shaping such converged solutions.

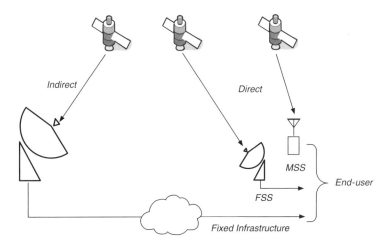

Figure 1.1   The personal (end-user) satellite applications domain.

The unifying force of the Internet offers unprecedented connectivity and tailored solutions such as Internet Protocol (IP) telephony, e-mail, e-learning instruments and audio/video streaming. The evolution in processing capability of personal computers continues unabated. Furthermore, cellular radio technology, based on the concept of radio spectrum multiplication through spatial reuse, now provides instant connectivity across continents. Within a span of just two decades, three generations of cellular systems have been fielded, and research for the introduction of the fourth and even the fifth generation is well under way. The unprecedented success of personal mobile systems has laid the foundations for the commercial viability of WLAN, which enriches the lives of millions through wireless accessibility to the Internet – not only at home and in the office but also in public areas such as cafes and airports.

The extent and speed of introduction of satellite-enabled solutions into the personal domain has surpassed expectations. In broad terms, such applications fall in the areas of personal communications, navigation, broadcast, distress–safety, Earth observation and amateur radio.

Figure 1.1 illustrates conceptually the use of satellite systems for personal applications, indicating the wide scope covered by this book.

1.3   Background and Applications

1.3.1   Background

The space era began with the launch of Sputnik and Explorer by the former Soviet Union and the United States in 1957 and 1958 respectively. Following a series of innovative technical developments, the era of geostationary satellite communications dawned with the launch of Early Bird in 1965. Until the mid-1970s, these communication satellites were mainly used to interconnect large telephone exchanges on national or, more usually, international trunk routes – an application quite remote from individuals. For the individual, the only manifestation of the satellite routing was the propagation (and echo) delay. In parallel, satellite applications extended to numerous other disciplines, namely Earth observation, navigation and radio amateur communications, etc. Monitoring of the Doppler frequency shift of radio signals from the first Sputnik satellite led to the concept of using satellites for navigation, and the first TRANSIT navigation satellite was subsequently launched in 1959 by the US Navy.

Space-enabled technology was furthered by space agencies, manufacturers and operators, leading to a wide range of applications. Direct broadcasts and mobile communications were demonstrated in the 1970s. The well-known Navigation System for Timing and Ranging (NAVSTAR), commonly known as the Global Positioning System (GPS), was launched in 1978 by the US Department of Defense (DoD). A competing system known as the Global Navigation System (GLONASS) was launched by the former Soviet Union in 1986. Yet another system known as the Galileo Positioning System, or simply Galileo, initiated by the European Union and the European Space Agency, is due for launch in early 2014.

Earth observation is a generic term used for a variety of satellite monitoring or, more precisely, remote sensing functions related to environment, meteorology, map-making, forestry, agriculture, etc. Vanguard-2 (launched 1959) was the first earth observation satellite, although TIROS-1 (Television and Infrared Observation Satellites – launched 1960) is widely regarded as the first successful Earth observation (weather) satellite, owing to a malfunction on Vanguard-2. Today, several countries and international bodies own and operate Earth observation satellites. This book encompasses applications such as weather monitoring and map-making where they are directly perceived by individuals. Some existing Earth observation satellites are:

- GMS (Geosynchronous Meteorological Satellite) – these satellites are placed in a geostationary orbit for meteorological sensing;
- Landsat – These satellites are placed in 700 km polar orbit for monitoring mainly land areas;
- NOAA (National Oceanic and Atmospheric Administration) – these satellites are placed in 850 km in polar orbit for meteorological observation and vegetation monitoring.

Amateur radio operators (affectionately known as 'hams') share an interest in construction and communication through non-commercial amateur radio satellites. Ham satellites are known generically as Orbiting Satellite Carrying Amateur Radio (OSCAR), the first of which, OSCAR 1, was launched into a low Earth orbit in 1961. There were almost 20 of these satellites operational in 2006 with plans of numerous additional launches. The Radio Amateur Satellite Corporation (AMSAT) was formed in 1969 as a non-profit educational organization, chartered in the United States to foster amateur radio's participation in space research and communication. Similar groups were formed throughout the world with affiliation to each other. These individuals have pioneered several breakthroughs and continue to do so.

As an aside, we present a few interesting observations that reveal some of the less obvious strengths of satellite systems and position them favourably in a modern context (Robson, 2006/2007).

- A typical Ariane 5 satellite launch emits about half the carbon dioxide emission of a transatlantic jumbo flight.
- Satellites are solar powered and hence environmentally friendly.
- By eliminating or reducing the need for terrestrial infrastructure where possible, it is feasible to reduce environmental load and costs (e.g. through lower use of electricity).
- Satellites are the most cost-effective delivery method for television broadcasts over a wide area.
- Terrestrial TV is heavily dependent on satellites for programme distribution.
- Personal broadband service in remote areas is more cost-effective via satellite than terrestrial techniques.
- Satellites can sometimes offer higher maximum speeds for broadband Internet access for individuals than terrestrial wireless mobile systems (albeit at a higher cost).
- Free satellite broadcast channels are available to users, much as their terrestrial counterpart; hence, the notion that satellite broadcasts are unaffordable to the less well off is debatable.
- The space economy is growing at a rapid rate, proportionately benefiting companies and individuals associated with the industry.

1.3.2   Applications

A wide range of personal applications has been enabled through the collective effort, encourage-ment and financial support of the satellite industry and various governments, complemented by the assistance of the regulatory authorities and an innovative research community. The recent trend in liberalization and privatization has introduced considerable motivation for an enhanced commer-cialization of the satellite industry. A notable feature of the changed environment is that industry's attention is likely to be favourable towards personal applications that promise a mass market. This trend is likely to result in a wider portfolio of personal satellite services and solutions in conjunction with cost benefits due to economies of scale.

When dealing with progress in technology, it is convenient to group applications by their ser-vice class owing to their inherent commonality. Typical applications of personal satellite systems categorized by their services are listed in Table 1.1, and an evolution timeline is summarized in Table 1.2. Appendix A lists a more comprehensive set of personal satellite applications.

Table 1.1   Personal applications by service category

Service category	Applications
Telecommuni-cations: (fixed and mobile)	*Social*: Mobile communications from remote locations (e.g. a remote holiday destination) or while travelling (e.g. on a ship, in a car, or an aircraft)
	Business: Broadband communications from small offices or remote larger offices
	Emergency: Communications from an individual in distress (e.g. during a mountaineering expedition or a maritime rally)
	Entertainment: Interactive Internet gaming, live television and radio during flight
	Military: Command and control; Situation awareness; Welfare communication
Broadcast	*Television*: Direct-to-home broadcasts
	Radio: Direct broadcasts for long-distance car travel, expatriate listening, live broadcast to aircrafts, etc.
	Multicast: Broadcast to a group/region (e.g. weather forecast, sports results)
	Unicast: Broadcast to individuals – financial/stock exchange update
Navigation	*Location dependent:* (e.g. road traffic conditions)
	Route guidance: (e.g. SATNAV)
	Distress
	Trekking
	Agriculture: (e.g. crop spraying)
	Military
Earth observation	*Weather*: Daily TV broadcasts
	Photographs/maps: Education, city maps
Distress and safety	*Internationally approved system*: GMDSS
	Local or regional service
	Ad hoc arrangements
Amateur	*Amateur communication*
	School projects
	Distress and safety
	Innovation

Table 1.2 Evolution timeline of personal satellite applications

Personal system	Approximate year of entry
Amateur radio	1961
Low-speed data land /maritime	Late 1980
Maritime phone	Early 1980
Direct-to-home broadcasts	1989 (Europe)
Fixed broadband	Early 1990
Personal navigation aid	Early 1990
Aeronautical phone	Early 1990
Maritime medium-speed data	Early 1990
Remote pay booth	Mid-1990
Desktop portable phones	1997
Handheld phone	1999
Affordable satellite imagery	Late 1990
Satellite radio	2001
Digital video broadcasting – satellite handheld	2004
Portable multimedia	2005
Satellite digital multimedia broadcast	2005
Mobile multimedia (ships, aircraft, land vehicles)	2007–2008

1.3.2.1 Telecommunications

Personal satellite telecommunication applications are most effective in remote regions without adequate terrestrial infrastructure, as well as in a mobile environment. The low penetration of satellite communication systems in areas lying within a terrestrial coverage is attributed to the relatively high end-user costs of satellite systems. However, satellite-enabled solutions are becoming increasingly synergistic and cost effective.

1.3.2.2 Fixed Satellite Service

In the Fixed Satellite Service (FSS) arena, steady inroads into the fixed personal broadband have continued, beginning in the early 1990s. The uptake of personal satellite broadband service has increased steadily, particularly in rural and remote areas of developed countries, because of an increasing reliance on Internet-delivered services and applications. There were around 2 million Very-Small-Aperture Terminals (VSATs) dispersed around the world in 2010 (Source: David Hartshorn, Global VSAT forum, 2009). VSAT networks are suited for content distribution, interactive services or services for interconnected mesh networks. In addition to well-entrenched applications, Internet-enabled applications such as TV over IP protocol (IPTV) and Voice over IP (VoIP) are increasing in popularity. Many enterprises have widely dispersed offices that are often inaccessible using only terrestrial networks. Such enterprises typically exploit Virtual Private Networks (VPNs) over satellite because these ensure the desired connectivity tagged with security at an attractive cost. Other applications where fixed satellite solutions are proving beneficial include both one-way and two-way interactive distance learning and telemedicine.

Today's typical high-end VSAT system includes a user terminal capable of supporting multiple telephone channels and Personal Computer (PC) networks, connected to a host network capable of delivering toll-quality voice and IP transmission. These solutions particularly appeal to small office/home office (SOHO) users, Internet cafe owners, etc.

VSAT networks are based on both proprietary technology and open standards. The latter allow economies of scale owing to competition. A case in point is the widely used Digital Video

Figure 1.2 A broadband personal terminal. Reproduced by permission of © Pearson Education. All rights reserved.

Broadcast–Return Channel by Satellite (DVB-RCS) standard developed in Europe with international participation. By providing an asymmetric data rate return channel from the users, it offers interactivity useful in applications such as interactive TV, the Internet and distance education.

Various refinements are under way to enhance the viability of the personal VSAT service. Migration to Ka band (30 GHz uplink/20 GHz downlink) is predicted to lower the user terminal cost and service charge through smaller-diameter antennas and higher space segment capacity (resulting from increased power and frequency reuse through smaller coverage patterns). IP acceleration and network optimization solutions improve the throughput, thereby enhancing quality of service and reducing service cost.

Figure 1.2 shows a typical VSAT terminal for providing broadband Internet at home.

1.3.2.3 Mobile Satellite Service

The Mobile Satellite Service (MSS) era dawned in the late 1970s with the launch of the Marisat satellite by COMSAT in the United States and the successful demonstration of MSS technology. Subsequent formation of the International Maritime Satellite Organization (Inmarsat) at the initiation of the International Mobile Organization (IMO) began the era of a public mobile satellite service. Beginning with large portable maritime user sets weighing hundreds of kilograms and capable of supporting only a single telephone channel, technology has evolved to a point where the smallest modern MSS terminals with an identical capability resemble a cellular phone. The data throughput has increased from a few kilobits per second to half megabit per second (Mb/s), and the services extend to the aeronautical and land sectors.

Figure 1.3 (left) illustrates a dual-mode satellite phone capable of operating either via terrestrial or via a low-Earth-orbit satellite infrastructure, as desired. Figure 1.3 (right) shows a phone with a similar capability but operating via a geostationary satellite system.

An interesting development in this sector is the migration of VSAT (broadband) and direct television broadcast services, traditionally associated with fixed services, to the mobile domain, transcending the service distinction formalized by the ITU.

In-flight real-time audio/video and the Internet MSS facilities are now available via L or Ku band systems. Ku-band systems (14/12 GHz) have an edge in throughput owing to increased spectrum allocation, whereas the L-band systems (~1.6/1.5 GHz) lead in terms of wider and more robust coverage and lower terminal and service migration costs. Trials have also shown the viability of

Figure 1.3 Handheld dual-mode satellite phones (Not to scale) used in: (a) a low-Earth-orbit satellite system. Courtesy © Globalstar; (b) a geostationary satellite system. Courtesy © Thuraya.

using cellular phones during a flight, where the aircraft acts as a mobile picocell connected to the terrestrial systems via a satellite terminal – leading to the introduction of commercial systems.

There are ambitious service and business plans to exploit for communication a technique known as Ancillary Terrestrial Component (ATC), where the satellite signals are retransmitted terrestrially in areas of poor satellite coverage to enhance coverage reliability. Other emerging mobile technologies are mobile TV and multimedia services.

Figure 1.4 illustrates a broadband personal portable device capable of supporting a data rate up to 0.5 Mb/s.

1.3.2.4 Direct Broadcast Service

The earliest interest in direct satellite television broadcast reception is attributable to enthusiasts who intercepted TV programme distribution transmissions (via satellite) for personal viewing.

Figure 1.4 A broadband mobile user terminal for packet or circuit-mode operation. Courtesy © Nera Satcom.

An industry grew around this mode of (unauthorized) viewing by the mid-1970s to the extent that programme distributors began encrypting transmissions. The Satellite Instructional Television Experiment (SITE), conducted by the Indian Space Research Organization (ISRO) in India in collaboration with NASA via Application Test Satellite-6, demonstrated the powerful potential and viability of direct broadcasts. Direct-to-home broadcasts were first introduced in Europe in the late 1980s. Currently, dozens of DBS systems and tens of millions of users are receiving the service throughout the world. The majority of these transmissions are subscription television, but large numbers of free broadcasts are also available. Considerable regulatory participation and decisions are necessary in bringing direct broadcast to the public domain, and the timing and the complexity of such decisions vary by country and region. To this day, the direct broadcast service is not permitted in some countries.

Satellite broadcast systems are both complementary as well as competitive to their terrestrial counterparts; however, in remote regions, direct broadcast systems are the only viable solution. With the recent introduction of satellite-delivered high-definition television (HDTV), it would appear that the era of home cinema has truly arrived. Figure 1.5 depicts a personal satellite 'dish' (antenna) that folds into a suitcase ready for easy transportation – perhaps to a remote holiday destination. Direct broadcast services to ships and aircrafts are available commercially, enriching the quality of life of thousands of crew and passengers alike.

Satellite Digital Multimedia Broadcasting

Satellite Digital Multimedia Broadcasting (S-DMB) refers to a recent standard for the transmission of multimedia television, radio and data to mobile devices. It has a hybrid satellite–terrestrial architecture where terrestrial repeaters retransmit the signal in areas of poor satellite coverage. The service was trialed in several countries, including China for a possible service roll-out to cater for the 2008 Olympics. A commercial service in Korea already provides television, radio and data, as well as a short message service, to mobile receivers integrated with various types of personal device such as laptop computers and cell phones.

Figure 1.5 Left: A portable dish with a low-noise front end to receive Sky broadcasts in Europe; the dish folds into a briefcase. Right: A satellite receiver. Reproduced from © Maplin Electronics Ltd.

Digital Video Broadcasting

Digital Video Broadcasting DVB is a suite of international video broadcasting standards that caters for numerous transmission media while ensuring equipment compatibility. The widespread adoption of these standards has enabled the cost of broadcast equipment and receivers to be lowered dramatically through economies of scale.

The DVB-S (DVB-Satellite) standard for satellite television was introduced in 1995. The multimedia transport scheme is based on the Motion Picture Expert Group (MPEG)-2 standard. It is a commonly used format for broadcast feed and services such as Sky TV (Europe) and Dish Network (United States).

DVB-S2 (DVB-S, second generation), ratified in 2005, replaces the DVB-S standard. This standard deploys a more advanced transmission technique than DVB-S, allowing change and adaptation of code rate in real time (in response to changing propagation conditions), and provides a throughput gain of about 30% over DVB-S, together with more flexible data encapsulation (with backwards compatibility). Its main current application is the distribution of high-definition television (HDTV). It is suitable for television broadcasts and interactive services with access to the Internet. The return message sent by a user can be channelled through a telephone, an optical fibre or a satellite medium. The DVB-S2 standard also permits professional applications such as content distribution and Internet trunking.

Digital Video Broadcast to Satellite Handheld (DVB-SH), proposed by Alcatel, is yet another potential satellite handheld solution comprising a hybrid satellite–terrestrial architecture at S-band (2–3 GHz) similar to that of S-DMB but using a more powerful geostationary satellite. Alcatel proposed to introduce a DVB-SH service in Europe in 2009. The DVB technical module, called Satellite Services to Portable Devices (SSP), has started to develop a standard for satellite handheld along these lines.

1.3.2.5 Satellite Radio

Commercially introduced around 2001, satellite radio – by which high-fidelity, specialist radio channels are wide-area broadcast directly to users – is growing rapidly in terms of subscriber base. This service holds a niche in the developed world, targeting individuals or businesses such as hotels wanting specialist audio channels – uninterrupted music, sport or news – on fixed sets, long-distance car travellers desiring uninterrupted high-quality broadcasts throughout a journey, people/businesses wanting regular weather forecasts, commercial airliners desiring live music or news, expatriates aspiring for a rebroadcast of their home channel, etc. In developing regions, direct transmissions are the only source of a wide listening choice. A variety of English and regional language news, entertainment, sports and popular music channels are available in far-flung regions of over a 130 countries around the world. Figure 1.6 shows a typical satellite radio used in the Asian and European regions.

1.3.2.6 Navigation

There are two global navigation satellite systems currently available to the general public, GPS and GLONASS, although use of the GPS system is more prevalent (GLONASS having fallen into disrepair). Numerous personalized location services are available around these systems. Many personal devices – such as cellular phones – can now integrate GPS functionality, allowing integrated location-based applications. Navigation aids for car owners, trekkers, mountaineers and other adventure activities are in regular use. In addition, there are other existing and planned regional systems, as discussed in Chapter 14.

Owing to their proven merits and truly global coverage, satellite navigation systems are useful even in areas where terrestrial communication systems dominate, although satellite navigation

Figure 1.6   A home satellite radio set. Reproduced from © Worldspace.

systems are unreliable indoors because of the low penetration and non-line-of-sight path of navigation signals within buildings or heavily shadowed locations. Figure 1.7 illustrates a navigation receiver integrated with a mobile phone and a Personal Digital Assistant (PDA), illustrating the trend in converged handheld solutions.

1.3.2.7   Distress and Safety

Satellites are a powerful means of supporting distress and safety services because of their ubiquitous and unbiased coverage. In a satellite-enabled distress and safety system, a navigation receiver determines the position of the affected party while a transmitter delivers the distress message, along with the position fix, to a rescue and coordination centre. Alternatively, the fixed component of the satellite distress system itself determines the transmitter location. Hence, satellite distress

Figure 1.7   A navigation receiver integrated with a mobile phone and a PDA. Reproduced from © Mio Technology.

and safety systems are vital for the international maritime and aeronautical transport industries, as well as for facilitating the management of distress situations in remote locations. There are at least two safety and distress application categories:

- delivery of distress messages;
- communication support from a disaster zone.

The former application is formalized internationally for maritime applications through the Safety of Life at Sea (SOLAS) treaty. For inland events, local authorities generally offer the requisite service. The latter application is improvised; here, individuals and/or aid agencies utilize an ad hoc satellite infrastructure until the terrestrial infrastructure is set up, which may take from days to months or even years in extreme cases such as the aftermath of a war. A common practice to establish the desired connectivity is to establish a mobile satellite solution initially, followed by deployment of a fixed satellite system. The MSS requires little set-up time because the terminals are ready to use and free from regulatory formalities, however, the service provides a limited bandwidth. An FSS (VSAT) arrangement offers a much wider bandwidth but may require a few days to activate owing to the installation effort and sometimes the need to obtain a regulatory clearance. In developed regions such as the United States, disaster and emergency management vehicles and systems replenished with a variety of fixed and mobile satellite communications units are finding favour with the welfare and disaster management agencies.

   Figure 1.8 shows a personal Emergency Position Indication Radio Locator (EPIRB), which operates via a low-orbit satellite distress and safety system known as COSPAS-SARSAT (Cosmicheskaya Sistyema Poiska Avariynich Sudov or Space System for the Search of Vessels in Distress – Search and Rescue Satellite-Aided Tracking) (Cospas-Sarsat, 2010).

1.3.2.8   Radio Amateurs

Radio amateurs are an active group, establishing contacts with each other through the Orbiting Satellites Carrying Amateur Radio (OSCAR) satellites since 1961. The OSCAR series was originally

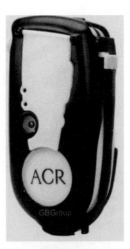

Figure 1.8   An EPIRB for personal maritime distress. When activated, signals are transmitted to a registered maritime agency for search and rescue. Reproduced by permission of © ACR.

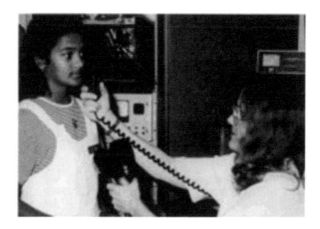

Figure 1.9   A ham radio transreceiver. Reproduced courtesy of NASA.

meant for amateurs to experience satellite tracking and participate in radio propagation experiments; however, these satellites have become increasingly advanced over the years. They are often used to support school science groups, provide emergency communications for disaster relief and serve as technology demonstrators. The first slow-scan TV reception in space was demonstrated aboard the Challenger STS-51 mission through an amateur transmission. The first satellite voice transponder and some advanced digital 'store-and-forward' messaging transponder techniques are also accredited to this group.

Figure 1.9 (Peck and White, 1998) illustrates radio amateur equipment during a school experiment. The Space Amateur Radio EXperiment (SAREX) is a notable example of youth participation. It facilitates communication between astronauts and school students. Through SAREX, astronauts make both scheduled and unscheduled amateur radio contacts from the Shuttle orbiter with schools selected through a proposal process. These contacts serve to enthuse students and families about science, technology and learning (Peck and White, 1998).

1.3.2.9   Military

Satellites are today indispensable to the military throughout the world, during peace, war or the reconstruction phase at the end of a war. The military of many advanced countries have dedicated military satellite systems for both strategic and tactical purposes. Indeed, such is the demand for such services that the military tend to utilize the communication capacity available on civilian satellite systems in order to satisfy their total requirements. In addition, the introduction of assured *service-based* military satellite provision (for example, Skynet 5 and Xtar) has allowed other nations to benefit from satellite capability without the associated expense of asset purchase.

During periods of peace, satellites assist in reconnaissance, communication between headquarters, logistics, welfare, etc. Military navigation systems have become indispensable at every phase of military operations. Both GPS and GLONASS are military systems. During a war, satellites play an indispensable part in tactical communication, surveillance and navigation. During the reconstruction phase, satellites can be used for communications between the military and other government and non-government organizations, and for distance learning to impart training in the affected regions and auxiliary services such as asset tracking and management (Mitsis, 2004).

Figure 1.10 Satellite imagery showing the stadium at the University of Texas. Reproduced by permission of © GeoEye Inc.

1.3.2.10   Earth Observation

Earth observation satellites, also referred to as remote sensing satellites, are attractive to assist in weather forecasts, agriculture yield estimates, forest management, commercial fishing, urban planning, etc. Such activities are normally associated with governments, welfare agencies, etc. Significantly, in recent years the data gathered by Earth observation satellites have become available to the general public and businesses.

Weather forecasts based on remote sensing satellite data embellished with satellite pictures are a regular part of television transmissions. Satellite imagery superimposed on route maps is quite a powerful route and location guidance tool for the general public. High-quality satellite images are widely available commercially. As an illustration, Figure 1.10 shows an image of the stadium at the University of Texas taken by a GeoEye satellite.

1.4   Trends

The topic of technology forecasting interests academics, researchers, strategists and planners and commercial companies, although each for different reasons. On one hand the likes of academics, the government and intergovernment bodies aim to further the frontiers of knowledge to pave the way for long-term economic and social benefit, while on the other hand a commercial company judges the short-to-medium trends in targeted areas to develop its business. Satellite manufacturers are interested in the medium-to-long term (5–15 years), and satellite carriers, service providers, application developers, etc., have a short-to-medium-term interest. We have seen the disastrous consequence of incorrect market forecasts in the MSS sector towards the end of the last century, when several operators had to file for bankruptcy owing to misreading the market trend in the early 1990s.

This section outlines the general trends in the topics discussed so far without trying to project a synthesized scenario of the future. The reader is referred to Chapter 17 for a more detailed exposition. Suffice to mention here that, with the steady convergence of telecommunication, broadcast, position location systems and computing and an increasing integration of functionality in personal devices, the information society will expect a wider accessibility to information (anywhere), greater mobility (everywhere), large bandwidth (broadband) and 'converged' personal devices – all at a lower cost. Judging from the present situation, it is evident that satellite systems will have a fair role but will have to evolve rapidly to be in step.

Handheld

A notable trend in recent years is the integration of an increasing number of functionalities in the personal handheld, facilitated by user demand and the refinements of low-cost semiconductor processes such as CMOS and SiGe Bi-CMOS. The enormous processing capability of modern programmable chips enables a variety of embedded functionality on the same platform. From the user's perspective, it is convenient to have a unified package of multimedia, communications and computation. Thus, a handheld device may potentially incorporate some or all of the following units – a camera, a multiple standard cellular/satellite phone, a mobile TV, a satellite positioning device, a Wireless Local Area Network (WLAN) interface, personal music/video systems, gaming, etc. Consumer demands for data *pushed* to a handheld device are on the rise. Examples are satellite imagery and maps, GPS coordinates, location-specific and general news, sports results, etc. Wireless technologies of interest here are satellite, cellular, WLAN and WiMAX (Worldwide Interoperability for Microwave Access). Using GPS as the location detection engine, handheld location-based services are gaining in popularity, riding on applications such as fleet management and proof-of-delivery signatures.

Owing to the availability of satellite imagery through Internet search engines and thereon to handheld devices, this type of data is whetting the appetite for more advanced remote imaging applications. Specialist companies are gearing up their space segment to provide higher-resolution imagery. These data, combined with local information, provide effective applications. For example, location knowledge can enable the downloading of desired location-specific traffic or weather conditions.

An interesting example of a future personal application (Sharples, 2000), in line with the UK government's Green Paper on life-long learning (Secretary of State, 1998) is the concept of a framework of educational technology constituting a personal handheld (or wearable) computer system that supports learning from any location throughout a lifetime. Such a concept would require an all-encompassing wireless network covering the majority, if not all, of the populated area, consisting of a hybrid, terrestrial–satellite architecture.

VSAT

It is recognized that personal VSAT systems benefit from high bandwidth and smaller terminal size. VSAT migration to Ka band provides a much larger available bandwidth ($\sim$1 GHz) and enables a smaller terminal size through the use of a greater number of smaller high power beams. It is therefore anticipated that Ka-band VSAT systems will be more attractive for many applications. Other notable areas of interest are as follows:

- Considerable effort is under way to maximize spectral efficiency and enhance transport protocol robustness to the inherent signal latency of satellite systems.
- Several operators offer a mobile VSAT solution for moving platforms such as ships. One limitation of a VSAT, when considered in a mobile application, is that, because it belongs to the fixed satellite service, there are mobility restrictions on these services in various parts of the world.

MSS

A major limitation of MSS systems is their relatively poor coverage in shadowed (especially, urban) areas. Following a liberalization of the spectrum regulations, the Auxiliary Terrestrial Component (ATC), also known as the complementary terrestrial component, has drawn considerable interest in the MSS and satellite radio broadcast communities, where satellite signals are rebroadcast terrestrially in difficult coverage areas to fill in such gaps (see S-DMB and DVB-SH above). In effect, the satellite spectrum is made available to terrestrial transmissions. The potential of a massive increase in the user mobile satellite base with consequent revenue gains and the possibility of a lucrative collaboration with terrestrial operators are strong incentives to the satellite operators.

The majority of modern MSS systems deploy an integrated architecture. Users can thus roam between these networks, identified by their subscriber identity module (SIM) card. This feature, along with a multimode receiver software-defined architecture, will allow a seamless connectivity while roaming across multiple services.

With the users' insatiable desire for larger bandwidth, the MSS industry is striving hard to provide higher throughput in spite of the harshness of the propagation channel, cost constraints and meagre spectrum availability in the favoured L and S bands. In the area of the L-band broadband MSS arena, 0.5 Mb/s portable mobile technology was introduced by Inmarsat in 2005, and the service was extended to all types of mobile environment by the year 2008. However, the scarcity of the L and S-band spectrum has prompted the industry to investigate higher-frequency bands – the Ka band to an extent but the Ku band in particular – where adequate bandwidth exists in spite of the problems of rain attenuation, stringent antenna pointing needs and low coverage of sea and air corridors. (Arcidiacono, Finocchiaro and Grazzini, 2010). High data rates of up to 2 Mb/s are available to ships in Ku band via Eutlesat satellites, while in the aeronautical sector Boeing Service Company provides a Ku-band broadband network service comprising high-speed Internet communications and direct broadcast television to the US Air Force Air Mobility Command following the closure of its commercial operations called Connexion to civil airlines. The forward link to the aircraft provides 20–40 Mb/s of shared capacity. Mobile platforms may transmit at up to 1.5 Mb/s (Jones and de la Chapelle, 2001). A number of commercial aeronautical Ku band services also exist.

Broadcast

Satellite radio technology has reached a mature stage within a rather short timespan, riding on the back of the most rapid service uptake of all the satellite products to date. Companies in the United States, in particular, have been at the forefront of its commercial introduction. The user base in the United States and other parts of the world is growing steadily despite competition from terrestrial and Internet radio, indicating the long-term commercial viability of the technology, particularly following a consolidation between US operators in order to achieve a critical mass, and introduction of novel commercial applications.

Terrestrial mobile TV systems are being promoted as the next commercial broadcasting success. By the same criteria, satellite mobile TV technology also holds promise. Handheld simulated-satellite TV trials are under way in Europe based on the DVB-SH technology. An operational system will use high-power spot-beam satellites and Orthogonal Frequency Division Multiplexing (OFDM) and will be compatible with DVB-H and 3G mobile standard. A similar system resident on another technology platform was in service for a while in Japan via MBSat since 2004, offering eight channels along with 30 audio channels. It also broadcasted maritime data such as sea currents and sea surface heights. Korea launched a similar service in 2005, and China intended to introduce the service prior to the 2008 Olympics, while India plans to deploy a multimedia mobile satellite for S-DMB IP-based services to individuals and vehicles.

High-Altitude Platforms

High-Altitude Platforms (HAPs) have been studied as surrogate satellites for well over a decade. They potentially offer an effective alternative to ground base stations by eliminating the logistics and costs associated with the installation of terrestrial base stations. Being closer to the Earth than satellites, they do not require an expensive launch, and nor do they need space-qualified hardware; their equipment can be readily repaired; they can provide considerably higher downlink received powers and uplink sensitivity than satellites and a smaller coverage area for a given antenna size owing to their proximity to the Earth (<60 km) and hence can support a variety of wireless technologies, for example, mobile and fixed broadband, WiMax, etc. Use of HAPs as ATC platforms is also an interesting concept. A number of initiatives are being pursued actively, and a commercial launch is awaited.

Network

Following the introduction of the 3G systems, the focus of the research community is towards the next generation of mobile system, 4G and beyond. While the specific details are under active research, some of the requirements have become reasonably well entrenched. It is believed that the 4G network will comprise a multitude of interoperable wireless technologies and interfaces. It is also clear that satellites will be one of the constituents of such an all-encompassing network. Techniques under investigation include software-defined radio, which should permit users to migrate across a heterogeneous mixture of air interfaces and protocols, an ad hoc network to facilitate mobility and allow formation of decentralized network management, efficient resource management techniques, handover between heterogeneous networks, etc.

Region

The United States continues to lead in the exploitation of personal satellite services, but there are indications that consumer demand for satellite communications products will increase considerably in the Middle East, the Far-East, India and South Asia. The areas of expected growth there include IP-based VSAT services, Virtual Private Network (VPN) services, combined data/video/voice services and robust handheld MSS solutions.

1.5   Overview of this Book

This chapter has provided a sample of the use of satellite technology for personal applications. The remainder of this book is divided into two parts. Part I, comprising Chapters 2 to 8, introduces fundamental principles and concepts of the applicable satellite systems, including some novel concepts of future interest. While some of the concepts presented in Part I are also documented elsewhere, it is the intention here that these topics be treated specifically in the context at hand. This material is therefore presented somewhat succinctly except where the subject is relatively new or, in the authors' view, has not received adequate attention in other texts. Part II, comprising the remaining chapters, introduces practical system techniques and architectures, with illustrative examples, concluding with a chapter on trends and evolution. For use with relevant educational courses, or to assist home study, each chapter includes a set of revision questions.

Chapter 2 describes the main attributes of satellite and high-altitude systems. Topics include satellite orbits, geometrical relationships and characteristics of high-altitude platforms. Chapter 3 introduces relevant aspects of spectrum regulation and electromagnetic wave propagation, including the effects of the atmosphere and the operational environment, as well as remote sensing windows. An understanding of system aspects of antennas, the transmission equation and the origins and characteristics of noise is fundamental to a radio communication system. Chapter 4 introduces these concepts. Next, Chapter 5 discusses the related topics of modulation and error control coding, including highly efficient modulation schemes. Coding topics include various block and convolution codes as well as newer turbo and LDPC codes. This leads to Chapter 6, which covers link budget and satellite access methods – particularly multiple access – and generic aspects of satellite networking. Chapter 7 covers the concepts of Doppler and ranging-based navigation systems, which are widely used in contemporary satellite navigation systems. In the final chapter of Part I, we discuss information entropy and concepts of data compression as well as speech, audio and video encoding.

With the reader thus prepared, the second part of the book goes on to discuss techniques and architectures – illustrating them with selected case studies and examples. Such system concepts are best dealt with in conjunction with the overall architecture, and hence these concepts are included in this part rather than in Part I. In Chapter 9, we address digital broadcasting techniques and architectures, including a review of MPEG multimedia standards, the concept of multiplexing and transporting, direct-to-home broadcast system architectures and prevalent transmission standards. In the next chapter, we introduce numerous contemporary state-of-the-art broadcast systems of each category – satellite radio, direct multimedia broadcasts, direct-to-home satellite television systems used in various parts of the world (with a focus on Europe and the United States). In the final section we introduce a multimedia broadcast system used by the US military. In Chapter 11 we address various types of network topology and connectivity, illustrating them with typical satellite and tentative HAP systems. In particular, we use network concepts for a hypothetical mobile satellite network as a baseline. These topics prepare us for chapter 12, where we present examples and case studies of various types of satellite communication system (mobile, fixed, amateur and portable military systems) including a few prominent HAP communications research initiatives undertaken in Europe. The next chapter, Chapter 13, introduces the techniques prevalent in navigation systems, including system aspects of Doppler and ranging navigation, the satellite augmentation system and hybrid communication–navigation system architecture, including issues related to navigation receivers. In the final part of the chapter we introduce the concepts of distress, safety and location-based services. Chapter 14 follows with examples of prominent global and regional navigation systems, satellite augmentation systems and distress and safety systems, concluding with an example of a system to provide a location-based service. Chapters 15 and 16 respectively introduce system-level remote sensing techniques/architectures, and prominent remote sensing systems and optical imaging systems.

What of the future? In the final chapter we address the evolution of personal satellite applications and systems, taking into consideration trends in user expectations, technology, regulatory and standardization efforts and the inherent strengths and limitations of satellite systems.

The authors hope that the reader will find this book to be a useful reference. They firmly believe that, with the passage of time, satellite systems will enrich the quality of people's lives far more than is visualized today.

References

Arcidiacono, A., Finocchiaro, D. and Grazzini, S. (2010) *Broadband Mobile Satellite Services: the Ku-Band Revolution*. Available: http://www.eutelsat.com/products/pdf/broadbandMSS.pdf [accessed January 2010].

Cospasat-Sarsat (2010) http://www.cospas-sarsat.org/ [accessed February 2010].

Global VSAT Forum (2009) http://www.gvf.org [accessed February 2009].

Jones, W.H. and de la Chapelle, M. (2001) Connexion by Boeing/sup SM/-broadband satellite communication system for mobile platforms. MILCOM 2001. Communications for Network-Centric Operations: Creating the Information Force, 28–31 October, Washington, DC, IEEE, Vol. 2, pp. 755–758.

Mitsis, N. (2004) Rebuilding Iraq: Satellite-enabled Opportunities. *Via Satellite*, April, S6–S8.

Peck, S. and White, R. (1998) *Amateur Radio in Space – A Teacher's Guide with activities in science, Mathematics and Technology*. NASA, Office of Human Resources and Education Division.

Raman, B. and Chebrolu, K. (2007) Experience in using W-Fi for rural Internet in India. *IEEE Communications Magazine*, **45**(1), 104–110.

Robson, D. (2006/2007) The need for space. *IET Communications Engineer*, **4**, (December/January), 26–29.

Secretary of State (1998) Secretary of State for Education and Employment, Green Paper on lifelong learning, UK Government.

Sharples, M. (2000) The design of personal mobile technologies for lifelong learning, *Computers and Education*, **34**, 177–193. Available: http://www.elsevier.com/locate/compedu [accessed May 2007].

Part I

Basic Concepts

Part II

Basic Concepts

2

Satellites and High-Altitude Platforms

2.1 Introduction

At the risk perhaps of stating the obvious, a vital aspect of any satellite-based service used for personal applications is the satellite itself. In this chapter, we focus on the satellite as a platform for provision of such services. In subsequent chapters we shall go on to discuss the concepts behind the payloads carried on these platforms, which permit us to use these satellites to provide broadcast, communications, navigation and remote sensing services.

Fundamental to the use of satellites for different applications is an appreciation of their orbits, and in this chapter we introduce the key parameters that describe the motion of these satellites around the Earth, and allow us to determine their orbital period and instantaneous position. Armed with this basic understanding, we go on to focus on the principal types of orbit in use, of which the geostationary orbit is, perhaps, the most widely known.

A key consideration in the selection of a particular satellite orbit type, for a particular application, is its potential ground coverage (both of individual satellites and the combined coverage of a constellation of satellites). Coverage is generally taken to mean that region of the Earth's surface visible from the satellite at any given time and its 'track' as the locus of the point on the Earth nearest to the satellite (the sub-satellite point) as it moves with respect to the Earth. The coverage effectively determines the available market for satellite services, or, conversely, the number of satellites needed to cover the desired region, and also the distance at which the spectrum can be reused by other satellites without interference.

Having gained an understanding of orbits and coverage, we go on to relate the motion of a satellite in the plane of orbit of the satellite to key spatial reference frames, and ultimately in the local coordinate system that permits the user to determine the look angles for pointing his/her terminal apparatus at the satellite (if using a directional antenna). We round off our discussion of satellites by briefly discussing the key features of the satellite bus itself – the host platform to which service payloads are added.

Our notion of satellites relates primarily to objects orbiting the Earth sustained by their momentum, located well above the drag of the Earth's atmosphere. However, at lower altitudes, the use of 'surrogate' satellites, in the form of High-Altitude Platforms (HAPs), is gaining significant interest for servicing smaller coverage areas at lower cost and shorter timescales, and potentially with higher performance. The altitude of these high-altitude platforms typically lies in the $\sim$16-22 km range. By contrast to conventional satellites, these platforms rely on aerodynamics and/or buoyancy to maintain their position relative to the Earth. Although they are still in their relative infancy, we cover the basic concepts of these HAPs, in anticipation of things to come, and contrast them with their satellite rivals.

2.2 Satellites

2.2.1 Orbital Motion

The motion of man-made satellites around the Earth follows the same laws of physics as those that govern the motion of the planets around the Sun. The laws that govern their motion were postulated almost 400 years ago by Johannes Kepler and subsequently proved by Sir Isaac Newton. We shall give only a basic treatment of such topics here – sufficient only to gain a basic understanding of the key concepts involved in determining a satellite's position, orbital period, coverage, etc. For a more detailed exposition of the topic, the reader is directed to the literature (e.g. Richharia (1999) and Maral and Bousquet (2002) for a system level treatment, and Bate, Mueller and White (1971) for a more classical treatment).

2.2.1.1 Kepler's Orbital Laws

The starting point for our discussion of orbital motion is Kepler's three laws of planetary motion, originally defined to describe the motion of the planets around the Sun. Rephrasing these laws to describe the motion of a satellite around the Earth, we have:

1. The orbit of every satellite is an ellipse with the Earth at one focus.
2. A line joining a satellite and the Earth sweeps out equal areas during equal intervals of time.
3. The square of the orbital period of a satellite is directly proportional to the cube of the semi-major axis of its orbit (its maximum radius).

Orbital Mechanics

Kepler's laws were subsequently proved by Newton using his law of gravitation. The gravitational force attracting a satellite to the Earth at radius r is

$$F = \frac{G m_E m_s}{r^2} \tag{2.1}$$

where m_E is the mass of the Earth, m_s is the mass of the satellite and G is the gravitational constant. This gravitational force governs the acceleration of the satellite (in the direction of the Earth) as it orbits. The motion of satellites under this gravitational pull is subsequently governed by Newton's laws of motion.

An idealized model can be used to explain the behaviour of satellite orbits broadly, and may subsequently be enhanced to represent the real system more accurately. Key simplifications used are:

- The influence of all other bodies (the Sun, Moon and so on) is ignored.
- The mass of the Earth is assumed to be much greater than that of the satellite (i.e. $m_E \gg m_s$).
- The Earth and satellite are assumed to be spherically symmetrical objects and may thus be considered as having masses that act at their centres.

Under these assumptions – the so-called two-body problem – the motion of the satellite may be described by (Richharia, 1999; Maral and Bousquet, 2002)

$$\ddot{\mathbf{r}} + \mu_G \frac{\mathbf{r}}{r^3} = \mathbf{0} \tag{2.2}$$

where $\mathbf{r}$ is the radial vector from the centre of the Earth to the centre of the satellite, $\ddot{\mathbf{r}}$ is the acceleration vector (second time derivative of $\mathbf{r}$) and μ_G ($\equiv Gm_E$) is known as the gravitational parameter.

Orbital Shape
The solution to equation (2.2) is a *conic section* (circle, ellipse, parabola or hyperbola). Our primary interest here is with circular and elliptical orbits (a circular orbit may be considered to be a type of elliptical orbit – one with zero ellipticity) and Kepler's laws for such orbits.

Prior to discussing these orbits in more detail, it is useful to define some of the key terms used, by convention, to describe them:

- *Perigee.* This is the closest point of the orbit from the Earth.
- *Apogee.* This is the furthest point of the orbit from the Earth.
- *True anomaly.* This is the angle between the satellite and perigee measured at the Earth in the orbital plane.
- *Subsatellite point.* This is the point on the Earth directly below the satellite. Specifically, it is the point on the Earth through which the surface normal passes through the satellite position.
- *Semi-major axis and semi-minor axis.* These describe the size of the orbital conic section. They correspond to the maximum and minimum orbital radii respectively.

Keplerian Elements
For the two-body problem, the satellite motion is described by six orbital parameters – the Keplerian elements:

- *Semi-major axis a_K.* This distance describes the size of the orbit.
- *Eccentricity e_K.* This ratio describes the eccentricity of the orbit ($e_K \rightarrow 0$ for a circular orbit, $0 < e_K < 1$ for an elliptical orbit, $e_K \rightarrow 1$ for a parabolic orbit and $e_K > 1$ for a hyperbolic orbit). We are interested here only in circular and elliptical orbits.
- *Mean anomaly M_K (0).* This angle is related to the time since the satellite last passed the perigee at some specified date (epoch).
- *Inclination i_K.* This angle describes the inclination of the orbit relative to the Earth's equatorial plane.
- *Right ascension Ω_K.* This angle describes the angle between the vernal equinox (a celestial reference direction specified by the position of the Sun at the spring equinox) and the *ascending node*. The right ascension of the ascending node is often abbreviated as RAAN. The ascending node represents the point where a satellite crosses the equatorial plane, moving from south to north; correspondingly, the *descending node* represents the point where a satellite crosses the equatorial plane, moving from north to south.
- *Argument of perigee ω_K.* This describes the angle between the ascending node (intersection with the equatorial plane) and the perigee (closest point) in the orbital plane.

The first three Keplerian elements describe the motion of the satellite *within* the orbital plane, while the remaining parameters describe the *orientation* of the orbital plane relative to the equatorial plane. Figure 2.1 illustrates how the Kepler orbital parameter set is used to specify an orbit in three dimensions.

Epoch
In addition to specifying the six Keplerian elements, it is also necessary to specify an epoch (a specific moment in time) at which the parameters are specified. It is relatively common to use the so-called *J2000* epoch, which is defined as noon on 1 January 2000, terrestrial time (an astronomical

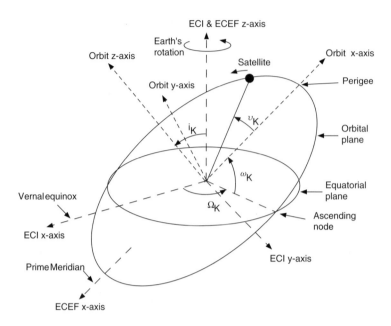

Figure 2.1   Kepler orbital parameters.

time) – equivalent to 11:58:55.816 on 1 January 2000, Universal Coordinated Time (UTC). UTC is equivalent to Greenwich Mean Time (GMT) – also known as Zulu (Z) time. UTC is derived from International Atomic Time (TAI), a statistical combination of the global ensemble of atomic clocks, but, unlike TAI, UTC incorporates leap seconds needed to synchronize with solar time (so as to compensate for the Earth's rotation not being exactly 24 h or 86 400 s). As of January 2009, UTC lagged behind TIA by 34 leap seconds.

The 'J' in J2000 refers to the use of a decimal calendar system in which the year is *exactly* 365.25 days and hence related to the average year of the calendar introduced by Julius Caesar in 45 BC. The convenience of the Julian year is that it is easy to calculate the time difference between two Julian dates – hence the Julian date is used to determine the time (in decimal days) since epoch. The Julian date is actually the number of days (and fractions of a day) since noon on 1 January 4713 BC (Julian dates start at noon), and the Julian date for the J2000 epoch is 24 51 545.0.

Orbital Radius and Period
The orbital radius r may be expressed as

$$r = \frac{a_K \left(1 - e_K^2\right)}{1 + e_K \cos\left(\nu_K\right)} \tag{2.3}$$

where ν_K is the true anomaly. From Kepler's third law, the orbital period (the time to complete one orbit) T_p is given by

$$T_p = 2\pi \sqrt{\frac{a_K^3}{\mu_G}} \tag{2.4}$$

Table 2.1 Orbital period and coverage for various circular orbits

Orbit	Altitude (km)	Period	Velocity	$\theta°$	$\psi°$	Coverage radius (km)	Coverage Area (km²)	Fractional coverage (%)
HAP	20	–	–	79.0	0.97	108	36 563	0.0072
LEO	800	1 h 40 min	7450 m/s	61.0	19.0	2106	1.38×10^7	2.7
MEO	20 000	11 h 50 min	3887 m/s	13.7	66.3	7351	1.52×10^8	29
GEO	35 786	23 h 56' 4.09''	3 074.7 m/s	8.54	71.5	7928	1.73×10^8	34

For the particular case of a circular orbit of a satellite at altitude h_s, the orbital period is found by substituting $a_K \rightarrow (R_E + h_s)$, where R_E is the mean Earth radius. Table 2.1 summarizes nominal orbital periods for the circular orbit types.

True Anomaly

Having determined the orbital period, the mean anomaly M_K at some specified time t may be determined from that at epoch via

$$M_K(t) = M_K(0) + 2\pi \frac{(t - t_o)}{T_p} \tag{2.5}$$

where t_o is the epoch time. The value of M_K is constrained to the range 0–2π rad by subtracting integer multiples of 2π.

The mean anomaly is essentially a parameter of convenience giving the equivalent angle from perigee for the satellite if it were orbiting with constant *average* orbital velocity. In order to determine the instantaneous satellite position, it is necessary to convert between the mean anomaly and the *true anomaly*. In general, this conversion must be carried out numerically, usually by first finding yet another angle – the *eccentric anomaly* (Richharia, 1999). Fortunately, for small values of ellipticity, the following approximation may be used (Green, 1985):

$$\nu_K \simeq M_K + 2e_K \sin(M_K) + \frac{5}{4} e_K^2 \sin(2M_K) \tag{2.6}$$

Clearly, for a circular orbit, the ellipticity is zero and the true anomaly is just the same as the mean anomaly.

2.2.1.2 Orbit Perturbations

In reality, this idealized two-body system forms an incomplete picture. Satellite motion is disturbed by external forces exerted by the non-uniform gravitational field created by the Earth's non-uniform mass (equatorial bulge) and the gravitational forces of other heavenly bodies – particularly the Sun and Moon. In the general case, the Keplerian elements (which are specified at a particular epoch) will vary with time.

The gravitational field of the non-spherical Earth causes variations in the argument of perigee ω_K, right ascension Ω_K and mean anomaly M_K (Maral and Bousquet, 2002). Therefore, in order accurately to predict the satellite position some time after epoch, additional parameters are used that permit calculation of these drift terms to an appropriate accuracy for the intended application (Maral and Bousquet, 2002; Kaplan and Hegarty, 2006). It can be shown that the rate of change of the argument of perigee $\dot{\omega}$ and rate of change in right ascension $\dot{\Omega}_K$ may be given as (Piscane, 1994)

$$\dot{\omega}_K = \frac{3}{4}\sqrt{\frac{\mu_G}{a_K^3}} J_2 \frac{R_E^2}{a_K^2 \left(1 - e_K^2\right)^2} \left[5\cos^2\left(i_K\right) - 1\right] \tag{2.7}$$

$$\dot{\Omega}_K = -\frac{3}{2}\sqrt{\frac{\mu_G}{a_K^3}} J_2 \frac{R_E^2}{a_K^2 \left(1 - e_K^2\right)^2} \cos\left(i_K\right) \tag{2.8}$$

where $J_2 = 1.0826 \times 10^{-3}$ is a coefficient of the spherical harmonic description of the oblateness of the Earth.

Significantly, at an inclination of $63.4°$, the term $\left(5\cos^2\left(i_K\right) - 1\right)$ in the expression for the rate of change in argument of perigee reduces to zero, and the precession of perigee is zero, and hence the apogee and perigee of the orbit become invariant. This inclination is therefore sometimes referred to as the *critical inclination*.

Furthermore it is possible to engineer the rate of change of right ascension to match the rotation of the Earth relative to the Sun (in a sidereal day) by appropriate choice of orbit altitude and inclination. This results in the so-called Sun-synchronous orbits, which are attractive for visible satellite imaging because the attitude of the Sun is the same at each pass.

Additional contributions result from the influence of the Sun and Moon. The gravitational influence of the Sun and Moon are generally negligable for lower Earth orbits, but they are important for geostationary satellites. Conversely, atmospheric drag due to the residual atmosphere is not significant for geostationary satellites but can be important for low-orbit satellites (Roddy, 2006).

2.2.2 Principal Types of Orbit

2.2.2.1 Overview

General Considerations

In 1958, a team led by James Van Allen discovered that the Earth is surrounded by radiation belts of varying intensity, caused by the interaction of electrically charged space-borne particles and the Earth's magnetic field. Figure 2.2 illustrates these radiation belts, generally known as the Van Allen belts. There are two main radiation belts, an inner belt of high-energy protons and electrons at around 1.5 Earth radii and an outer belt of energetic electrons at around four Earth radii. The significance of these belts to satellite applications is that these charged particles can damage spacecraft electronics and software. Consequently, with very few exceptions, satellites are preferably placed in benign regions of low radiation intensity, inside, in between or beyond the Van Allen belts.

The majority of the more useful of the available benign orbits may be categorized into three basic types:

- *Low Earth Orbit* (LEO): those located in the altitude region range $\sim$750–1500 km (below the inner Van Allen belt).
- *Medium Earth Orbit* (MEO): those located in the altitude range $\sim$10 000–20 000 km (between the Van Allen belts).
- *Geostationary/Geosynchronous Orbit* (GEO): those located in a circular or near-circular orbit at an altitude of 35 786 km above the equator (beyond the Van Allen belts).

Note, however, that the ranges are not demarked formally. For example, some authors define LEO and MEO altitude ranges respectively as 500–2000 km and 8000–20 000 km (Intelsat Online. 2010). A majority of existing or planned communication, navigation and remote sensing systems lie within these regions.

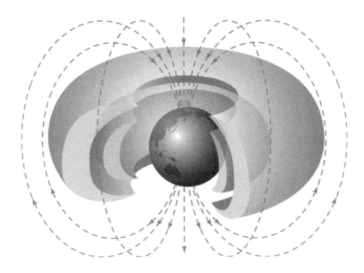

Figure 2.2   Van Allen belts and magnetic flux lines. Reproduced courtesy of NASA.

2.2.2.2   Coverage

Of significant interest is the amount of the Earth's surface visible from a satellite in a particular orbit (altitude). The region within the visibility of a satellite is known as the coverage area (often called the footprint). Note that it is necessary to distinguish between the coverage determined by the limit of visibility and the nominal coverage provided by design for a particular beam of a satellite payload system – which will be less than or equal to the visibility coverage according to the design objectives. A satellite need not service the entire coverage and hence in such instances the coverage area differs from the service area.

It is useful to be able to visualize a particular orbit, its coverage and the track of its subsatellite point using an appropriate map projection. The reader will appreciate that the apparent shape of the coverage (and track) will depend on the projection used. A non-stationary satellite coverage pattern, on the contrary, is dynamic owing to the movement of the satellite and the Earth's rotation. The Mercator projection is in common use. The longitudes are spaced uniformly by vertical lines and the latitude is represented by horizontal lines placed closer to the equator. Satellicentric projection is a view of the Earth from the satellite and is useful for assessing the satellite antenna pattern shape without the distortion caused by projection (as would happen, for instance, in a Mercator projection). The polar projection, used by radio amateurs, provides a simple way of generating ground plots. In a rectangular projection, longitude and latitude are respectively the X- and Y-axes.

For reliable broadcasting, communication and navigation, the elevation angle from the edge of coverage must be >5–$10°$, which reduces the coverage somewhat. The visible coverage with specified minimum elevation angle is defined by the geometry indicated in Figure 2.3, in which a spherical Earth is assumed with radius R_E, satellite altitude h_s and user altitude h_p.

Nadir and Coverage Angle and Range
The half (cone) angle subtended by the coverage area at the satellite is known as the *nadir angle*. Applying the sine rule to the geometry in Figure 2.3, we obtain for the nadir angle θ for a specified

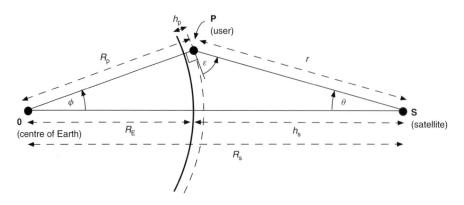

Figure 2.3 Coverage geometry.

elevation angle ϵ

$$\theta = \sin^{-1}\left(\frac{R_p}{R_s}\cos(\epsilon)\right) \tag{2.9}$$

where $R_p = (R_E + h_p)$ is the geocentric radius of the observation point (e.g. the position of the user terminal) and $R_s = (R_E + h_s)$ is the geocentric radius of the satellite.

Similarly, the half-angle subtended by the coverage at the centre of the Earth is known as the *coverage angle* or *Earth central angle*. The coverage angle ϕ is obtained from

$$\phi = \frac{\pi}{2} - \theta - \epsilon \tag{2.10}$$

Finally, applying the cosine rule gives the range from the user to the satellite:

$$r = \sqrt{R_p^2 + R_s^2 - 2R_p R_s \cos(\phi)} \tag{2.11}$$

Earth Coverage Radius and Area
The coverage 'radius' R_C (defined here as the radius from the sub-satellite point to the edge of coverage – measured along the Earth's surface) is given by:

$$R_C = R_E\phi \tag{2.12}$$

For a cone of apex angle 2ϕ the cone solid angle Ω_C is given by

$$\Omega_C = 2\pi(1 - \cos(\phi)) \tag{2.13}$$

Thus, the coverage (surface) *area* may be obtained from the product of the Earth solid angle Ω_C and the Earth radius squared

$$A_C = \Omega_C R_E^2 \tag{2.14}$$

Table 2.1 illustrates the nadir and coverage angles and coverage radius and area for low, medium and geostationary circular orbits. assuming a minimum elevation angle of $\epsilon \to 10°$. The table also gives the fraction of the Earth's surface area covered by the satellite. Note that HAPs are included in the table, in anticipation of the discussion of these platforms later in the chapter.

It is apparent from the table that the fractional coverage varies by more than four orders of magnitude between that of a HAP and a geostationary satellite.

2.2.2.3 Geostationary and Geosynchronous Orbits

At an altitude of $h_s = 35\,786\,000$ m (equivalent to an orbital radius of $R_s = 42\,164\,000$ m), the orbital period exactly equals one *sidereal* day. A sidereal day is the time taken for the Earth to rotate by 360° with reference to an inertial reference frame, and differs from a *solar* day due to the Earth's simultaneous rotation and orbit around the Sun. The length of a sidereal day is 23 h 56 mins 4.1 s. Orbits with periods equal to a sidereal day are called *geosynchronous*, and the subsatellite point returns to the same place at the same time each day. Furthermore, if the orbit is circular and oriented in the equatorial plane, the satellite appears, to an observer on the Earth, to be stationary, located above the equator, making services provided by a single satellite available on a continuous basis, with fixed antennas. This is the *geostationary* orbit, and in practice the term 'geosynchronous' is most commonly used to describe orbits that are very nearly geostationary. The track of the subsatellite point of a geosynchronous satellite in inclined orbit typically describes a figure-of-eight pattern.

The coverage contour of a geostationary satellite is static, and hence design and interpretation of the service area are straightforward. Figure 2.4 illustrates the coverage of the Intelsat IS 801 geostationary satellite. (Note: satellite visibility coverage and orbital tracks in this chapter were produced using the free JSatTrack satellite tracking software available at http:// www.gano.name/ shawn/JSatTrak).

Three geostationary satellites can cover the Earth between latitudes of approximately ±80° (depending on elevation angle); the World Meteorological Organization recommends six satellites for meteorological purposes to ensure full redundancy.

The extraneous forces along the orbit become noticeable at the geostationary orbit when a satellite's inclination deviates from the ideal (i.e. inclination is non-zero), causing the satellite to drift along the orbit. This drift must be compensated for regularly by firing thrusters as a part of east–west station-keeping. Disturbance is also induced by the gravitational forces of the Sun and

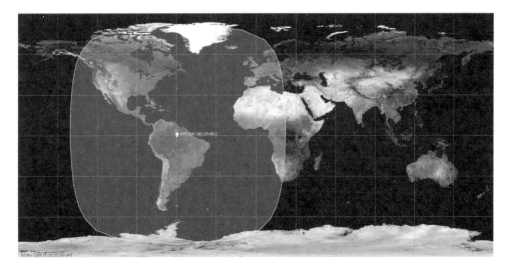

Figure 2.4   Visibility coverage for a geostationary satellite.

Moon, the impact of which is quite noticeable at geostationary orbit. Satellites receive a stronger gravitational pull towards these heavenly bodies when near to them, resulting in a change in orbit inclination. The change due to the Moon varies cyclically (18.6 year period) between ~0.48 and 0.67 deg/year, while the change in inclination due to the Sun is steady at ~0.27 deg/year. The net variation is therefore 0.75–0.94 deg/year over the cycle (Richharia, 1999).

2.2.2.4   Low and Medium Earth Orbits

We have noted that a geostationary satellite provides continuous visibility from the ground with fixed pointing angle and hence is preferred for a majority of communication and broadcast systems. However, under some conditions, non-geostationary systems are preferred. Geostationary satellite systems exhibit an inherent latency (~250 ms) due to the propagation time of electromagnetic waves for this range. Large deployable antennas and powerful on-board transmitters are typically necessary to provide high-quality services to mobile and handheld user terminals. To minimize these limitations, several mobile service operators prefer to use non-geostationary orbits. For global navigation systems, non-geostationary constellations are preferred because they can provide multiple visibility of satellites from any location on the Earth. In such applications it becomes necessary to use a constellation of satellites in order to maintain real-time service continuity. Design and optimization of such constellations are complex, involving cost-benefit analysis, coverage optimization, constellation launching and maintenance considerations, etc. (Richharia, 1999).

Purely from coverage considerations, the equatorial, inclined and polar orbits favour the equatorial, mid-latitude and polar regions respectively. Orbit eccentricity determines the dwell time over a specific region, with the region visible from apogee exhibiting the maximum dwell time.

Global Constellation

Both path delay and coverage area decrease with decreasing altitude. Consequently, lowering orbital altitudes increases the number of satellites needed in a global constellation. Various authors have provided quantitative relationships for continuous single or multiple visibility for global coverage. Polar constellations consist of planes intersecting at the equator, thereby favouring coverage to high-latitude regions (e.g. Adams and Rider, 1987). Others have provided solutions of a more uniformly spread constellation (e.g. Ballard, 1980). It is relatively common to describe such constellations in terms of the orbital patterns described by Walker (1984). *Walker Star* constellations utilize near-polar circular orbits to provide 'streets of coverage' (Adams and Rider, 1987) and gain their name from the star pattern formed by the orbital tracks at the poles. A notable feature of Walker Star constellations is that there exist one or more 'seems' between adjacent planes of satellites with ascending and descending nodes. *Walker Delta* constellations use less inclined orbits forming a pattern, also known as a *Ballard Rosette* (Ballard, 1980). Both constellations may be described in Walker's notation by

$$i_k : N_s/N_p/p$$

where i_k is the inclination, N_s is the total number of satellites in the constellation, N_p is the number of planes (RAAN values) and p is a parameter that describes the phasing between planes – the interpretation of which differs somewhat for Walker Star and Delta constellations.

A lower bound on the minimum number of satellites for total coverage of the Earth's surface for non-stationary orbits is given by Beste in terms of the coverage angle (Beste, 1978)

$$N_s \geq \frac{4}{1 - \cos(\phi)} \tag{2.15}$$

$$\rightarrow 4 \left(\frac{R_E}{h_s} + 1 \right) \text{ (for } \epsilon = 0) \tag{2.16}$$

where h_s is the satellite altitude. This assumes an optimum non-stationary constellation geometry. By way of example, for an LEO constellation with an altitude of 800 km, the minimum number of satellites needed to ensure that one satellite is always visible is approximately 36. Higher numbers of satellites will be needed for particular constellation types and to provide the coverage overlap needed to ensure effective handover, and for applications that require multiple satellites to be in view at once – such as satellite navigation.

A major constraint of an LEO or MEO constellation with regard to communication arises owing to the perpetual motion of the satellite – and satellite footprint. Therefore, satellites have to be tracked from the ground for those service links requiring a directional antenna, and the service handed over from one satellite to the next.

Notable examples of LEO constellation systems include Iridium and Globalstar. Iridium comprises 66 satellites in circular orbit distributed in six planes of 86° inclination at an altitude of 780 km. Globalstar consists of 48 satellites distributed in eight circular orbit planes each inclined at 52° at 1414 km altitude. Figure 2.5 illustrates the differences in footprint and orbital tracks for a satellite from each of these constellations (note that tracks for three orbits are shown for each satellite), while Figure 2.6 illustrates the difference between the Iridium and Globalstar orbital patterns.

The GPS navigation satellite system is an example of an MEO constellation. It comprises 24 satellites distributed in six near-circular orbit planes at an altitude close to 26 559.7 km and an inclination of 55°.

2.2.2.5 Highly Inclined Orbits

It was previously stated that the rate of change in the argument of perigee owing to the non-spherical Earth gravity potential is zero at an inclination of 63.4°. This desirable attribute, together with an elliptical orbit, may be used to provide quasi-stationary satellite services to high-latitude regions, where the GEO arc appears at a very low elevation angle ($<$ 10 deg), by operating satellites in the apogee region of a highly elliptical orbit. A satellite slows in the apogee region because velocity is inversely proportional to the altitude – thus it remains visible for several hours over high-latitude regions. Numerous variants of this class of orbit are in use, such as Molniya and Tundra. Molniya orbits have a period of half a sidereal day, while Tundra orbits have a period of one sidereal day

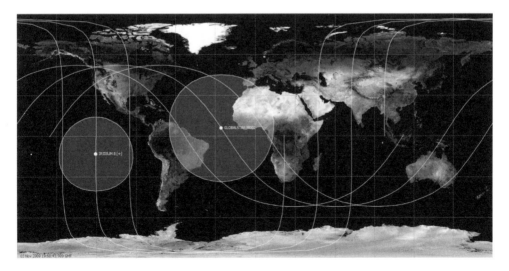

Figure 2.5 Snapshot of coverage and orbital tracks for Iridium and Globalstar LEO satellites.

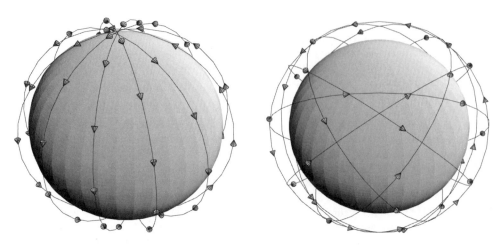

Figure 2.6 Iridium Walker Star Constellation (left) and Globalstar Walker Delta (right). Arrowheads indicate the instantaneous position and direction of motion of satellites.

(and are therefore geosynchronous). For instance, such orbits are used for communications and broadcasting in Canada and Russia which have significant land masses at high latitudes.

Examples of systems in this type of orbit are the Sirius radio broadcast system serving the US and Canadian regions and the Molniya satellite system which services the Russian Orbita Television network in Russia. Figure 2.7 depicts the satellite ground track (locus of the subsatellite points) for three Sirius RadioSat satellites located in highly elliptical orbits at an inclination of 63.4° (Tundra orbit). The constellation of three satellites, phased 120° apart, provides seamless broadcasting to the North American region (and beyond). Notice that the satellites appear virtually overhead in these northerly regions, whereas geostationary satellites would appear close to or below the horizon.

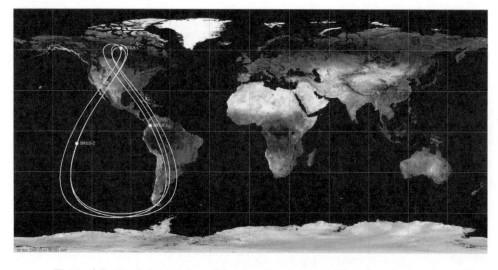

Figure 2.7 Footprints of three Sirius satellites in tundra highly elliptical orbit.

2.2.2.6   Sun-Synchronous Orbits

The Earth moves around the Sun by 0.986° per day, and, if the satellite orbital parameters are adjusted such that the orbital precession synchronizes to the Earth's motion, then the orbit retains the same relationship to the Sun throughout the year. This class of orbit is known as the Sun-synchronous orbit. Satellites in this orbit view each point on the Earth at about the same time each day and hence at similar illumination levels. This property is utilized for remote sensing, where it is useful to view each scene repeatedly at a similar illumination level to assess a change – for instance, to judge the progress of a crop or urban growth.

Table 2.2 gives orbital details for the Landsat-7 Sun-synchronous satellite orbit (NASA, 2009). The Landsat satellite series has provided extensive Earth-observing land resource data since the early 1970s. Its near-polar Sun-synchronous orbit has 705 km nominal altitude at the equator and a period of nearly 99 min. The spacecraft crosses the equator from north to south at around 10:00 a.m. on each pass and covers the region between 81° north and south latitude every 16 days.

2.2.2.7   Utilization and Orbit Selection

Utilization

From an applications perspective, the duration of a pass for a non-geostationary satellite depends on the elevation angle above which the satellites can provide a useful service. For instance, communications from a satellite phone would be reliable at an elevation angle of $> \sim 10\text{--}15°$. Thus, the duration of a pass is of the order of tens of minutes for an LEO satellite system, extending to a few hours for an MEO system. By contrast, satellites in GEO orbit remain stationary when viewed from the Earth because the velocity of a satellite in GEO synchronizes to that of the Earth. GEO systems are in widespread use owing to numerous desirable features, such as: simpler ground segment transceivers, easier radio link design and management, easily predictable inter/intrasystem interference, simpler operation, etc. Table 2.3 summarizes typical orbits used for different services.

Orbit Selection

Orbit selection for any new type of system will typically involve a comprehensive study of the trade-offs between various orbit types with regard to coverage, performance (e.g. capacity, information rate, latency, scene illumination, etc.), availability, reliability and cost. By way of illustration, consider the rationale in selection of a constellation for the European navigation system known as Galileo (Benedicto *et al.*, 2000). The system had to provide global coverage and be compatible with existing satellite navigation systems (GPS and GLONASS), which operate in medium earth orbit. Thus, suitability of MEO has been established in terms of multiple satellite visibility, visibility duration, resistance to radiation, constellation deployment and management, etc. MEO was therefore selected because of its proven track record and the necessity for compatibility with existing systems. Considering the wide area coverage and robustness of GEO technology, a *hybrid*

Table 2.2   Parameters for NASA Landsat-7 Sun-synchronous orbit

Parameter	Value
Nominal altitude	705 km
Inclination	98.2°
Orbital period	98.8 min
Mean solar time at equator	9.45 a.m.

Table 2.3 Commonly used orbits for each service

System type	Commonly used orbit	Comments
Fixed telecommunication	GEO	
Mobile communication	GEO, LEO and MEO	Polar or inclined LEO or MEO constellations necessary for truly global coverage
Television broadcasting	GEO	
Radio broadcasting	GEO, highly elliptical (fixed perigee)	Generally, three-satellite constellation necessary for highly elliptical orbit
Navigation	LEO and MEO	Constellation necessary to ensure global coverage
Remote sensing	LEO (polar and Sun synchronous), GEO	A single satellite in near-polar orbit can revisit each point regularly
Amateur radio	LEO	

MEO–geostationary combination was another option. Availability and time to replace a failed satellite were chosen as the primary evaluation criteria. A GEO system failure would affect large parts of the service area, whereas an MEO failure would be less catastrophic as the service gaps would be time and space variant – causing outages of tens of minutes until a replacement could be activated. In the final reckoning, a 27-satellite MEO constellation with three back-up satellites was selected, the final constellation utilizing an MEO orbit with an altitude of 23 222 km, with three orbital planes (right ascension), each inclined at 56° and populated with equally spaced operational satellites, together with in-orbit spares.

2.2.2.8 Eclipse and Conjunction

Eclipse
Spacecrafts generate electricity from solar cells backed up by rechargeable batteries to cover periods when the satellite is shadowed by the Earth. For geostationary satellites, eclipses occur for about ±22 days around the spring and autumn equinox, when the Sun, equatorial plane and Earth lie in the same plane. Full eclipse occurs for about 69.4 min on the day of the equinox (21 June and 21 September) at local midnight, while partial eclipse starts and ends 22 days before and after (Richharia, 2001).

For other orbits, the eclipse can occur at any time depending on the Sun–satellite vector in relationship to the Earth. LEO satellites could have several thousands of eclipses per year. A satellite in 780 km equatorial orbit is shadowed for 35% of the orbit. The eclipses may occur up to 14 times a day during the equinox, and a satellite could be shadowed for upto 8 h.

In an MEO system, under similar conditions the maximum duration of an eclipse would occur for 12.5% of the orbital period; eclipses would occur 4 times a day, and the total duration would be about 3 h.

A satellite can also be eclipsed by the Moon. The occurrence of such eclipses is variable owing to the time variance in the satellite–Moon vector in relationship to the Earth.

Conjunction
In situations when the Sun appears behind a satellite, a received satellite radio signal may be corrupted by intense thermal noise emanating from the Sun. The occurrence of such conjunctions

depends on the receiver location and the satellite's orbital characteristics. For geostationary satellites, Sun interference can occur within ± 22 days around the equinox. Fortunately, for small terminal antennas used in personal systems, the impact is usually relatively insignificant because the receiver's inherent noise in itself is much higher than the noise power collected by the antenna from the Sun.

2.2.3 Position in Different Reference Frames

The orbit of a satellite is described by the six Keplerian elements in terms of the orbital plane, and its relation to an inertial frame (i.e. one that does not spin with the Earth). However, we need to be able to convert the satellite position to other reference frames of more utility to a user located at a fixed position on the Earth (and spinning with it).

If the satellite position is given in the orbital plane by radius r and true anomaly v_K, the Cartesian coordinate vector s_{ORF} for the satellite in the *orbital reference frame*, that is, with the z-axis normal to the orbital plane, the x-axis passing through the perigee and the y-axis forming a right-handed set, is

$$s_{ORF} = \begin{bmatrix} r\cos(v_K) \\ r\sin(v_K) \\ 0 \end{bmatrix} \qquad (2.17)$$

Rotation Matrix Transformation

The conversion between different reference frames generally requires rotation (and occasional translation) of a coordinate system. A rotation of the reference frame about the x-, y- or z-axis by angle α is described by multiplying the vector by the matrices $\mathbf{M}_x$, $\mathbf{M}_y$, or $\mathbf{M}_z$ respectively[1]

$$\mathbf{M}_x(\alpha) = \begin{bmatrix} 1 & 0 & 0 \\ 0 & \cos(\alpha) & \sin(\alpha) \\ 0 & -\sin(\alpha) & \cos(\alpha) \end{bmatrix} \qquad (2.18)$$

$$\mathbf{M}_y(\alpha) = \begin{bmatrix} \cos(\alpha) & 0 & -\sin(\alpha) \\ 0 & 1 & 0 \\ \sin(\alpha) & 0 & \cos(\alpha) \end{bmatrix} \qquad (2.19)$$

$$\mathbf{M}_z(\alpha) = \begin{bmatrix} \cos(\alpha) & \sin(\alpha) & 0 \\ -\sin(\alpha) & \cos(\alpha) & 0 \\ 0 & 0 & 1 \end{bmatrix} \qquad (2.20)$$

Therefore, coordinate transformations between different reference frames may be effected by multiplying the coordinates by the appropriate sequence of matrix rotations. Conversely, the inverse transform is obtained with the inverse sequence (with opposite angles).

2.2.3.1 Earth-Centred Inertial and Earth-Centred Earth-Fixed Reference Frames

Earth-Centred Inertial

Three of the Keplerian elements describe the orientation of the orbital plane with respect to an intertial reference frame (i.e. one that points to the same points in space). It is convenient to use the Earth-Centred Intertial (ECI) reference frame for this purpose, in which the z-axis is aligned with the axis of rotation of the Earth (polar axis), the x-axis points to the vernal equinox (in the equatorial plane) and the y-axis is chosen to obtain a right-handed set (also in the equatorial plane).

---

[1] These matrices effect a clockwise rotation around the appropriate axis as seen from the origin. Note that the rotation matrices for the coordinate axes and for vectors are of opposite sense.

Hence, with reference to Figure 2.1, in order to convert from orbital plane to ECI, one must rotate around z by the negative angle of perigee, rotate about x by the negative inclination and rotate about z by the negative right ascension. Hence, the satellite coordinates in ECI, s_{ECI}, are obtained from

$$s_{ECI} = \left[\mathbf{M}_z \left(-\Omega_K \right) \mathbf{M}_x \left(-i_K \right) \mathbf{M}_z \left(-\omega_K \right) \right] s_{ORF} \tag{2.21}$$

Earth-Centred Earth-Fixed

It is often desirable to evaluate the satellite position in an Earth-Centred Earth-Fixed (ECEF) reference frame, which rotates with the Earth. This may permit the conversion to and from geocentric latitude, longitude and altitude. In ECEF, the z-axis is again aligned with the polar axis (as in ECI), but the x-axis is aligned with the prime meridian (zero longitude), and again the y-axis is chosen to form a right-handed set. To convert from ECI to ECEF, one needs only to rotate about the z-axis by the angle Ω_M appropriate to the sidereal time difference between the vernal equinox and sidereal time at the prime meridian. The satellite position in ECEF, s_{ECEF}, is thus

$$s_{ECEF} = \mathbf{M}_z \left(\Omega_M \right) s_{ECI} \tag{2.22}$$

To convert coordinates back from ECEF to ECI, one just applies the inverse trasform (negative angle).

2.2.3.2 Geodetic Latitude and Altitude

Thus far, in considering coordinate transformations, we have assumed that the Earth is a perfect sphere. In fact, the Earth radius varies with latitude, bulging at the equator. For this reason, reference is commonly made to a geodetic model or *reference ellipsoid* – an oblate (flattened) spheroid. Such a geodetic model is essential for accuracy in navigation applications.

WGS 84

By far the most common geodetic model in current use is that used by the GPS, the World Geodetic System 1984 (WGS84), the parameters of which are given in Table 2.4. The WGS-84 reference spheroid radius varies from 6 378 137 m at the equator (semi-major axis) to 6 356 752 m at the poles (semi-minor axis). The WGS84 prime meridian is that of the International Earth Rotation and Reference Systems Service (IERS). However, as a result of its historical relation to the former US TRANSIT satellite navigation system, the IERS meridian is actually offset slightly from the Greenwich prime meridian – by a few arc seconds (about 100 m). Note that the WGS-84 (and GPS) altitude does not exactly correspond to that which would be measured from local mean sea level – the difference being due to local variations in the Earth's *geoid* (the notional equipotenial surface where force of gravity is everywhere perpendicular).

Table 2.4   WGS reference ellipsoid parameters (ICAO, 1998)

Parameter	Value
Semi-major (equatorial) axis a_G	6 378 137 m
Inverse flattening ($\frac{1}{f_G}$)	298.257223563
Geocentric gravitational constant μ_G	398 600.5 km^3/s^2

Key derived parameters for the reference ellipsoid include the semi-minor (polar) axis b_G and the first and second elipsoid eccentricities e_G and e'_G respectively

$$b_G = a_G (1 - f_G) \tag{2.23}$$

$$e_G = \frac{a_G^2 - b_G^2}{a_G^2} \tag{2.24}$$

$$e'_G = \frac{a_G^2 - b_G^2}{b_G^2} \tag{2.25}$$

With reference to Figure 2.8, the user's position in the ECEF reference frame $\mathbf{p}_{\text{ECEF}}$ may be obtained from geodetic latitude Θ and altitude h via (Farrell and Barth, 1998)

$$\mathbf{p}_{\text{ECEF}} = \begin{bmatrix} (R_N + h)\cos(\Theta)\cos(\Phi) \\ (R_N + h)\cos(\Theta)\sin(\Phi) \\ \left((1 - e_G^2) R_N + h\right)\sin(\Theta) \end{bmatrix} \tag{2.26}$$

where R_N is the length of the radius of curvature normal to the point on the geodetic surface, given by

$$R_N = \frac{a_G}{\sqrt{1 - e_G^2 \sin^2(\Theta)}} \tag{2.27}$$

Note that, except at the poles and equator, the normal does not pass through the centre of the Earth.

In general, an iterative numerical solution is required, used to convert from geocentric vector coordinates to geodetic latitude, longitude and altitude (Kaplan and Hegarty, 2006; Kelso, 1996). However, Bowring (1976) gives the following approximation which is highly accurate for most practical altitudes:

$$\Theta \simeq \tan^{-1}\left(\frac{z + b_G e'^2_G \sin^3(\Theta')}{\sqrt{x^2 + y^2} - a_G e_G^2 \cos^3(\Theta')} \right) \tag{2.28}$$

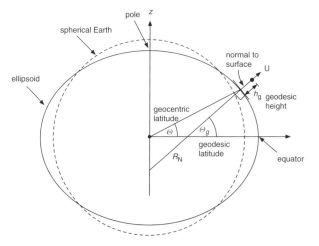

Figure 2.8 Geodetic latitude and normal.

where Θ' is the reduced (or parametric) latitude given by

$$\Theta' = \tan^{-1}\left(\frac{a_G}{b_G}\frac{z}{\sqrt{x^2 + y^2}}\right) \tag{2.29}$$

where $\{x, y, z\}$ are the ECEF vector coordinates.

The geodetic longitude is the same as the geocentric longitude

$$\Phi = \tan^{-1}\left(\frac{y}{x}\right) \tag{2.30}$$

where care must be taken when evaluating the arctan function in order to ensure that the angle is given in the correct quadrant.

The geodetic altitude is given (except at the poles) by

$$h = \frac{\sqrt{x^2 + y^2}}{\cos(\Theta)} - R_N \tag{2.31}$$

The maximum difference between geodetic and geocentric latitude is less than $0.2°$. Although this sounds quite small, it equates to a maximum positional error of about 22 km. Therefore, for satellite *navigation* applications, it is essential to use an appropriate geodetic model of the Earth.

2.2.3.3 Topocentric Horizon

The direction of the satellite at the user's position is of significant interest with regard to finding the satellite and pointing an antenna at it. The direction is normally given in terms of the *azimuth* and *elevation* angles in a topocentric horizon system. Azimuth is normally specified relative to true North. Elevation angle is normally specified relative to the local horizontal.

In the topocentric horizon reference frame, the z-axis is aligned to be normal to the Earth's surface (zenith), the x-axis is generally oriented to be along the horizontal towards the South and the y-axis is aligned along the horizontal towards the East (forming a right-handed set) as illustrated in Figure 2.9 (Kelso, 1995).

To effect the necessary transformation, first we require the vector from the user's position to the satellite given by the difference in the two position vectors. To convert from ECEF to topocentric horizon, we rotate this vector about the z (polar) axis by the user longitude Φ_p and then about the y-axis by $90°$ minus the user geocentric latitude Θ_p

$$\mathbf{ps}_{TH} = \left[\mathbf{M}_y\left(\frac{\pi}{2} - \Theta_p\right)\mathbf{M}_z\left(\Phi_p\right)\right](\mathbf{s}_{ECI} - \mathbf{p}_{ECI}) \tag{2.32}$$

Finally, the satellite range r, azimuth α and elevation ϵ may be obtained as follows:

$$r = \sqrt{x^2 + y^2 + z^2} \tag{2.33}$$

$$\alpha = \tan^{-1}\left(\frac{-y}{x}\right) \tag{2.34}$$

$$\epsilon = \sin^{-1}\left(\frac{z}{r}\right) \tag{2.35}$$

where x, y and z are the south, east and up vector components, respectively, for the topocentric horizon range vector $\mathbf{ps}_{TH}$.

Note that the azimuth angle is specified relative to true North (hence the minus sign). Again, care must be taken when evaluating the arctan function in order to ensure that the azimuth angle is given in the correct quadrant.

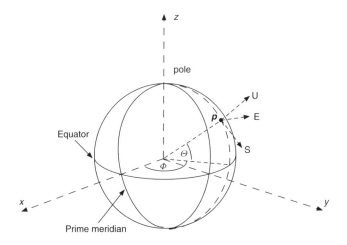

Figure 2.9   Topocentric horizon reference frame.

Geostationary Case

For the important case of a geostationary orbit, the coverage angle is given by

$$\phi = \cos^{-1}\left(\cos\left(\Phi_p - \Phi_s\right)\cos\left(\Theta_p\right)\right) \tag{2.36}$$

If we assume a spherical Earth, the resultant expressions for the satellite elevation and azimuth angles simplify to (Richharia, 1999)

$$\epsilon \rightarrow \tan^{-1}\left(\frac{\cos\left(\phi\right) - \frac{R_p}{R_s}}{\sin\left(\phi\right)}\right) \tag{2.37}$$

$$\alpha \rightarrow \tan^{-1}\left(\frac{\tan\left(\Phi_p - \Phi_s\right)}{\sin\left(\Theta_p\right)}\right) \tag{2.38}$$

where, again, care must be taken to ensure that the azimuth angle is in the correct quadrant.

SES-ASTRA example. Consider by way of example, the satellite look angles for a user terminal in London, UK (latitude 51.5° N, longitude 0° E) to the ASTRA satellites with the subsatellite point at 19.2° E. Using equations (2.37) and (2.38), the elevation and azimuth angles for these satellites are 28.4° and 156.0° respectively (the subsatellite point is south-east of London).

2.2.4   Satellite Bus

Having examined satellite orbits and their coverage, we now briefly turn our attention to the satellites themselves. The primary aim of a satellite is to provide a platform with a specified performance throughout the design lifetime of the system, and preferably exceeding it. The specific design and capabilities will clearly depend on the particular mission goal – telecommunication, broadcast, navigation, remote sensing, etc. Influencing factors are: services (communication, remote sensing, etc.), user terminals (size, cost), spacecraft capacity (power, bandwidth), service region, orbital characteristics, radio frequency, space environment, state of technology, manufacturers' infrastructure and expertise, etc.

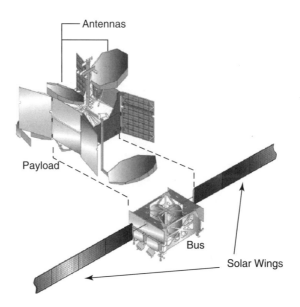

Figure 2.10   Boeing 702 geostationary satellite bus. Reproduced by permission of © Boeing.

A satellite comprises a support subsystem known as a *bus* and a subsystem responsible for pro-viding the core functionality, known as the *payload*. Figure 2.10 illustrates the general relationship between bus and payload on a Boeing 702 geostationary satellite, which employs a standard bus with a payload tailored to the specific application. Meanwhile, Figure 2.11 shows the relationship between bus and payload for smaller Iridium LEO satellites, which have been mass produced for one specific application. The bus provides support to the primary functions of the satellite. These support func-tions tend to have considerable commonality between mission types, whereas the payload architec-ture is heavily influenced by the applications (e.g. telecommunication, broadcast, navigation, etc.).

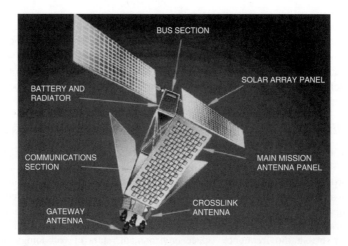

Figure 2.11   Iridium satellite (Schuss *et al.*, 1999). Reproduced by permission of © 1999 IEEE.

The bus comprises a variety of subsystems:

- the satellite structure itself, which provides mechanical support throughout the life of a satellite, beginning from satellite launch;
- the attitude and control subsystem (AOCS), which stabilizes the spacecraft in the presence of numerous disturbances, and controls satellite orbit;
- the propulsion system, which provides the necessary velocity and torque to the AOCS;
- the telemetry, tracking and command (TT&C) subsystem, which monitors and sends the status of various spacecraft subsystems and accepts commands from the ground to perform the desired manoeuvres, and also assists in ground station tracking and orbit determination aspects of a satetellite;
- the electrical power supply subsystem, which provides power for functioning of a satellite.

Because of the secondary role of the satellite bus in our general discussion of satellite applications, and keeping to the intended scope of this book, we will focus in subsequent chapters almost exclusively on the salient features of the payload. Nevertheless, the interested reader may want to consult the wide-ranging literature on the subject of satellite technology for more information on bus design and also satellite launch systems (Richharia, 1999; Maral and Bousquet, 2002).

2.3   High-Altitude Platforms

2.3.1   Overview

Thus far we have considered only satellites located well beyond the atmosphere, and that rely on *momentum* to sustain their orbits, as platforms, for carrying payloads suited to services supporting personal applications. There is, however, growing interest in the use of aerial platforms located at much lower altitudes – within the atmosphere – and that are sustained by aerodynamics and hydrostatics. These are generally referred to as *High-Altitude Platforms* (HAPs). In effect, HAPs provide a low-altitude quasi-geostationary orbit (albeit one with significantly reduced coverage) and thus HAPs combine the low latency and small coverage of a LEO with the simplified antenna pointing of a GEO. HAPs have been proposed for continuous services, but a particular advantage of this type of platform is the speed with which it may be launched – for example, to allow rapid introduction of a new service or, alternatively, to provide a temporary service for disaster recovery.

HAPs are based on a familiar concept (i.e. aerial platforms), but their practical use will require much greater endurance. Indeed, HAPs are variously referred to as High-Altitude Long-Endurance (HALE) and even High-Altitude Long-Operation (HALO) platforms. Chapter 11 explores HAP network architecture, benefits and applications, while Chapter 12 introduces examples of representative research programmes and commercial initiatives. In this section, we focus on the HAP platform itself as an alternative or adjunct to satellites.

Several factors have given impetus to the development of HAPs. These include: Unmanned Aerial Vehicle (UAV) technology evolution driven by military applications; a renaissance in large balloon construction; advances in composite materials technology and in low-speed, high-altitude aerodynamics and propulsion systems; and growing commercial interests in alternative methods of delivering mass-market services (Tozer and Grace, 2001a; Tozer and Grace, 2001b).

Optimum Altitude

There is currently an undefined region above controlled airspace (between about 20 km and near space) that is attractive for use by HAPs. Indeed, the ITU-R now considers the altitude range 20–50 km as the region of HAP operation when dealing with spectrum assignment. A significant factor with regard to selection of operating altitude is (peak) wind velocity. Consider the instantaneous power, P, necessary to maintain stability of an aerial platform. This can be modelled as (Djuknic, Freidenfelds and Okunev, 1997)

$$P = \frac{1}{2}\rho\, C_d S_c v^3 \tag{2.39}$$

where ρ is the air density, C_d is the aerodynamic drag coefficient, S_c is the airship cross-sectional area and v is the instantaneous wind velocity.

It is apparent that the instantaneous wind velocity has a significant influence on the power needed to maintain an HAP on station for long periods with high availability, and wind gusts can exert considerable load on the propulsion system. The stratospheric region in the altitude range $\sim$17–22 km appears to exhibit low wind turbulence and reduced wind velocity by comparison with lower (or higher) altitudes, and is reasonably accessible. Depending on latitude, the long-term average wind velocity at this altitude ranges between 10 and 40 m/s (with seasonal variations).

The stratospheric region is clearly attractive for future deployment of HAPs, although lower altitudes are attractive for platforms with reduced endurance, which for practical reasons, cannot operate at this altitude. It was previously established that coverage area depends on altitude, and, according to Table 2.3, for an altitude of 20 km, an HAP potentially offers a maximum visibility radius of around 100 km for a 10° elevation angle, decreasing to 75 km for 15°. To put this into perspective, Figure 2.12 illustrates how a constellation of six stratospheric HAPs might be deployed to cover an area the size of the United Kingdom (Tozer and Grace, 2001a).

Types of HAP

There are essentially two main classes of HAP:

- *Heavier-than-Air (HTA) platforms.* This class includes manned fixed-wing aircraft and unmanned aerial vehicles (UAVs), including solar-powered 'gliders'. HTA platforms rely on propulsion and aerodynamics to maintain their altitude – generally flying in a controlled circular path;

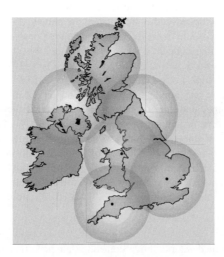

Figure 2.12 Proposed coverage of the United Kingdom from a constellation of six HAPs, generally allowing line-of-sight visibility (Tozer and Grace, 2001a). Reproduced with permission of © IET.

- *Lighter-than-Air (LTA) platforms.* This class includes tethered aerostats and unmanned balloons. LTA platforms are effectively density neutral, floating at the required altitude, with propulsion used only to maintain their position (and for transit to and from the ground station), if required.

Each class (and subclass) clearly has a different requirement for ground facilities and air crew, with larger platforms requiring a normal landing strip and a rotating crew.

2.3.2   Heavier-than-Air Platforms

2.3.2.1   Manned HTA Platforms

In the heavier-than-air category, stratospheric *manned aircraft* have found favour in some commercial endeavours owing to the relatively low technology risk of suitable aircraft (Global HAPS Network, 2009). An operational aircraft-based HAP system would potentially comprise a succession of aircraft flying over the service area in a small circle ($\sim$4–6 km diameter) for a reasonable duration ($\sim$6–8 h), with each handing over the operation to the next aircraft before descending to refuel and change pilots.

Figure 2.13 shows an example of a Proteus piloted fixed-wing stratospheric aircraft developed by Scaled Composites, a version of which was proposed for use by Angel Technologies for 'super-metropolitan area' broadband communication services (Angel Technologies, 2009). The intention is to deploy HALO-based networks on a city-by-city basis, around the world. The HALO Proteus aircraft can operate in the altitude range $\sim$15.5–18.3 km, with the aircraft utilizing airports anywhere within a 300 mile radius of the city.

2.3.2.2   Unmanned Aerial Vehicles

There are a wide range of Unmanned Aerial Vehicles (UAVs) currently under development around the globe, with the main drivers being for use in military applications. UAV solutions are used in

Figure 2.13   A high-altitude long-duration Proteus piloted fixed-wing aircraft. Reproduced courtesy of NASA.

surveillance, remote sensing and other military applications – mostly in short-duration low-altitude flights. An exception is the Northrop Grumman RQ4 Global Hawk, with its wing span in excess of 35 m and a payload of more than 1000 kg – to altitudes of 20 km. At the time of writing, a NASA Global Hawk holds the official world UAV endurance record of 35 h.

Although the military have been rapid adopters of UAV technology, their use for civilian applications has been relatively slow to emerge, in part owing to their military heritage, and in part owing to residual concerns over the safety of flying these machines within controlled airspace. The capabilities of most of these systems in terms of size, power and altitude are generally insufficient to support viable large commercial programmes.

2.3.2.3   Solar-Powered Heavier-than-Air Platforms

For high-availability services, the need to land and refuel (and transit to and from the operating zone) is a significant constraint on the endurance of HTA platforms. Multiple platforms are then needed to sustain a continuous service. Ultralightweight aircraft using solar power have been developed, thereby avoiding the need to land and refuel. Day operation of the aircraft is supported by solar power generated by amorphous silicon cells that cover the aircraft's wings, while rechargeable lithium–sulphur batteries provide power at night (although storage batteries essential for night flying increase an aircraft's weight and thereby claw away the payload weight).

In July 2008, one such platform, the QinetiQ Zephyr HALE aircraft, achieved an unofficial UAV endurance record of about 3.5 days, at an altitude of about 18.3 km (QinetiQ Online, 2009). Zephyr weighs 30 kg and has a wingspan of up to 18 m. At the present time, the maximum payload mass for such solar-powered platforms is extremely restricted. Nevertheless, the US Defense Advanced Research Projects Agency (DARPA) is funding the development of an advanced solar-powered HALE platform – known as 'Vulture'. This platform has a target endurance of 5 years (with 99% availability), carrying a 1000 lb, 5 kW payload, with an interim platform offering 12 months endurance planned by 2012.

2.3.3   Lighter-than-Air Platforms

2.3.3.1   Tethered Aerostats

A tethered aerostat HAP is essentially a lighter-than-air balloon tethered to the ground. Although the tether ultimately limits the maximum altitude that can be attained, its significant advantage is that it may be used as a means of feeding DC power and communications to the aerostat from the ground station. Tethered systems can be set up rapidly and hence are particularly suitable for applications where a rapid response is desirable. A tether can reach altitudes of up to ~5 km. However, it is a potential hazard to the controlled aerospace and hence is better suited to remote areas where there is little air traffic or where the airspace is authorized for this type of system, say in a disaster zone.

One of the larger commercially available aerostats is shown in Figure 2.14. Table 2.5 contains additional details of this type of aerostat, made by TCOM. An operational system from Platform Wireless International uses a 1–5 km tether, fixed to either a permanent or a mobile base station capable of servicing up to 80 000 users in an area that is claimed can extend to 225 km in diameter – that is, extending over the horizon (Platform Wireless International, 2009). Tethered aerostats are in operational service with the US Air Force Tethered Aerostat Radar System (TARS), which is used to detect illegal drug trafficking along the US border with Mexico and in Florida. The TARS aerostats located at Cudjoe Key, Florida, are also used to broadcast US 'news and current affairs' TV to Cuba (known as TV 'Marti').

The limited maximum altitude for this type of platform means that it is affected by bad weather – indeed, the Cudjoe Key aerostat has twice broken away from its mooring in bad weather.

Figure 2.14   Large tethered aerostat with payload radome. Reproduced by permission of © TCOM.

Table 2.5   Key characteristics for the TCOM 71M tethered aerostat

Characteristic	Value
Length	71 m
Maximum altitude	4.8 km
Payload mass	1600 kg
Payload power	22 kW
Endurance	30 days
Wind speed (operational)	70 knots

Because of the tether, the airspace around these aerostats has to be restricted. The endurance of aerostats is ultimately limited by the loss of helium. Consequently, two aerostats are typically used, in alternation.

2.3.3.2   Unmanned Airships

Airships are powered, untethered balloons with a means of propulsion and steering. These helium-filled platforms – typically 100–200 m in length and ∼50 m in diameter – can operate at higher altitude and over greater distances than aerostats – ideally above the bulk of the weather and strong winds. The current endurance record for an unmanned balloon of 42 days is held by the NASA Cosmic Ray Energetics And Mass (CREAM) scientific balloon – used to circle the South Pole.

Because of their large size, airships can potentially carry a payload of ∼1000 kg. One High-Altitude Long-Endurance (HALE) platform designed under ESA's sponsorship was specified as 220 m in length and 55 m in diameter to carry payloads of up to about 1000 kg (ESA, 2009). Payload power must be generated on-board, and therefore, in order to achieve long endurance, solar cells are usually installed on the upper surface facing the Sun. Solar cells supplemented with a rechargeable battery for operation in the absence of Sunlight provide both the primary and payload power.

Figure 2.15 shows an artist's impression of a High-Altitude Airship (HAA) prototype currently being developed by Lockheed Martin to facilitate the hosting of a variety of applications – surveillance, telecommunications, weather observation, etc. Power will be generated by an array

Figure 2.15 An artist's impression of a High-Altitude Airship (HAA) being developed by Lockheed Martin, showing the solar cells on the upper surface. Reproduced by permission of © Lockheed Martin Corp.

of solar cells on the balloon upper surface. The performance goals include sustained operation for at least 2 weeks at about 18 km (60 000 ft) altitude and 500 W payload power generated by thin-film solar cells capable of yielding about 15 kW supported by rechargeable lithium ion–polymer batteries of 40 kW h capacity for night operation. The payload weight is targeted at about 22.7 kg (50 lb). Lockheed Martin is currently building a scaled-down version of the HAA, called the High-Altitude Long-Endurance Demonstrator (HAL-D), and the specified characteristics of the HALE-D are given in Table 2.6. The prototype will demonstrate long-endurance station-keeping and flight control capabilities (Lockheed Martin, 2009). Another major HAP airship project, known as SkyNet, was sponsored by the Japanese government and led by Yokosuka Communications Research Laboratory and aims to produce an integrated network of about 15 airships to provide broadband communication and broadcasting services in Japan.

2.3.3.3 Bus

Conceptually, HAP payloads have much in common with satellite payloads. However, payload power and weight are critical issues in selection of an HAP system. Aircraft systems can utilize

Table 2.6 Key characteristics of the Lockheed Martin HALE-D airship

Characteristic	Value
Length	73 m
Maximum altitude	20 km
Payload mass	23 kg
Payload power	500 W
Endurance	15 days

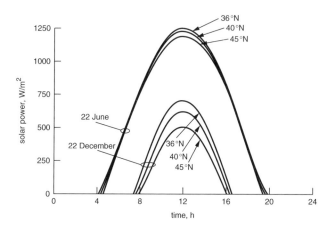

Figure 2.16 Seasonal and daily variations in solar power in mid-latitude regions of the earth (Tozer and Grace, 2001a). Reproduced by permission of © IET.

conventional technology (e.g. a generator), while airships, owing to their large surface area, can potentially generate significant solar power (up to ~20 kW) but must be backed up by rechargeable batteries for night operation. Tethered systems are easier to manage because they are fed from the ground. Solar-powered HTA systems have the stringent task of supporting the flight as well as the payload even in the absence of Sunlight. Solar-powered systems must contend, in addition to a nightly back-up system, with the variation in the solar flux density throughout the year.

Figure 2.16 demonstrates that the variation in the flux density between winter and summer months from mid-latitude regions when the Sun is respectively the nearest (June) and the furthest from the Earth (December) amounts to almost 60% reduction in the flux density (Tozer and Grace, 2001a). A further reduction in power occurs owing to a gradual loss in solar cell efficiency over the lifetime of the system (several years). The power variations must be compensated for by a voltage stabilization subsystem and an appropriate sizing of the payload. A noteworthy advantage of the HAP platform is that it can be retrofitted in the case of a malfunction or even during the operational phase to allow a technology refresh.

2.3.4 Viability and Trade-offs

It must be stressed that most of the applications of high-altitude platforms discussed in this section are *future* prospects. They are currently in the developmental phase. As a result, their designs are evolving to meet the numerous challenges involved.

Orbital drift and attitude perturbations of a platform can cause a jitter in the beam pattern – particularly at the periphery of coverage – resulting in signal fluctuations at user terminals and unwarranted intercell handover; the movement of the platform typically necessitates tracking at directional ground terminals.

LTA HAP systems, in particular, require an on-board attitude and control system to maintain the platform on station (i.e. within a specific three-dimensional space), leading to power and weight overheads. The design of the platform itself and the antenna configuration are among the most crucial technologies. In a spot beam system, a trade-off between the number of beams used (and hence the system capacity) against the size–weight–cost of the antenna becomes crucial. For an aircraft system, the antenna platform has to remain constantly pointed to the ground to avoid beam movements as the plane moves in tight circles. In addition to alongside and lateral movements, rotation around pitch, roll and yaw axes also have to be compensated for. Manned aircraft must contend with regular take-off and landing, involving a complex communication handover between aircraft when switching planes.

The electrical payload power capacity for HAPs necessitates a trade-off between platform sizes, payload weight and power, both for solar-powered and for fuel-driven platforms. For airships, the power, size and weight constraints are lessened because of the large size of airships and their weight-carrying capability. Further, airships can be brought back to the ground for replenishment on a regular basis if endurance capability is limited. Other issues affecting their deployment include:

- *Security.* These include malicious jamming and terrorist threat.
- *Regulatory.* These include: spectrum clearance; authorization from civil aviation and other authorities for operating an aerial system; obtaining airworthiness licensing and certificates for equipment and software.
- *Operational.* These include: developing contingency procedures; payload design cost-benefit trade-off; selection of air interface and frequency band, etc.
- *Commercial.* These include general concerns regarding an unproven commercial model and hence difficulty in obtaining investment; competition from terrestrial and satellite systems, etc. Risk mitigation strategies adapted in business plans include multi-functional solutions – remote sensing, disaster communication fixed and mobile solutions with maximum synergy with partner satellite and terrestrial systems. The economics of a commercial system has yet to be proven. Various start-ups have failed owing to inadequate funding, lack of infrastructure, etc.

Revision Questions

1. Describe the six Keplerian elements and indicate on a diagram their relation to the ECEF coodinate system. What is the difference between the true anomaly and the mean anomaly? What is an epoch?
2. Estimate the nadir angle, coverage angle, coverage radius and coverage area for a satellite in the Galileo constellation (altitude 23 200 km), assuming a minimum elevation angle of 5°. What is the orbital period?
3. Discuss the benefits of a Sun-synchonous orbit and the types of applications that use it.
4. Satellites are perturbed in their orbit by the non-uniform gravitational force and the gravitational pull of the Sun and the Moon. (a) What are the effects of these extraneous forces on satellites orbiting in a low Earth orbit and the geostationary orbit. (b) Explain how these effects are used to advantage in highly elliptical orbits.
5. Calculate the geodetic latitude and altitude for a point at mean sea level with geocentric latitude 35° N and geocentric altitude 20 km. How great is the difference between the two latitudes, and what is the positional error? Plot the difference in geocentric latitude versus geodetic latitude for a point on the Earth's surface. What is the maximum difference and at what latitude?
6. In spite of the significant research and development effort expended in evolving HAP technology, a viable operational system seems elusive. Discuss the main issues that are impeding the development of HAP systems.
7. Calculate the ratio of the maximum to minimum range from the Earth's surface to a HAP and to LEO, MEO and GEO satellites, assuming minimum elevations of 10° and 5°, and plot this ratio as a function of altitude.

References

Adams, W.S. and Rider, L. (1987) Circular polar constellations providing continuous single or multiple coverage above a specified latitude. *Journal of Astronautics Science*, **35**(2).

Angel Technologies (2009) http://www.angeltechnologies.com/ [accessed July 2009].

Ballard, A.H. (1980) Rosette constellations of earth satellites. *IEEE Trans. Aerospace and Electronic Systems*, **AES-16**(5), 656–673.

Bate, R.R., Mueller, D.D. and White, J.E. (1971) *Fundamentals of Astrodynamics*, Dover Publications, Inc., New York, NY.

Benedicto, J., Dinwiddy, S.E.. Gatti, G., Lucas, R. and Lugert, M. (2000). *GALILEO: Satellite System Design and Technology Developments*. European Space Agency Publication, November. Available: http://esamultimedia.esa.int/docs/galileo_world_paper_Dec_2000.pdf [accessed July 2009].

Beste, D.C. (1978) Design of satellite constellations for optimal continuous coverage. *IEEE Trans. Aerospace and Electronic Systems*, **AES-14**(3), 466–473.

Bowring, B.R. (1976) Transformation from spatial to geographical coordinates. *Survey Review*, **XXIII**(181), July.

Djuknic, G.M., Freidenfelds, J. and Okunev, Y. (1997) Establishing wireless communications services via high altitude aeronautical platforms: a concept whose time has come? *IEEE Communication Magazine*, September, 128–135.

ESA (2009) http://esamultimedia.esa.int/docs/gsp/completed/comp_i_98_N50.pdf, [accessed July 2009].

Farrell, J. and Barth, M. (1998) The Global Positioning System and Inertial Navigation. McGraw-Hill, New York, NY.

Global HAPS Network (2009) http://globalhapsnetwork.co.uk/page.php?domain_name=globalhapsnetwork.co.uk&viewpage=home [accessed July 2009].

Green, R.M. (1985) *Spherical Astronomy*, Cambridge University Press, Cambridge, UK.

ICAO (1998) EUROCONTROL, *WGS 84 Implementation Manual*. Available: http://www2.icao.int/en/pbn/ICAO%20Documentation/GNSS%20and%20WGS%2084/Eurocontrol%20WGS%2084.pdf [accessed February 2010].

Kaplan, E.D. and Hegarty, C.J. (2006) *Understanding GPS: Principles and Applications*. Artech house, Norwood, MA.

Kelso, T.S. (1995) Orbital coordinate systems, Part II. *Satellite Times*, November/December. Available: http://celestrak.com/columns/v02n02/ [accessed February 2010].

Kelso, T.S. (1996) Orbital coordinate systems, Part III. *Satellite Times* January/February. Available: http://celestrak.com/columns/v02n03/ [accessed February 2010].

Lockheed Martin (2009) *HALE-D factsheet*. Available: http://www.lockheedmartin.com/data/assets/ms2/HALE-D_factsheet.pdf [accessed February 2010].

Maral, G. and Bousquet, M. (2002) *Satellite Communications Systems*, John Wiley & Sons, Ltd, Chichester, UK.

NASA (2009) Orbit and coverage. *Landsat Data User's Handbook*, Ch. 5. Available: http://landsathandbook.gsfc.nasa.gov/handbook/handbook_htmls/chapter5/chapter5.html [accessed July 2009].

Piscane, V.L. (ed) (1994) *Fundamentals of Space Systems*, Oxford University Press, Oxford, UK.

Platform Wireless International (2009) http://www.plfm.net/pdf/plfm/arc_web_site.pdf [accessed August 2009].

QinetiQ Online (2009) http://www.qinetiq.com/home/newsroom/news_releases_homepage/2008/3rd_quarter/qinetiq_s_zephyr_uav.html [accessed July 2009].

Richharia, M. (1999) *Satellite Communication Systems: Design Principles*, 2nd edition. Macmillan Press Ltd, Basingstoke, UK.

Richharia, M. (2001) *Mobile Satellite Communications: Principles and Trends*. Pearson Education Ltd, London, UK.

Roddy, D. (2006) *Satellite Communications*, McGraw-Hill, New York, NY.

Schuss, J.J., Upton, J., Myers, B., Sikina, T., Rohwer, A., Makridakas, P., Francois, R., Wardle, L. and Smith, R. (1999) The IRIDIUM main mission antenna concept. *IEEE Trans. Antennas and Propagation*, **47**(3).

Tozer, T.C. and Grace, D. (2001a) High-altitude platforms for wireless communications. *Electronics and Communications*, **13**(3), 127–137.

Tozer, T.C. and Grace, D. (2001b) HeliNet – the European Solar-Powered HAP Project. Unmanned Vehicle Systems Technology Conference, 6–7 December. Available: http://www.elec.york.ac.uk/comms/pdfs/20030506163224.pdf [accessed February 2010].

Walker, J.G. (1984) Satellite constellation's. *Journal of the British Inerplanetary Society*, **37**, 559–571.

3

Spectrum and Propagation

3.1 Introduction

In the previous chapter we considered satellites as platforms used to support the provision of various personal satellite applications. In this chapter we turn our attention to the mechanism of transfer of useful information to and/or from the satellite. Without exception, this is achieved via electromagnetic radiation (which extends from radio waves, through infrared, visible light, ultraviolet and X-rays, to gamma rays).

A fundamental consideration with regard to the choice of the region of electromagnetic spectrum used for a particular service is that the Earth's atmosphere must be reasonably transparent over the chosen frequency range because satellites orbit well beyond the bulk of the atmosphere. A further consideration is the ease and efficiency with which electromagnetic signals may be generated and/or detected. In this chapter we focus primarily on the use of *radio waves* for satellite broadcasting, communications and navigation, but briefly consider also the use of infrared and visible light for remote sensing applications.

The radio spectrum is a finite resource that must be shared by the myriad of potential spectrum users, with the significant likelihood of interference between systems. In particular, regions of the radio spectrum are already heavily utilized (not just for satellite applications), and we discuss how regulation and coordination of the use of the radio spectrum is an essential feature of satellite system provision. As electromagnetic waves do not respect political borders, such regulation has an international aspect, embodied in the work of the International Telecommunications Union – Radio Communications Bureau (ITU-R).

Having briefly explored the major constraints on selection of operating frequency, the bulk of the chapter focuses on the various impairments to propagation through the atmosphere of radio waves, and shows that the most significant atmospheric layers from the point of view of atmospheric propagation are the *troposphere* (the lowest ∼7–18 km, which contains most of the atmospheric mass and weather systems) and the *ionosphere* (the upper regions of the atmosphere, ∼50–1 000 km altitude, where the thin residual atmospheric gases are ionized by the Sun). Propagation impairments in the troposphere include gaseous absorption, scattering by rain and ice and scintillation, while those in the ionosphere include absorption, scintillation and polarization rotation. Significantly, future high-altitude platforms are likely to be located in the upper troposphere or just above in the lower stratosphere, and therefore only tropospheric impairments will be significant to those platforms.

Finally, an important consideration for mobile satellite services is the influence of the local electromagnetic environment produced by proximity to nearby buildings, trees, moving cars, etc. We shall describe how reflection and diffraction of radio waves can result in multiple delayed signals from multiple ray paths (multipath), which can cause fading of the received signal.

Satellite Systems for Personal Applications: Concepts and Technology   Madhavendra Richharia and Leslie David Westbrook
© 2010 John Wiley & Sons, Ltd

3.2   Spectrum

The use of electromagnetic waves is fundamental to any exploitation of satellites or high-altitude platforms. These waves provide the means for communications or broadcasting, for radio ranging or for remote sensing of the Earth's surface and atmosphere.

3.2.1   Atmospheric Windows

It is well known that electromagnetic waves extend from low-frequency radio waves, through infrared, visible and ultraviolet light, to X-rays and gamma-rays, and the region of spectrum utilized by satellite services will generally depend on the application. For applications of interest here, however, a primary consideration is that the Earth's atmosphere is essentially transparent to these waves (for a dry atmosphere with no clouds). Figure 3.1 shows the transmission of the Earth's atmosphere (on a dry day) for a vertical path for frequencies up to 10^{16} Hz (near ultraviolet). As we shall see later, the added effects of clouds and rain can significantly reduce the availability of a satellite-based service.

3.2.1.1   Radio Windows

In communications, broadcasting and navigation applications, the ability efficiently (and cost-effectively) to generate, manipulate and detect electromagnetic waves is an important consideration, and as a result the radio-frequency portion of the spectrum is by far the most highly used.

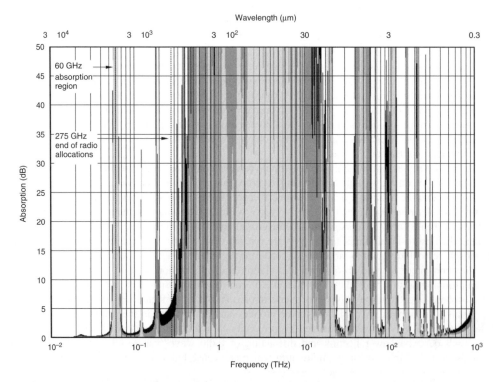

Figure 3.1   Atmospheric transmission as a function of frequency (or wavelength) for a vertical path (ITU-R P.1621-1,2001). Reproduced by permission of © ITU.

Table 3.1 ITU frequency range designations relevant to satellite applications

Acronym	Designation	Frequency range
VHF	Very High Frequency	30–300 MHz
UHF	Ultra High Frequency	300 MHz–3 GHz
SHF	Super High Frequency	3–30 GHz
EHF	Extremely High Frequency	30–300 GHz

Frequency Band Designations

Within the radio-frequency region of the electromagnetic spectrum, the International Telecommunications Union (ITU) designates a number of broad radio-frequency bands, and the generic bands relevant to this book are indicated in Table 3.1.

These ITU designations are rather broad (each band covering a frequency range of 10:1), and it is common practice to refer instead to the frequency ranges embodied by the Institute of Electrical and Electronic Engineers (IEEE) Radar bands (IEEE, 1984), set out in Table 3.2, which have a frequency range closer to an octave (2:1). These IEEE frequency bands have their origins in World War II Radar (where use of letters to indicate frequency ranges was a way of circumventing classification restrictions), but their use for classifying frequency ranges in other disciplines is now commonplace.

Within these generic bands, the spectrum is further divided up by the ITU and national regulatory authorities and apportioned to different services. As one would expect, larger spectrum allocations (larger blocks of frequency spectrum) tend to be available at higher frequencies because, for any block of spectrum equivalent to some fraction of its centre frequency, the absolute allocation will increase with frequency. Inspection of the frequency allocations confirms this to be the general case. Newer wideband satellite applications requiring significant additional signal bandwidth – such as HDTV – will in general be required to operate at higher frequencies, and, at the time of writing there is significant activity in the provision of Ka-band satellite capacity. As we shall see later in this chapter, however, the use of higher frequencies generally incurs increased variability in atmospheric losses.

3.2.1.2 Remote Sensing Windows

For remote sensing, the choice of spectral region is generally dictated by some physical property of the object(s) being measured. Certain segments of the electromagnetic spectrum are better suited

Table 3.2 IEEE (radar) band designations

Band	Frequency range
L	1–2 GHz
S	2–4 GHz
C	4–8 GHz
X	8–12 GHz
Ku	12–18 GHz
K	18–27 GHz
Ka	27–40 GHz
V	40–75 GHz
W	75–110 GHz

Table 3.3 Electromagnetic spectrum range of interest for remote sensing

Band	Wavelength	Comments
Ultraviolet	3–400 nm	Generally not used
Visible (VIS)	0.4–0.7 μm	Earth's reflected solar energy detection by passive sensor
Near Infrared (NIR)	0.7–1.5 μm	Earth's reflected solar energy detection by passive sensor
Short Wavelength Infrared (SWIR)	1.5–3 μm	Earth's reflected solar energy detection by passive sensor
Mid Wavelength Infrared (MWIR)	3–8 μm	Earth's emitted energy detection
Long Wavelength Infrared (LWIR)	8–15 μm	Earth's emitted energy detection
Far Infrared (FIR)	>15 μm	
Microwave	1 mm–~1 m	Useful for many applications

for remote sensing as indicated in Table 3.3. The wavelength of the visible band is approximately 0.4–0.7 μm. The infrared region 0.7–3.0 μm is reflective in nature, whereas lower-wavelength regions are useful for detecting thermal, radiating heat. Microwave remote sensing (of heat and reflected heat) is a relatively recent introduction. The main interest in the context of personal applications is confined to the visual and meteorological bands.

The Sun is an important illuminating source for visible and near-visible bands. Aerial photography – one of the most useful applications for personal applications – is totally dependent on it. The ultraviolet region is not widely used owing to the absorption and scattering properties of the band, while the infrared (IR) region is useful in agriculture and forestry applications because of the sensitivity of vegetation to IR *reflectance*.

3.2.2 Regulation and the ITU-R

The useful radio spectrum is a finite resource that today must be shared among a wide range of users with very different applications. At one extreme, users may be broadcasting powerful TV signals using kilowatts of power, while at the other extreme, users may be using sensitive measuring apparatus to monitor extremely weak signals from natural phenomena (radio astronomers, for instance), or using safety-critical communications in which lives are at stake. Clearly, significant potential exists for services to interfere with each other.

Furthermore, radio signals transcend national boundaries, and consequently international coordination of their use is needed. Indeed, it was not long after the first successful wireless telegraphy transmissions in 1896 that the need to regulate the use of the radio spectrum was identified, with the first International Radiotelegraph Conference being held in 1906, under the auspices of the International Telegraph Union (which had itself been formed in 1865 to regulate international telegraph services).

The 1927 International Radiotelegraph Conference established the Consultative Committee on International Radio (CCIR) which began allocating frequency bands to the different radio services, so as to minimize the potential for interference, and in 1932 the International Telecommunication Union (ITU) was created to reflect the wider remit of the union (i.e. not just telegraphy). Later, in 1947, the ITU was adopted as an agency of the United Nations (UN), and, somewhat more recently, 1992 saw the creation of the ITU Radiocommunications (ITU-R) bureau which superseded the CCIR (ITU Online, 2010).

Following the launch of the first geostationary satellite in the mid-1960s, the ITU mandate was widened to include the regulation of satellite frequency assignments, orbital positions and other satellite parameters, in order to avoid interference to/from satellites.

Today, the ITU-R defines its current role (ITU-R Mission Statement, 2010) as being to:

- effect allocation of bands of the radio-frequency spectrum, the allotment of radio frequencies and the registration of radio-frequency assignments and of any associated orbital position in the geostationary satellite orbit in order to avoid harmful interference between radio stations of different countries;
- coordinate efforts to eliminate harmful interference between radio stations of different countries and to improve the use made of radio frequencies and of the geostationary satellite orbit for radiocommunication services.

ITU-R Radio Regulations

Changes to frequency allocations are discussed at the World Radiocommunication Conference (WRC),[1] which is held every 2–4 years, and conference decisions are encapsulated in the ITU-R Radio Regulations (ITU-R Radio Regulations, 2008), which form an international agreement between the member countries of the ITU, governing the use of the radio frequency spectrum and both geostationary and non-geostationary satellite orbits.

The ITU has adopted three regions worldwide, and some frequency band allocations differ between these regions:

- *ITU Region 1:* including Europe, Africa, the Middle East and the Russian Federation;
- *ITU Region 2:* including North and South America and Greenland;
- *ITU Region 3:* including Australia, New Zealand, Japan, China, India and the Far East.

ITU member states are afforded some flexibility in implementing the radio regulations in accordance with their national interests, as long as they do not permit unnecessary interference to services in other ITU-R member countries. Most countries have a national body responsible for spectrum management/coordination – for instance in the United Kingdom this is the role of the Office of Communications (OFCOM). In addition, there exist regional bodies that also play a part in regulating the radio spectrum. For example, the European Radiocommunications Committee (ERC) – part of the European Conference of Postal and Telecommunications Administrations (CEPT) – is the regulatory body responsible for spectrum regulation and harmonization within the European Union (EU).

Frequency Allocations

A complete list of ITU frequency allocations may be found in Article S5 of the ITU-R Radio Regulations, and currently fills over 130 pages (ITU-R Radio Regulations, 2008). Within the frequency allocation tables, the same spectrum may be allocated to several services, on either a primary or secondary basis. Primary services are those protected from interference. Secondary services are not protected from interference from primary services but allowed to share the spectrum.

In addition, individual member nations have some discretion with regard to licensing these bands. Nevertheless, by way of general illustration of the typical frequency ranges used worldwide for

---

[1] The World Radiocommunication Conference was formerly known as the World Administrative Radiocommunication Conference (WARC).

Table 3.4   Principal Satellite and HAP bands for the United Kingdom (for indication only)

Frequency range	Service	Comment
144.0–146.0 MHz	Amateur satellite	
243–270 MHz	MSS (government)	D
292–317 MHz	MSS (government)	U
399.9–400.05 MHz	RNSS	D
406.0–406.1 MHz	MSS	U (EPIRB beacons)
420.0–450.0 MHz	Amateur satellite	
1164–1240 MHz	RNSS	D
1215–1240 MHz	RNSS	D
1260–1350 MHz	Amateur satellite	U
1452–1492 MHz	BSS	D (1worldspace)
1525–1559 MHz	MSS	D (Inmarsat)
1559–1610 MHz	RNSS	D
1610–1626.5 MHz	MSS	U/D (Globalstar, Iridium)
1626.5–1660.5 MHz	MSS	U (Inmarsat)
1885–1980 MHz	HAP (IMT 2000)	U
1980–2010 MHz	MSS (IMT 2000)	U
2110–2160 MHz	HAP (IMT 2000)	D
2170–2200 MHz	MSS (IMT 2000)	D
2483.5–2500 MHz	MSS	D (Globalstar)
2520–2670 MHz	BSS	D
3600–4200 MHz	FSS	D (paired with 5850–6450 MHz)
5725–7075 MHz	FSS	U
7250–7750 MHz	FSS + MSS (government)	D (paired with 7900–8400 MHz)
7900–8400 MHz	FSS + MSS (government)	U (paired with 7250–7750 MHz)
10.7–11.7 GHz	FSS/BSS	D
11.7–12.5 GHz	BSS	D
12.5–12.75 GHz	FSS	U/D
12.75–13.25 GHz	BSS	U (feeder links)
13.75–14 GHz	FSS	U
14–14.25 GHz	FSS	U (paired with 12.5–12.75 GHz)
14.25–14.5 GHz	FSS	U
17.3–17.7 GHz	FSS/BSS	U (feeder links)
17.7–18.4 GHz	FSS/BSS	U (feeder links)
17.7–19.7 GHz	FSS	D (paired with 29.5–30 GHz)
19.7–20.2 GHz	FSS + MSS	D
20.2–21.2 GHz	FSS (government)	D
27.5–29.5 GHz	FSS	U (paired with 17.7–19.7 GHz)
29.5–30 GHz	FSS	U (paired with 19.7–20.2 GHz)
30.0–31.0 GHz	FSS + MSS (government)	U (downlink in 20.2–21.2 GHz)
43.5–45.5 GHz	MSS (government)	U (downlink in 20.2–21.2 GHz)
47.2–47.5 GHz	HAP	D
47.9–48.2 GHz	HAP	U

satellite applications, Table 3.4 lists the principal frequency bands allocated to satellite and HAP services in the UK. These services are generally classified as one of:

- *Fixed Satellite Service* (FSS) communications;
- *Mobile Satellite Service* (MSS) communications;[2]
- *Broadcast Satellite Service* (BSS) broadcasting;
- *Direct Broadcast Service* (DBS) broadcasting;
- *Radio Navigation Satellite Service* (RNSS) navigation;
- *Aeronautical Radio Navigation Service* (ARNS);
- *Amateur Satellite Service* (ASS);
- *Inter Satellite Service* (ISS).

Table 3.4 is intended for indication only and is not exhaustive. For a definitive list of frequency allocations, the reader is directed to the most recent version of the radio regulations (ITU-R Radio Regulations, 2008) and to their national regulatory authority. The spectrum above 300 GHz is currently unregulated.

3.3   Propagation

Exploitation of space-based or high-altitude platforms depends on the ability of electromagnetic radiation to propagate through the Earth's atmosphere with reasonably low loss with high availability. In the main, use is therefore limited to a number of atmospheric transmission 'windows' (regions of the electromagnetic spectrum with relatively low attenuation). However, even in these atmospheric transmission windows there are a number of natural phenomena that affect – to varying degrees – the ability to detect electromagnetic waves after propagation through the atmosphere (and its weather). These phenomena include:

- molecular absorption by atmospheric gases;
- scattering and molecular absorption by hydrometeors (rain, ice, cloud, fog, etc.) and to a lesser extent by airborne particles (e.g. smoke and dust);
- scintillation due to turbulence in the troposphere;
- Faraday rotation (of electromagnetic wave orientation) due to ionized gases in the ionosphere in the presence of the Earth's magnetic field;
- attenuation by ionized gases in the ionosphere;
- scintillation due to inhomogeneities in the ionsophere.

In addition, for mobile and hand-held satellite applications, a number of additional propagation phenomena are associated with the proximity of nearby objects (or terrain) to the user terminal:

- reflections from buildings and/or surrounding terrain;
- diffraction around building edges and corners;
- shadowing by trees, tall buildings and other objects;
- Doppler frequency shift due to the relative motion of the satellite and user terminal.

---

[2] Note that the L-band MSS is further subdivided into Maritime Mobile Satellite System (MMSS), Land Mobile Satellite System (LMSS) and Aeronautical Mobile Satellite System (AMSS).

3.3.1   Impact of Propagation on Service Availability

Satellite service providers usually talk about the *availability* of a particular satellite service, that is, the fraction of time that the service is available at the supplier's stated level of Quality of Service (QoS). Service availability is related to the probability P that the service is unavailable

$$\text{Availability} = (1 - P) \tag{3.1}$$

We may define the cumulative service *outage* as the cumulative time that a service is unavailable – usually given as the total 'down' time over a calendar year. The annual service outage times for various probabilities (and availabilities) are given in Table 3.5, and it may surprise the reader to learn that a service quoting 99% availability will still result in a total of almost 4 days cumulative unavailability within a given year! Availabilities of 99.99% are fairly typical of toll-quality commercial satellite communications; however, achieving such high availability is expensive to maintain (particularly at the higher frequency bands), and lower values of availability are not uncommon in personal satellite services.

In many cases, the limiting contributing factor on service availability is due to the prevailing atmospheric conditions. As propagation in the troposphere is affected by climate and that in the ionosphere by solar conditions, it is not possible to predict the propagation impairments for a particular location for a given time interval. Instead, we must consider a statistical description of these effects, and we shall thus be interested in the probability that a certain condition – such as a particular attenuation value (signal fade depth) – is exceeded.

The probability that the value of some random variable *exceeds* some specified threshold is given by the Complementary Cumulative Distribution Function (CCDF), defined as

$$P(x > X) = 1 - \int_{-\infty}^{X} \rho(x)\, dx \tag{3.2}$$

where $\rho(x)$ is the Probability Density Function (PDF)[3] of the random process.

General Approach

There are two quite different approaches to estimating the impairments in propagation resulting from the phenomena listed above. At one extreme, a *physical* model is used to describe

Table 3.5   Service availability, probability and cumulative annual outage for different probabilities

Availability (%)	Probability (%)	Outage (h/year)
90	10	876.6
95	5	438.3
99	1	87.66
99.5	0.5	43.83
99.9	0.1	8.766
99.95	0.5	4.383
99.99	0.01	0.8766

[3] The Probability Density Function (PDF) $\rho(x)$ describes the probability that some random variable lies in the range between x and $x + dx$ (where dx is some small increment). The PDF is normalized in such a way that the total probability (for all x) is unity.

the phenomena affecting propagation, with model parameters determined by a mixture of analysis and experiment. Such methods are sometimes constrained by our limited understanding or, indeed, the sheer complexity of the natural phenomena involved. At the other extreme is a wholly *empirical* approach in which empirical formulae are obtained by curve-fitting the available experimental data, often with little reference to the underlying physics. The latter approach can be highly accurate within defined ranges, but the lack of reference to a physical model tends to obscure any intuitive understanding of the underlying phenomena, and the limits of applicability of the resultant formulae. Consequently, in this section, we shall lean towards the physics-based approach.

3.3.2 Wave Propagation Fundamentals

3.3.2.1 Propagation of Plane Waves

One-Dimensional Wave Equation

The starting point for any theoretical consideration of electromagnetic wave propagation is a wave solution of Maxwell's set of phenomenological equations (Ramo, 1965) for the electric field vector[4] **E** and magnetic field vector **H**. Instantaneous power flow *per unit area* is given by the product of **E** and **H**. From Maxwell's equations, one may derive a general equation that describes electromagnetic wave propagation. For simplicity, we shall limit our discussion to a plane wave propagating in the $+z$-direction (i.e. a wave with no spatial variation in either x or y).

Phasor Notation

It is mathematically convenient to employ complex phasor notation[5] when describing the evolution of the amplitude and phase of sinusoidal signals, and we shall adopt this notation occasionally. The use of phasor notation greatly simplifies mathematical manipulation by permitting the factoring out of common time-dependent terms (thereby transforming differential equations into simpler algebraic equations). A *phasor* is just a complex number that describes both the magnitude and phase of a sinusoidal wave. A phasor *vector* is a vector of such phasors.

Field Solution

In complex phasor notation, the solution to the 1D wave equation for waves travelling in the $+z$-direction is (Ramo, 1965)

$$\mathbf{E}\,(z,t) = \Re\left\{\bar{\mathcal{E}}_{o} \exp\left[i\,\omega\left(t - \frac{\bar{n}}{c}z\right)\right]\right\} \tag{3.3}$$

where $\bar{\mathcal{E}}_{o}$ is a phasor vector[6] that describes the magnitude and relative phase of the electric field components in the chosen coordinate system, $i \equiv \sqrt{-1}$ is the imaginary constant and $\Re\{...\}$ signifies use of the *real* part of an expression taking a complex value.[7]

The constant $c = 2.99792458 \times 10^{8}$ m/s is the velocity of electromagnetic waves in a vacuum (often called 'free space') – more commonly known as the speed of light in a vacuum. The complex quantity $\bar{n}$ describes the electromagnetic properties of the medium, and its physical interpretation

---

[4] Note: we use heavy type to denote symbols for vectors and matrices.

[5] Complex phasor notation makes use of Eulers, equation: $e^{i\omega t} \equiv \cos(\omega t) + i \sin(\omega t)$.

[6] We use a bar to denote a complex scalar, vector or matrix – such as $\bar{a}$ or $\bar{\mathbf{A}}$, and an asterisk to denote its complex conjugate – hence $\bar{a}*$.

[7] It is common practice to omit the $\Re\{\}$ symbol in equations employing phasor notation, but it is always implicitly assumed for all physical quantities.

may be more easily explained by expressing it in terms of its real and imaginary parts[8]

$$\bar{n} \equiv n - i\,k \qquad (3.4)$$

Where n is the (real) refractive index of the medium, defined as the ratio of the phase velocity[9] of electromagnetic waves in a vacuum to that in the medium itself (hence, by definition, $n \to 1$ in a vacuum), and k is known as the extinction coefficient, as it determines the rate of decay (extinction) of the wave.

Equation 3.3 describes the electric field of a sinusoidal wave of frequency f (where $\omega = 2\pi f$ is known as the *angular frequency*) and peak amplitude (vector) $|\bar{\mathcal{E}}_o|$ propagating in the $+z$-direction with (limiting) velocity $\frac{c}{n}$, and which *decays* exponentially with distance at a rate determined by $\frac{\omega k}{c}$. The distance travelled by the wave in oscillation period $T = \frac{1}{f}$ (the distance between successive wave peaks) is called the *wavelength* λ, given by $\lambda = \frac{c}{f}$.

Power Flow and Attenuation

The attenuation of power flow in a medium is of significant importance in quantifying the effects of the atmosphere on propagation. It can be shown that the magnitude of the time-averaged power per unit area is proportional to the square of the electric field strength and decays exponentially with distance (Beer–Lambert law) with specific attenuation coefficient

$$\alpha = \frac{2\omega k}{c} \qquad (3.5)$$

where α is in Nepers per metre (Np/m), although in everyday engineering practice, it is usual to quote power ratios in Decibels (dB).[10]

The slant path attenuation A for some path through the atmosphere may be obtained by integrating the specific attenuation constant over the specified path

$$A = \int_{\text{path}} \alpha \; dl \qquad (3.6)$$

3.3.2.2 Phase and Group Velocity

The velocity $\frac{c}{n}$ at a specific frequency is known as the *phase velocity*. It describes the velocity of a single sinusoid. Signals that carry useful information occupy a spread of frequencies, and, as the refractive index is generally frequency dependent (as a result of frequency-dependent absorption), experience dispersion. By way of example, Figure 3.2 illustrates the propagation of a wave envelope, or packet. The velocity of the signal *envelope* is normally associated with the flow of real *information* and generally differs from the phase velocity.

---

[8] $\bar{n}$ is generally known as the complex refractive index. The reader may encounter an alternative definition of the complex refractive index in the literature, in which the sign of the imaginary part is reversed. This is a consequence of the somewhat arbitrary nature of the definition of the sign of k.

[9] The phase velocity at a given frequency is the velocity of individual wave peaks and troughs.

[10] The conversion factor from Np to dB is

$$1 \; \text{Np} = \frac{10}{\ln(10)} = 4.343 \, \text{dB} \qquad (3.7)$$

The reader should be aware that an additional factor of 2 results if the ratio is given in terms of voltage or electric field and the result required is a power ratio in dB.

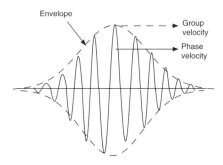

Figure 3.2   Propagation of a wave packet.

It is reasonable to assume that the peak of such an envelope occurs where the phases of the component sinusoids all add in-phase (yielding the maximum signal). This condition implies that the propagation phase shift is the same for all frequencies (Lipson, 1995), from which it is easily shown that the velocity of the signal envelope (known as the group velocity) may be expressed in terms of n_g, the group refractive index (or simply the group index), given by

$$n_g = n + \omega \frac{\partial n}{\partial \omega} \tag{3.8}$$

3.3.2.3   Excess Delay

For some applications we are interested in estimating very accurately the time it takes electromagnetic waves to travel through the atmosphere (or more specifically, how much *extra* time it takes, compared with propagation in a vacuum).

We may define the excess delay Δt – the difference between the true delay (the time taken to propagate over the specified path) and the equivalent delay in a vacuum. The excess phase and group delays are given by

$$\Delta t_p = \frac{1}{c} \int_{path} (n - 1) \cdot dl \tag{3.9}$$

$$\Delta t_g = \frac{1}{c} \int_{path} \left(n_g - 1\right) \cdot dl \tag{3.10}$$

3.3.2.4   Estimation of the Atmospheric Slant Path

When estimating propagation impairments for a satellite path, it is necessary to be able to predict the path taken by the wave. In the general case, this is complicated by bending of the ray[11] path as a result of refraction caused by the variation in refractive index n with altitude, with the curvature increasing the nearer the ray is to the ground, for low elevation angles (Robertson, 1980).

Flat Earth Approximation

Fortunately, in the vast majority of satellite applications, the elevation angle is usually sufficiently large ($\epsilon > 5°$) to allow the use of a flat Earth approximation, which ignores the refractivity variation

---

[11] It is convenient to talk in terms of the propagation of *rays*. A ray is essentially the locus of the propagation of the normal to the wavefront of a plane wave.

with altitude and curvature of the Earth. Under the flat Earth approximation, the incremental path is related to the incremental altitude via the cosecant rule

$$\mathrm{d}l \rightarrow \frac{1}{\sin{(\epsilon)}}\mathrm{d}h \tag{3.11}$$

and the integral of those media properties (attenuation, delay, etc.) that depend only on altitude are approximately equal to their values for a zenith (vertical) path times the cosecant of the elevation angle.

3.3.2.5 Polarization

The orientation of the electromagnetic field vectors is significant for several reasons. Efficient reception of electromagnetic energy depends on appropriate alignment of the receiving antenna with the fields of the incoming wave. Furthermore, the strength of certain propagation phenomena varies according to the field orientation with respect to the stratified layers of the atmosphere, the Earth's magnetic field, and even the orientation of non-spherical raindrops. The orientation of the electric field vector is known as the *polarization* of the electromagnetic wave.

Linear Polarization
Linear polarized waves (also called plane polarized waves) have the property that the electric field **E** oscillates entirely in a single dimension perpendicular to the direction of propagation. For example, in Cartesian coordinates, a plane wave propagating in the $+z$-direction and having its electric field vector directed along the x-direction is considered to be linearly polarized in x. In fact, the orientation of the electric field may be at any angle τ relative to the x-axis in the $x-y$ plane, but can always be resolved into a linear sum of an x-polarized wave and a y-polarized wave.

Circular and Elliptical Polarization
In circular polarization, the magnitude of the electric field is constant, but the electric field orientation rotates in a plane normal to the direction of propagation, tracing a circle (when viewed along the direction of propagation), and making one revolution every $\frac{1}{f}$ seconds, where f is the frequency. By definition (IEEE, 1993), if the field rotates clockwise, when looking in the direction of propagation, then the wave is said to be Right-Hand Circular Polarized (RHCP). Conversely, if the field rotates anticlockwise, the wave is said to be Left-Hand Circular Polarized (LHCP).

The most general state of polarization is *elliptical* polarization. Elliptical polarization is a state between linear and circular polarization in which the direction of the electric field rotates as with circular polarization but the magnitude of the field also varies as it rotates, tracing an ellipse (when viewed along the direction of propagation). Both linear and circular polarizations may be considered as special cases of elliptical polarization.

Polarization Ellipse
There are numerous different methods employed for defining the polarization state. One of the simplest to visualize is the polarization ellipse, whereby polarization is characterized in terms of the parameters of an ellipse, formed by the locus of the peak of the normalized electric field vector, normal to the direction of propagation. Referring to Figure 3.3, the polarization ellipse is characterized in terms of the orientation angle ψ_p of its major axis (diagonal at the widest part) relative to one of the coordinate axes (x) and by the ellipticity (the ratio of the major to minor axes) equal to the arctangent of angle χ_p. For circular polarized waves, the residual ellipticity is known as the *axial ratio* (usually expressed in dB).

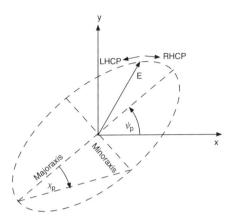

Figure 3.3 Polarization ellipse.

Polarization Vector

When using phasor (vector) notation, a more convenient method of specifying polarization is obtained through the introduction of a polarization vector $\bar{\mathbf{p}}$–a (phasor) unit vector that defines the polarization state. Whereupon, the electric field vector phasor $\bar{\mathcal{E}}$ can be described in terms of its magnitude $|\bar{\mathcal{E}}|$ and the unit polarization vector $\bar{\mathbf{p}}$

$$\bar{\mathcal{E}} = \bar{\mathbf{p}} \, |\bar{\mathcal{E}}| \tag{3.12}$$

The polarization vector is particularly convenient when we need to take into account arbitrary antenna orientation. Some example polarization vectors are given in Table 3.6.

Polarization Misalignment

Misalignment of the receiver antenna polarization with that of the incoming electromagnetic wave will cause some of the input wave to be rejected (reflected or absorbed), thereby reducing the received signal strength, and we may associate a polarization mismatch 'loss' with this misalignment. The polarization loss L_{pol} is given by (Balanis, 1997)

$$\frac{1}{L_{\text{pol}}} = |\bar{\mathbf{p}}_t \cdot \bar{\mathbf{p}}_r|^2 \tag{3.13}$$

Where the subscripts 't' and 'r' indicate transmitter and receiver respectively.

Table 3.6 Some example polarization vectors

Polarization state	Polarization Vector $\bar{\mathbf{p}}$
Linear polarization along x-axis	$\{1, 0, 0\}$
Linear polarization at $45°$ to x-axis	$\frac{1}{\sqrt{2}} \{1, 1, 0\}$
RHCP	$\frac{1}{\sqrt{2}} \{1, i, 0\}$
LHCP	$\frac{1}{\sqrt{2}} \{1, -i, 0\}$

For the particular case of a linear polarized wave, the polarization loss (ratio) is simply given in terms of the polarization mismatch angle $\Delta\varphi$ by $L_{pol} \rightarrow \cos^2(\Delta\varphi)$.

Cross-Polarization

Cross-Polar Discrimination (XPD) is the ratio of the power in the wanted polarization to that in the unwanted polarization (usually given in dB). For the particular case where the receiver is configured to detect a specific linear polarization, and an orthogonal linear polarization is present, misalignment of the polarization vector by an angle $\Delta\varphi$ will result in a cross-polar discrimination of XPD $\rightarrow 1/\tan^2(\Delta\varphi)$.

3.3.3   Tropospheric Effects

We are now ready to address specific propagation impairments. The principal influences on the propagation of electromagnetic radiation between satellites and user terminals located on or near the Earth's surface result from the presence of the Earth's atmosphere, and we begin with those due to the troposphere – the lowest layer of the atmosphere. As a general rule of thumb, tropospheric effects dominate for frequencies above ~3 GHz (Ippolito, 1999).

3.3.3.1   Structure of the Atmosphere

An elementary understanding of our atmosphere is clearly desirable in order to understand the origins and scope of these effects. As illustrated in Figure 3.4, the vertical structure of the Earth's atmosphere is generally considered to comprise several distinct regions (Barry and Chorley, 1998):

- *Troposphere.* The lowest atmospheric region, of the order of 8–16 km thick (depending on latitude and season – 11 km being a typical average value), this region encompasses most of the clouds, water vapour and precipitation and is characterized by turbulent mixing of the atmospheric gases. 75% of the mass of the atmospheric gases is contained in the troposphere (Barry and Chorley 1998). This layer is heated primarily by convection from the Earth's surface, and, as the air rises, it experiences lower pressure and cools, leading to an average rate of change in temperature with altitude (known as the lapse rate) of around −6.5 K/km. At the top of the troposphere, the temperature profile reaches a plateau – known as the tropopause.
- *Stratosphere.* This atmospheric region exists between altitudes of around 11–50 km (the exact value being dependent on latitude) and comprises many strata (layers), with only limited mixing between these layers. Most of the atmosphere's ozone is formed in this region, and, in contrast to the troposphere, the stratosphere is heated primarily from above by the absorption of ultraviolet radiation from the Sun by atmospheric ozone, and this results in a positive average rate of change in temperature with altitude of between around 1 and 2.8 K/km. At the top of the stratosphere, at around 50 km, the temperature reaches another plateau – called the stratopause.
- *Mesosphere.* This atmospheric region exists between altitudes of about 50–80 km (again, the exact values being dependent on latitude). In this region the temperature once again falls with increasing altitude, with an average rate of change of around −2.8 K/km. At the top of the mesosphere there is another layer where the temperature is constant – the mesopause.
- *Thermosphere.* This atmospheric region – the largest – exists from about 100 km altitude to above 500 km, and is characterized by an increase in temperature with altitude owing to absorption of high-energy solar radiation by the residual atmospheric gases. In contrast to the lower atmospheric layers, where the dry gases that comprise the atmosphere exist in fixed proportions owing to constant mixing, in the thermosphere these gases separate according to their different

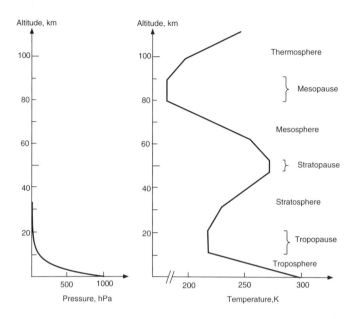

Figure 3.4   Characterization of the atmospheric regions.

molecular weight. Significantly, the thermosphere includes the ionosphere (see Section 3.3.4), where the residual gases are ionized by solar radiation into electrons and ions.

As the properties of the vertical structure of the lower atmosphere vary with both latitude and season, it is generally useful to consider an *averaged* atmospheric profile when estimating the effect of the atmosphere on propagation. The most widely-used averaged atmospheric profile is the *US Standard Atmosphere* (1976) which represents an averaged, mid-latitude vertical profile of the Earth's atmosphere. Derived from measurements in the United States, it comprises an idealized model of temperature, pressure and density profiles up to 100 km.

Concept of Scale Height
We expect the effect of the atmospheric gases to depend – at least to first order – on gas density, which varies approximately exponentially with altitude, with exponential decay characterized by a parameter known as the *scale height*. One way to visualize the scale height is the height of an equivalent hypothetical atmosphere containing the same amount of gas (i.e. the same number of molecules) but where the pressure and density remain constant at their surface values up to the scale height and are zero above this.

It turns out that the scale heights are quite different for water vapour (the 'wet' atmospheric gas) and the remaining 'dry' atmospheric gases. For the dry atmospheric gases, the effect of the variation in atmospheric temperature with altitude (Figure 3.4) is only of secondary effect (the variation in *absolute* temperature with altitude in the first 100 km being only around 15%), and the scale height is determined by the variation in pressure owing to the weight of the gases themselves. The resulting scale height for the dry atmospheric gases is around 6–8 km.

By contrast, the profile of water vapour density with altitude is very sensitive to the temperature variation, as the partial pressure of water vapour is constrained by saturation – with additional water vapour falling as precipitation. The dominant influence on the density and partial pressure of atmospheric water vapour is the temperature dependence of the saturated water vapour pressure (which is approximately exponential), and the resultant scale height for water vapour is of the order of 2 km. In addition, the surface water vapour concentration varies significantly – both geographically and seasonally.

3.3.3.2   Climatic Regions

The impact of propagation impairments involving the troposphere for a given level of service (probability) clearly depends on local climate, and design engineers typically estimate such impairments for any specified location (latitude and longitude) and probability using global climatic datasets, such as those available from the ITU-R (ITU-R P.836-3, 2001). Clearly it is not feasible to reflect the scope and complexity of such datasets in this book. Instead, our aim has been to give the reader a flavour of the likely impairments for different climate types, and to indicate, where appropriate, the relevant procedures/datasets that will allow the reader to calculate the impairments for him/herself for any given location and probability.

   To that end, it is instructive to focus on the propagation impairments for some example climate types, and the most widely used climate classification scheme is that due to Köppen. Köppen divided the climates of the world into five general categories: *tropical*, *arid*, *temperate*, *continental* and *polar*, and further subdivided these according to their precipitation and temperature patterns. In the interests of brevity, we shall limit our illustrative examples in this book to 10 example locations for the principal climatic types for the main populated areas of the world, as indicated in Table 3.7.

3.3.3.3   Gaseous Absorption

The atmosphere attenuates electromagnetic waves as a result of molecular resonances of its constituent gases. These resonances are referred to as *absorption lines* (a term originally used to describe vertical dark lines in photographs of the emission spectra). For frequencies below 1000 GHz the dominant gaseous absorption lines are those due to oxygen and water vapour, with approximately 44 oxygen and 34 water vapour absorption lines in the spectrum below 1000 GHz; the lowest

Table 3.7   Köppen climate regions, with some example locations

Climate	Köppen classes	Example	Key
Tropical, rainforest	Af	Singapore	S
Tropical, monsoonal	Am	Mumbai, India	M
Tropical, savanna	Aw	Darwin, Australia	D
Arid, desert	BWk, BWh	Yuma, Arizona, USA	Y
Arid, steppe	BSk, BSh	Denver, Colorado, USA	d
Temperate, Mediterranean	Csa, Csb	Rome, Italy	R
Temperate, humid subtropical	Cfa, Cwa	New Orleans, Louisiana, USA	N
Temperate, maritime	Cfb, Cwb	London, UK	L
Temperate, maritime subarctic	Cfc	–	–
Continental, hot summer	Dfa, Dsa, Dwa	Beijing, China	B
Continental, humid	Dfb, Dsb, Dwb	–	–
Continental, subartic	Dfc, Dsc, Dwc	Quebec, Canada	Q
Continental, extreme subartic	Dfd, Dwd	–	–
Polar, tundra	ET	–	–
Polar, ice cap	EF	–	–

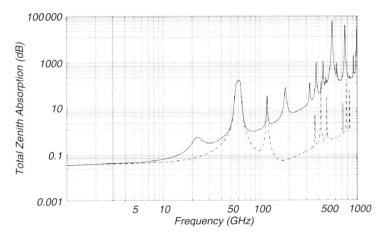

Figure 3.5 Sea level zenith attenuation due to atmospheric gases for dry air (dashed line) and an atmosphere with 7.5 g/m^3 surface water vapour density (solid line) using the method of ITU-R Recommendation P.676-6 Annex 1 (ITU-R P.676-6, 2005).

lines are around 57 GHz and 22 GHz for oxygen and water vapour respectively. Gasous absorption depends on the gas density and is therefore strongly altitude dependent.

Accurate calculations of gaseous absorption generally rely on line-by-line calculations of the absorption, involving a separate calculation for each gaseous absorption line (Liebe, 1981), such as that outlined in ITU-R Recommendation P.676-6 Annex 1 (ITU-R P.676-6, 2005). An example of the frequency dependence of the sea level zenith attenuation for both dry air and water vapour concentration of 7.5 g/m^3, obtained using the line-by-line method, is shown in Figure 3.5 (attenuation at other elevations may be estimated using the cosecant law).

Approximate Estimation Method

Line-by-line calculations of gaseous absorption are computationally intensive and unsuited to everyday use. Fortunately, a number of approximate formulae have been devised for estimating gaseous absorption at radio frequencies, based on estimating the specific absorption coefficients for oxygen and water vapour at sea level and invoking a scale height. Annex 2 of ITU-R Recommendation P.676-6 (ITU-R P.676-6, 2005) provides one such method.

For a rapid estimation of loss due to gaseous absorption up to 57 GHz, the simpler approximate formulae of Rogers (Rogers, 1985) are useful. The specific gaseous attenuation due to oxygen $\alpha_o(0)$ and that due to water vapour $\alpha_w(0)$ at sea level are approximately given by

$$\alpha_o(0) \simeq \left(\frac{7.1}{f_{\text{GHz}}^2 + 0.36} + \frac{4.5}{(f_{\text{GHz}} - 57)^2 + 0.98} \right) 10^{-3} f_{\text{GHz}}^2 \quad (\text{dB/km}) \tag{3.14}$$

$$\alpha_w(0) \simeq \left(0.067 + \frac{3}{(f_{\text{GHz}} - 22.3)^2 + 7.3} \right) 10^{-4} \rho_w f_{\text{GHz}}^2 \quad (\text{dB/km}) \tag{3.15}$$

where ρ_w is the surface water vapour density[12] (for a given probability P) and f_{GHz} is the frequency in GHz. The variation in surface water vapour density with probability is illustrated in Figure 3.6 for the example climates of Table 3.7.

---

[12] The surface water vapour density is also known as the absolute humidity (Barry and Chorley, 1998).

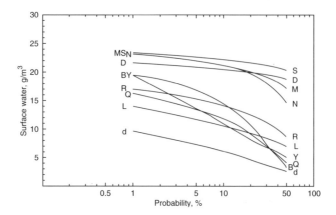

Figure 3.6 Surface water vapour density ρ_w versus probability P for the example climates (see Table 3.7 for key).

The approximate zenith attenuation is obtained by multiplying the sea level attenuation by an equivalent height for each gas. Typically, we are interested in a path from sea level up to a point well beyond the scale height. The zenith gaseous attenuation A_{gz} (in dB) is given by

$$A_{gz} \rightarrow \alpha_o(0)\,h_o + \alpha_w(0)\,h_w \tag{3.16}$$

Rogers gives the following approximate values for the equivalent heights (Rogers, 1985):

$$h_o \simeq 6\,\text{km} \tag{3.17}$$

$$h_w \simeq 2.2 + \frac{3}{(f_{GHz} - 22.3)^2 + 3}\,\text{km} \tag{3.18}$$

The physical justification for the frequency dependence introduced for the equivalent height for water vapour near to the molecular resonance frequency at 22 GHz is that the (constant) scale height approximation becomes less accurate as the specific attenuation constant increases (introducing significant attenuation even where the atmosphere is thin).

Finally, the total zenith slant path gaseous attenuation is obtained from the zenith attenuation using the cosecant rule.

3.3.3.4   Radiation Scattering by Water Droplets and Ice Particles

Attenuation due to electromagnetic scattering is caused by various types of particle: gas molecules in the upper atmosphere; pollen, smoke and water in the lower portions of the atmosphere. Scattering by rain is typically the dominant source of attenuation above 10 GHz for high-availability satellite services. Less severe, but still significant, is attenuation due to water droplets in clouds and fog. These loss processes are typically characterized, on a per raindrop (or water droplet) basis, by an equivalent cross-section (equal to the equivalent cross-sectional area of a hypothetical solid particle removing energy from the wave on a purely geometrical basis), called the extinction cross section.

Mie Scattering
The scattering of a plane wave by a sphere of arbitrary size is a classical problem in physics, the solution to which was given by Mie (Born and Wolf, 1997). Mie scattering theory is somewhat

complex; however, there are useful limiting cases. When the size of the scattering sphere is much smaller than the wavelength, Mie scattering is well described by the simpler Rayleigh scattering approximation, derived from electrostatic considerations (van de Hulst, 1981), and the degree of scattering is proportional to the cube of the drop(let) cross-sectional area. At the other extreme, when the drop size is large compared with the wavelength, the normalized cross-section tends to a value of 2 – sometimes known as the Fraunhoffer approximation.

Since both cloud and rain consist of a range of varying droplet sizes, the specific attenuation coefficient (in dB/unit distance) is obtained by integrating the extinction coefficient σ_{ext} over all droplet diameters:

$$\alpha_r = 4.3434 \int_0^\infty \sigma_{ext}(D) N(D) \, dD \quad \text{(dB/unit distance)} \tag{3.19}$$

where $N(D)$ is the density of drop size with diameter D.

3.3.3.5   Attenuation due to Cloud and Fog

Cloud, Mist and Fog

Non-precipitating clouds, mist and fog[13] contain minute, suspended water droplets, with typical water droplet sizes of $<100\,\mu m$.

In the particular case where the signal wavelength is much larger than the diameter of the scatterer (for example, lower radio frequencies and/or very small droplets sizes), Mie scattering tends to the Rayleigh approximation. Specifically, in the case of Rayleigh scattering at radio frequencies by water droplets, the extinction cross-section is dominated by the absorption cross-section (Garace and Smith, 1990; van de Hulst, 1981)

$$\sigma_{ext} \simeq \sigma_{abs} = \frac{\pi^2 D^3}{\lambda} \Im\left\{ -\frac{\bar{n}_w^2 - 1}{\bar{n}_w^2 + 2} \right\} \tag{3.20}$$

where $\bar{n}_w$ is the complex refractive index of water and $\Im\{...\}$ signifies the use of the imaginary part of a complex expression.

As the extinction is proportional to the *volume* of the water droplet (and thereby on its mass), the integral contained in equation (3.19) can be related to the density of water contained in a given volume. Under the Rayleigh approximation, therefore, the attenuation due to cloud for a zenith path A_{cz} may then be expressed as

$$A_{cz} = K_w W \tag{3.21}$$

where W is the total cloud columnar water content integrated over the zenith path (the total amount of water per square metre in a vertical column) and K_w is a frequency dependent constant.

ITU-R Recommendation P.840-3 (ITU-R P.840-3, 1999) provides a number of geographical maps of total cloud columnar water (i.e. from the ground to the top of the cloud) for probabilities from 0.1 to 50% derived from the same ITU-R dataset. The variation in cloud columnar water with probability is shown in Figure 3.7 for the example climates of Table 3.7. For frequencies up to 30 GHz, K_w may be approximated by (Maral and Bousquet, 2002)

$$K_w \simeq 1.2 \times 10^{-3} \, (f_{GHz})^{1.9} \tag{3.22}$$

It is apparent that the attenuation due to cloud increases with increasing RF frequency. Table 3.10 contains an example calculation of the loss due to scattering from clouds.

---

[13] The distinction between a fog and a mist is the point where the visibility is less than or equal to 1 km.

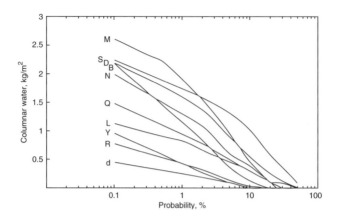

Figure 3.7 Cloud columnar water (normalized to $0°$ C) versus probability P for the example climates (see Table 3.7 for key).

Effect of Cloud Cover at Optical Frequencies

The Rayleigh scattering approximation breaks down at high frequencies even for small droplets, and cloud attenuation becomes very significant, particularly so at optical frequencies – both for Earth observation and for future very high-data-rate free-space optical satellite communications. To a first approximation[14] one may estimate the availability of a particular optical service from the observed amount of cloud of any type (including fog) obtained from summaries of surface synoptic weather reports. Table 3.8 (taken from Hahn and Warren, 2007) indicates the average cloud amount for our example climates. The global average probability of cloud cover is 64.8%, indicating a significant constraint on high availability optical systems.

3.3.3.6 Attenuation due to Rain

Distribution of Rain Drop Sizes

When considering attenuation due to rain, the full Mie scattering theory must be used, together with a knowledge of the distribution of rain drop sizes. The distribution of raindrop sizes for a given rain rate (in mm/h) was measured by Laws and Parsons (1943) and others, using various ingenious methods (such as using flour to trap the raindrops for subsequent measurement), and Marshall and Palmer (1948) derived an empirical function that accurately approximates the measured distribution under most conditions. The Marshall–Palmer function for the density N of drops with diameter D in terms of the measured rain rate R is given by

$$N(D) = N_0 \, \exp(-\alpha R^{-\beta} D) \tag{3.23}$$

where N_0, α and β are fitting parameters.

Specific Attenuation due to Rain

Owing to the practical difficulties in evaluating equation (3.19), it is common practice to use an empirical fit to the attenuation due to rain (in dB/km) for a giver rain rate R (in mm/h), given by

$$\alpha_r = aR^b \quad \text{(dB/km)} \tag{3.24}$$

---

[14] Clearly this is a gross simplification, and a small degree of attenuation/scattering may be tolerable.

Table 3.8 Average cloud amount for the example climates of Table 3.7

Location	Average cloud amount
Singapore	80%
Mumbai	44%
Darwin	41%
Yuma	29%
Denver	50%
Rome	43%
New Orleans	57%
London	71%
Beijing	48%
Quebec	66%
Global average	64.8%

where a and b are frequency-dependent parameters. Although the origin of equation (3.24) is empirical, Olsen, Rogers and Hodge (1978) have shown that this is, in fact, an exact solution of equation (3.19) for the Laws and Parsons raindrop size distribution, in both the low-frequency and optical limits (and also as $R \rightarrow 0$).

Values of the parameters a and b for vertical polarizations given in ITU-R Recommendation P.838-3 (ITU-R P.838-3, 2005) are tabulated in Table 3.9 for a number of common frequency bands used by satellite applications. Recommendation P.838-3 provides formulae to obtain the values of a and b for other frequencies and polarizations (in particular, circular polarization), although these corrections are relatively small.

Table 3.9 Rain attenuation coefficients for vertical polarization (adapted from ITU-R P.838-3, 2005)

Frequency (GHz)	a_v	b_v
1	0.0000308	0.8592
1.5	0.0000574	0.8957
2	0.0000998	0.9490
4	0.0002461	1.2476
6	0.0004878	1.5728
7	0.001425	1.4745
8	0.003450	1.3797
10	0.01129	1.2156
12	0.02455	1.1216
14	0.04126	1.0646
18	0.07708	1.0025
20	0.09611	0.9847
27	0.1813	0.9349
30	0.2291	0.9129
40	0.4274	0.8421
45	0.5375	0.8123

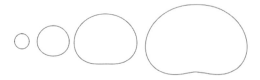

Figure 3.8   Increasingly non-spherical raindrop shapes with drop size (after Pruppacher and Pitter, 1971) – the drop size on the left is 1 mm diameter.

Shape of Rain Drops

The specific attenuation coefficients due to rain differ slightly depending on polarization. Why is this? After all, raindrops are round, aren't they? In fact, the larger the raindrop, the less spherical is its shape as a result of the aerodynamic forces acting on it as it falls.

Pruppacher and Pitter (1971) have derived the *average* shapes of raindrops, based on aerodynamic considerations. As illustrated in Figure 3.8, the shape of raindrops becomes increasingly non-spherical for very large drop sizes. The shape of the larger raindrops also becomes less *stable*. Furthermore, wind and atmospheric turbulence cause these flattened raindrop shapes to become tilted with respect to the vertical – a process known as *canting*.

The impact of these non-spherical raindrops is to produce polarization-dependent specific attenuation coefficients, and also depolarization of the wave, as the scattering coefficient becomes orientation dependent with respect to the raindrop. One might anticipate some correlation between the attenuation due to rain and the degree of depolarization (quantified by the cross-polarization discrimination). Indeed, an empirical relation has been observed between the cross-polarization due to rain XPD_r and the attenuation due to rain A_r for a given probability (Ippolito, 1981).

Rain Rate

In order to estimate the attenuation due to rain, we need to know the rain rate R (in mm/h) for a specified probability. If such information is plotted on a map, then the user can read off the value appropriate to his or her location (or region of interest). ITU-R recommends that, for the purposes of estimating rain attenuation using equation (3.24), the rain rate integration (measurement) time should be 1 min. In practice, the rain rate is rarely measured on such a short timescale (it is more commonly measured on a half-hourly or hourly scale). Rice and Holmberg (1973) therefore devised a method of estimating the 1 min rain rate from the available weather reports using available information on the ratio of thunderstorm rain to stratiform rain.

Crane (1980) introduced a rain attenuation model based around the concept of *rain zones* – based in part on the Köppen climate classification scheme – by way of a method of estimating rainfall for given probabilities without local measurement data (i.e. by reference to the rainfall characteristics of the local rain zone). More recently, processed global rainfall data have become widely available. ITU-R Recommendation P.837-4 (ITU-R P.837-4, 2003) provides global maps of rainfall rate, and Figure 3.9 shows the variation in rain rate with probability value, derived from the ITU-R rainfall data, for the example climates of Table 3.7.

Effective Height for Rain

We also require an estimate of the maximum height of the rain. To a first approximation one can assume that rain exists – at the same rainfall rate – from ground (or sea) level up to the altitude corresponding to the 0° C isotherm, which is also called the freezing height. In thermal equilibrium, this would be where falling ice crystals melt to form water droplets, and therefore any water above this height should be in the form of ice crystals. In fact, the assumption of strict

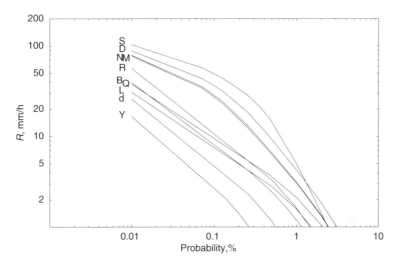

Figure 3.9 Rain rate R versus probability P for the example climates (see Table 3.7 for key).

thermodynamic equilibrium does not strictly hold owing to supercooling (non-adiabatic cooling) of raindrops lifted up rapidly above the $0°$ C isotherm inside convective clouds, and the rain height needs to be corrected in such cases.

ITU-R Recommendation P.839-3 (ITU-R P.839-3, 2001) provides maps of the yearly average rain height (in km) as a function of latitude and longitude; however, we use a simple empirical relationship for the variation in rain height with latitude.

Rain Cells and Effective Rainy Path Length

Attenuation due to rain is affected by the horizontal spatial variation in rate that occurs within heavy rainstorms. It is therefore not appropriate to assume a constant rain rate over the *entire* slant path (up to the rain height). Whereas the rain associated with stratiform (horizontally layered) clouds, for example Nimbostratus clouds, is relatively uniform (and relatively light), rain associated with convective (heaped) clouds, for example Cumulonimbus clouds, is more localized and can be extremely heavy. It is observed that heavy rain, associated with *rainstorms*, appears to be concentrated in rain 'cells'. The average radius of these cells is observed to decrease with increasing rain rate. Furthermore, measurements of the rain rate across a cell suggest an approximately *exponential* variation in rain rate with distance from the centre of a cell.

Rain Attenuation Spatial Models

There are a great many models that attempt to account for the spatial variation in rain rate along the slant path, and these models are constantly being improved and updated. A detailed review of such models has been undertaken under the European Union COST-255 study, and includes a comparison of the *accuracy* for European climates (COST-255, 2010). Feldhake and Ailes-Sengers (2002) have compared the various models for a US climate, using NASA ACTS satellite data. It is, however, important to note that all of these rain attenuation models have a significant error margin in the estimated attenuation (in dB) over the wide range of probabilities of interest to satellite system designers, with the error for the better rain models typically being in the range 35–45% (COST-255, 2010).

A common feature of many models is the introduction of an effective path length through the rain through the use of a horizontal correction factor

$$A_r = \left(a\, R^b\right) L_s\, r_h \tag{3.25}$$

where r_h is a horizontal correction factor, which typically depends on rain rate and ground range L_g (the horizontal extent of the path through the rain). Under the flat Earth approximation, the ground path length is related to the slant path length (bounded by the rain height) by

$$L_g \rightarrow L_s \cos(\epsilon). \tag{3.26}$$

Simple Attenuation Model (SAM)

The majority of the available spatial models are empirical, derived from best fits to measure attenuation data, with ITU-R Recommendation P.618-8 being one of the more accurate (ITU-R P.618-8, 2003). Unfortunately, the empirical nature of this recommendation causes some of the underlying physics to be lost. A few rain attenuation models are based on physical rain cell models. For example, the Excell model (Capsoli and Paraboni, 1987) is based on a distribution of exponential rain cells with circular symmetry, and typically ranks in the top three (COST-255; Feldhake and Ailes-Senger, 2002). An exposition of the Excell rain model is beyond the scope of this book, and the slant path rain attenuation model chosen for further discussion here is the Simple Attenuation Model (SAM) proposed by Stutzmann and Dishman (1984). It is not the most accurate, but it is one of the simplest retaining a physical picture of the approximately exponential variation in rain rate in storm cells. In a European climate, this model has been found to have an error of around 40% (COST-255, 2010).

In the simple attenuation model, the rain rate along the slant path is assumed to vary exponentially, if the peak rain rate is above a certain threshold value (used to distinguish stratiform rain from thunderstorm rain), according to

$$R = \begin{cases} R_o & \text{if } R_o \leq R_{min} \\ R_o \exp\left(-\Gamma \ln\left(\frac{R_o}{R_{min}}\right) l \cos(\epsilon)\right) & \text{if } R_o > R_{min} \end{cases} \tag{3.27}$$

where R_o is the local rain rate and l is the horizontal path distance. $R_{min} = 10\,\text{mm/h}$ is the rain rate below which the rain rate profile (assumed to be due to stratiform rain) may be assumed to be flat, and $\Gamma \simeq \frac{1}{14}$ is a best-fit parameter. The slant path attenuation is then obtained by integrating the specific attenuation (from equation (3.24)) over the path, yielding the equivalent horizontal path reduction factor used in equation (3.25):

$$r_h = \begin{cases} 1 & \text{if } R_o \leq R_{min} \\ \dfrac{1-\exp\left(-b\,\Gamma \ln\left(\frac{R_o}{R_{min}}\right) L_g\right)}{b\,\Gamma \ln\left(\frac{R_o}{R_{min}}\right) L_g} & \text{if } R_o > R_{min} \end{cases} \tag{3.28}$$

where b is the rain attenuation coefficient discussed previously. In the SAM model, rain height is assumed to be equal to the freezing height for stratiform rain and a rain-rate-dependent height for thunderstorm rain, determined from

$$h_r \simeq \begin{cases} h_f & \text{if } R_o \leq R_{min} \\ h_f + \log\left(\frac{R_o}{R_{min}}\right) & \text{if } R_o > R_{min} \end{cases} \tag{3.29}$$

where h_f is the height of the $0°\,C$ isotherm. The SAM model employs a simple empirical approximation for the variation in $0°\,C$ isotherm height h_f with latitude Θ:

$$h_f \simeq \begin{cases} 4.8 & \text{if } |\Theta| \leq 30° \\ 7.8 - 0.1\,|\Theta| & \text{if } |\Theta| > 30° \end{cases} \tag{3.30}$$

An example of a rain attenuation calculation using SAM is given in Table 3.10. The combined attenuation (in dB) due to rain and gaseous attenuation is illustrated in Figure 3.10 for various rain rates (for mid-latitudes). It is apparent that attenuation due to rain is particularly significant for frequencies above 10 GHz.

3.3.3.7 Tropospheric Scintillation

Random fluctuations in the refractive index of the troposphere cause the phase of a propagating plane wave to vary spatially, resulting in scintillation at the receiver, as different signal contributions

Table 3.10 Example tropospheric loss calculation (* indicates value for 1% probability used)

Quantity	Value			
Probability of exceedance (%)	10	1	0.1	0.01
Location/climate		London/temperate, maritime		
Altitude (km)		0		
Elevation angle (deg)		20		
Frequency (GHz)		12		
Rain rate (mm/h)	$\sim$0	1.6	8	28
Water vapour density (g/m^3)	10.5	13.9	13.9*	13.9*
Columnar water (kg/m^2)	0.3	0.83	1.1	1.1
Oxygen specific attenuation (dB/km)		0.00037		
Water vapour specific attenuation (dB/km)	0.014	0.019	0.019	0.024
Oxygen effective height water vapour (km)		6		
Water vapour effective height (km)		2.03		
Zenith gaseous attenuation (dB)	0.031	0.040	0.040	0.040
Slant path gaseous attenuation (dB)		0.12		
Cloud specific attenuation ((dB/km)/(g/m^3))		0.13		
Zenith cloud attenuation (dB)	0.040	0.11	0.15	0.15
Slant path cloud attenuation (dB)	0.12	0.33	0.43	0.43
Freezing height (km)		2.65		
Rain height (km)	2.65	2.65	2.65	3.1
Rainy slant path (km)	7.75	7.75	7.75	9.06
Ground path (km)	7.28	7.28	7.28	8.51
Rain a coefficient		0.024		
Rain b coefficient		1.18		
Reduction factor	1	1	1	0.707
Rain attenuation (dB)	$\sim$0	0.319	2.171	7.83
Water vapour partial pressure (hPa)	14.5	19.2	19.2	19.2
Wet refractivity	60.3	79.8	79.8	79.8
ITU-R P.618-8 scintillation fade depth (dB)	0.21	0.54	0.85	1.1
Total tropospheric attenuation (dB)	0.33	0.96	2.9	8.5

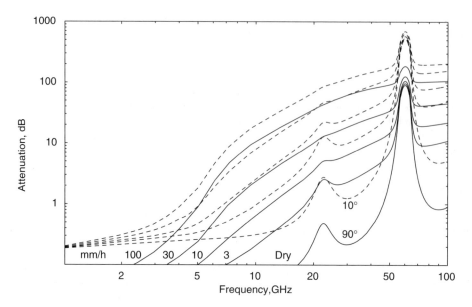

Figure 3.10 Combined attenuation due to scatter by rain and gaseous absorption as a function of frequency for various rain rates at zenith (solid lines) and 10° elevation (dashed lines).

add either constructively or destructively, depending on the instantaneous phase variation. Scintillation may cause temporal fluctuations in signal amplitude, relative phase and even angle of arrival. Such scintillation is observed to exhibit both short-term and long-term variations. Short-term fluctuations are attributed to turbulence in the atmosphere, while long-term variations in the short-term mean level are attributed to variations in meteorological variables (temperature, water vapour density, etc.).

The theory of tropospheric scintillation is somewhat complex and beyond the scope of this book; however, it is informative to note that theory predicts that the standard deviation of the signal power (in dB) due to scintillation will vary with frequency f and slant path length to the turbulence L_s according to $f^{\frac{7}{12}} L_s^{\frac{11}{12}}$ (Karasawa, Yamada and Allnut, 1988). In practice, the dependence on slant path length is not exactly as predicted by theory, which makes certain assumptions regarding the horizontal extent of the refractive index perturbations compared with the wavelength, and it is usual to use empirical formulae to estimate the attenuation due to tropospheric scintillation. For a method of estimating tropospheric scintillation for frequencies between 4 and 20 GHz, the reader is directed to ITU-R Recommendation P.618 (ITU-R P.618-8, 2003). Tropheric scintillation is mainly significant for systems with low service availability (corresponding to a high probability of scintillation).

3.3.3.8   Combining Tropospheric Attenuation Contributions

Various approaches have been proposed to estimate the net effect of the various contributions to the attenuation of electromagnetic signals from gaseous attenuation, rain, cloud and scintillation. ITU-R Recommendation P.618-8 (ITU-R P.618-8, 2003) gives the following method of combining these effects for a given probability P of the attenuation being exceeded

$$A_{\text{total}}(P) = A_g(P) + \sqrt{(A_r(P) + A_c(P))^2 + A_s(P)^2} \text{ (dB)} \qquad (3.31)$$

where $A_g(P)$ is the gaseous attenuation (using the appropriate water vapour content corresponding to P), $A_r(P)$ is the attenuation due to rain, $A_c(P)$ is the attenuation due to cloud and $A_s(P)$ is the fading due to scintillation. The form of equation (3.31) reflects a degree of correlation between the effects of rain, cloud and scintillation (COST-255, 2010).

As a large part of the cloud attenuation and gaseous attenuation is already included in the rain attenuation prediction for time percentages below 1%, the gaseous and cloud attenuation figures are constrained for $P < 1\%$ to their 1% values. An example of a complete tropospheric link impairment calculation is illustrated in Table 3.10 for availabilities of 90–99.99%.

3.3.3.9 Tropospheric Excess Delay

In certain applications (such as navigation), the excess time delay for propagation is a potential source of error that, given knowledge of the path, may be estimated. The excess phase and group delays are given by equations (3.9) and (3.10). The real part of the atmospheric refractive index n_{atm} depends on altitude and water vapour concentration (ITU-R P.453-8, 2001). Because the refractive index differences are small ($n \sim 1.00026$ at sea level), it is usual to express the atmospheric refractive index in terms of the *refractivity* N_{atm}, which is found to be well approximated by (Bean, 1962)

$$N_{atm} \equiv 10^6 (n_{atm} - 1) \simeq 77.6 \frac{P}{T} + 3.73 \times 10^5 \frac{\rho_w}{216.7\,T} \tag{3.32}$$

where P is the atmospheric pressure and T is the temperature (at the specified altitude). The first term is that due to the dry gases, while the second term is that due to the water vapour. These refractivity components vary approximately exponentially with altitude according to the different scale heights of these gases. The wet term also varies geographically and seasonally. Fortunately, the dominant effect on tropospheric delay is that due to the dry gases, and an average sea level temperature and pressure may be used, together with an average scale height and elevation angle, to estimate the excess delay.

3.3.4 Ionospheric Effects

We turn our attention next to the effect of the ionosphere on electromagnetic wave propagation. The strength of all ionospheric propagation impairments decreases rapidly with frequency, and, as a rule of thumb, ionospheric effects are significant only for operating frequencies below $\sim 3\,GHz$ (Ippolito, 1999).

3.3.4.1 Ionosphere Electron Density

Above about 50–100 km, the residual atmospheric gases are ionized by high-energy solar radiation, causing their separation into electrons and ions – forming the ionosphere. The maximum free electron concentration occurs at altitudes of about 200–400 km.

A key parameter in determining the strength of ionospheric phenomena is the total electron content (TEC) – the total electron count along the designated propagation path per square metre of cross-section. The total electron content N_t is found by integrating the electron density n_e along the path taken by the signal of interest.

$$N_t = \int_{path} n_e \cdot dl \tag{3.33}$$

Since the TEC for a given path will depend on elevation angle, it is generally more convenient to quote the Vertical TEC (VTEC) N_v, that is, the TEC for a zenith (vertical) path. Figure 3.11

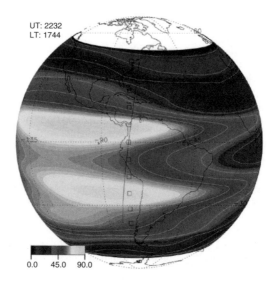

UT: 2232
LT: 1744

0.0   45.0   90.0

Figure 3.11 Map of Ionospheric Vertical Total Electron Content (VTEC). Reproduced by permission of © Naval Research Laboratory.

illustrates the global variation in VTEC derived from the US Naval Research Laboratory (NRL) SAMI3 ionosphere model. Since the Earth rotates with respect to the Sun at an inclination of 23.4° and also orbits the Sun, there is both a diurnal and a seasonal variation in the amount of solar ionization of the thermosphere for a particular latitude and longitude. In addition, there is a longer-term dependence on sunspot activity (which has an ~11 year cycle) and a dependence on solar flares. The TEC for any given location thus varies significantly with time, date and latitude, with VTEC typically varying between about 10^{16} and 10^{18} electrons/m^2, the higher figure relating to daytime at low latitudes during the peak of a solar cycle (Ippolito, 1999).

3.3.4.2 Propagation in a Plasma in a Static Magnetic Field

Although the ionosphere is composed of both electrons and ionized atoms (together with unionized gases), the dominant effect on the propagation of electromagnetic waves is due to the motion of electrons in the Earth's magnetic field. The motion of ions is much smaller owing to their much greater mass.

The electromagnetic characteristics of the ionosphere are determined by the equation of motion for electrons in a magnetic field. Of particular interest is the effect of the component of the Earth's magnetic field that is parallel to the direction of propagation. This field induces a circular electron motion around the direction of propagation, and the principal effect on propagation is an effective rotation of the wave polarization as the wave travels through the ionosphere.

As polarization rotation occurs, it is convenient to think in terms of the propagation of circular polarized waves.[15] Under the assumption of propagation nearly parallel to the static magnetic field – ignoring relativistic effects due to the contribution of the magnetic field of the propagating wave itself—one obtains the following expression for the refractive index n_p and extinction coefficient k_p for left- and right-hand circular polarized waves in an ionized medium under the influence

---

[15] We may decompose linear polarized waves into a pair of orthogonal circular polarized waves.

of a static magnetic field:

$$n_{p\pm} \simeq 1 - \frac{\omega_p^2}{2\omega^2}\left(1 \mp \frac{\Omega_p}{\omega}\right) \tag{3.34}$$

$$k_{p\pm} \simeq \frac{\nu\,\omega_p^2}{2\omega^3} \tag{3.35}$$

where the '$\pm$' sign signifies either LHCP (-) or RHCP (+) waves.

The parameter ω_p is known as the *plasma frequency*. Below the plasma frequency, waves are rapidly attenuated by the plasma. The parameter Ω_p is the cyclotron frequency – the frequency at which electrons would rotate around the magnetic field lines. The parameter ν is the plasma recombination rate. The plasma and cyclotron frequencies are related to the electron density and magnetic field via

$$\omega_p \equiv \sqrt{\frac{n_e e^2}{m_e \epsilon_0}} \tag{3.36}$$

$$\Omega_p \equiv \frac{e B_\parallel}{m_e} \tag{3.37}$$

where $\epsilon_o = 8.8542 \times 10^{-12}$ F/m is the permittivity of free space and $B_\parallel$ is the component of the (static) magnetic field parallel to the direction of wave propagation. It is apparent that the influence of the ionosphere varies as $1/f^2$, and therefore is generally significant only at lower frequencies.

Excess Delay

For frequencies of interest here, the wave frequency is much higher than both the recombination rate and the cyclotron frequency (i.e. $\omega \gg \nu,\ \Omega$). Under these conditions, the dependences of the excess ionospheric phase and group delays on frequency and TEC are (ignoring small polarization differences)

$$\Delta t_p \simeq -\frac{e^2 N_t}{2 c m_e \varepsilon_o \omega^2} = -\frac{1.34 \times 10^{-7}}{f^2} N_t \tag{3.38}$$

$$\Delta t_g \simeq +\frac{e^2 N_t}{2 c m_e \varepsilon_o \omega^2} = +\frac{1.34 \times 10^{-7}}{f^2} N_t \tag{3.39}$$

3.3.4.3 Faraday Rotation

Referring to equation (3.34), it is apparent that left-hand and right-hand circular polarized waves experience slightly different refractive indices in an ionized medium under the influence of the Earth's magnetic field, with the result that the propagation of the left-hand polarized component is slightly *slower* than that of the right-hand polarized component. If the input polarization is linear (which may be decomposed into equal amplitudes of left- and right-hand polarizations), then the result will be to effect a rotation in the polarisation vector at the output – an effect known as Faraday rotation. For this reason, the use of linear polarization is generally avoided for frequencies below $\sim$3 GHz. Table 3.11 gives example Faraday rotations at various frequencies (Goodman and Aarons, 1990; ITU-R P.531-9, 2007). A first estimate of the impairments at other frequencies and elevation angles may be obtained assuming a $\sec(\epsilon)/f^2$ scaling relationship.

Table 3.11 Estimated maximum ionospheric effects for a 30° elevation path (adapted from Goodman and Aarons, 1990; ITU-R 531-9, 2007)

Effect	100 MHz	1 GHz	10 GHz
Faraday rotation	30 rotations	108°	1.08°
Propagation delay	25 μs	250 ns	2.5 ns
Refraction	<1°	<0.6'	<0.36"
AOA	20'	12"	0.12"
Absorption (auroral/polar)	5 dB	0.05 dB	0.0005 dB
Absorption (mid-latitude)	<1 dB	<0.01 dB	<0.0001 dB
Scintillation		(refer to text)	

3.3.4.4 Ionospheric Absorption

Ionospheric absorption occurs owing to the recombination of electrons and ions, and enhanced ionospheric attenuation is observed at both low and high latitudes. At the polar caps, periods of excess absorption, lasting a few days at a time, occur during peaks of high solar activity owing to the effect of energetic particles trapped by the Earth's magnetic field near the poles. Typically, there may be of the order of a dozen such events per year (Ippolito, 1999).

Around that part of the ionosphere nearest to the Sun, energetic particles cause excess ionization that creates an aurora. This phenomenon takes the form of an elongated oval-shaped region of enhanced ionization around the point closest to the Sun (which moves during the day). Enhanced ionospheric absorption occurs with durations of the order of hours in the auroral oval region, associated with solar activity. Both auroral and polar cap absorption are generally important only for low operating frequencies.

3.3.4.5 Ionospheric Scintillation

Similar to tropospheric scintillation, ionospheric scintillation is the result of spatial non-uniformities in the refractive index of the ionosphere. There are two intense zones of scintillations, one at high latitudes, near the poles, and the other centred within approximately 20° of the magnetic equator. The scintillation fade depth varies with latitude, time of day, season, solar activity (e.g sunspot number) and also magnetic disturbances. Scintillation is most significant in the equatorial regions (within ~20° of the geomagnetic equator), approximately 1–2 h after sunset (Ippolito, 1999) (rather like a wake following the sunset), as illustrated by the dark patches on the map in Figure 3.12 (NWRA Online, 2010).

The modelling of ionospheric scintillation has proved difficult, with the result that prediction methods still incur a large degree of error. However, observations suggest that, at mid-latitudes, fading due to ionospheric scintillation is generally important only below ~1 GHz, although it can also be significant at higher frequencies during periods of high solar activity.

3.3.5 Multipath

We turn next to the effect on propagation of the local electromagnetic environment. When a user terminal – especially one with a non-directional antenna – is located near other objects, such as buildings, cars, etc., the potential exists for the wanted signal to take more than one path to the receiver. Electromagnetic waves may be reflected by nearby surfaces and diffracted by nearby edges (corners), with the result that the total signal may comprise contributions from multiple

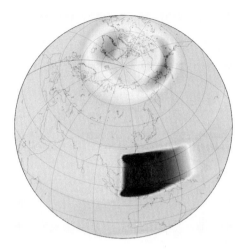

Figure 3.12 Global distribution of ionospheric plasma irregularities for 1300 UT 21 March 2000. Reproduced by permission of © North West Research Associates, Inc.

paths – hence the term *multipath*. One impact on the signal is that the resultant signal strength may vary (fade) as a result of the (position-dependent) destructive interference between the various paths. The amplitude of the signal can also increase slightly if these paths add constructively. The integrity of the signal may also be corrupted if the variation in propagation delay for the different paths exceeds the shortest period over which the data signal changes. Lastly, any motion of either transmitter or receiver will potentially cause a different shift in frequency due to the Doppler effect for each path. Propagation in the multipath environment is dependent on the application area (i.e. land, air or sea) and satellite type (LEO or GEO). A generic treatment is introduced here, but for a detailed treatment of these environments the reader is directed to the wider literature.

3.3.5.1 Fading

In the mobile user environment, signal fading is commonly classified according to how rapidly the signal level varies with user motion:

- *Fast fading* is commonly used to describe fading mechanisms where the amplitude varies rapidly with user motion. It is generally caused by destructive interference between signals taking different paths as a result of reflection and diffraction effects in the local environment, and thus occurs on a scale equivalent to the signal wavelength. Fast fading is typically most significant in urban environments.
- *Slow fading* is commonly used to describe fading mechanisms where the amplitude varies slowly with user motion (on a scale much greater than the wavelength). Examples of slow fading are due to shadowing by trees, bridges, street furniture, tall buildings, etc.

Fading may also be characterized according to its frequency dependence:

- *Flat fading* results when the coherence bandwidth (inversely related to spread in time delay for the various propagation paths) of the propagation channel is wide compared with the bandwidth of interest, and all frequency components suffer similar fading.

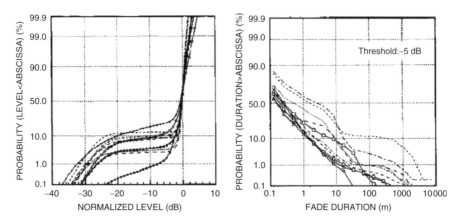

Figure 3.13 CDFs (left) for the normalized received L-band satellite signal power and fade duration (right) for −5 dB fade threshold (Tanaka, *et al.*, 1994). Reproduced by permission of © 1994 IEEE.

- *Frequency-selective fading* results when the coherence bandwidth is narrow compared with the bandwidth of interest, and different frequency components in the signal suffer very different fading.

3.3.5.2 Fading Models for Mobile Satellite Applications

As, in general, we have no information regarding the number, direction, amplitude or phase of the multipath contributions, or shadowing of the direct path, it is generally useful to consider statistical distributions derived from a physical model whose parameters can be adjusted to reflect

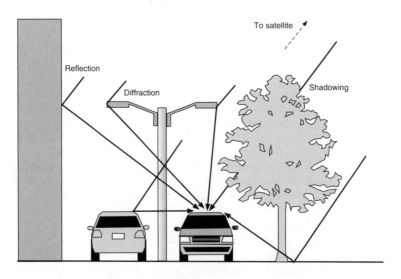

Figure 3.14 The multipath environment.

different generic environments – such as urban, rural, highway, etc. Figure 3.13 serves to illustrate the spread in statistical characteristics of fade depth and fade duration for a number of mobile satellite environments in Japan (Tanaka *et al.*, 1994). Each plot represents the statistics of fade depth and duration experienced along different roads in Japan.

Consider the case, illustrated in Figure 3.14, where the received signal comprises the sum of a direct signal, together with multiple indirect-path signals (from multiple directions and with varying amplitudes and delays/phases), resulting from reflection and/or diffraction from nearby objects.

For convenience, we shall continue to assume a sinusoidal wanted signal, whereupon the received narrowband signal may be expressed in terms of its instantaneous envelope r and relative phase φ (i.e. compared with the direct path). If the amplitude of the direct signal contribution is v, and the amplitude and phase of the ith indirect signal contribution is a_i and φ_i respectively, the received signal will be the summation of the various contributions

$$r \cos(\omega t + \varphi) = v \cos(\omega t) + \sum_{i}^{N} a_i \cos(\omega t + \varphi_i) \tag{3.40}$$

and the received signal power will be proportional to the signal envelope r squared.

The phases φ_i of the indirect contribution are assumed to be random and uniformly distributed between 0 and 2π rad, whereupon the in-phase and quadrature indirect signal contributions are considered to have uncorrelated but identical statistical distributions.

We can estimate the *probability* that the envelope takes a particular value by determining the probability distribution for the envelope r.

3.3.5.3 Rician Fading

Consider the case where the received signal comprises a mixture of contributions from both direct (line-of-sight) and indirect paths. This scenario is particularly relevant to satellite services, as the low received power in many satellite applications often (but not always) dictates direct line-of-sight visibility of the satellite by the user terminal in order to achieve acceptable received power. By contrast, in terrestrial communications, it may be common for no direct signal to be present.

In this case, the in-phase signal component comprises a constant-amplitude direct-path signal plus a Gaussian distributed indirect signal, while the quadrature phase component contains a Gaussian distributed indirect signal only. The PDF of the resultant envelope r is given by the Rice distribution (Bennett, 1956). It is common to express the Rice PDF in terms of the power ratio of the direct to multipath signal contributions $K \equiv \frac{v^2}{2\sigma_R^2}$ and mean signal amplitude $s = \sqrt{v^2 + 2\sigma^2}$ (Norton *et al.*, 1955), whereupon the Rice probability density function can be expressed as (Corazza and Vatalaro, 1994)

$$\rho(r)_{\text{Rice}} = 2(K+1)\frac{r}{s^2}\exp\left(-(K+1)\frac{r^2}{s^2} - K\right) I_0\left(2\frac{r}{s}\sqrt{(1+K)\,K}\right) \tag{3.41}$$

Expressing Rician fading in this way allows us to characterize the local fast-fading multipath environment in terms of a single parameter, the Rice K factor. The CCDF under Rician fading is given by the Marcum Q-function (a tabulated function) (Proakis, 2000). In the special case where there is no direct path ($K \to 0$), the Rice distribution simplifies to the Rayleigh distribution.

3.3.5.4 Log-Normal Fading

Ricean/Rayleigh fading results from signal contributions from multiple indirect paths. In addition, we must take account of variations in signal strength owing to intermittent shadowing – partial obstruction of the propagation path by objects such as trees lining a highway, or street furniture (such as streetlights), roadsigns, bridges, building superstructures, etc.

Fading due to shadowing is observed to approximate a log-normal distribution, given by

$$\rho\,(r)_{\text{LN}} = \frac{1}{\sqrt{2\pi}\,\sigma_{\text{LN}}\,r}\,\exp\left(-\frac{(\ln\,(r) - \mu_{\text{LN}})^2}{2\,\sigma_{\text{LN}}^2}\right) \tag{3.42}$$

where μ_{LN} and σ_{LN} are the mean and standard deviation of the log signal envelope respectively.

3.3.5.5 Composite Fading Models

Real-life mobile satellite environments display a mixture of multipath and shadowing, and the basis on which these contributions should be combined will depend on the particular range of environments being modelled (Loo, 1985; Lutz et al., 1991; Vucetic and Du, 1992; Corazza and Vatalaro, 1994). A *composite* fading model is thus needed to describe the fading characteristics illustrated in Figure 3.13.

If we have a probability distribution function of two variables, rather than just one, and we know the PDF of one of these variables on its own, then, from the theorem of total probability, the total PDF can be expressed in terms of the *conditional PDF*

$$\rho\,(x) = \int_{-\infty}^{\infty} \rho(x \mid z)\,\rho\,(z)\,\mathrm{d}z \tag{3.43}$$

where $\rho\,(x \mid z)$ is the conditional PDF, that is the probability of x, given a specific value of z.

In the model due to Loo (1985), the received signal is assumed to comprise Rician fading with the direct-path signal amplitude subject to log-normal fading but with a constant indirect-path contribution. The model of Loo is intended for the situation where the direct signal path is available for most of the time, and is generally considered to be most appropriate for rural environments. Loo's model tends to a log-normal distribution for large envelope amplitudes r and to a Rayleigh distribution for small r.

In the model due to Corazza and Vatalaro (1994), the mobile satellite channel fading is also assumed to be Rician, but in this model *both* the direct and indirect paths are assumed to be simultaneously and equally affected by log-normal fading via the signal level. The Rice factor is assumed to be constant. The model of Corazza and Vatalaro is reasonably successful at describing observed fading for different environments, using just three parameters: K (the Rice factor), μ_{LN} (log-normal mean) and σ_{LN} (log-normal standard deviation).

In order to account for the user mobility between different environments, various multistate models have been proposed. For example, the two-state, fading model due to Lutz et al., (1991) assumes that the propagation channel is alternately 'good' and 'bad' as a result of intermittent shadowing, where the 'good' propagation channel statistics apply for those areas with an unobstructed view of the satellite. Lutz's model assumes Rician fading for the 'good' channel case and Rayleigh fading with log-normal distributed power in the 'bad' channel case (in other words, it is assumed that no direct signal is present in the 'bad' channel case). The proportions of 'good' and 'bad' channels are determined by an additional parameter that quantifies the fractional time for which the channel quality may be classed as 'bad', that is, when the direct path is shadowed by obstacles.

3.3.5.6 Multipath Doppler Spectrum

The Doppler shift in received frequency owing to relative motion between the satellite and user terminal is discussed in Chapter 7, with emphasis on the motion of the satellite. For the multipath environment, the received signal spectrum will be the sum of a Doppler-shifted direct-path signal, as discussed above, and multiple indirect-path contributions, which will have a spread of Doppler shifts because of the different paths taken to the receiver (and hence relative velocities). Thus, in addition to an offset in centre frequency, the received signal bandwidth is also broadened.

The Doppler spectrum for the Rayleigh fading component (i.e. no direct path) has been estimated by Jakes (1994) for an unmodulated signal and omnidirectional user terminal antenna, assuming equal probability of indirect-path contributions from all directions in the plane of motion. The Jakes model assumes multipath contributions in the plane of motion only, and was extended by Aulin (1979) to include the effects of non-planar scattering (i.e. three-dimensional multipath), and by Kasparis, King and Evans (2007) for the effect of non-omnidirectional antennas.

In practice, the observed spectrum is likely to be broadened by modulation of the emitted signal (to allow it to carry useful information), with the result that the observed spectrum will be a convolution (smearing) of the modulated spectrum and Doppler spectrum. In such cases, the significance of the Doppler broadening will depend on the ratio of the maximum Doppler shift to the modulated bandwidth of the signal.

Revision Questions

1. Outline the mission of the ITU-R. How often does the World Radio Conference meet?
2. Identify the IEEE frequency bands associated with: (a) 1.575 GHz, (b) 3.5 GHz, (c) 11 GHz, (d) 30 GHz, (e) 47 GHz.
3. Indicate which frequency band might be used to provide a direct broadcast service.
4. Which frequency ranges are currently allocated for future services from high altitude platforms?
5. List the various propagation impairments that might affect a satellite service. Which effect may be expected to dominate for high-availability services operating at frequencies above 10 GHz?
6. Outline the difference between linear and circular polarization. What are the parameters of the polarization ellipse for circular polarization? Give two reasons why one would use circular polarization for mobile satellite services operating at L-band.
7. Estimate the gaseous attenuation, attenuation due to cloud, attenuation due to rain, and their total, exceeded for 0.01% of the time for the 20 GHz downlink of a K-band geostationary broadcast satellite link to a terminal located at sea level in Rome, Italy (latitude 41.9° N), assuming an elevation angle of 30°.
8. Repeat the previous question for the case of a 1.6 GHz mobile satellite service. Describe the impact of the shape and orientation of raindrops on propagation.
9. Outline the potential impairments for a *mobile* satellite service. Which statistical propagation models are most appropriate where (a) there is a strong direct signal, (b) there is no direct signal, (c) there is shadowing and (d) the environment varies?
10. What is the polarization loss for a RHCP signal and linear polarization antenna? Using the polarization vector notation, demonstrate that the polarization loss for an RHCP signal with an LHCP signal is infinite.

References

Aulin, T. (1979) A modified model for the fading signal at a mobile radio channel. *IEEE Trans. Vehicular Technology*, **28**(3), 182–203.

Balanis, C.A. (1997) *Antenna Theory, Analysis and Design*, 2nd edition. John Wiley & sons, Inc., New York, NY.

Barry, R.G. and Chorley, R.J. *Atmosphere, Weather and Climate*, Routledge, New York, NY.

Bean, B.R. (1962) The radio refractive index of air. *Proceedings of the IRE*, March, 260–273.

Bennett, W.R. (1956) Methods of solving noise problems. *Proceedings of the IRE*, **44**(5), 609–638.

Born, M. and Wolf, E. (1997) *Principles of Optics: Electromagnetic Theory of Propagation, Interference and Diffraction of Light*, 6th edition, Cambridge Univerity Press, Cambridge, UK.

Capsoli, C., Fedi, F. and Paraboni, A. (1987) A comprehensive meteorologically orientated methodology for the prediction of wave propagation parameters in telecommunication applications beyond 10 GHz. *Radio Science*, **22**(3), 387–389

Corazza, G.E. and Vatalaro, F. (1994) A statistical model for land mobile satellite channels and its application to nongeostatonary orbit systems. *IEEE Trans. on Vehicular Technology*, **43**(3), 738–742.

COST-255 (2010) *Radiowave Propagation Modelling for SatCom Services at Ku-band and above*. European Cooperation in the Field of Scientific and Technical Research (COST) 255. Available: http://www.cost255 .rl.ac.uk [accessed February 2010].

Crane, R.K. (1980) Prediction of attenuation by rain. *IEEE Trans. on Communications*, **28**, 1717–1723.

Feldhake, G.S. and Ailes-Sengers, L. (2002) Comparison of multiple rain attenuation models with three years of Ka band propagation data concurrently taken at eight different locations. *Online Journal of Space Communication*, (2). Available: http://satjournal.tcom.ohiou.edu/issue02/pdf/paper_feldhake.pdf [accessed February 2010].

Garace, G.C. and Smith, E.K. (1990) A comparison of cloud models. *IEEEE Antennas and Propagation Magazine*, October. 32–38.

Goodman, J.M. and Aarons, J. (1990) Ionospheric effects on modern electronic systems. *Proceedings of the IEEE*, **78**(3), 512–528.

Hahn, C.J. and Warren, S.G. (2007) A gridded climatology of clouds over land (1971–96) and ocean (1954–97) from surface observations worldwide, 2007, Carbon Dioxide Information Analysis Centre, Oak Ridge National Laboratory, Dataset NDP-026E.

IEEE (1993) Standard definitions of terms for antennas, June.

IEEE (1984) Standard letter designations for radar-frequency bands, IEEE 521-1984, 22 March.

Ippolito, L.J. (1981) Radio propagation for space communications systems. *Proceedings of the IEEE*, **69**(6), 697–727.

Ippolito, L.J. (1999) *NASA Handbook: Propagation Effects Handbook for Satellite Systems Design*.

ITU Mission Statement (2010) http://www.itu.int/ITU-R/index.asp?category=information&rlink=mission-statement&lang=en [accessed February 2010].

ITU Online (2010) http://www.itu.int [accessed February 2010].

ITU-R Radio Regulations (2008) Vol. 1, Article 5.

ITU-R P.453-8 (2001) Recommendation P.453-8, The radio refractive index: its formula and refractivity data.

ITU-R P.531-9 (2007) Recommendation P.531-9, Ionospheric propagation data and prediction methods.

ITU-R P.618-8 (2003) Recommendation P.618-8, Propagation data and prediction methods required for the design of Earth–space telecommunication systems.

ITU-R P.676-6 (2005) Recommendation P.676-6, Attenuation by atmospheric gases.

ITU-R P.836-3 (2001) Recommendation P.836-3, Water vapour: surface density and total columnar content.

ITU-R P.837-4 (2003) Recommendation P.837-4, Characteristics of precipitation for propagation modelling.

ITU-R P.838-3 (2005) Recommendation P.838-3, Specific attenuation model for rain for use in prediction methods.

ITU-R P.839-3 (2001) Recommendation P.839-3, Rain height model for prediction methods.

ITU-R P.840-3 (1999) Recommendation P.840-3, Attenuation due to clouds and fog.

ITU-R P.1621-1 (2001) Recommendation P. 1621, Propagation data required for the design of Earth–space systems operating between 20 THz and 375 THz.

Jakes, W.C. (1994) *Microwave Mobile Communications*, Wiley-IEEE Press.

Karasawa, Y., Yamada, M. and Allnut, J.E. (1988) A new prediction method for tropospheric scintillation on Earth–space paths. *IEEE Trans. on Antennas and Propagation*, **36**(11), 1608–1614.

Kasparis, C., King, P. and Evans, B.G. (2007) Doppler spectrum of the multipath fading channel in mobile satellite systems with directional terminal antennas. *IET Communications*, **1**(6), 1089–1094

Laws, J.O. and Parsons, D.A. (1943) The relation of raindrop size to intensity. *Trans. American Geophysical Union*, **24**, 452–460.

Liebe, H.J. (1981) Modelling attenuation and phase of radio waves in air at frequencies below 1000 GHz. *Radio Science*, **16**, 1183–1199.

Lipson, S.G., Lipson. H. and Tannhauser, D.S. (1995) *Optical Physics*, Cambridge University Press, Cambridge, UK.

Loo, C. (1985) A statistical model for a land mobile satellite link. *IEEE Trans. on Vehicular Technology*, **VT-34**(3), 122–127.

Lutz, E., Cygan, D., Dippold, M. and Papke, W. (1991) The land mobile satellite communication channel – recording, statistics and channel model. *IEEE Trans. on Vehicular Technology*, **40**(2), 375–386.

Maral, G. and Bousquet, M. (2002) *Satellite Communications Systems*. John Wiley & Sons, Ltd, Chichester, UK.

Marshall, J. and Palmer, W. (1948) The distribution of raindrops with size. *Journal of Meteorology*, **5**, 165–166.

Norton, K.A., Vogler, L.E. Mansfield, W.V. and Short, P.J. (1955) The probability distribution of the amplitude of a constant vector plus a Rayleigh-distributed vector. *Proceedings of the IRE*, October, 1354–1361.

NWRA Online (2010) North West Research Associates. Available: http://www.nwra.com/ [accessed February 2010].

Olsen, R.L., Rogers, D.V. and Hodge, D.B. (1978) The aR^b relation in the calculation of rain attenuation. *IEEE Trans. on Antennas and Propagation*, **AP-26**, 318–329.

Proakis, J.G. (2000) *Digital Communications*. McGraw-Hill, New York, NY.

Pruppacher, H.R. and Pitter, R.L. (1971) A semiempirical determination of the shape of cloud and rain drops. *Journal of Atmospheric Science*, **28**, 86–94.

Ramo, S., Whinnery, J.R., and Van Duzer, T. (1965) *Fields and Waves in Communication Electronics*, John Wiley & Sons Inc., New York, NY.

Rice, P.L. and Holmberg, N.R. (1973) Cumulative time statistics of surface-point rainfall rates. *IEEE Trans. On Communications*, **COM-21**(10).

Robertson, G. (1980) Effective radius for refractivity of radio waves at altitudes above 1 km. *IEEEE Trans. on Antennas and Propagation*, **34**(9), 1080–1105.

Rogers, D.V. (1985) Propagation considerations for satellite broadcasting at frequencies above 10 GHz. *IEEE Journal on Selected Areas in Communicatons*, **3**(1), 100–110.

Stutzmann, W.L. and Dishman, W. (1982) A simple model for the estimation of rain-induced attenuation along Earth-space paths at millimetre wavelengths. *Radio Science*, Vol **17**(6), pp. 946.

Suzuki, H. (1977) A statistical model for urban radio propagation. *IEEE Trans. on Communications*, **25**(7), 673–680.

Tanaka, K., Obara, N., Yamamoto, S. and Wakana, H. (1994) Propagation characteristics of land mobile satellite communications in Japan using ETS-V satellite. *IEEE Vehicular Technology Conference*, June 1994, pp. 929–933.

van de Hulst, H.C. (1981) *Light Scattering by Small Particles*, Dover, New York, NY.

Vucetic, B. and Du, J. (1992) Channel modelling and simulation in satellite mobile communication systems. *IEEE Journal on Selected Areas in Communications*, **10**(8), 1209–1218.

4

Antennas and Noise

4.1 Introduction

In the previous chapter we considered the use of electromagnetic waves – in particular, radio waves – to convey useful information to/from the satellite (or HAP), and discussed, in some detail, the various impairments to propagation that result from the presence of the Earth's atmosphere. In this chapter, we consider how we launch and collect these electromagnetic waves efficiently, and the key device in this respect is the *antenna*.

Our principal aim is to provide the reader with a basic understanding of the key system properties of antennas. Such an understanding allows us to determine the transmission (loss) between two antennas, such as might occur where a signal is broadcast from the satellite to a user terminal (or vice versa). This is an important precursor in the determination of received signal strength.

There are a bewildering number of different antenna designs in use for satellite applications, and probably several times this number in use for other (i.e. terrestrial) applications. Our intention here is to give the reader an appreciation of the *generic* antenna types used for various types of satellite application (fixed, mobile, handheld, etc.), together with some illustrative examples. In this respect we have categorized antennas according to their gain (their ability to focus signals in specified angular ranges) – low medium or high.

Our focus in this chapter is primarily on radio waves. We shall, however, briefly discuss optical antennas (i.e. telescopes) used for remote sensing at infrared and visible light frequencies. and future very high-data-rate communications to satellites and HAPs. The reader may be surprised to learn of a close resemblance between the designs of high-performance space telescopes and those of certain high-gain radio antennas. The explanation is that those same antenna designs were originally derived from optical telescopes (which predate antennas by several hundreds of years).

Our discussion of antennas will be essential as we next turn our attention within this chapter to the sources of noise in a satellite receiver (which may be either in the user's satellite terminal or in the satellite itself). It is well known that all *real-world* signals are corrupted to some degree by noise and the reader will no doubt be familiar with the concept of noise generated within electronic devices (such as those within the satellite receiver). However, received noise is also collected by the antenna – from terrestrial, atmospheric, galactic and even cosmic sources. Quantifying the dominant noise sources is an important step in determining the received *signal-to-noise* ratio, which, as subsequent chapters will show, ultimately determines the maximum data rate for satellite broadcast and communications links (or, for that matter, the sensitivity of remote sensors).

Satellite Systems for Personal Applications: Concepts and Technology   Madhavendra Richharia and Leslie David Westbrook
© 2010 John Wiley & Sons, Ltd

4.2   Antennas

Thus far, we have discussed the propagation of electromagnetic waves without consideration of how we launch and recover unguided electromagnetic waves between satellites and the Earth. Antennas[1] perform this vital function. They are the interface between unguided waves and the guided waves contained within a satellite or user terminal. Note that, although our primary focus in this section will be antennas for use with radio waves, this definition of an antenna is sufficiently general to allow the term to be used for other types of electromagnetic wave – including optical 'antennas' (more usually known as telescopes).

4.2.1   General Concepts

4.2.1.1   Guided versus Unguided Waves

Electromagnetic waves passing between satellite and user terminals on or near the ground are examples of *unguided* waves. Antennas effect the transition between these unguided electromagnetic waves and *guided* waves which can be routed within the satellite or user terminal to the appropriate electronics. Such guided waves are contained within *waveguides*. Familiar forms of waveguide include coaxial cable, rectangular and circular metal waveguide (a hollow metal tube) and microstrip – the latter being particularly suited to Printed Circuit Board (PCB) manufacturing techniques and the uses of surface-mount electronic components.

A transmitting antenna thus takes an output electrical signal from a transmitter waveguide and launches it as an unguided wave (into space or the atmosphere), while a receiving antenna captures some fraction of an incoming unguided wave and channels it into the receiver waveguide input.

4.2.2   Antenna Properties

A full exposition of the properties of antennas is beyond the scope of this book, and interested readers are directed to the relevant texts contained in Balanis (1997) and Kraus (1988). Nevertheless, it will be useful, within the context of this book, to have an appreciation of some of the key properties and parameters of antennas, in order that we can develop our understanding of the constraints on satellite service performance.

The vast majority of antennas are passive, reciprocal devices – that is, they function equally when used to transmit (guided to unguided wave) and receive (unguided to guided wave) – exceptions to this generalization being specialized antennas that contain non-reciprocal active electronic elements (such as RF amplifiers, isolators, etc). It is convenient when introducing antenna properties to consider first the antenna as being transmitting (radiating) and thereafter to consider the antenna as receiving, and finally to demonstrate the relationship between these properties resulting from reciprocity.

A fundamental attribute of a transmitting antenna is the degree to which the radiated unguided energy is distributed in angle. The equivalent property for receiving antennas would be the degree to which antenna sensitivity to incoming radiation is distributed in angle. In both cases we shall assume that we are sufficiently far away from the antenna to be in its far-field (the region where the electromagnetic field is effectively a spherical wave with field components normal to the direction of propagation).

---

[1] Note that the Oxford English Dictionary gives the plural of antenna (as in radio aerials) as *antennas* – in contrast to antennae, which are the sensory appendages on the heads of insects, etc.

Spherical Polar Coordinate System

When considering antenna radiation properties, where we are generally interested in the variation in radiation strength with angle and distance, it is usual practice to use the spherical polar coordinate system $\{R, \theta, \phi\}$ rather than the normal Cartesian coordinates $\{x, y, z\}$. Here, R is the radial distance from the origin, θ is the zenith angle ('elevation' angle measured downwards from the z-axis – typically oriented along the antenna boresight) and ϕ is the azimuth angle (in the $x-y$ plane).

4.2.2.1   Antenna Directivity

Consider first a transmitting antenna. We define the radiation intensity $I(\theta, \phi)$ resulting from this antenna in a given direction defined by the spherical polar angles θ and ϕ as the power radiated in that direction per unit solid angle (IEEE, 1993).

Much reference is made in the field of antennas to an *isotropic radiator* – a hypothetical antenna that radiates equally in all directions – as, for an isotropic radiator, the radiation intensity will be the same in all directions and equal to the power accepted by the antenna divided by 4π (since there are 4π sr in a sphere).

We may quantify the degree to which a real antenna discriminates in favour of some directions at the expense of others in terms of its *directivity*. Antenna directivity $D(\theta, \phi)$ is defined as the ratio of the radiation intensity in a given direction divided by the radiation intensity averaged over all directions. In terms of the total radiated power, its *directivity* is thus

$$D(\theta, \phi) = 4\pi \frac{I(\theta, \phi)}{P_{\text{rad}}} \tag{4.1}$$

Clearly, for the hypothetical isotropic radiator mentioned earlier, the directivity is unity in all directions. Figure 4.1 illustrates antenna directivity polar plots for example low-gain (turnstile antenna), medium-gain (horn antenna) and high-gain reflector (prime focus reflector) antennas. Note that the radial scale is logarithmic (in dBi).

A directional antenna radiation pattern typically comprises a single main lobe, together with a number of lesser *sidelobes* (and/or backlobes). Much effort is expended in designing antennas to maximize the efficiency of the main lobe and minimize the average level of sidelobes – which affect received noise and interference from unwanted directions. We are typically interested in the

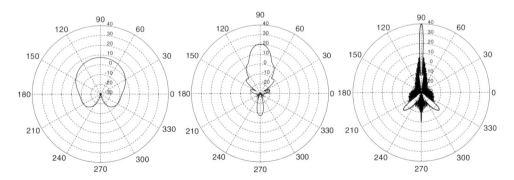

Figure 4.1   Example polar diagrams of antenna directivity (in dBi) for (left to right): crossed-dipole on a ground plane (turnstile antenna), horn and horn-fed reflector antennas.

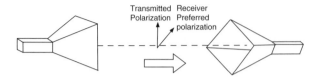

Figure 4.2   Antenna polarization mismatch.

maximum directivity D_{max} (for the antenna's preferred polarization), as this defines the maximum degree to which the antenna can direct radiated power in a single direction (usually the antenna boresight in a unidirectional antenna).

4.2.2.2   Antenna Polarization

Antennas are designed to radiate waves with a particular polarization. This is usually either vertical or horizontal linear polarization or either left- or right-handed circular polarization. An antenna will therefore have preferred polarization, and the directivity may be resolved into directivities for two orthogonal polarizations (generally referred to as copolarization and cross-polarization). In general, directivities for the preferred and orthogonal polarizations are significantly different. The polarization mismatch loss factor resulting from non-optimum alignment of the receive antenna (refer to Figure 4.2) was defined previously in Chapter 3.

4.2.2.3   Antenna Gain

Although a fundamental antenna characteristic, the directivity of an antenna is rarely quoted, as it relates the radiation intensity to the radiated power rather than to the *input* power. The radiated power will always be less than the input power in practical antennas, owing to dissipative (ohmic) losses in the antenna construction and feed network, and a more useful property of antennas is antenna *gain*.

   Antenna gain for a transmitting antenna is defined as the radiation intensity in a particular direction divided by the average radiation intensity that would result if all of the accepted input power were radiated, the latter being equal to the accepted input power P_{in} divided by 4π sr (we shall explore the precise meaning of *accepted* input power later in Section 4.2.2.5. Antenna gain is thus defined as

$$G(\theta, \phi) \equiv 4\pi \frac{I(\theta, \phi)}{P_{in}} \tag{4.2}$$

   Gain and directivity may be related by introducing the antenna radiation efficiency η_{rad}, defined as the ratio of the radiated power to the accepted input power, where upon

$$G = \eta_{rad} D \tag{4.3}$$

   A typical value for the radiation efficiency for a commercial antenna is in the region of 60–80%.
   Antenna gain is usually quoted relative to the hypothetical lossless isotropic radiator, in dBi (meaning dB relative to isotropic). Gain of circular polarization antennas is quoted in dBiC (the C indicating circular polarization). The *maximum* gain G_{max} (for the antenna's preferred polarization) is often of most interest.

4.2.2.4   Beamwidth

The Full-Width Half-Maximum (FWHM) beamwidth of an antenna is an important parameter that needs to be optimized for the particular application. The beamwidth of a satellite antenna defines the beam coverage area on the Earth. The beamwidth of a user terminal defines the degree to which the antenna needs to be pointed at the satellite, and the scope for interference to/from other satellites.

Relation Between Maximum Gain and Beamwidth

We have previously stated that antennas with high maximum gain have narrow beamwidths, and vice versa. As radiation intensity is defined as the power radiated per unit solid angle, we can relate the antenna maximum directivity to an equivalent beam solid angle Ω_B, defined as the solid angle through which all the radiated power would stream if the power per unit solid angle were constant at the maximum value of the radiation intensity through this solid angle and zero elsewhere. Maximum directivity and beam solid angle are related via

$$D_{\max} = \frac{4\pi}{\Omega_B} \qquad (4.4)$$

Gain Estimate

Equation (4.4) provides a useful way to estimate the maximum directivity for high-gain antennas. The beam solid angle for a high-gain 'pencil beam' antenna having antenna beamwiths $\Delta\theta_x$ and $\Delta\theta_y$ in the two principal axes can usefully be approximated[2] by $\Omega_B \approx \Delta\theta_x \, \Delta\theta_y$, from which the approximate maximum directivity (and potentially maximum gain) for specified antenna beamwidths may be estimated. Maximum antenna gain may thus be estimated using a suitable estimate for the radiation efficiency. Balanis (1997) gives the following approximate formula for maximum antenna gain 'for many practical antennas':

$$G_{\max} \approx \frac{30\,000}{(\Delta\theta_x)_{\deg} \left(\Delta\theta_y\right)_{\deg}} \qquad (4.5)$$

where $(\Delta\theta_x)_{\deg}$ and $\left(\Delta\theta_y\right)_{\deg}$ are the FWHM beamwidths, in degrees, for the two principal planes.

Beamwidth of Circular-Aperture Antennas

The FWHM beamwidth $\Delta\theta$ (in radians) for the important case of a circular-aperture antenna of diameter D may be calculated analytically for wavelength λ

$$\Delta\theta = k\frac{\lambda}{D} \qquad (4.6)$$

where the value of k depends on the distribution of illumination across the radiating aperture. For a uniformly illuminated aperture, $k \to 1.02$ rad (58.4°). However, the aperture illumination of practical antennas is normally tapered towards the edges (from a maximum at the centre) in order to optimize efficiency while controlling sidelobe levels. A more typical illumination taper results in $k \to 1.22$ rad (70°).

---

[2] More accurate approximations for maximum directivity as functions of beamwidth are available (Stutzmann, 1998; Kraus, 1988).

Table 4.1 Approximate antenna
pointing loss versus pointing error
(normalized to 3 dB beamwidth)

Pointing loss (beamwidths)	Pointing error
0.01 dB	0.029
0.02 dB	0.041
0.05 dB	0.065
0.1 dB	0.091
0.2 dB	0.13
0.5 dB	0.20
1.0 dB	0.29

Gain Approximation Near Boresight

The variation in antenna gain with offset angle for the main lobe near boresight is important when considering the accuracy required for antenna pointing and may be approximated (for each axis) by a parabola[3]

$$G\left(\theta\right) \approx G_{max}\left(1 - 2\left(\frac{\delta\theta}{\Delta\theta}\right)^2\right)$$ (4.7)

where $\delta\theta$ is the mispointing angle – the angle offset from the peak gain direction (boresight) for the axis under consideration. We may use this approximation to estimate the antenna mispointing loss for different pointing errors, expressed as a fraction of the FWHM (3 dB) antenna beamwidth for that axis, as indicated in Table 4.1.

4.2.2.5   Impedance and Bandwidth

The impedance of an antenna is the ratio of the voltage across the antenna terminals to the current flowing into the antenna terminals. In general, the impedance of an antenna is complex (in both senses of the word), the voltage and current generally being offset in phase. The fraction of the real part of the antenna impedance that corresponds to power radiated from the antenna is known as the *radiation resistance*. It must be noted that there is a fundamental relation between the volume of an antenna and its impedance bandwidth (Wheeler, 1947; Chu, 1948), which ultimately limits their size reduction.

Typically, an antenna is only useful over a limited bandwidth, and the useful bandwidth of an antenna may be defined in a number of ways, but it is common to describe the bandwidth as the frequency range over which the antenna impedance presents an acceptable *impedance mismatch loss*.

Impedance Mismatch

Our definition of antenna gain relates the radiation efficiency to the *accepted* input power. Let us now explore the meaning of this phrase. In electronic circuits and transmission lines, efficient

---

[3] When expressed in dBi, an alternative approximation for the main lobe gain is (for small angles)

$$(G\left(\theta\right))_{dB} \approx (G_{max})_{dB} - 12\left(\frac{\delta\theta}{\Delta\theta}\right)^2$$ (4.8)

power transfer depends on matching the load impedance to that of the source. In general, a fraction of the input signal will be reflected at the antenna input terminals back along the transmission line towards the source (where it is dissipated) and the ratio $\tilde{\Gamma}$ of the amplitude of the reflected wave to that of the input wave (in complex phasor notation) is related to the mismatch in source and antenna impedances (Ramo, Whinnery and Van Duzer, 1965). The power reflection coefficient is $\left|\tilde{\Gamma}\right|^2$.

We define the accepted input power to be that which is not reflected back towards the source. The ratio of the accepted input power to the maximum available input (source) power (i.e. the input power if the antenna were properly matched) is known as the impedance mismatch loss (ratio) L_Z given by

$$\frac{1}{L_Z} = \left(1 - \left|\tilde{\Gamma}\right|^2\right) \qquad (4.9)$$

An equivalent situation occurs for the receiving antenna, in which case the ratio of the power delivered from the antenna to the load via the transmission line relative to the maximum output power that would be delivered into a matched load is also given by equation (4.9).

4.2.2.6 Transmitter Figure of Merit

The radiation intensity in a particular direction is proportional to the product of the accepted power and the antenna gain times. This product is the Effective Isotropic Radiated Power (EIRP), defined as

$$\text{EIRP} \equiv P_t G_t \qquad (4.10)$$

where P_t is the accepted power in the transmitter antenna (after any transmission line losses), and G_t its gain. Physically, EIRP is the equivalent power that would have to be radiated by an, isotropic radiator to achieve the same intensity as the actual antenna in the direction of interest (usually in the direction of maximum antenna gain). The EIRP is thus an important figure of merit for a transmitting antenna. EIRP is typically given in dBW (dB relative to 1 W) or dBmW (dB relative to 1 mW).

4.2.2.7 Effective Aperture Area

Power Flux Density
We may relate the power flux density (PFD – the power per unit area) at some radius R from a transmitting antenna to the transmitted EIRP, ignoring any propagation losses, via

$$\text{PFD} = \frac{\text{EIRP}}{4\pi R^2} \qquad (4.11)$$

So far we have focused on the properties of a transmitting antenna in terms of the radiation intensity for a given input power. For a receiving antenna we may conveniently introduce an *effective area* (or effective aperture), defined as the ratio of the available power at the output terminals of a receiving antenna to the power flux density incident on the antenna from a specified direction (under the assumption that the input wave polarization is matched to that of the antenna). The concept of effective area is particularly intuitive for aperture antennas which have a defined physical aperture area, although, in general, the effective area is somewhat less than the physical area owing to inefficiencies.

Consider the arrangement of Figure 4.3. Using the definition for effective area, the received power delivered to the antenna terminals is given by

$$P_r = \text{PFD} \cdot A_e \qquad (4.12)$$

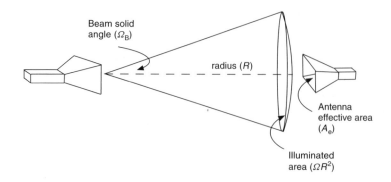

Figure 4.3   Beam solid angle, PFD and effective aperture.

where A_e is the effective aperture. As with directivity and gain, we are usually interested in the maximum effective area (the effective area in the direction of maximum antenna gain).

As a result of reciprocity, it should not matter whether the power flows from antenna 1 to antenna 2 or vice versa. As a consequence, we note that the product of the transmitter antenna gain and the receiver effective area is constant (i.e. $G_1 A_{e,2} = G_2 A_{e,1}$), which implies that, for *any* antenna, the ratio of effective area to maximum gain is a constant. Using this together with analytical expressions derived for the gain and effective area for a particular antenna – the short dipole – Friis (1946) derived a fundamental relationship between gain and effective aperture, valid for any antenna:

$$G = \frac{4\pi}{\lambda^2} A_e \qquad (4.13)$$

Approximate Gain of Aperture Antennas

Equation (4.13) provides a useful means to relate antenna maximum gain of an aperture antenna to its physical aperture area. We shall define the antenna aperture efficiency η_a to be the ratio of the maximum effective aperture area to the physical aperture area. For well-designed reflector antennas, aperture efficiency is typically in the region 60–80%.

Hence, for a circular aperture antenna (such as a circular reflector antenna) of diameter D, the maximum antenna gain is

$$G_{\max} \rightarrow \eta_a \left(\frac{\pi D}{\lambda} \right)^2 \qquad (4.14)$$

4.2.2.8   Antenna Arrays

Individual antennas may be connected together to create antenna arrays with desirable properties. In antenna arrays, the component antenna elements are made to radiate together with specified amplitude and phase relationships (for each frequency). For this reason they are also referred to as *phased arrays*. Antenna arrays permit the construction of antenna systems with high maximum gains from low- or medium-gain element antennas. Such arrays can also offer advantages in terms of their physical geometry – for example, the antenna may take the form of a (thin) flat panel or be made *conformal* to some surface. Phased arrays also provide additional degrees of flexibility with regard to control over radiation patterns – in principle, an array with N elements has $(N - 1)$ degrees of freedom available in the control of its radiation pattern. Lastly, phased arrays may be constructed

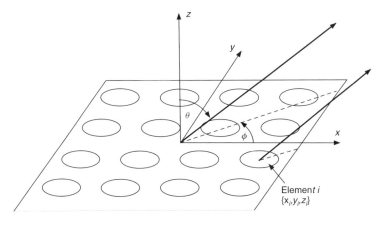

Figure 4.4 Planar phased array.

such that the phase relationship between the elements can be adjusted *electronically*, allowing the radiation pattern to be modified, at will – facilitating beam scanning and beam steering (although this flexibility usually comes at a significant additional cost, weight and power consumption).

Although antenna arrays may in principle take almost any shape, linear, circular and planar geometry arrays are the most common forms, with linear and planar arrays being the most relevant array configurations for personal satellite applications.

It will be useful to have a limited understanding of phased-array antennas; however, for a more detailed understanding of their behavior, the reader is directed to the many excellent textbooks in this field (Balanis, 1997; Kraus, 1988).

With reference to the planar array in Figure 4.4, consider the contribution to the radiated far-field of an antenna array owing to the ith element of the array, located at $\{x_i, y_i, z_i\}$. For a given far-field direction (θ, ϕ), the propagation phase advance (or delay) $\Delta\psi_i$ of this element, relative to the array reference origin, is given by

$$\Delta\psi_i = \frac{2\pi}{\lambda}\left(x_i\,\cos(\phi)\,\sin(\theta) + y_i\,\sin(\phi)\,\sin(\theta) + z_i\,\cos(\theta)\right) \qquad (4.15)$$

Now, if $\bar{\mathcal{E}}_i$ is the (complex phasor) electric field due to the ith element (referenced to the element coordinates), then the total farfield of the contributions from each array element (referenced to the array *origin*) is given by the sum of all field contributions

$$\bar{\mathcal{E}}_{\text{array}}(\theta, \phi) = \sum_{1}^{N} \tilde{w}_i\,\bar{\mathcal{E}}_i(\theta, \phi)\,\exp(i\,\Delta\psi_i) \qquad (4.16)$$

where $\tilde{w}_i$ is a (complex) weight for the ith element, introduced to control the radiation pattern. The phase and amplitude of the individual element weights may be used to control (or scan) the beam maximum gain direction and control beam shape and sidelobe levels.

Array Factor

Under the assumption that all of the antenna elements are identical and have the same relative orientation, and that electromagnetic coupling between elements can be neglected, the farfield patterns of all of the array elements (referenced to the elements themselves) will also be identical, and the total farfield may be expressed as the product of the element farfield and a (complex) *array factor* $\bar{AF}$ given by

$$\bar{AF} \equiv \sum_{1}^{N} \tilde{w}_i \, \exp\left(i \, \Delta \psi_i\right) \tag{4.17}$$

The array factor is essentially the radiation pattern of an equivalent array of isotropic elements. The array antenna gain G_{array} may thus be simply expressed as the product of the element gain G_{elem} and the modulus squared of the array factor

$$G_{\text{array}}\left(\theta, \phi\right) = G_{\text{elem}}\left(\theta, \phi\right) \left|\bar{AF}\left(\theta, \phi\right)\right|^2 \tag{4.18}$$

Array Factor Under Cophasal Excitation

A particularly simple method to direct maximum array gain in a particular direction is to set the phase of each element weight equal and opposite to the relative phase delay resulting from the element position in the array. This is known as *cophasal excitation*. For an array with a fixed beam pattern, this may be implemented using a passive feed network.

As the array factor is a summation over N elements, and as the array gain is proportional to the square of the array factor, it may be inferred that the maximum array gain can be enhanced by a factor of up to N^2 compared with that for a single element. However, in a passively fed phased array there is a requirement to split (for a transmit array) or combine (for a receive array) the signal N ways, in order to form a single feed point for the antenna. This splitting/combining results in a power loss factor of $\frac{1}{N}$. Consequently, in this type of phased array, the maximum gain enhancement is only a factor of N.

Linear Array

This result provides a useful first estimate of the number of elements needed to obtain a particular gain (or the gain given the number of elements). It may be used, for example, to estimate the maximum gain of a Yagi–Uda array antenna. The Yagi–Uda array is essentially a medium-gain linear array of elemental dipoles in which only one of the array elements is actually driven; the rest are excited *parasitically*.

Planar Array

The other common form of antenna array is the planar array. Typically, this comprises low-gain antenna elements (such as patch antennas) or sometimes medium-gain antennas (such as horn antennas) arranged in a rectangular or hexagonal grid. If the array comprises N closely packed aperture antenna elements, the maximum effective aperture of the array will tend to N times the effective aperture of a single element. Furthermore, if the effective area of the individual antenna elements is comparable with its physical aperture, then, for densely packed planar arrays, the maximum array gain will approach that of a single-aperture antenna with the equivalent total effective aperture.

Antenna array performance is sensitive to the interelement array spacing: too close and excessive mutual coupling degrades the beamwidth and impedance match; too far apart (more than half a wavelength) and a nominally unidirectional pattern will instead exhibit multiple (main) lobes – so-called grating lobes.

4.2.3   Transmission Between Two Antennas

We can now consider the transmission factor between two antennas – the ratio of output power at a receiving antenna to the accepted input power at a transmitting antenna (neglecting any transmission line losses and assuming that impedance, antenna orientation and polarization have all been optimized so as to maximize output). This may usually be written as (Friis, 1946)

$$P_r = P_t G_t G_r \left(\frac{\lambda}{4\pi R}\right)^2 \frac{1}{L_T} \tag{4.19}$$

where G_r is the receiver antenna gain and L_T is the total loss (ratio) comprising contributions from atmospheric propagation loss, polarization mismatch loss and transmitter and receiver impedance mismatch loss:

$$L_T \equiv \left(L_{Z,t}\, L_{atm}\, L_{pol}\, L_{Z,r}\right) \tag{4.20}$$

Equation (4.19) includes the effects of non-optimum antenna alignment if the antenna gains G_t and G_r are the gain of the antennas in the direction of the other antenna.

Free-Space 'Loss' Term

The term $\left(\frac{\lambda}{4\pi R}\right)^2$ is commonly referred to as the *free-space loss* (FSL). It is a 'loss' in the sense that; it results from the geometric dilution factor (inverse square law) according to $\frac{1}{4\pi R^2}$; it also reflects the variation in effective aperture of an isotropic antenna (relative to which all antenna gains are defined). The apparent frequency dependence associated with this free-space transmission thus arises purely from our use of a hypothetical isotropic antenna as a reference. The concept of frequency dependence of a term ostensibly associated with propagation in free space (with no physical rationale for any frequency dependence) offers significant potential for confusion, and the reader should exercise some care when using this form of the Friis equation to compare transmission at different frequencies.

Friis Equation for Point-to-Multipoint Links

Equation (4.19) is the most commonly used form of Friis's transmission equation and is appropriate when estimating the transmission between two antennas *at a specific frequency* in terms of available antenna gains. An alternative form of the Friis equation that provides a better physical picture of transmission, and one that is particularly appropriate when considering the point-to-multipoint transmission between a satellite antenna having *fixed coverage* on the Earth (and hence fixed beam solid angle and fixed gain) and a terminal antenna of a given size is

$$P_r = \left(\frac{P_t G_t}{4\pi R^2}\right) \frac{1}{L_T} A_{e,r} \tag{4.21}$$

where the first two terms combined represent the power flux density at the receiving antenna, including losses. It is apparent that, in this more physical description, there is no frequency dependence of transmission.

Friis Equation for Point-to-Point Links

Lastly, when considering point-to-point transmission between two antennas of *fixed size*, a third form of the Friis equation may sometimes be appropriate:

$$P_r = P_t A_{e,t} A_{e,r} \left(\frac{1}{\lambda R}\right)^2 \frac{1}{L_T} \tag{4.22}$$

Table 4.2   Effect of constraints on the frequency
dependence of the Friis transmission equation

Antenna A	Antenna B	f Dependence
Fixed beamwidth	Fixed beamwidth	$1/f^2$
Fixed beamwidth	Fixed aperture	None
Fixed aperture	Fixed aperture	f^2

In this particular case, and neglecting any frequency-dependent propagation impairments (contained in L_T), transmission loss is potentially improved by using higher frequencies owing to the improved beamwidth for a given antenna size at higher frequencies.

The different frequency dependences of these three perspectives of the same transmission equation are summarized in Table 4.2.

4.2.4   Antennas for Personal Satellite Applications

4.2.4.1   Types of Antenna

Antennas come in a bewildering range of shapes and sizes, depending on their intended application and design heritage. Indeed, there are probably more different antenna designs than pages in this book. Nevertheless, most of these designs are adaptations of a relatively small number of generic types. Furthermore, antennas for use in satellite applications represent a relatively small subset of the available antenna types.

To facilitate our discussion, we shall initially categorize antennas for use in satellite applications according to the general type of *radiation pattern* they produce. These patterns reflect the relative distribution of radiation as a function of angle for transmitting antennas and the relative distribution of sensitivity with angle for a receiving antenna. The following generic pattern types are considered here:

- torus-shaped, omnidirectional pattern (e.g. that for a common dipole antenna);
- low-gain (wide-beamwidth) unidirectional pattern (e.g. that for a patch antenna);
- medium-gain (moderate-beamwidth) unidirectional pattern (e.g. that for a horn antenna);
- high-gain (narrow-beamwidth) unidirectional pattern (e.g. that for a reflector antenna).

4.2.4.2   Torus-Shaped Omnidirectional Antennas

The torus-shaped omnidirectional antenna pattern is exemplified by the common *dipole* antenna and its cousin, the *monopole* (in effect, one-half of a dipole above a reflecting surface). This type of antenna, and its derivatives, is used in everyday devices such as mobile phones and wireless network devices. The familiar 'doughnut'-shaped radiation pattern is particularly attractive for use in terrestrial mobile applications where transmissions emanate from near the horizon and where the azimuth angle of the signal with respect to the antenna is not predictable in advance (and where the cost of implementing a signal tracking scheme would be prohibitive). However, the usefulness of such antennas for satellite applications (for which the elevation angle may be significant) is limited by their toroidal radiation pattern which has maximum gain at the horizon (normal orientation) and zero gain at the zenith. Furthermore, their linear polarization would result in angle-dependent polarization loss. Antennas with torus-shaped omnidirectional patterns therefore find limited application in satellite applications.

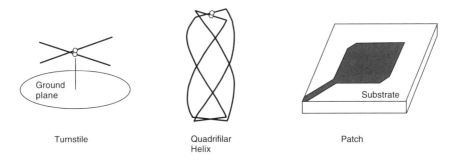

Figure 4.5   Low-gain antennas for circular polarization.

4.2.4.3   Low Gain (Near-Hemispherical-Coverage) Unidirectional Antennas

A more useful type of antenna pattern for use in mobile and handheld satellite applications is the low-gain unidirectional antenna with hemispherical (or near-hemispherical) coverage, as illustrated in Figure 4.1 (left). Correctly oriented, these antennas provide coverage of the upper hemisphere (where all visible satellites are located) and are useful where the azimuth of the satellite signal is unknown. No pointing of the antenna is needed in this case. Such antennas are widely used in mobile satellite applications, for example in GPS or other satellite navigation system receivers.

Examples of low-gain circular polarized antennas having nominally hemispherical radiation patterns are illustrated in Figure 4.5. These include: the crossed dipole or *turnstile* antenna (a pair of dipoles above a reflecting surface); a quadrifilar helix (a helix comprising four arms) and patch antenna.

4.2.4.4   Medium-Gain Directional Antennas

The nominally hemispherical radiation patterns of low-gain antennas, while attractive for mobile applications where antenna pointing is a problem, have limited application where increased sensitivity is required and some degree of antenna pointing is acceptable, or where coverage needs to be reduced. An example radiation pattern for a medium-gain antenna (a conical horn) is illustrated in Figure 4.1 (centre).

Medium-gain antennas, some examples of which are illustrated in Figure 4.6, provide more angular selectivity, which results in greater sensitivity. In the figure, examples are shown of a Yagi–Uda array of crossed dipoles (a variant of the familiar Yagi–Uda antenna[4] used for terrestrial television reception), an axial-mode helix antenna and a horn antenna. The operation of the horn antenna (of which both pyramidal and conical variants exist) is perhaps easiest to understand; the antenna forms a taper between its aperture (opening) and the waveguide section at the other end; to a first approximation, the slower the taper, the narrower is the beam. Typical gains of these antennas are in the region of 10-20 dB better than for the low-gain antennas.

4.2.4.5   High-Gain Directional Antennas

Still higher EIRP and sensitivities (or smaller beamwidths) require still higher gain antennas. Typically, very high gains are produced by further focusing the patterns from a medium-gain antenna (the *feed antenna*) using techniques originally developed in optics, in particular using lenses and

---

[4] The Yagi–Uda antenna is also commonly referred to as a Yagi antenna.

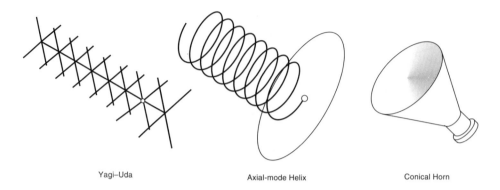

Yagi–Uda   Axial-mode Helix   Conical Horn

Figure 4.6   Medium-gain antennas.

telescopes. Alternatively, high-gain antennas may be produced using two-dimensional arrays (grids) of lower medium-gain antennas (see Section 4.2.2.8).

High-gain *reflector* antennas transform the radiation patterns of a medium-gain antenna – such as a horn antenna – using reflecting lenses. Referring to Figure 4.7, the simplest reflector antenna has a feed antenna at the focus of a parabolic reflector (the point where parallel rays incident on the reflector converge). Such an antenna is known as a prime-focus antenna (top left).

A drawback of the prime-focus antenna is that the feed (antenna) is located some distance in front of the reflector, and must be supported (and fed) with minimal obstruction to the main beam. In the Cassegrain reflector antenna (named after the seventeenth-century telescope inventor), the feed is instead located at a small opening in the main reflector, and a small convex hyperbolic subreflector is used to deflect the beam from the feed onto the main parabolic reflector (Figure 4.7, top right). The use of a subreflector provides an additional degree of freedom in the antenna design,

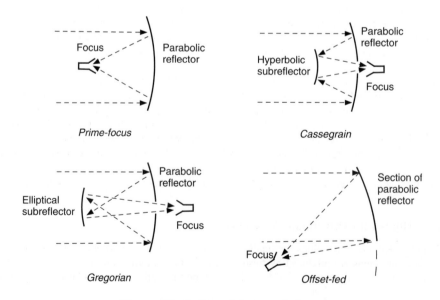

Figure 4.7   Reflector/telescope designs.

which allows Cassegrain antennas with higher gain than prime-focus reflectors, but, in order to be effective, the subreflector must itself be several wavelengths across (Balanis, 1997), which increases blockage. As a result, Cassegrain antennas are generally only used for very high-gain antennas.

A third main catagory of reflector antenna is the Gregorian reflector (also named after a seventeenth-century telescope inventor), in which a concave elipsoid subreflector is used instead of the convex hyperberloid of the Cassegrain design. An advantage of the Gregorian reflector is that the separation of the feed and subreflector is increased and the feed antenna may be located behind the main reflector (Figure 4.7, bottom left). Again, Gregorian antennas are generally only used for very high-gain antennas.

Feeding the antenna from behind the reflector is attractive in many situations, and, for relatively small reflector antennas, where the subreflector is small (in wavelengths) and almost flat, this type of antenna is referred to as a 'splash-plate' reflector.

A common drawback of all three reflector antenna types is that some part of the main beam is obstructed by the feed or subreflector and their support structures. This reduces antenna efficiency. In order to overcome this problem, the reflectors can be fed asymmetrically, as illustrated in Figure 4.7 (bottom right). In effect, these offset-fed reflector antennas have reflectors and subreflectors that are off-centre portions of larger reflectors. Such offset-fed reflectors can achieve higher efficiencies than centre-fed equivalents, albeit with slight degradation of polarization purity.

Maximum Gain of a Reflector Antenna

As the aperture size of an antenna increases (in wavelengths), the beamwidth decreases and the gain increases. In principle one might expect that the gain of an aperture antenna, and in particular that of a reflector antenna, can be increased *ad infinitum* simply by increasing the aperture size (for a given operating frequency). In practice, the maximum achievable gain will ultimately be limited by the departure of the antenna reflector geometry from the ideal. The effect of a given RMS surface roughness on surface efficiency has been estimated by Ruze (1966).

An RMS surface roughness of $\frac{\lambda}{32}$ will result in a surface efficiency of $\sim$85%, while an RMS roughness of $\frac{\lambda}{16}$ yields a surface efficiency of just $\sim$50%. Thus, at a frequency of 30 GHz (Ka-band), we would require an RMS surface roughness better than 0.3 mm in order to achieve a high antenna efficiency.

For mass-produced reflector antennas used in personal satellite applications, where the maximum antenna aperture is typically constrained by the required portability, pointing accuracy and/or mounting arrangements, this limit determines the required manufacturing tolerance and reflector mechanical strength (to avoid flexure) for these low-cost, mass-produced antennas. For the larger Earth station antennas used to anchor such services, additional strengthening is necessary, and ultimately the maximum achievable gain of extremely large-aperture (in wavelengths) Earth station antennas will be limited. For large satellite antennas, which must be very light and may be required to be unfurled in space, the impact of this limit on maximum antenna size can be significant, and is one limit on the exploitation of very narrow beamwidths (less than 1°) in satellite applications.

4.2.4.6   Satellite Antennas

A variety of satellite antenna types are used to provide personal satellite services, depending on the satellite altitude, the desired beam coverage and the desired frequency and polarization. Intelsat, one of the largest fixed satellite service providers, which operates some 50 satellites spread worldwide, defines its satellite coverage by four types of beam:

- the global beam, which covers roughly one-third of the globe;
- the hemispherical beam, which covers roughly one-sixth of the disk;

- the zonal beam, which covers a large landmass like Europe;
- the spot beam, which covers a 'specific geographical area'.

Dual polarized antenna systems support two orthogonal polarizations – effectively doubling the satellite capacity.

Antennas for Earth Cover Beam

Clearly, the beamwidth of a satellite antenna to provide coverage of the visible Earth will depend on satellite altitude. For a geostationary satellite, we find from Chapter 2 that the satellite nadir angle for zero minimum elevation angle is approximately $8.75°$ (i.e. a FWHM satellite antenna beamwidth of $\sim 17.5°$). The solid angle of a cone with this angle is ~ 0.073 sr. Therefore, the required antenna directivity is approximately 172 (22 dBi). In practice, the antenna pattern is not a simple conical shape, and the exact directivity of an Earth cover antenna will depend on the allowed gain reduction at the edge of cover. Earth coverage for a geostationary satellite is thus typically achieved by an antenna with a gain in the region of 13–22 dBi.

At microwave and millimetric frequencies, this value of gain is conveniently achieved using conical horn antennas, while at lower frequencies, for circular polarization, it is typically achieved using axial-mode helix antennas. Of course, where available, phased-array antennas may also be employed for Earth cover beams (using the appropriate phase weights).

Antennas for Zonal/Spot Beams

Spot beams – that is beams that are smaller than those that cover the visible Earth – are typically used to provide services requiring higher power/sensitivity. The shape and beamwidth of such spot beams depends on a number of factors and represent a compromise between maximum gain and coverage area. Radiation pattern for satellite antennas are often represented as a contour plot on the Earth to facilitate estimation of received signal quality.

The simplest spot beam antenna pattern is circular in shape, although the illuminated area appears non-circular on the ground owing to the Earth's curvature (essentially an elongation towards high latitude and/or longitude difference). To improve the transmission efficiency, radiation patterns (beams) may be shaped to best fit the service zone.

Spot beams are typically provided using high-gain reflector antennas – typically, offset-fed reflector antennas. By way of example, Figure 4.8 shows the complement of reflector antennas on the Anik F-2 spacecraft, which provides a range of telecommunication services.

Phased-arrays antenna may also be used to provide spot beams for certain advanced applications, notably where the beam shape or location is altered dynamically.

At lower frequencies, the size of the satellite antenna reflector needed to obtain useful spot beams can present a problem, and typically these reflectors comprise a very light wire mesh, which is stowed away at launch and unfurled only when the satellite is in orbit. Notable examples of unfurlable reflectors are the Astromesh reflectors from Northrop Grumman (2010) used on Inmarsat-4, Thuraya and MBSAT satellites. Figure 4.9 shows an artist's impression of the 9 m diameter Astromesh reflector deployed on the Inmarsat-4 satellite. Another example of an unfurlable reflector is the Harris Folding Rib reflector, as deployed on the ACeS (Garuda) and ICO G1 satellites, which opens up rather like an umbrella (Harris Online, 2010). The maximum size of operational antenna at present is 18 m launched recently by Terrestar in the United States to support personal applications and other services (Terrestar Online, 2010).

Antennas for Contoured Beams

For some applications it is desirable to synthesize the beam coverage to match particular geographic or political boundaries.

There are two general approaches to forming complex contoured beams. One method is to use a standard reflector antenna but to deform the main reflector (or subreflector in a Cassegrain

Figure 4.8   Anik F2 satellite antenna farm. Reproduced by permission of © Boeing.

arrangement) so as to achieve the desired pattern. This approach is cost-effective but does not offer the flexibility to adjust the coverage once the satellite is launched.

Until recently, beam shapes of satellite antennas usually tended to remain fixed throughout a satellite's lifetime; more recently, and particularly for personal communication systems, satellites have incorporated the capability to alter beams shapes, size or numbers, which allows for flexibility in allocation of satellite resources. For instance, an operator may decide to reduce or reshape beam size and illuminate uncovered areas in response to an event, or to introduce new products, or readjust coverage owing to a change in traffic pattern. The most flexible (but more expensive) approach is to use a phased array to feed a standard parabolic reflector. By adjusting the phase weights of the array elements, the beam shape can be optimized (and reoptimized) for the required coverage.

Figure 4.9   Artist's impression of the Inmarsat-4 satellite with its 9 m Astromesh reflector unfurled. Reproduced by permission of © Northrup Grumman.

Figure 4.10 A 120-element Inmarsat-4 satellite L-band feed array being assembled (Stirland and Brain, 2006). Reproduced by permission of © 2006 IEEE.

Antennas for Cellular, Multiple-Spot-Beam Coverage

It is often desirable to produce multiple spot beams, and it is increasingly common practice to maximize system capacity by synthesizing multiple beams in a hexagonal cellular-type pattern so as to maximize the potential for reuse of the limited available spectrum to a particular operator between non-adjacent 'cells', and to provide high EIRP. This type of cellular pattern of spot beams is generally formed by feeding a reflector antenna from a phased array at or near its focus. Figure 4.10 illustrates a 120-element L-band feed array being assembled for the Inmarsat-4 satellite.

Inmarsat-4 (I-4) satellites can synthesize over 200 beams in the L-band service link, as illustrated in Figure 4.11. The satellite generates three types of beam – a global beam to cover the entire disc, 19 wide beams and 200 narrow spot beams to service portable user terminals. Each satellite deploys a 9 m furlable gold-plated molybdenum mesh reflector antenna which is offset fed by 120 helical elements. Each beam is formed from about 20 feed elements fed with the appropriate amplitude and phase coefficients from an on-board digital beam former. Over 220 beams of different shapes and sizes can be formed by altering the beam coefficients. I-4 beam patterns have been altered on a number of occasions to satisfy the operational requirements.

4.2.4.7 Ground Station Antennas

Ground station antennas used as a hub to anchor satellite services will typically comprise high-gain reflector antennas at microwave frequencies. By way of illustration, Figure 4.12 shows an 9 m Cassegrain antenna from ViaSat. At lower frequencies, for circular polarization, arrays of axial-mode helices or crossed-dipole Yagi–Uda antennas are often used.

4.2.4.8 User Equipment Antennas

Antennas for Fixed Satellite Applications

For applications where the orientation of the satellite is fixed with respect to the user, high-gain antennas may be used. For wideband applications, offset-fed reflector antennas are generally employed. A typical example of a fixed reflector antenna from Kathrein is illustrated in Figure 4.13 (left). Planar arrays of patch antennas may also be used for medium-gain antennas for fixed services,

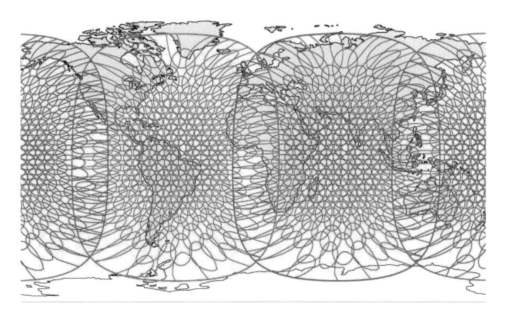

Figure 4.11 Example of cellular spot beam coverage produced by three Inmarsat fourth-generation satellites, located at 53° W, 64° E and 178° E. Reproduced by permission of © Inmarsat Global Limited.

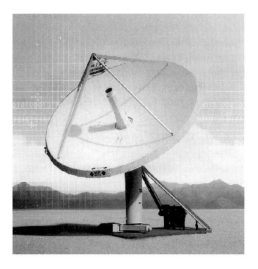

Figure 4.12 An example 9 m cassegrain ground station antenna. Reproduced by permission of © Viasat.

Figure 4.13 Example Ku-band satellite TV antennas: offset-fed reflector (left); motorized square planar antenna array (right). Reproduced by permission of © Kathrein-Werke KG.

and Figure 4.13 (right) shows a modern example of a mass-market square planar array mounted on a motorized turntable (auto-pointing antenna for use on recreational vehicles), also from Kathrein (Kathrein Online, 2010).

Antennas for Portable Satellite Applications

A number of medium-gain directional antenna designs are available for portable applications where the user is required to erect the antenna and manually point it at the satellite. At UHF, foldable Yagi–Uda antennas are available that can be packed into a small carrying case, while medium-gain planar-phased arrays offer a compact form factor for medium-gain portable satellite applications.

Tracking Antennas for Mobile Satellite Applications

Antennas for applications where the user is mobile (for example, in a car, train or boat) generally fall into two categories. Less demanding applications may be satisfied using low-gain antennas with hemispherical radiation patterns, such as patches and quadrifilar helices. More demanding applications require a medium–high-gain directional antenna which needs continually to track the satellite as the platform moves.

The majority of tracking antennas employ an arrangement with the antenna (and often the RF electronics) mounted on a mechanical gimbal, which tracks (and compensates for) the motion of the platform. The antenna must track the satellite in at least two orthogonal axes: usually azimuth and elevation. In addition, those systems employing linear polarization generally require the antenna feed to track the polarization relative to the moving platform. The use of electronic beam steering using phased-array antennas instead of mechanical tracking is relatively rare owing to the generally higher cost and somewhat lower efficiency of electronic phased arrays – although a few designs employ electronically steered phased arrays for one axis (resulting in hybrid mechanical/electronic beam steering). Nevertheless, the use of mechanically-steered staring arrays (i.e. fixed boresight phased arrays) is increasingly common, as these can result in a reduced swept volume while tracking in comparison with normal reflectors.

The control mechanism for such tracking antennas may take one of two forms:

- *Open loop.* The direction of the satellite is predicted using only the motion of the platform. Platform motion is typically obtained using electromechanical sensors. Typically, angular motion is obtained from gyroscopes, while heading is obtained from a flux gate compass (to compensate for gyroscope drift). A more sophisticated option is to detect platform motion electronically using an attitude sensor comprising an array of satellite navigation antennas.
- *Closed loop.* After the satellite has been found using open-loop control, the satellite signal is continuously monitored and the antenna is repointed so as to maximize the received signal. The tracking mechanism used will depend on the tracking speed required. A conical scan is typically employed, as it is relatively simple and cost effective to implement. In this scheme, the antenna continuously rotates around the nominal pointing direction, a fraction of a beamwidth away. Any signal variation is attributed to pointing error and is corrected, therefore conical scan can be confused by blockage/fading. The alternative is a monopulse arrangement, where the signal is received simultaneously in two or more directional beams with slightly different boresights, with the relative beam power used to monitor pointing.

High-gain tracking reflector antennas have been available for use on marine platforms for some time, and a splash-plate reflector antenna is often used on a small marine platform. The use of reduced-profile offset-fed high-gain antennas on Recreational Vehicles (RVs – motorized caravans) is a more recent development.

The height of tracking antennas on road vehicles is a significant issue. At Ku-band, a number of tracking antenna systems employ mechanically scanned planar arrays (so-called staring arrays). An example of such an antenna is that from Raysat (Raysat Online, 2010), illustrated in Figure 4.14. Although still mechanically scanned, these array antennas present a smaller swept volume than reflector antennas. Mechanically steered staring arrays are also used at L-band for mobile satellite services; typically, the array comprises a small number of axial-mode helices.

Figure 4.15 illustrates a design for an ultralow-profile tracking antenna for Ku-band TV reception, from RaySat, suited to mid-latitudes (RaySat Online, 2010). This antenna employs a number of flat, planar, staring, phased arrays that are mechanically steered in both azimuth and elevation to point at the satellite. The aim is to limit the overall height of the antenna by using multiple subarrays, the outputs of which are combined electronically.

Ultimately, however, such low-profile antennas are limited by Lambert's cosine law of illumination, with the result that many low-profile designs have lower limits on satellite elevation range

Figure 4.14   Raysat T5 planar-phased array antenna. Reproduced by permission of © Raysat.

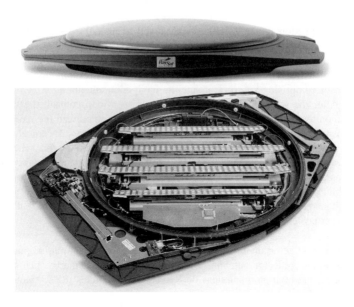

Figure 4.15 Raysat Speedway 1000 hybrid electronically/mechanically steered ultralow profile Ku-band antenna array for mobile TV reception. Reproduced by permisson of © Raysat.

because, in order to achieve a constant effective area (and hence constant antenna gain), as the satellite elevation decreases, the antenna will inevitably extend higher. For this reason, a common approach to low profile is to trade antenna height for width, thereby maintaining a constant area.

Antennas for Handheld Satellite Applications

For handheld devices, satellite antennas must be compact and accommodate the unknown orientation of the satellite relative to the handset. The antenna pattern will therefore be omnidirectional in azimuth and preferably provide coverage of the whole upper hemisphere. Typical antenna types used for these applications are patch and quadrifilar helix antennas. In order to avoid the need to align the antenna polarization, circular polarization is almost universally employed for handheld services.

Clearly, the size of the antenna is a significant constraint in modern personal electronic devices. Increasingly high dielectric materials – such as ceramics – are used to reduce the antenna size by reducing the wavelength in the dielectric.

Figure 4.16 illustrates a range of GPS patch antenna from Cti International that use a ceramic dielectric insulator. The patch design to achieve circular polarization is a square with chamfered corners. For maximum sensitivity, GPS patch antennas are sometimes integrated with low-noise preamplifier to overcome the effects of cable loss.

Figure 4.17 shows a GeoHelix antenna designed for use in a variety of handheld satellite devices from Sarantel that employs a miniature quadrifilar helix wrapped around a high dielectric core (Sarantel Online, 2010). This type of antenna is employed in a number of handheld devices, including the latest Iridium mobile satellite handset.

Figure 4.16   A selection of dielectric-loaded circular-polarized GPS patch antennas, ranging from 12.5 × 12.5 mm. Reproduced by permission of © CTi Ltd.

4.2.5   Optical Antennas

Although the primary focus in our discussion is on antennas for use at radio frequencies, we shall briefly widen our scope to discuss optical antennas (i.e. telescopes). Such telescopes are used for remote sensing applications, and in future they may also be used to provide very high-data-rate links between satellites, and between satellites and HAPs. Ultimately, they may even be used to provide very high-data-rate links to specialized ground terminals (weather permitting).

Although telescopes can be manufactured using either refractive (e.g. glass) or reflective lenses (and hybrids thereof), the use of reflector telescope designs is most common owing in part to the absence of chromatic distortion. As high-gain reflector antennas are themselves derived from seventeenth-century optical reflector telescope designs, the basic designs of such telescopes are already familiar to us.

Figure 4.17   Sarantel GeoHelix dielectric-filled miniature quadrifilar helix L-band antenna. Reproduced by permission of © Sarantel Ltd.

Telescopes for Free-Space Optical Communications

With regard to the use of optical communications for space–space and space–ground communications links, distinguishing features of optical telescopes are their ultranarrow beamwidth and ultrahigh antenna gain. Such telescopes are typically of the Cassegrain design (and its derivatives), however, as the tolerance on reflector accuracy scales with the wavelength, large optical apertures are usually fabricated on a glass or silicon carbide substrate and are therefore more substantial than the microwave equivalent of the same size.

Although high-gain reflector antennas and optical telescopes share a common heritage, let us highlight differences in scale by comparing antennas with the same aperture size at a Ku-band frequency of 12 GHz, where the wavelength is 25 mm, with an optical telescope at a typical near-infrared frequency used in terrestrial fibre-optical communications of $\sim 2 \times 10^{14}$ Hz, where the wavelength is 1.5 µm (i.e. a factor of 16 667 times different). Let us consider a telescope with a modest 0.5 m diameter aperture. The approximate antenna gain for a circular-aperture antenna was given by equation (4.14), and the respective values are $\sim$33 dB at Ku-band and $\sim$118 dB for the optical link – that is a massive 85 dB difference!

The size of optical telescopes for use in free-space optical communications is constrained in part by their mass and in part by the difficulty in pointing their ultranarrow beams. The beamwith of a circular-aperture antenna was given by equation (4.6), and the antenna pointing (angle) tolerance scales with the wavelength. Continuing with our comparative example, the respective antenna beamwidths are $\sim$3.5° at Ku-band and $\sim$0.0002° for the optical telescope. It is thus not hard to see the enormous practical difficulties that must be overcome to permit the exploitation of free-space optical links between satellites – particularly between non-geostationary satellites – and between satellites and the ground. It is also apparent that future optical links between satellites and ground terminals will have very limited beam coverage areas, and will thus primarily be restricted to single point-to-point communications.

Telescopes for Imaging

A variety of optical designs are utilized for imaging applications owing to the need to balance different constraints for each application. An example optical telescope configuration adapted for use on IKONOS Earth observation applications is the three-reflector anastigmat telescope. A third mirror is added compared to the Cassegrain configuration, and two additional flat 'fold mirrors' are used to fold the beam path to make the telescope more compact. Benefits of the three-reflector arrangement are a relatively wide field of view and greater control over aberrations.

Optical Resolution

Of significant interest in imaging telescopes is the ability to resolve spatially adjacent objects in the farfield. The ability to discern separate objects depends on a number of factors; however, a common metric is the *Rayleigh criterion*, which relates to the limits on resolving two adjacent objects imposed by diffraction from the antenna aperture.

For a fully illuminated (unobscured) circular aperture of diameter D, the Rayleigh criterion is the separation at which the peak intensity due to one infinitely small light source is collocated with the first null of an adjacent point source. This condition is illustrated in Figure 4.18.

The Rayleigh limit on angular resolution θ_R is given by

$$\sin(\theta_R) = 1.22 \frac{\lambda}{D} \tag{4.23}$$

$$\simeq \theta_R \text{ (for small } \theta_R) \tag{4.24}$$

This value must be modified somewhat for the case where the aperture is partially obscured (e.g. by the second mirror in a Cassegrain configuration).

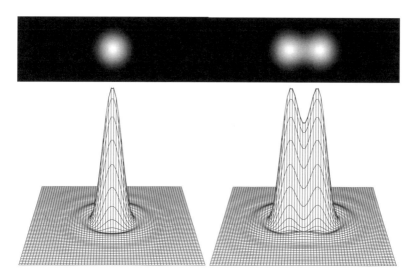

Figure 4.18 Top: Image intensity for point source (left) and for two point sources at the Rayleigh criterion (right). Bottom: Corresponding 3D intensity maps.

GeoEye-1 Example. Consider, by way of an example, the panchromatic resolution of the GeoEye-1 satellite. According to the specifications for these satellites, the unobscured telescope aperture diameter is 1.1 m. At a wavelength of 550 nm, this gives an angular resolution at the Rayleigh limit of approximately 6×10^{-7} rad. Multiplying this value by the satellite altitude of 684 km gives a tangential spatial resolution on the Earth (at the subsatellite point) of approximately 0.41 m.

4.3 Noise

4.3.1 Overview

We turn our attention next to the subject of noise. The received signal strength of the *wanted* signal is only part of the picture, since all real physical signals are degraded to some degree by random noise. A key parameter in the performance of all satellite applications is therefore the received signal-to-noise ratio S/N.

Received noise is partly that produced by the electronic components within the receiver itself, partly received noise from the transmitter and partly unwanted electromagnetic radiation collected by the receiver antenna. We shall briefly discuss the various contributions to the receiver noise and factors that influence these contributions, and introduce figures of merit for the performance of a receiver.

4.3.1.1 Noise Sources

Noise sources of significance in satellite applications include:

- *Antenna noise* comprising contributions from:
  - *sky noise*, comprising blackbody radiation emanating from the sky (and ground) and collected by the antenna, including *cosmic noise* (background cosmic blackbody radiation), blackbody radiation from the atmosphere and radiation from the Earth;

- *noise from the Sun*, that is, electromagnetic radiation from the Sun when it passes into the beam of an antenna;
- *galactic noise*, that is, electromagnetic radiation emanating from other stars of our Galaxy (the Milky Way), and consequently limited to antenna beam angles aligned with the galactic plane;
- *atmospheric noise*, such as that due to electrical storms;
- *man-made noise*, such as noise from electrical machinery.
- *Electronic noise* comprising contributions within the system electronics from:
  - *thermal noise*, a noise source typically attributed to the random 'Brownian' motion of electrons in resistive electronic components;
  - *shot noise*, also referred to as quantum noise, a noise source that results from the discrete, particle nature of electrons and photons;
  - *flicker noise*, that is, low-frequency electronic device noise associated with conduction which has a power spectrum frequency dependence of $\frac{1}{f}$;
  - *excess noise*, a noise source associated with electronic noise in certain electronic devices, over and above thermal and shot noise, that typically results, for example, from avalanche multiplication.

4.3.2   Antenna Noise

4.3.2.1   Sky Noise

We begin our consideration of antenna noise with sky noise. Definitions of exactly what constitutes sky noise is a matter of opinion, and in this book we shall define sky noise as noise due to blackbody radiation having a fixed relationship to the antenna, comprising cosmic background radiation, radiation from the atmosphere and hydrometeors (e.g. rain) and also radiation from the Earth. We choose this definition in order to distinguish sky noise from other sources of noise that do not have a fixed relationship to the antenna and thus tend to be more transient.

Blackbody Radiation
We are familiar with the notion of radiation by warm (and hot) objects, generally referred to as blackbody radiation. Thermal radiation from such objects may be collected by a receiving antenna and will take the form of unwanted random noise in the antenna output.

The power radiated per unit bandwidth, per unit surface area and per unit solid angle, B_f–called the *brightness*–of a blackbody at temperature T is given by Planck's blackbody radiation law

$$B_f = \frac{2hf^3}{c^2} \frac{1}{\left(e^{\frac{hf}{kT}} - 1\right)} \tag{4.25}$$

where c is the speed of light in a vacuum, $k = 1.3806 \times 10^{-23}$ J/K is Boltzmann's constant and $h = 6.626 \times 10^{-34}$ Js is Planck's constant.

Now, for radio frequencies we generally have a photon energy much lower than the thermal energy (i.e. $hf \ll kT$), and, for these frequencies, Planck's formula reduces to the Rayleigh–Jeans approximation. As these formulae give the brightness for *all* polarizations, the brightness for a single polarization will be half this value.

Consider a hypothetical lossless directional receiving antenna pointing at this blackbody source with beam solid angle Ω_B and maximum effective aperture A_e. From equation (4.4) and equation (4.12), the received blackbody noise power per unit bandwidth for a single polarization N_{bb}

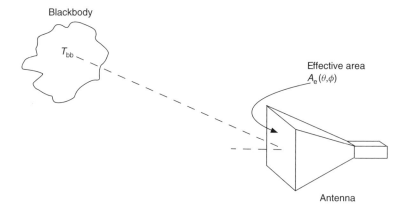

Figure 4.19 Antenna noise.

will be

$$N_{bb} = \frac{1}{2} B_f \Omega_B A_e \approx kT \qquad (4.26)$$

The blackbody radiation noise may therefore be characterized completely by its temperature T.

4.3.2.2 Antenna Temperature

In the general case, with reference to Figure 4.19 (ignoring noise from the Sun and Moon, galactic noise and terrestrial noise), the antenna noise temperature is obtained by integrating the sky noise contributions multiplied by antenna gain in that direction, over all angles. Intuitively, for an ideal narrow-beamwidth antenna, with a single mainlobe, one expects that the antenna temperature tends to the sky noise temperature for the direction of peak gain.

In the general case, the antenna noise is (in spherical polar coordinates)

$$T_{ant} = \frac{1}{4\pi} \int_0^{2\pi} \int_0^{\pi} G(\theta, \phi) T_{sky}(\theta, \phi) \sin(\theta) \, d\theta \, d\phi \qquad (4.27)$$

Evaluation of this integral is inconvenient, and for many practical cases the antenna noise temperature of a moderately high-gain antenna with a single mainlobe with boresight elevation angle ϵ may be approximated by (Boculic, 1991)

$$T_{ant}(\epsilon) \approx \eta_o \left(\eta_b T_{sky}(\epsilon) + (1 - \eta_b) T_{sky\text{-}av} \right) + (1 - \eta_o) T_o \qquad (4.28)$$

where η_o is the antenna efficiency due to ohmic losses, η_b is the antenna beam efficiency (fraction of radiated power in the mainlobe), $T_{sky\text{-}av}$ is the sky noise temperature average over all of the sidelobes and T_o is the antenna physical temperature (usually taken to be 290 K). As approximately half of the antenna sidelobes collect noise from the Earth at around 300 K, the value of $T_{sky\text{-}av}$ lies in the approximate range 150–300 K (depending on atmospheric conditions).

A high-gain ground terminal antenna may be expected to be directed at some point in the sky, and will only collect a small amount of thermal radiation noise from the Earth – from its radiation sidelobes, and should therefore have a relatively low antenna noise temperature (depending on atmospheric conditions). By contrast, a low-gain ground terminal antenna is likely to collect a significant amount of radiation from the Earth, and the antenna temperature in this case may be in the region of >150 K.

Relation Between Sky Noise and Atmospheric Attenuation

There is a fundamental relationship between the sky noise temperature and atmospheric attenuation for a particular slant path. If we may assume that the atmosphere is in local thermodynamic *equilibrium*, then any power absorbed by atmospheric attenuation must be balanced by power radiated by blackbody radiation – otherwise the local temperature would not be stable. This concept is known as the *principle of detailed balance*.

In the absence of any atmospheric attenuation, the sky noise temperature seen by an idealized pencil beam antenna would be that of the cosmic background radiation, $T_{cos} = 2.725$ K. If, however, the atmospheric loss (ratio) is actually L_a, then the power absorbed (or scattered) in the atmosphere, in bandwidth B, is $\left(1 - \frac{1}{L_a}\right) kT_cB$. In order to maintain equilibrium, this power must be balanced by radiation.

The resulting sky noise temperature seen by the antenna for that slant path is thus

$$T_{sky} \approx \left(\frac{1}{L_{atm}}\right) T_{cos} + \left(1 - \frac{1}{L_{atm}}\right) T_{atm} \tag{4.29}$$

where T_{atm} is the nominal atmospheric temperature.

ITU-R Recommendation P.618-8 recommends appropriate values of T_{atm} of 260 K if raining and 280 K if not (ITU-R P.618-8, 2003). Figure 4.20 shows the dependence of sky noise temperature on atmospheric attenuation.

As the atmospheric attenuation depends on elevation angle, so too does the sky noise temperature, and the sky noise will be greater at low elevations. It should be noted that sky noise temperature will be significantly affected by the presence of attenuation due to rain.

The dependence of sky noise on atmospheric attenuation means that the received signal-to-noise power ratio will increase faster than it would otherwise owing to attenuation of the signal alone, as the signal is reduced and the sky noise increased, simultaneously.

Noise Temperature of the Earth

The Earth typically has an apparent blackbody radiation temperature of around 290 K, although the observed temperature will depend on the local surface emissivity (which results in the brightness being less for the sea than for land). Any radiation lobes of an antenna that are directed at the Earth's surface will thus collect noise at this temperature. A satellite antenna providing coverage

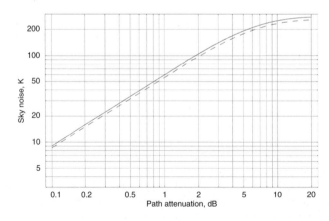

Figure 4.20 Sky noise temperature versus tropospheric path attenuation for the cases of no rain (solid line) and rain (dashed line).

of the Earth will have seen a temperature of around 200–300 K, depending on the particular coverage.

4.3.2.3 Solar and Galactic Noise

Effective Noise Temperature of the Sun
Very occasionally, the Sun or the Moon may pass through the beam of a satellite or ground terminal antenna. The impact will depend on their radiation temperature and the solid angle subtended by these objects compared to the beams solid angle. As one would expect, the Sun has a very high noise temperature. At very high frequencies the sun has the radiation characteristics of a blackbody with an approximate temperature of 6 000 K (Ho et al., 2008). However, the brightness at radio frequencies is higher than one would expect from an ideal blackbody, and, to a first approximation, the mean effective noise temperature of the quiet Sun (i.e. during a period of low solar activity) at GHz frequencies may be expressed as (Shimabukuro and Tracey, 1968)

$$T_{sun} \approx 120000\, f^{3/4}\ \text{(K)} \tag{4.30}$$

Galactic Noise
Galactic noise can be significant at low frequencies in the galactic plane. The frequency dependence of the mean galactic noise temperature is illustrated in Figure 4.21, which shows the average effective noise temperature due to galactic sources, as well as the nominal sky noise for a dry atmosphere and photon noise (due to the quantum nature of electromagnetic radiation) – see section 4.3.5.1.

4.3.3 Electronic Noise

4.3.3.1 Electronic Thermal Noise

The dominant noise source in radio receivers is thermal noise. Thermal noise was first characterized by Johnson (1928) and subsequently explained by Nyquist (1928), who considered thermal noise as a one-dimensional case of blackbody radiation, using the principle of detailed balance. For frequencies for which the Rayleigh–Jeans approximation holds, the noise power per unit bandwidth is kT – as for the blackbody radiator, and thus, for most practical purposes, thermal noise at radio frequencies may be considered *'white'* – that is, the noise per unit bandwidth is the same for (almost) all frequencies.

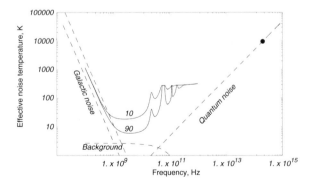

Figure 4.21 Effective sky noise temperature versus frequency (log scale) up to optical frequencies.

4.3.3.2   Noise Factor and Noise Figure

Noise in a two-port electronic device – such as an amplifier – is often characterized by its *noise factor F* (sometimes called the excess noise ratio), defined as the ratio of the input signal-to-noise divided by the output signal-to-noise (Friis, 1944).

If the input signal power is S_{in}, the output signal power will be $S_{out} = GS_{in}$, where G is the amplifier (power) gain. The input noise power in bandwidth B is kT_oB, where T_o is the ambient/system reference temperature (usually taken as 290 K), whereupon the output noise power is equal to the sum of the amplified input noise kGT_oB and the noise power added by the amplifier itself $GkT_{eff}B$, where T_{eff} is the effective noise temperature of the amplifier, referred to its input port.

The noise factor F is thus

$$F = 1 + \frac{T_{eff}}{T_o} \tag{4.31}$$

It is generally more convenient to refer to the *noise figure* NF of a network element. The noise figure is just the noise factor expressed in decibels

$$NF = 10\log_{10}(F) \tag{4.32}$$

Noise Figure for an Attenuator
We have already seen that, in equilibrium, the principle of detailed balance requires that the noise power generated in an attenuator must balance the input noise power absorbed in the attenuator. It is readily shown that the noise figure for an attenuator is equal to its loss factor (in dB).

4.3.4   System Noise

4.3.4.1   Noise Temperature for Concatenated Network Elements

In general, a satellite receiver (or sensor) comprises several concatenated stages of amplification and possibly frequency translation. Each concatenated device contributes its own noise figure. However, it is generally convenient to determine an effective input noise temperature, noise factor or noise figure for the concatenated chain of devices making up the receiver system.

It is easily shown that the effective noise temperature at the input of a cascade of network elements having gains G_1, G_2, etc., is

$$T_{eff} = T_1 + \frac{T_2}{G_1} + \frac{T_3}{G_1 \cdot G_2} + \frac{T_4}{G_1 \cdot G_2 \cdot G_2} + \cdots \tag{4.33}$$

where T_1, T_2, etc., are the effective noise temperatures of each stage. The corresponding system noise factor and noise figure may be determined by application of equations (4.31) and (4.32).

4.3.4.2   System Noise Temperature

We are finally in a position to determine the noise temperature for the complete receiver *system*, comprising contributions from the antenna, the receiver and any lossy feeder connecting the antenna and receiver. We shall assume that we have previously determined the effective noise temperature for the cascade of electronic amplifiers – the effective temperature of the complete receiver system – referred to the first amplifier in the chain, using equation (4.33), and that this temperature is T_r. We shall further assume a feeder loss (ratio) of L_f. The total system noise temperature is then

$$T_{\text{sys}} = \left(\frac{1}{L_{\text{f}}}\right) T_{\text{ant}} + \left(1 - \frac{1}{L_{\text{f}}}\right) T_{\text{o}} + T_{\text{r}} \tag{4.34}$$

4.3.5 Signal-to-Noise Density and Receiver Figure of Merit

Combining equations (4.19) and (4.10) and our expression for the noise, we find that the actual received signal to one-sided noise spectral density may be expressed as[5]

$$\frac{C}{N_{\text{o}}} = \text{EIRP} \left(\frac{G}{T}\right) \frac{1}{k} \left(\frac{\lambda}{4\pi R}\right)^2 \frac{1}{L_{\text{T}}} \tag{4.35}$$

Note that, by convention, the unmodulated carrier (power)-to-noise density is normally used to represent the signal-to-noise density – hence C/N_{o} is used in equation (4.35) rather than S/N_{o}, which is mainly used when considering the demodulated baseband signal. Subsequent chapters will show that the information throughput of a satellite link is directly related to C/N_{o}.

Signal-to-Noise Plus Interference

We can extend the concept of carrier-to-noise spectral density to include interference in a simple way by using an effective value of C/N_{o}, comprising the signal-to-noise plus interference spectral density, given by

$$\frac{1}{\left(\frac{C}{N_{\text{o}}}\right)_{\text{eff}}} = \frac{1}{\left(\frac{C}{N_{\text{o}}}\right)} + \Sigma_i \left[\frac{1}{\left(\frac{C}{I_{\text{o}}}\right)_i}\right] \tag{4.36}$$

where I_{o} is the effective interferer power spectral density averaged over the (wanted) signal bandwidth. The contribution from all of the potential interferers is determined in the same manner as for the wanted signal using an appropriate received power calculation (equation (4.19)), together with an estimate of the spectral overlap between the interfering signal and the receiver bandwidth (used to convert the interferer power into an equivalent mean power spectral density).

As interferers are typically located at different pointing angles to the wanted signal, the interference calculations must take into account the effect of different antenna pointing errors (which are usually to the users' advantage).

Concatenated Links

Satellite transmission typically occurs using two concatenated links: an uplink (from a ground station to the satellite) and a downlink (from the satellite to the second ground station), except where the signals are regenerated on-board the spacecraft (i.e. demodulated and remodulated).

For concatenated, unregenerated transmission, the noise introduced in each link adds linearly, and consequently the received signal-to-one-sided noise spectral density is obtained using

$$\frac{1}{\left(\frac{C}{N_{\text{o}}}\right)} = \frac{1}{\left(\frac{C}{N_{\text{o}}}\right)_1} + \frac{1}{\left(\frac{C}{N_{\text{o}}}\right)_2} + \cdots \tag{4.37}$$

---

[5] The same formula can also be expressed (in dB) as

$$\left(\frac{C}{N_{\text{o}}}\right)_{\text{dB}} = (\text{EIRP})_{\text{dB}} + \left(\frac{G}{T}\right)_{\text{dB}} + 376.1 - 20\log(f) - 20\log(R) - (L_{\text{T}})_{\text{dB}}$$

where $\left(\frac{C}{N_o}\right)_i$ is the signal-to-one-sided noise spectral density for the ith transmission link.

It is readily verified that the total C/N_o is typically dominated by (and approximately equal to) that of the constituent link with the lowest C/N_o contribution.

Receiver Figure of Merit

As the received signal power depends on the antenna gain G_r and the receiver noise power depends on the system noise temperature T_{sys}, we may define a figure of merit for a receiver that is reflected in the signal-to-noise ratio – the ratio of the receive antenna gain to system noise temperature

$$\left(\frac{G}{T}\right) \equiv \frac{G_r}{T_{sys}} \tag{4.38}$$

The receive figure of merit is typically given in units of dB/K. The equivalent figure of merit for the transmitter was the EIRP.

4.3.5.1 Signal-to-Noise Density in Free-Space Optical Receivers

We mentioned previously the potential use of optical frequencies for very high-data-rate free-space communications – for intersatellite links, for satellite–HAP links and potentially for specialized satellite–ground links (availability permitting). At very high frequencies the particle nature of electromagnetic radiation becomes significant and noise results from the random arrival of photons. This noise is variously referred to as photon noise or quantum noise, but is directly related to electronic shot noise in the optical receiver.

The variance of the received shot noise current $\langle i_{SN}^2 \rangle$ is related to the mean current I

$$\langle i_{SN}^2 \rangle = 2eIB \tag{4.39}$$

where $e = 1.602 \times 10^{-19}$ C is the electronic charge and B is the receiver bandwidth.

Comparison of RF and optical receivers is generally not straightforward, except for some special cases. Fortunately, we can compare the performance for a *quantum-limited* coherent optical receiver, which represents a fundamental limit on sensitivity of any optical communications receiver (although it is often more convenient to use less sensitive techniques). In a coherent optical receiver, incoming radiation is mixed with a local (optical) oscillator (LO) in a photodetector (or more typically in a pair of photodiodes). Mixing produces either an intermediate frequency (heterodyne) or a baseband (homodyne) signal, which may subsequently be demodulated in the same manner as a modulated RF signal.

It can be shown that the instantaneous IF carrier signal from an optical mixer is (Barry and Lee, 1990)

$$i_{IF}(t) = \frac{2\eta_d e}{hf} d(t) \sqrt{P_s P_{LO}} \cos(2\pi \Delta f\, t) + \text{noise} \tag{4.40}$$

where P_s and P_{LO} are the (weak) incident optical signal and (strong) local oscillator power respectively, Δf is the difference frequency between the input and LO and η_d is the photodetector quantum efficiency (the number of electrons produced per incident photon). $d(t)$ is the data modulation (see Chapter 5).

In the quantum limit, the mean photocurrent is dominated by the LO power $I \simeq \frac{\eta_d e}{hf} P_{LO}$, and, if all other noise sources may be neglected, the resultant carrier-to-noise density is given by

$$\frac{C}{N_o} = \eta_d \frac{P_s}{hf} \tag{4.41}$$

Comparison with the equivalent expression for an RF receiver (with an appropriate definition of received power) shows that quantum-limited optical signals have an effective noise temperature of

$$T_{\text{eff}} = \frac{hf}{k} \qquad (4.42)$$

and Figure 4.21 includes this equivalent noise temperature for very high frequencies. At a typical optical communications frequency of 2×10^{14} Hz ($\lambda = 1.5\,\mu m$), the effective noise temperature is of the order of $19\,000$ K! One is thus prompted to ask, what then is the advantage of free-space optical links if their receiver noise is several orders of magnitude higher than that of microwave receivers? The answer – for point-to-point links – lies in the even larger increase in antenna gain (which scales as the frequency squared).

Revision Questions

1. Estimate the beamwidth and maximum gain of a uniformly illuminated circular-aperture antenna of 1.2 m diameter at a frequency of 12 GHz.
2. Which types of satellite antenna are typically used for the following coverage situations:

   - earth cover beams;
   - spot beams;
   - contoured beams;
   - multiple satellite spot beams?

3. Estimate the pointing loss for an 0.5 m antenna at 12 GHz, for pointing errors of 0.5, 1 and 2°. What is the antenna beamwidth?
4. What types of antenna would you expect to find in a handset used for satellite applications at L-band? Why is circular polarization used for most handheld terminal antennas?
5. Up to what frequency can the unmodified Rayleigh–Jeans formula be used at/near room temperature? Estimate the effective noise temperature of an optical signal at $\lambda = 1.06\,\mu m$.
6. What noise temperatures correspond to noise figures of 1, 2 and 5 dB?
7. Estimate the figure of merit for a low-gain antenna with hemispherical coverage, and explain your reasoning.
8. The diameter of the Sun is approximately 1.4 million km and mean distance from the Earth is ~149.6 million km. Estimate the solid angle subtended by the Sun and use this to estimate its maximum contribution to antenna noise for a 1 m antenna at 12 GHz.

References

Balanis, C.A. (1997) *Antenna Theory, Analysis and Design*, 2nd edition, John Wiley & Sons, Inc., New York, NY.

Barry, J.R. and Lee, E.A. (1990) Performance of coherent optical receivers. *Proceedings of the IEEE*, **78**(8), 1369–1394.

Boculic, R. (1991) Use basic concepts to determine antenna noise temperature. *Microwaves and RF*, March.

Chu, L.J. (1948) Physical limitations in omnidirectional antennas. *Journal of Applied Physics*, **19**, 1163–1175.

Friis, H.T. (1944) Noise figures of radio receivers. *Proceedings fo the IRE*, July, 419–422.

Friis, H.T. (1946) A note on a simple transmission formula. *Proceedings fo the IRE*, **34**, 254.

Harris online (2010) *Commercial Deployable Antenna Reflectors*. Available: http://download.harris.com/app/public_download.asp?fid=463 [accessed February 2010].

Ho, C., Slobin, S., Kantak, A. and Asmar, S. (2008) Solar brightness temperature and corresponding antenna noise temperature at microwave frequencies. *NASA InterPlanetary Network Progress Report 42-175*, 15 November. Available: http://ipnpr.jpl.nasa.gov/ progress_report/42-175/175E.pdf [accessed February 2010].

IEEE (1993) *IEEE Standard Definitions of Terms for Antennas*, June.

ITU-R P.618-8 (2003) Recommendation P.618-8, Propagation data and prediction methods required for the design of Earth–space telecommunication systems.

Johnson, J. (1928) Thermal agitation of electricity in conductors. *Physical Review*. **32**, 97.

Kathrein Online (2010) http://www.kathrein.com/ [accessed February 2010].

Kraus, J.D. (1988) *Antennas*. McGraw-Hill, New York, NY.

Northrup Grumman (2010) *Astromesh*. Available: http://www.as.northropgrumman.com/businessventures/ astroaerospace/ [accessed Feruary 2010].

Nyquist, H. (1928) Thermal agitation of electric charge in conductors. *Physical Review*, **32**, 110–113.

Ramo, S., Whinnery, J.R. and Van Duzer, T. (1965) *Fields and Waves in Communication Electronics*, John Wiley & Sons, Inc., New York, NY.

Raysat Online (2010) http://www.raysat.com [accessed February 2010].

Ruze, J. (1966) Antenna tolerance theory: a review. *Proceedings of the IEEE*, **54**(4), 633–640.

Sarantel Online (2010) http://www.sarantel.com/ [accessed February 2010].

Shimabukuro, F. and Tracey, J.M. (1968) Brightness temperature of quiet sun at centimeter and millimeter wavelengths. *The Astrophysical Journal*, **152**(6), 777–782.

Stirland, S.J. and Brain, J.R. (2006) Mobile antenna developments in EADS astrium. First European Conference on Antennas and Propagation. EuCAP 2006. Volume, Issue, 6-10 Nov. 2006, pp. 1–5.

Stutzmann, W.L. (1998) Estimating directivity and gain of antennas. *IEEE Antennas and Propagation Magazine*, **40**(4), 7–11.

Terrestar Online (2010) http://www.terrestar.com/satellite.php [accessed February 2010].

Wheeler, H.A. (1947) Fundamental limitations of small antennas. *Proceedings of the IRE*, December, 1479–1488.

5

Modulation and Coding

5.1 Introduction

Previous chapters have laid the foundations of understanding of the concepts of transmission of information via electromagnetic radiation between a satellite and a user terminal, the latter being located on or near the Earth's surface. Concepts discussed so far allow us to estimate the received signal-to-noise (power) ratio at a receiver of a particular satellite service. In this chapter, we turn our attention to the transmission of *digital* signals and explore the relationship between signal-to-noise, occupied bandwidth and maximum data rate. Ultimately, we are interested in the *capacity* of these systems (the maximum achievable bit rate with minimal error) – either for a single link or for the total system capacity via Shannon's channel capacity theorem.

Practical digital satellite signals are examples of *passband* signals – that is, they occupy a finite spectral bandwidth, centred around some nominal, allocated frequency. However, in order to explore the fundamental limitations on digital transmission, it is initially convenient to consider *baseband* digital signals – that is, those centred on zero frequency. There is a fundamental relationship between the capacity of a digital link, the signal-to-noise ratio and the occupied bandwidth, and we illustrate the relationship between the probability of a transmission bit error and normalized signal-to-noise ratio. This knowledge provides a basis with which we may compare practical modulation schemes.

We next consider the three main *binary* modulation formats (amplitude, frequency and phase shift keying), with particular emphasis on phase shift keying owing to its constant signal amplitude and superior sensitivity. We go on to consider how, if the signal-to-noise ratio is sufficiently high, the modulated spectral efficiency, that is, the number of transmitted bits per second divided by the occupied spectrum, may be further improved by utilizing non-binary modulation schemes – in particular, non-binary (*M*-ary) phase shift keying and (combined) amplitude and phase shift keying. We also briefly discuss alternative modulation schemes used to reduce the practical occupied spectral width (in order to increase capacity), or to mitigate frequency elective fading due to multipath.

A consequence of Shannon's famous channel capacity theorem is that there must exist a method of coding the transmitted data that permits the channel capacity limit to be approached. In this chapter, therefore, we shall also look at how coding prior to transmission can be used to detect and even to correct errors, without resort to retransmission. We discuss the two main types of error control codes: *block* codes (in which a block of data are coded at a time) and *convolutional* codes (in which the data are coded as a continuous stream).

Surprisingly perhaps, in spite of the fact that the fundamental limits on channel capacity and error control coding have been known for decades, it is only very recently that practical codes

Satellite Systems for Personal Applications: Concepts and Technology   Madhavendra Richharia and Leslie David Westbrook
© 2010 John Wiley & Sons, Ltd

have been produced that can approach these limits. These are turbo codes and Low-Density Parity Check (LDPC) codes, and we discuss how these codes differ from their predecessors.

5.2 Modulation

Figure 5.1 serves to illustrate the principal functions in a simplex (one-way) communications link between transmitter apparatus (top) and receiver apparatus (bottom), via a physical channel (e.g. the ether). Where the original information source is analogue, such as a sound recoding or image, this must first be converted into an electrical signal via some form of transducer, such as a microphone or camera sensor. The resultant analogue electrical signal is then encoded into a digital signal as efficiently as possible (efficient source encoding is covered in Chapter 8). The digitized source will typically be further encoded so as to optimize it with regard to the particular properties of the physical channel, and in particular to make it more robust against noise and interference (channel coding is covered in the next section). The fully encoded baseband signal is then modulated onto a high-frequency carrier (a great many such carriers being simultaneously present on the ether).

At the receiver, the corresponding inverse functions are invoked. The received waveform is first demodulated to obtain the coded baseband signal. After this, channel decoding and source decoding produce a replica of the input electrical signal at the transmitter (after allowing for the transmission and encoding/decoding delays). Any analogue signals present may then be output, as necessary, via an appropriate transducer – such as a loudspeaker or video display unit.

Within a transmitter or receiver apparatus, communications signals cover various distinct frequency ranges. Generally, we may refer to *baseband* signals, *intermediate-frequency* (IF) signals and *radio-frequency* (RF) signals. Baseband signals are essentially those signals having spectral content extending down to zero frequency, while the modulated bandpass signal will be centred around some carrier frequency. In practice, it is generally convenient to carry out this process in two (or more) steps, and the output of the modulator is typically an IF signal. This is subsequently converted into the required RF signal (dictated by the system spectral allocation) via frequency translation (also known as up-conversion), prior to transmission. The remainder of this section focuses on the primary methods employed for modulation and demodulation of digital signals.

5.2.1 Modulation/Demodulation

Modulation is the process of imparting the data symbols on a carrier wave to form a useful bandpass (IF or RF) signal. In fact, there are a great many modulation techniques, and no single

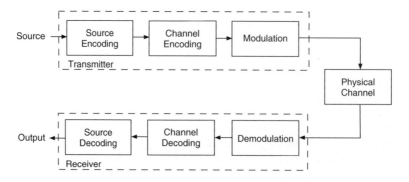

Figure 5.1   Major functional blocks in a digital communications system.

scheme is dominant (although some are more common than others). For example, we shall see that some schemes are more spectrally efficient than others, while others necessitate a greater receiver sensitivity for the same mean power (we shall see also that these qualities are essentially mutually exclusive).

Important factors to consider when selecting a modulation scheme include:

- *Sensitivity.* Transmissions may be restricted by battery power in the user's terminal equipment, by the required robustness to multipath, radiological protection levels for humans, by mutual interference when many users are collocated or by regulatory restrictions.
- *Spectral efficiency.* The need for spectral efficiency is important when sharing the channel with others. The greater the spectral efficiency, the more services/users can coexist in a given amount of spectrum.
- *Channel linearity.* Nonlinearity of any equipment used in the communications channel could distort the signals. In transponded satellite communications (i.e. where the received uplink signal is amplified, filtered and frequency translated), the linearity of the transponder is a critical issue when operated at or near maximum power (the most power efficient operating mode).
- *Cost and complexity.* Some modulation schemes are significantly more complex and hence more expensive to implement than others.

Modulation, and its inverse demodulation, are achieved by means of a MOdulator/DEModulator (MODEM) subsystem. Since, in general, it becomes more difficult to implement accurate modulation directly at higher RF carrier frequencies, an intermediate frequency (IF) – typically around 70 MHz or in L-band (950–1450 MHz) – is generally used for convenience. Block up and down (frequency) converters are then used between this IF and the actual carrier frequency.

5.2.2 Baseband Digital Signals

We shall consider initially baseband digital signals, that is, those where the digital signals are restricted to frequencies between zero and some (moderate) upper limit, such as might be found *within* the source coding communications equipment. Discussion of such signals is simpler, yet with care the properties of baseband signals may be applied to many aspects of bandpass signals.

5.2.2.1 Symbol Rate

The reader will be reasonably comfortable with the concept of the transmission and storage of digital information in the form of *bits* (Binary unITs), although we shall leave the precise definition of a bit until later on in this chapter. A bit is essentially the useful information associated with a digital system that has only two states (e.g. 'one' or 'zero'). In the general case, however, we must allow for the possibility of both binary and non-binary transmission systems. We therefore assume that the transmission comprises a sequence of *symbols* (modulation states), with each symbol taken from a finite *alphabet* (i.e. the set of allowed symbols). Of course, for the specific case where the transmission alphabet contains just two symbols, we have the familiar binary case, and each symbol is equivalent to 1 bit.

Let the time taken to transmit each symbol (the symbol period – in seconds) be T_s; then the symbol rate[1] R_s (in symbols per second) is given by

$$R_s = \frac{1}{T_s} \tag{5.1}$$

---

[1] The symbol rate is sometimes referred to as the Baud rate.

5.2.2.2   Maximum Symbol Rate

Intuitively, for a given channel bandwidth there ought to be an upper limit on the number of symbols that can be transmitted per second and still recovered at the receiver. At extreme symbol rates, a finite channel bandwidth (including any filters) will result in smearing (broadening) of the received symbol pulses, as illustrated in Figure 5.2 (top), as the channel is not able to respond at the required rate. This smearing causes adjacent symbols in a transmitted sequence to overlap significantly, and the receiver is no longer able to distinguish successive symbols correctly owing to this InterSymbol Interference (ISI).

Nyquist Criteria for Zero Intersymbol Interference

The limit on symbol rate because of finite bandwidth was analysed by Nyquist (1924), who determined that, in the absence of noise, successive symbols may be successfully distinguished if the received amplitude of all the other symbols is zero at the point where the wanted pulse is sampled (measured), as illustrated in Figure 5.2 (bottom).

Nyquist considered a baseband channel in which transmission is possible up to some maximum frequency only, corresponding to a bandwidth B. Such a (hypothetical) transmission channel may be described as having a rectangular, or 'brick-wall', filter characteristic. The normalised impulse response (the response at the output of the channel to an impulse at the input) corresponding to this rectangular spectrum is[2]

$$f(t) \rightarrow \text{sinc}(2\pi t B) \tag{5.2}$$

where sinc (x) is the unnormalized sinc function.[3] The sinc function has the property that there are periodic nulls (i.e. zero amplitude), with a null occurring at time offset equal to multiples of $1/2B$,

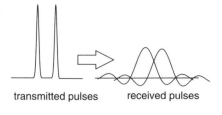

transmitted pulses   received pulses

received pulses

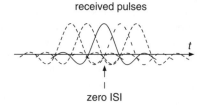

zero ISI

Figure 5.2   ISI due to pulse broadening in a finite bandwidth (top) and the Nyquist condition for pulses having zero ISI (bottom).

---

[2] Time and frequency domain signal descriptions are interrelated by the Fourier transform.

[3] The unnormalized sinc function is defined as

$$\text{sinc}(x) \equiv \frac{\sin(x)}{x}$$

either side of the peak. Nyquist thus reasoned that the maximum symbol rate with zero ISI for such a bandlimited channel is given by

$$R_s \leq 2B \tag{5.3}$$

that is the symbol rate cannot exceed twice the baseband bandwidth.

Raised Cosine Filter

The Nyquist 'brick-wall' channel characteristic is impractical (and unphysical), and would result in infinite 'ringing' (oscillation) of the output pulses. There are, however, practical channel filter shapes that also satisfy the Nyquist condition for zero ISI. An important example in digital communications is the Raised-Cosine (RC) spectral shape, the normalised form of which is given by

$$H(f) = \begin{cases} 1 & \text{if } 2T_s |f| \leq (1-\beta) \\ \cos^2\left(\frac{\pi T_s}{2\beta}\left(|f| - \frac{1-\beta}{2T_s}\right)\right) & \text{if } (1-\beta) < 2T_s |f| \leq (1+\beta) \\ 0 & \text{if } (1-\beta) > 2T_s |f| \end{cases} \tag{5.4}$$

where the parameter β is usually referred to as the *roll-off factor*, as the maximum occupied bandwidth (i.e. the maximum extent of non-zero power spectral density) for the raised-cosine pulse is a factor $(1+\beta)$ greater than the equivalent Nyquist 'brick-wall' channel filter characteristic. The raised-cosine filter reduces to the brick-wall filter in the limit $\beta \to 0$. The raised-cosine channel filter characteristic and resultant impulse shape are illustrated in Figure 5.3. Typical practical values for the RC roll-off factor are between 0.2 and 0.5.

Expressed in terms of the baseband bandwidth of an RC-filtered channel B_{RC}, the Nyquist limit may be expressed as

$$R_s \leq \frac{2}{(1+\beta)} B_{RC} \tag{5.5}$$

The raised-cosine filter is relatively easy to approximate in practice. Furthermore, as some form of filtering is generally required at both transmitter and receiver, RC filtering is commonly implemented by providing identical Root Raised-Cosine (RRC) filters at both the transmitter output and receiver input. As the name suggests, the RRC filter produces the *square root* of the desired overall RC filter characteristic (which then results from the product of the two RRC filters).

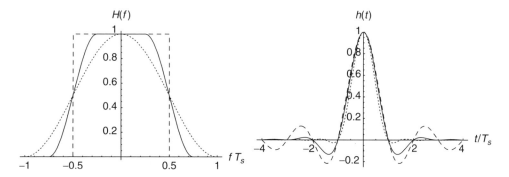

Figure 5.3 Normalised raised-cosine filter response (left: spectrum; right: pulses) for $\beta = 1$ (dotted line), $\beta = 0.5$ (solid line) and $\beta = 0$ (dashed line).

Nyquist Condition for Bandpass Channels

We may consider the pure translation from a baseband signal to a bandpass signal (often called *up-conversion*) as being achieved by multiplication of a continuous-wave (unmodulated) carrier signal at (centre) frequency f_c by the baseband signal.[4] The resulting RF (or IF) bandwidth is equal to *twice* the baseband bandwidth (in effect the RF bandpass spectrum incorporates the up-converted negative baseband frequencies as well as the positive frequencies).

5.2.2.3 Information Rate

How should we measure the amount of *information* passed between a satellite service and a user's terminal equipment (or the amount of information stored on some storage device)? Note that we must distinguish between information and *meaning*, as our interest is the measurement of information purely from an engineering point of view, rather than concerning ourselves with semantics and interpretation (Shannon, 1948).

The foundations for the measurement of information were laid down by Nyquist (1924) and Hartley (1928). Hartley reasoned that a symbol or character conveyed information only if there were alternative symbols that could be transmitted instead. He further argued that a *logarithmic* measure was appropriate for the information carried by a sequence of symbols chosen from a finite alphabet of symbols. Hartley's information measure was the logarithm of the number of possible symbol permutations.

Hartley argued that the information contained in a sequence of independent messages should add, and consequently the information contained in a sequence of M symbols should be the same as the information contained in M sequences of just one symbol. As the number of permutations in a message of M symbols is equal to the number of permutations in a message of just one symbol raised to the power of M, a logarithmic measure of information would have the correct property because $\log(k^M) = M \log(k)$.

The base b of the logarithm amounts to a choice of units. We are familiar with *binary units* (bits) having base $b = 2$.

5.2.2.4 Relation Between Bit Rate and Symbol Rate

In general, a symbol may contain the equivalent of several bits of information. We therefore pose the simple question: If we have symbol rate R_s, what is the equivalent bit rate?

If there are M symbols in our transmission alphabet (where $M \geq 2$), then the number of *bits per symbol* is m, where

$$m = \log_2(M) \tag{5.6}$$

since a group of m bits may take $M = 2^m$ different values.

The effective channel bit rate R_c – the rate at which bits are actually transmitted over the channel – is equal to the symbol rate times the number of bits per symbol. We shall allow for the eventual possibility that the effective user bit rate R_b (i.e. the information rate actually experienced by the user's application) is not the same as the effective channel bit rate R_c. Indeed, it is quite probable that the effective channel rate will be higher than the user bit rate (i.e. in general $R_c \geq R_b$), owing to the (deliberate) introduction of redundant bits to the user data bits (redundant from the user application's perspective) prior to transmission in order to facilitate the detection (and correction) of transmission errors (incorrectly detected bits), and thereby make the digital signal more robust to noise (and interference).

---

[4] Pure up/down conversion is a form of amplitude modulation.

The simultaneous use of symbol rate and user and channel bit rates can be somewhat confusing at times. The differences between these quantities may be summed up as follows:

- *User bit rate (R_b).* This is the rate at which useful (i.e. the user's application) bits are transmitted over the channel.
- *Channel bit rate (R_c).* This is the equivalent rate at which bits are actually transmitted over the channel, and typically differs from the channel rate owing to the inclusion of redundant bits in order to combat errors introduced during transmission.
- *Symbol rate (R_s).* This is the rate at which symbols (distinct digital values) are transmitted over the communications channel. Each symbol may contain one or more channel bits.

We shall discuss channel coding in the next section. For now, we shall simply account for it via the ratio r of the *user* bit rate to the *channel* bit rate (where $r \leq 1$). Hence, we have the following relationship between user bit rate, channel bit rate and symbol rate

$$R_b = r R_c; \quad R_c = m R_s \tag{5.7}$$

5.2.2.5 Communications in the Presence of Noise

Matched Filter

The ability to distinguish different symbols at the receiver is degraded by the presence of noise. Some form of filter is employed at the receiver input to limit noise and thereby maximize the received signal-to-noise ratio, and the signal-to-noise ratio achieved will clearly depend on the receiver filter response.[5] Clearly, if the receiver has an input filter bandwidth much wider than that of the input signal, a large amount of noise will be collected in addition to the signal and the signal-to-noise ratio will be suboptimal. If, on the other hand, the input filter is too narrow, not all of the signal will pass through the filter, and the signal may be corrupted.

In order to maximize the signal-to-noise ratio at the receiver detector stage, it can be shown that the optimum receiver filter arrangement is obtained when the receive filter impulse response has the same magnitude response as the normalized (amplitude) spectrum of the input signal itself – but with the opposite phase response.[6] In the time domain, the filter impulse response is the time-inverse of the input signal. Such a filter is known as a *matched filter*.

If the symbol pulse amplitude[7] is given by $s(t)$ and the added noise is given by $n(t)$, then the available energy in the symbol E_s may be obtained by integrating the signal power over all time, while the noise energy is equal to its statistical variance (the mean square of the noise amplitude)

$$E_s = \int_{-\infty}^{\infty} |s(t)|^2 \, dt \tag{5.8}$$

The maximum output signal-to-noise ratio from the receiver for this symbol is found to be

$$\gamma_s = \frac{2E_s}{N_o} \tag{5.9}$$

where N_o is the one-sided noise power spectral density at the receiver (also in units of energy).

---

[5] A filter is also generally employed at the transmitter output to limit interference to other communications channels.

[6] In phasor notation, the (complex) filter impulse response is the complex conjugate of the (complex) input signal amplitude spectrum.

[7] For mathematical convenience, it is common to describe signal amplitudes in units such that the amplitude squared gives the power directly. Physically, the amplitude would typically be measured in volts. This notation is therefore equivalent to assuming a 1 Ω reference impedance.

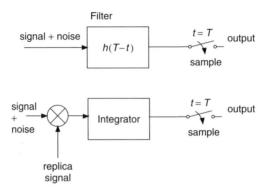

Figure 5.4 Alternate implementations for a matched filter (top: conjugate filter; bottom: multiplication by the expected signal and integration).

A matched filter is one that achieves the maximum signal-to-noise per symbol at its output. Figure 5.4 (top) illustrates the matched filter concept. It can be shown that a matched filter may also be realized (for time-limited signals) by multiplying the received signal (plus noise) by the expected signal and integrating. Consequently, for rectangular (non-Nyquist) pulses, the matched filter is equivalent to integrating the signal over the symbol period (with the integrator reset after each symbol period). A matched filter for nominally rectangular pulses is therefore typically implemented using a so-called 'integrate and dump' circuit. The matched filter for the case where the transmitted signal comprises rectangular (Nyquist) band-limited pulses at the transmitter is an identical rectangular filter. Similarly, the matched filter for RRC filtered pulses is an identical RRC – underlining the usefulness of the RRC filter.

It is generally more useful to talk in terms of the energy per symbol, averaged over *all* symbols. The average energies to one-sided noise for symbol, channel bit and user bit are related via

$$\frac{E_s}{N_o} = m\frac{E_c}{N_o}; \quad \frac{E_c}{N_o} = r\frac{E_b}{N_o} \tag{5.10}$$

In particular, the mean energy per bit to one-sided nose spectral density (E_b/N_o) is an invaluable 'yardstick' by which the sensitivity performance of various modulation schemes may be compared.

5.2.2.6 Shannon Channel Capacity Theorem

What is the maximum information that can reliably be passed over a communications link where there is noise present? Hartley (1928) established that the capacity (maximum rate of flow of information) C of a communication link is

$$C = 2B\log_b(M) \tag{5.11}$$

where M is the number of 'resolvable values' that a symbol can take (although Hartley did not define this further) and the base b of the logarithm is chosen to be appropriate to the information type. In a landmark paper, Shannon (1948 and 1949) used the mathematics of multidimensional space to determine that, in the presence of Additive White Gaussian Noise (AWGN) – that is, when the received signal is the linear sum of the uncorrupted signal plus noise whose spectral density is assumed to be constant with frequency (at least over the band of interest) and whose amplitude obeys a Gaussian distribution – the maximum number of discernable symbols for negligible error is given by $M \to \sqrt{1 + \frac{S}{N}}$, where S/N is the signal-to-noise ratio. Shannon's channel capacity theorem gives the capacity in bits per second of a channel with bandwidth B and signal-to-noise S/N to be (Shannon, 1948)

$$C = B \log_2 \left(1 + \frac{S}{N}\right) \tag{5.12}$$

Shannon's channel capacity theorem goes on to state that 'it is not possible by any encoding method to send at a higher rate and have an arbitrarily low frequency of errors'. The theorem defines the *fundamental* limit on channel capacity, although it does not imply that this limit is easily achievable!

Shannon Limit

The signal-to-noise ratio may be conveniently expressed in terms of the bit energy-to one-sided noise density by noting that the signal power is $R_b E_b$ and the noise power is $N_o B$, so that

$$\frac{S}{N} = \frac{R_b}{B} \frac{E_b}{N_o} \tag{5.13}$$

We now introduce the spectral efficiency for the channel ζ as being the ratio of the useful bit rate R_b to the channel bandwidth B, with units of (b/s)/Hz:

$$\zeta \equiv \frac{R_b}{B} \tag{5.14}$$

Clearly the spectral efficiency is very important where the electromagnetic spectrum is very limited. Noting that $R_b \leq C$, equation (5.12) may be rearranged so as to express the minimum energy per bit to one-sided noise spectral density as a function of the channel spectral efficiency ζ

$$\frac{E_b}{N_o} \geq \frac{2^\zeta - 1}{\zeta} \tag{5.15}$$

This fundamental relationship between sensitivity and spectral efficiency, known as the *Shannon limit*, is illustrated in Figure 5.9.

An important limiting case is that where the channel bandwidth is not a constraining factor (i.e. as $\zeta \to 0$). In this limit, the minimum energy per bit to one-sided noise ratio is given by

$$\frac{E_b}{N_o} \to \ln(2) = 0.169 \text{ (or } -1.6 \text{ dB)} \tag{5.16}$$

This value therefore represents the *absolute minimum* value of $\frac{E_b}{N_o}$ for nominal error-free transmission, which cannot be surpassed, regardless of channel bandwidth.

5.2.2.7 Binary NRZ Signals

It is useful to consider one particular form of baseband digital signal and to derive their performance by way of reference for later comparison with various modulation schemes. For simplicity, we shall restrict our discussion to the binary Non-Return to Zero (NRZ) signal format.

The bipolar NRZ format uses a positive signal of amplitude $+A$ to signify a logical 'one' and an (equal and opposite) negative amplitude $-A$ to signify a logical 'zero'. The symbol state is held constant for period T_s – the symbol period (here also equal to the bit period T_b), and therefore symbol changes occur only at intervals of multiples of the bit period. This baseband waveform is illustrated in Figure 5.5 (left).

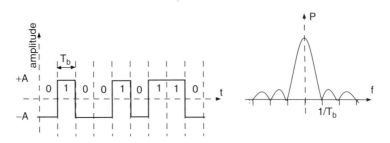

Figure 5.5 Uncoded NRZ baseband waveform (left: example time sequence; right: power spectrum).

At the output of a matched filter, the separation (difference) between the outputs for the 'one' and a 'zero' symbols is equal to

$$d_{\min} [\text{NRZ}] = 2\sqrt{E_b} \tag{5.17}$$

where $E_s, E_b \rightarrow A^2 T_b$. This difference is called the *minimum Euclidean distance* between the NRZ modulation symbol states.

For random sequences, the NRZ signal spectral envelope (envelope of the signal power spectral density) is (Proakis, 2001)

$$\text{PSD}(f) [\text{NRZ}] = A^2 T_b \, \text{sinc}^2 (\pi f T_b) \tag{5.18}$$

Although, in principle, the frequency spectrum of NRZ signals extends up to infinite frequency, albeit with rapidly diminishing amplitude, it is common to describe the bandwidth of such a signal as being given by the frequency of the first null in the power spectrum (i.e. $B \sim 1/T_b$).

5.2.2.8 Detection of Baseband Binary Signals

On reception of a digital communication signal, the demodulator must make a decision, for each symbol period, as to which of the permitted symbols was actually transmitted. Under the assumption of AWGN, both the input noise amplitude and matched filter output can be described in terms of the Gaussian (or normal) distribution PDF for variable x given by

$$\rho(x)_N = \frac{1}{\sqrt{2\pi}\,\sigma_N} \exp\left(-\frac{(x - \mu_N)^2}{2\sigma_N^2}\right) \tag{5.19}$$

where μ_N and σ_N are the distribution mean and standard deviation respectively. The noise variance is related to the noise power spectral density.

The distributions of the outputs from the matched filter for binary 'one' and binary 'zero' are illustrated in Figure 5.6.

Bit Error Ratio

It is important to be able to quantify the probability of a bit error – also known as the Bit Error Ratio (BER). The BER is the probability that an incorrect decision is made as to whether the received result is a 'one' or a 'zero'.

Hard Decision

In the simplest type of detector, a decision as to whether the received signal is a 'one' or a 'zero' is made by comparing the sample value with some threshold. Under the assumption of

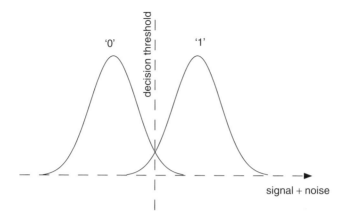

Figure 5.6   Binary detection threshold.

equal likelihood of the transmitted bit being a 'one' or a 'zero', the two symbol distributions at the matched filter output have identical variance (but different mean) and the optimum decision threshold is clearly located half-way between the mean signal amplitudes for logical 'one' and logical 'zero'; in the case of NRZ signals this will be at zero amplitude, and the 'tails' of these distributions – the parts that cross over the threshold – will result in an incorrect decision for those signal values.

If the probability of transmission of a 'one' and a 'zero' are P_1 and P_0 respectively, and the probability of an error when a 'one' is transmitted and when a 'zero' is transmitted are $P_{e,1}$ and $P_{e,0}$ respectively, then the probability of error of either kind is given by $(P_0 P_{e,0} + P_1 P_{e,1})$.

We assumed equal probability for logical 'zero' and 'one' (i.e. $P_0 = P_1$), and also that for AWGN the noise variance is the same for both logical 'zero' and 'one', whereupon the probability of uncoded channel bit error is

$$P_c = \frac{1}{2}\text{erfc}\left(\frac{|d_{min}|}{2\sqrt{N_o}}\right) \tag{5.20}$$

$$P_c\,[\text{NRZ}] = \frac{1}{2}\text{erfc}\left(\sqrt{\frac{E_b}{N_o}}\right) \tag{5.21}$$

where erfc (x) is the complementary error function (a tabulated function associated with the Gaussian distribution).[8]

Soft Decision

The threshold detector just described is known as a *hard-decision* decoder. It makes a hard (i.e. final) decision as to the likely symbol, using only the output from the matched filter. It turns out that, when using certain error-correcting codes (discussed in the next section), it can be advantageous to delay making a hard decision regarding the symbol until the error decoding process (post-demodulation).

---

[8] The complementary error function is defined as

$$\text{erfc}\,(x) \equiv \frac{2}{\sqrt{\pi}} \int_x^\infty e^{-z^2}\,\mathrm{d}z \tag{5.22}$$

In such cases, it is useful, instead, to make a (quantized) measurement of how *likely* a received symbol is to be a particular symbol. This is known as a *soft* decision, and the associated decoder is known as a soft-decision decoder. When combined with an appropriate error-correcting scheme, a soft-decision decoder will typically outperform its hard-decision equivalent – albeit at the expense of additional complexity.

There are many ways to implement a soft-decision detector. One approach is to quantize the output with a specified number of levels between the mean outputs for logical 'one' and logical 'zero'. Another approach is to output the Log-Likelihood Ratio (LLR) – the probability of the symbol being that indicated divided by the probability of it not being the symbol. For the binary case, the LLR is the natural logarithm of the ratio of the probability of transmission of a logical 'one', given the output value, to that of it being a logical 'zero'.

5.2.2.9   Probability of Error for Concatenated Regenerative Links

In some circumstances we may have two or more datalinks, one after the other, with regeneration (detection and retiming) between each link. This will be the case for modulated signals for an on-board processed satellite, where the uplink is regenerated (and potentially switched) before retransmission on the downlink.

The probability of bit error for two concatenated datalinks (with regeneration) is the probability of an error in link 1 times the probability of no error in link 2, plus the probability of an error in link 2 times the probability of no error in link 1:

$$P_c = P_{c,1}\left(1 - P_{c,2}\right) + P_{c,2}\left(1 - P_{c,1}\right) \tag{5.23}$$

$$\simeq P_{c,1} + P_{c,2} \tag{5.24}$$

where the assumption has been made that the probability of error is much smaller than 1 in both links.

5.2.3   Binary Digital Modulation Schemes

Having established limits on baseband communications, we move on to consider binary modulation schemes for bandpass communications. Binary modulation schemes (i.e. those with $m \to 1$) are generally the most robust against noise (and therefore provide the best sensitivity). The available options for binary modulation of an RF carrier are:

- *Amplitude Shift Keying (ASK).* The carrier amplitude is modulated according to the (normalized NRZ) data signal amplitude.
- *Frequency Shift Keying (FSK).* The carrier phase is modulated according to the (normalized NRZ) data signal amplitude.
- *Phase Shift Keying (PSK).* The carrier frequency is modulated according to the (normalized NRZ) data signal amplitude.

These basic modulation schemes are illustrated in Figure 5.7.

5.2.3.1   Binary ASK

In binary Amplitude Shift Keying (ASK), the amplitude of the RF carrier wave is alternately modulated on and off, depending on whether the data is a 'one' or a 'zero'. For this reason, this modulation scheme is also known as On-Off Keying (OOK).

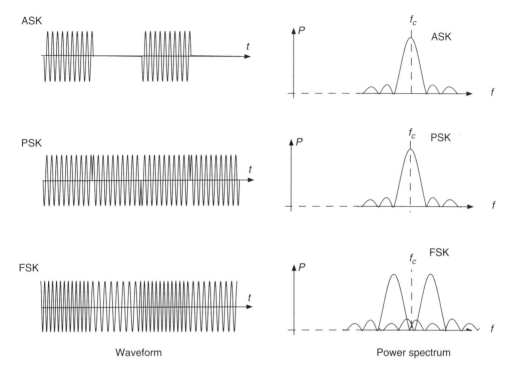

Figure 5.7   Binary modulation schemes: time domain (left); frequency domain (right).

The time-dependent signal amplitude corresponding to the binary amplitude shift keyed signal may be represented as

$$s(t) [\text{ASK}] = \frac{A}{2} (1 + d(t)) \cos(\omega t) \tag{5.25}$$

where $d(t)$ is the (normalized) NRZ binary input data sequence. The optimal demodulation method results from multiplying the received signal by an identical (unmodulated) carrier followed by low-pass filtering – a technique referred to as coherent demodulation.

Probability of Channel Bit Error

As no power is transmitted for the 'zero' state, the average energy per symbol E_s (and in this case, the energy per bit E_b) is equal to half the energy for the 'one' symbol, The minimum Euclidean distance d_{min} (i.e. the amplitude difference at the output of a matched receiver) between the two symbols is, for ASK,

$$d_{min} [\text{ASK}] = \sqrt{2E_s} \tag{5.26}$$

As the minimum distance for ASK is a factor of $\sqrt{2}$ smaller than for NRZ, it can be shown that the corresponding value of E_b/N_o for an uncoded (i.e. with no error coding) ASK signal for a defined BER is twice (i.e. 3 dB greater) that for NRZ signals with the same BER. The variation in BER for uncoded ASK with E_b/N_o is illustrated in Figure 5.8.

Although relatively simple to implement, binary ASK is generally inferior to alternative modulation schemes for personal satellite applications.

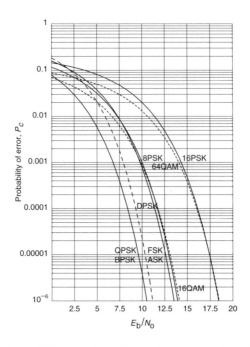

Figure 5.8   Variation in BER with E_b/N_o for various (uncoded) modulation schemes.

5.2.3.2   Binary FSK

In frequency shift keying (FSK), the signal amplitude is held constant but the carrier *frequency* is modulated according to the data value. In this case, the signal for an FSK signal may be represented as

$$s(t)\,[\text{FSK}] = A\cos\left((\omega_0 + k\,d\,(t))\,t\right) \tag{5.27}$$

where ω_0 is the mean angular frequency and k is a scaling constant that determines the frequency separation between the 'zero' and 'one' states.

The frequency separation between the on and off states in FSK is somewhat arbitrary; however, the minimum frequency separation between the two binary states for FSK for which the signals remain orthogonal (completely separable) is equal to one half of the symbol period T_s, and FSK with this frequency separation is known as Minimum Shift Keying (MSK).

Probability of Error

For binary FSK with the frequency separation of the two states such that the symbols are *just* orthogonal (i.e. MSK), the distance between the symbols is

$$d_{\min}\,[\text{MSK}] = \sqrt{2E_s} \tag{5.28}$$

which is the same as for ASK. The probability of error for coherent FSK is therefore the same as for coherent binary ASK.

5.2.3.3 Binary PSK

We now turn to a very important type of modulation for personal satellite applications – phase shift keying. In phase shift keying (PSK) the signal amplitude is held constant but the carrier *phase* is modulated according to the instantaneous baseband data value.

The signal for a phase-shift-keyed signal may be represented as

$$s(t)\,[\text{PSK}] = A\cos\left(\omega t + (1 + d(t))\,\frac{\pi}{2}\right) \tag{5.29}$$

Probability of Error

For Binary PSK (BPSK), the minimum distance between the symbols is

$$d_{\min}\,[\text{BPSK}] = 2\sqrt{E_s} \tag{5.30}$$

which is the same as for NRZ. Therefore, the probability of channel bit error for coherent demodulated BPSK is the same as that for NRZ.

Again, the variation in BER with E_b/N_o for uncoded BPSK is compared with other modulation schemes in Figure 5.8. The enhanced receiver sensitivity metric (E_b/N_o) for BPSK compared to that for both binary ASK and FSK is one reason for the popularity of phase shift keying.

5.2.3.4 Non-Coherent Detection

The discussion has so far focussed on coherent detection with an optimum receiver. Coherent detection requires that a replica carrier be generated at the receiver. This increases the complexity of the receiver circuitry, and so-called non-coherent demodulation can be attractive in certain scenarios.

Non-Coherent Detection of ASK and FSK

ASK and FSK may be detected incoherently by detecting the signal *envelope* using a so-called envelope detector. For non-coherently demodulated ASK and FSK, the probability of uncoded channel bit error is

$$P_c\,[\text{nc-ASK or FSK}] = \frac{1}{2}\exp\left(-\frac{1}{2}\frac{E_b}{N_o}\right) \tag{5.31}$$

The performance of non-coherent demodulation of ASK and FSK is marginally worse (about 1 dB worse at low BER) than for coherent detection.

Non-Coherent Detection of PSK

Non-coherent demodulation of Phase Shift Keying (PSK) may be achieved by using Differential Phase Shift Keying (DPSK) in which the phase of a symbol is compared with the phase of the previous (appropriately encoded) symbol, delayed by the symbol period. However the use of a relative noisy signal (the signal from the previous symbol period) as phase reference results in additional bit errors, and the BER for DPSK is given by

$$P_c\,[\text{DPSK}] = \frac{1}{2}\exp\left(-\frac{E_b}{N_o}\right) \tag{5.32}$$

Nevertheless, the performance of DPSK is still better than that of coherent demodulated ASK or FSK, and, for low error probabilities ($P_c < 10^{-5}$), the difference in E_b/N_0 values between uncoded DPSK and BPSK is only of the order of 1 dB. Given that DPSK is simpler to implement, the use of DPSK is clearly attractive from a cost perspective – except for those circumstances where every dB of sensitivity counts.

5.2.4 Modulation Schemes for High Spectral Efficiency

5.2.4.1 Spectral Efficiency of Binary Modulation Schemes

The modulated spectrum for ASK is essentially that of NRZ, with the appropriate frequency translation. Therefore, the spectral efficiency (bit rate over occupied bandwidth) for uncoded binary ASK with rectangular pulses is 0.5 (b/s)/Hz, while, for Nyquist (brickwall-filtered) pulses the spectral efficiency is increased to 1 (b/s)/Hz. The occupied spectrum for a binary FSK signal is determined by the chosen frequency spacing between the two states, with MSK being the most spectrally efficient form of FSK. The spectral width of binary PSK is also that of NRZ with appropriate frequency translation. Uncoded spectral efficiencies of various modulation schemes are tabulated in Table 5.1.

5.2.4.2 M-ary Modulation Schemes

So far, we have considered binary modulation schemes with uncoded spectral efficiencies less than or equal to just 1 (b/s)/Hz. Higher-order modulations, in which more than 1 bit is sent in a symbol period, are attractive for achieving significantly higher spectral efficiencies (in (b/s)/Hz), thereby maximizing use of the available spectrum.

The spectral efficiency of M-ary ASK and PSK modulation schemes (that is ASK and PSK schemes with M symbols, where $M > 2$) is increased by a factor $\log_2 (M)$. For example, the spectral efficiency for bandpass signal with raised-cosine-filtered M-ary PSK or ASK pulses with code rate r and m bits per symbol is

$$\zeta \equiv \frac{R_b}{B_{ch}} = \frac{rm}{(1 + \beta)} \tag{5.33}$$

Spectral efficiencies are summarized in Table 5.1 for various modulations, for the case of rectangular pulses and ideal (Nyquist brickwall-filtered) pulses.

Table 5.1 Uncoded spectral efficiencies of various modulation schemes (for both rectangular and Nyquist brickwall-filtered pulses) (Proakis, 2001)

Modulation type	m	Spectral efficiency ((b/s)/Hz)	
		(Rect. pulses)	(Nyquist pulses)
ASK; BPSK; DPSK	1	0.5	1
QPSK	2	1	2
8-PSK	3	1.5	3
16-PSK; 16-QAM	4	2	4
32-PSK	5	2.5	5
64-PSK; 64-QAM	6	3	6

The disadvantage of these higher-order modulation schemes is, in general, the reduced minimum distance between symbols; consequently, the probability for error is higher for a given energy per bit to noise, and, conversely, higher signal power is required to ensure error-free operations.

Constellation Diagram Representation

When considering higher-order ASK and PSK modulation schemes (and hybrids thereof), it is generally convenient to express the signal in terms of the In-phase (I) and Quadrature (Q) signal amplitudes, which together define a signal constellation (trajectory)

$$s(t) = I(t)\cos(\omega t) + Q(t)\sin(\omega t) \tag{5.34}$$

We note that this I and Q description also forms a convenient general scheme for implementing a wide range of modulation formats (including non-binary schemes) that rely on either amplitude or phase shift keying (or some combination of both). Furthermore, as the instantaneous signal frequency is just the rate of change in the instantaneous carrier phase, frequency modulation and phase modulation are intrinsically related to each other.

Relation between Symbol and Bit Error Rates

While the symbol error probability may be estimated for M-ary modulation schemes, the bit error ratio for M-ary signals ultimately depends on the *mapping* of bits to symbols. However, a lower bound on the bit error ratio for a given symbol error ratio occurs for (optimum) Gray-coded symbol mapping, because, for this coding scheme, adjacent symbols differ only by a single bit. As the most probable errors occur owing to the erroneous distinction between adjacent symbols in the constellation diagram, in the particular case of Gray coding, a single symbol error would normally be expected to cause only an m-bit error, and the bit error ratio is approximately the symbol error ratio divided by the number of bits per symbol. A lower bound on bit error ratio for M-ary modulation schemes is therefore

$$P_c \geq \frac{P_s}{m} \tag{5.35}$$

The performance of various modulation types in terms of energy per bit to one-sided noise spectral density $\frac{Eb}{N_0}$ and spectral efficiency η is compared with the Shannon capacity limit in Figure 5.9. It is apparent that, although the spectral efficiencies are higher for M-ary modulation schemes, the uncoded BER performance of all of the modulation schemes considered in this chapter still falls some way short of the theoretical Shannon-limited capacity. We shall show in the next section that the use of appropriate coding can achieve performance very close to the Shannon limit with many of these modulation schemes, thereby combining high spectral efficiency and sensitivity.

5.2.4.3   *M*-PSK

An important class of M-ary modulation for satellite communications is M-PSK (M-ary PSK). In M-PSK more than two phase states are used (4, 8, 16, etc). Figure 5.10 shows I and Q signal constellation diagrams for BPSK, Quaternary Phase Shift Keying (QPSK or 4-PSK), 8-PSK and 16-PSK. M-PSK retains the approximate constant amplitude of BPSK, which makes it less demanding on channel linearity than ASK.

The output of a matched filter for an M-PSK constellation is distributed uniformly around a circle of radius $\sqrt{E_s}$, so that the minimum distance between adjacent symbols for M-PSK is found, using simple geometry, to be

$$d_{\min}[\text{M-PSK}] = 2\sqrt{E_s}\sin\left(\frac{\pi}{M}\right) \tag{5.36}$$

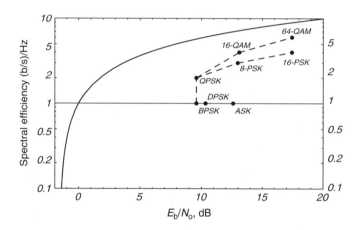

Figure 5.9   Uncoded modulation – performance against Shannon limit (solid line).

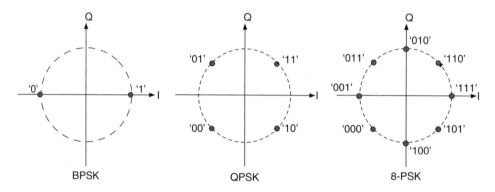

Figure 5.10 M-PSK modulation constellations (left to right: BPSK, QPSK, Gray-coded 8-PSK).

QPSK

A particularly important example of M-PSK is Quaternary Phase Shift Keying (QPSK). QPSK may be considered as two orthogonal BPSK signals (one each for I and Q), whereupon the probability of bit uncoded channel error may be shown to be the same as for BPSK. Of course, the symbol energy for QPSK is twice that for BPSK, and thus, for a given symbol rate, the transmitted power for a given BER will be twice that for BPSK; however, the bit rate will also be twice that of BPSK. Hence, for the same bit rate, QPSK offers the same sensitivity, but twice the spectral efficiency. This fact explains why QPSK and its derivatives are widely exploited in many satellite systems.

M-PSK with M > 4

For Gray-coded M-PSK with large M, it can be shown that the probability of channel bit error is (Proakis, 2001)

$$P_c\,[\text{M-PSK}] \approx \frac{1}{\log_2(M)}\text{erfc}\left(\sqrt{\left(\log_2(M)\sin^2\left(\frac{\pi}{M}\right)\right)\frac{E_b}{N_o}}\right) \qquad (5.37)$$

Non-Coherent Demodulation of M-PSK

As with binary modulation, M-ary modulation schemes may be demodulated non-coherently; however, the performance of M-ary differential PSK is relatively poor for $M > 2$. For example, E_b/N_o for Differential QPSK (DQPSK) is approximately 2.3 dB worse than for coherent demodulated QPSK (Proakis, 2001).

5.2.4.4 Combined Amplitude and Phase Shift Keying

Amplitude and phase shift keying may be combined in order to increase the density of modulation states in the constellation diagram. The result is usually referred to as Amplitude–Phase Shift Keying (APSK) or Quadrature Amplitude Modulation (QAM), depending on the particular constellation diagram shape, as illustrated in Figure 5.11 for $M = 16$. For the sake of clarity, in this book we shall henceforth refer to the rectangular grid constellation as M-QAM and the concentric circular constellation as non-rectangular M-APSK.

Rectangular Quadrature Amplitude Modulation

Rectangular M-QAM may be generated using two orthogonal $\sqrt{M}$-ary ASK signals (also known as pulse amplitude modulation – one for I and one for Q. The particular case of 4-QAM is the same as QPSK.

The probability of channel bit error for Gray coded rectangular M-QAM may be shown to be (Pahlavan and Levesque, 1995)

$$P_c\left[M\text{-QAM}\right] \approx \frac{2}{\log_2(M)}\left(1 - \frac{1}{\sqrt{M}}\right)\text{erfc}\left(\sqrt{\left(\frac{3\log_2(M)}{2(M-1)}\right)\frac{E_b}{N_o}}\right) \tag{5.38}$$

Referring to Figure 5.9, for $M > 4$, uncoded M-QAM has a lower E_b/N_o than the equivalent M-PSK modulation (both having the same spectral efficiency). This explains the use of QAM in terrestrial communications and broadcasting. However, the performance of M-QAM is sensitive to channel linearity, which can limit its use in satellite communications (where transponders are often operated near saturation – for high power efficiency).

Impact of Nonlinear Channel Distortion

Modulation schemes such as rectangular QAM, which incorporate significant variations in amplitude, are susceptible to distortion by any nonlinear elements in the communications chain. In satellite

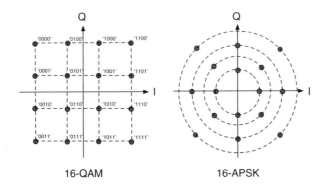

Figure 5.11   16-APSK constellations (left: rectangular 16-QAM; right: non-rectangular 16-APSK).

systems the dominant nonlinearity is caused by operation of a satellite transponder high-power amplifier (often a Travelling-Wave Tube Amplifier (TWTA)) near to saturation – for maximum efficiency.

A simple and convenient model of such nonlinear elements is that due to Saleh 1981. The input and output amplitudes for a bandpass (composite) signal $x(t)$ and $y(t)$ may be described by

$$x(t) = r(t) \cos(\omega_0 t + \psi(t)) \tag{5.39}$$

$$y(t) = A(r) \cos(\omega_0 t + \psi(t) + \Phi(r)) \tag{5.40}$$

where $r(t)$ and $\psi(t)$ are the instantaneous input signal magnitude and phase respectively (e.g. resulting from amplitude and phase modulation). The nonlinear function $A(r)$ describes the distortion of the output signal magnitude, while $\Phi(r)$ gives that of the output phase shift. These functions are dependent on input signal magnitude only and are conveniently modelled as

$$A(r) \simeq \frac{\alpha_A r}{\left(1 + \beta_A r^2\right)} \tag{5.41}$$

$$\Phi(r) \simeq \frac{\alpha_\Phi r^2}{\left(1 + \beta_\Phi r^2\right)} \tag{5.42}$$

where the four parameters α_A, β_A, α_Φ and β_Φ are usually determined by curve-fitting the actual nonlinear characteristic.

Figure 5.12 illustrates the effect of nonlinear distortion of a 64-QAM signal, for a hypothetical nonlinear amplifier, for various input levels. The signals levels were chosen to exaggerate the nonlinear effect. The effect is increasing distortion of the points furthest from the origin – ultimately forming a shape that has rounded edges and is rotated (Ngo, Barbulescu and Pietrobon, 2005).

Non-Rectangular Amplitude–Phase Shift Keying

The sensitivity of APSK to channel nonlinearities is reduced by adopting the non-rectangular QAM constellation. For this reason, non-rectangular APSK is used in some high-capacity satellite communications services, including those that employ the DVB-S2 waveform (Nemer, 2005).

Unfortunately, Gray coding cannot be used for this type of constellation and analytic expressions for the bit error probability are not available; nevertheless, the performance may be similar to that of QAM.

5.2.5 Modulation Schemes for High Spectral Purity

The number of similar communication channels that can be contained in a given allocation of spectrum is affected by the minimum permissible spacing between individual channels in order to

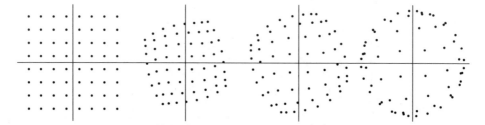

Figure 5.12 Nonlinear distortion of a 64-QAM signal constellation (left to right: input, 6 dB back-off, 3 dB back-off, no back-off).

minimize adjacent channel interference. The minimum channel spacing clearly depends on how rapidly the power spectrum decays with frequency offset. For rectangular pulses this decay is quite slow, and a significant guard band is required between similar channels.

In order to minimize the channel spacing and thereby maximize the number of channels in a particular spectrum allocation, there is a need for modulation schemes that exhibit rapid decay of the power spectra outside the Nyquist bandwidth. It is found that a cause of relatively slow decay in the spectral sidelobes in modulation schemes – such as QPSK – is the discontinuous phase 'jumps' when the input pulses are rectangular. A class of modulation known as Continuous Phase Modulation (CPM) is preferred, in which the amplitude of the signal is constant and the phase varies smoothly, with no sudden changes. Minimum shift keying, discussed earlier under FSK, is one example of CPM. As a result, MSK exhibits improved spectral purity over unfiltered QPSK, however in order to achieve the same sensitivity as BPSK, MSK must be detected not as normal FSK (e.g. envelope detector) but as a more complex variant of QPSK.

Further reduction in adjacent channel interference (for a fixed spacing) can be achieved (at the expense of some intersymbol interference) using Gaussian Minimum Shift Keying (GMSK). GMSK is equivalent to MSK where the input pulses are first shaped using a Gaussian low-pass filter. The Gaussian-shaped impulse response filter generates a modulated signal with low sidelobes, as illustrated in Figure 5.13 (Murota, 1985). The GMSK modulation scheme is characterized by the product of the bandwidth B of the Gaussian filter times the bit period T_b, a product of $BT_b = 1$ being equivalent to MSK. Smaller values of this product result in better spectral purity (reduced spectral sidelobes) but incur increased intersymbol interference. GMSK thus generally has a somewhat higher probability of error than QPSK.

5.2.6   Modulation Techniques for Frequency-Selective Channels

In Chapter 3 we indicated that, in the presence of multipath propagation, the channel experiences fading governed by statistics other than Gaussian. In addition, such fading may often be frequency

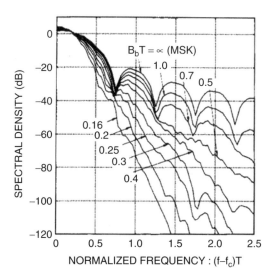

Figure 5.13   Spectral shapes for GMSK for various BT_b products (Murota, 1985). Reproduced by permission of © 1985 IEEE.

dependent (frequency-selective fading). The modulation and access techniques that may be deployed to counter multipath are primarily:

- *Spread Spectrum (SS)*. The user's signal is spread over a bandwidth substantially wider than the symbol rate. In the spread spectrum the signal is correlated with the spreading code. In a rake spread-spectrum receiver, a number of separate correlateoprs are used, each with a different symbol delay. When suitably combined, the effect of the multipath contributions may be minimized.
- *Orthogonal Frequency Division Multiplexing (OFDM)*. The data are divided between a number of smaller parallel low-data-rate channels in which the occupied bandwidth is smaller than the coherence bandwidth.

We shall cover these techniques later in Chapter 6 when discussing multiple access.

5.3   Error Control Coding

5.3.1   Overview

In this section we discuss how the performance of the communications and broadcasting systems employing the modulation schemes of this chapter can be improved through the use of *error control coding*. Such coding works through the introduction of redundant bits to the message stream, and may be classified as belonging to one of two types:

- codes that allow only the detection of transmission errors, with some method of acknowledgement used to request retransmission;
- codes that allow the detection and correction of errors without retransmission, known as Forward Error Correction (FEC) codes.

One of the simplest codes for error detection is the use of *parity* bits. An extra bit is added at the end of each block so as to make the sum of the bits (including the parity bit) even (or odd). If the sum of the bits in the received block is not also even (or odd), then an error has been detected. The detected error cannot be corrected, however, as there is no indication as to which bit has been changed during transmission and detection. Furthermore, if there is more than one error, parity may be restored, erroneously. More sophisticated checksum methods exists that can detect multiple errors. A popular method is the Cyclic Redundancy Check (CRC), which uses modulo (remainder) arithmetic to form the checksum bits and to validate the received data plus checksum. As we shall learn later, this type of arithmetic is relatively easy to implement in hardware using shift registers – hence its popularity. In this section, however, we shall focus predominantly on forward error correction codes.

In proving his capacity limit theorem, Shannon showed that there must exist a system of encoding that allows transmission at the capacity limit with an arbitrary small frequency of errors (Shannon 1948). Today, almost all satellite communications and broadcasting employ some form of FEC coding.

There are two principal types of FEC:

- *Block codes*. The message is partitioned into fixed-length blocks, redundant data are added to each block and the receiver decoder works on each received block in turn.
- *Convolutional codes*. Each bit is coded in a continuous stream, with the code for a particular data bit bearing some relation to a specified number of previous data bits.

In addition, there are important refinements to these two basic code types, which can significantly improve their performance:

- *Concatenated codes.* Code concatenation (the use of two different codes in series or parallel) is used to achieve improved performance over that of either single code.
- *Iterative decoding.* Some feedback mechanism is used at the decoder to improve the error performance with each iteration.

At this point, we must advise the reader that the mathematics of error-correcting codes is somewhat demanding, and therefore we shall only give a *flavour* of the concepts of error coding in this book. The interested reader is directed to the literature for further information on coding techniques (Sweeney, 2002; Proakis, 2001).

5.3.1.1 Code Rate

In this chapter, we have previously hinted at the use of error control codes by introducing the *code rate r* (the ratio of the user bit rate to the effective channel bit rate). If, for every k user bits, a total of n channel bits are transmitted, the code rate is

$$r = \frac{k}{n} \qquad (5.43)$$

and it is apparent that $(n - k)$ redundant bits have been added before transmission.

A consequence of the addition of redundant bits for error detection and correction is that the effective channel bit rate is *increased* by a factor $1/r$, as compared with the uncoded case. This has the effect of further constraining the capacity (for uncoded data bits). As a result, the optimum code rate represents a compromise between improved power-limited capacity on the one hand and reduced bandwidth-limited capacity on the other.

5.3.1.2 Coding Gain

Our overarching aim in the use of FEC codes is to reduce the probability of error (BER) experienced by the end-user application for a given received power, by detecting and correcting as many errors as possible in the decoder. Conversely, we also expect the minimum bit energy to one-sided noise spectral density E_b/N_o to be smaller for the same BER. The reduction in E_b/N_o for a given end-user BER performance, as compared with the case where no FEC is used, is known as the *coding gain*.

5.3.1.3 Hamming Distance

Before delving further into particular error-correcting codes, we shall require some means of quantifying by how much two different coded sequences differ, and therefore how likely it is that one will be converted into the other by corruption by added noise. Such a measure is the Hamming distance (Hamming, 1950). The Hamming distance d between two codewords of equal length is the number of symbol positions for which the corresponding symbols are different. For binary codewords, the Hamming distance is equal to the number of ones in the result of an exclusive OR (XOR) of the two codewords.

By way of example, consider the distance between the following code sequences c_1 and c_2, where

$$c_1 = [10010011]$$

$$c_2 = [00110011]$$

The sequences differ in the first and third columns, and the Hamming distance between these sequences is therefore $d = 2$.

The *minimum Hamming distance* d_{min} for any two code sequences in a particular coding scheme is a useful measure of the *robustness* of that code, as, the greater the minimum Hamming distance, the less likely it is that an error will convert one codeword into another. If we employ *minimum distance decoding* (i.e. choosing the valid codeword with the smallest Hamming distance from a valid codeword) we can relate the minimum Hamming distance, for a given code, to the maximum number of errors that may be detected and the maximum number that may be corrected.

For minimum distance decoding, the number of errors that may be detected e is one less than the minimum Hamming distance:

$$e = (d_{min} - 1) \tag{5.44}$$

while the number of errors that may be corrected t is half this:

$$t = \frac{(d_{min} - 1)}{2} \tag{5.45}$$

5.3.1.4   Errors and Erasures

The term *error* is fairly self-explanatory; however, we shall come across another term in the context of FEC – *erasure*. The term 'erasure' is used in coding to indicate a condition where the receiver has some information indicating that a symbol, or set of symbols, has been lost – for example as a result of fading (where the average received power drops). In contrast, the receiver has no prior information regarding an error. Therefore, the ability to correct erasures is generally better than the ability to correct errors.

5.3.2   Linear Block Codes

The first main type of error control code is the block code. As the name suggests, data are coded in fixed-size *blocks*.

We shall assume that each fixed-size data block contains k information bits, which we shall refer to as a *dataword*. There are thus a maximum of 2^k possible datawords. We shall further assume that we code each of these datawords into blocks of n bits by adding $(n - k)$ redundant bits – usually known as parity bits – formed by some combination of the dataword. We shall refer to the coded version of a dataword as a *codeword*. The maximum number of possible codewords is now 2^n, of which only 2^k are valid codewords. Hence, there may be a high probability that an error (or errors) will transform a valid codeword into an invalid codeword that the receiver can recognize as invalid.

5.3.2.1   Matrix Formulation of Linear Block Codes

It is convenient to describe the generation of linear block codewords and the subsequent detection of errors in terms of matrices.

Let the user dataword be represented by a row vector $\mathbf{u} = [u_0, u_1, \ldots, u_{k-1}]$, where u_i are the dataword bits. Similarly, let the codeword be represented by row vector $\mathbf{c} = [c_0, c_1, \ldots, c_{n-1}]$. The generation of the codeword from the dataword may then be represented as the multiplication of the dataword vector by a matrix

$$\mathbf{c} = \mathbf{u}\,\mathbf{G} \tag{5.46}$$

where the $n \times k$ matrix $\mathbf{G}$ is known as the *generator matrix*.

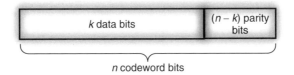

n codeword bits

Figure 5.14   Systematic block code showing data and parity bits.

Systematic Form

It is common to put the codeword into *systematic form*, as illustrated in Figure 5.14, wherein the codeword contains the unmodified dataword (usually at the beginning of the codeword). It can be shown that any linear block code can be put into systematic form.

For systematic block codes, the generator matrix may be described in terms of two submatrices

$$\mathbf{G} = [\mathbf{I}_k \mid \mathbf{P}] \tag{5.47}$$

where submatrix $\mathbf{I}_k$ is just the $k \times k$ identity matrix, and submatrix $\mathbf{P}$ is the part that generates the parity bits.

The validity of a received codeword is checked by multiplication by the *parity check matrix*. The parity check matrix is also formed from the same two submatrices: the transpose of the parity generation submatrix $\mathbf{P}^{\mathrm{T}}$ and an $n - k$ identity matrix $\mathbf{I}_{n-k}$

$$\mathbf{H} = \left[\mathbf{P}^{\mathrm{T}} \mid \mathbf{I}_{n-k} \right] \tag{5.48}$$

Syndrome

Now a fundamental property of the generator and parity check matrices is that the product of the generator matrix and the transpose of the parity check matrix is zero

$$\mathbf{G}\,\mathbf{H}^{\mathrm{T}} = \mathbf{0} \tag{5.49}$$

As the product of the transmitted (valid) codeword and the parity check matrix must thus be zero, the product of the received (corrupted) codeword and the parity check matrix yields a vector, known as the *syndrome*, which reflects the received codeword errors. The syndrome depends only on the errors, and not on the transmitted codeword itself.

If we assume that the received codeword $\mathbf{c}_{\mathrm{R}}$ is equal to the transmitted codeword plus some error vector $\mathbf{e}$ then the syndrome $\mathbf{s}$ vector is

$$\mathbf{s}\left(\equiv \mathbf{c}_{\mathrm{R}}\,\mathbf{H}^{\mathrm{T}} \right) = \mathbf{e}\,\mathbf{H}^{\mathrm{T}} \tag{5.50}$$

Thus, a syndrome comprising all zeros indicates no detected errors. If the syndrome is non-zero, one or more errors have been detected, and the value of the syndrome allows the determination of the position(s) of the error(s) – up to the maximum number of errors the particular code can detect/correct.

5.3.2.2   Hamming Codes

An early linear block code was the Hamming code (Hamming, 1950). Hamming codes have a minimum distance $d_{\min} \to 3$ (although, if an additional parity bit is added, the minimum distance is increased to 4). As such, Hamming codes can correct one error and potentially detect one other, per block. For binary Hamming codes, the codeword length is determined by (Proakis, 2001)

$$n \to 2^m - 1 \tag{5.51}$$

$$(n - k) \to m \tag{5.52}$$

$$t \to 1 \tag{5.53}$$

where m is a positive integer ($m \geq 2$).

The principal attraction of Hamming codes is their simplicity, and we shall use an example Hamming code to help illustrate the use of the generation and parity check matrices. Consider the following {7, 4} Hamming code (i.e. $n \to 7$ and $k \to 4$), where the parity check generator submatrix is given by (Proakis, 2001)

$$\mathbf{P} \to \begin{pmatrix} 1 & 0 & 1 \\ 1 & 1 & 1 \\ 1 & 1 & 0 \\ 0 & 1 & 1 \end{pmatrix}$$

From equation (5.47), the generator matrix for this block code will be given by

$$\mathbf{G} \to \begin{pmatrix} 1 & 0 & 0 & 0 & 1 & 0 & 1 \\ 0 & 1 & 0 & 0 & 1 & 1 & 1 \\ 0 & 0 & 1 & 0 & 1 & 1 & 0 \\ 0 & 0 & 0 & 1 & 0 & 1 & 1 \end{pmatrix}$$

and the parity check matrix from equation (5.48) is

$$\mathbf{H} \to \begin{pmatrix} 1 & 1 & 1 & 0 & 1 & 0 & 0 \\ 0 & 1 & 1 & 1 & 0 & 1 & 0 \\ 1 & 1 & 0 & 1 & 0 & 0 & 1 \end{pmatrix}$$

Let us consider the transmission of the dataword $\mathbf{u} \to [1010]$. From equation (5.46), the transmitted codeword will be $\mathbf{c_T} \to [1010011]$. In the absence of transmission errors, the received codeword $\mathbf{c_R}$ will be the same as the transmitted codeword, and it is easily verified that the syndrome is $\mathbf{s} \to [000]$, indicating that there are no errors to correct.

Suppose, however, that the third bit of the received codeword were changed during transmission, so that $\mathbf{c_R} \to [1000011]$. In this case, the syndrome becomes $\mathbf{s} \to [110]$, indicating an error. Referring to the relevant columns of the parity check matrix, we find that this sequence relates to bit 3, which is assumed to be in error.

5.3.2.3 Cyclic Linear Block Codes

Cyclic codes are an important subset of linear block codes that have the property whereby a *cyclic shift* of any codeword is also a codeword. As this cyclic shift is relatively easy to implement in hardware using shift registers, cyclic codes were convenient to implement prior to the availability of powerful digital signal processors. Cyclic codes also possess a certain structure that can be exploited during the coding and decoding processes.

Generator Polynomial

In a cyclic linear block code, if $[c_0, c_1, c_2, \ldots, c_{n-1}]$ is a codeword, then $[c_{n-1}, c_0, c_1, \ldots, c_{n-2}]$ is also a codeword. A useful feature of cyclic block codes is that they may be described in the form of a polynomial. Codewords are formed using a *generator polynomial* $g(x)$

$$g(X) = g_0 + g_1 X + \cdots + g_{n-k} X^{n-k} \tag{5.54}$$

in which multiplication by X in the polynomial represents a cyclic bit-shift (e.g. to the left), X^2 represents two bit-shifts, and so on.

Our k-bit user dataword may thus also be described in polynomial form

$$u(X) = u_0 + u_1 X + \cdots + u_{k-1} X^k \tag{5.55}$$

whereupon the n-bit codeword is obtained by multiplying the user dataword polynomial by the generator polynomial

$$c(X) = u(X) g(X) \tag{5.56}$$

$$= c_0 + c_1 X + \cdots + c_{n-1} X^{n-1} \tag{5.57}$$

As all valid codewords are exactly divisible by the generator polynomial, a syndrome polynomial is formed at the decoder by the remainder after dividing the received codeword by the generator polynomial. With some further manipulation, the cyclic block codeword can be put into the usual systematic form (Goldsmith, 2005).

5.3.2.4   BCH Codes

An important subclass of cyclic linear block codes are BCH codes, named after their discoverers Bose and Ray-Chaudhuri (1960) and Hocquenghem (1959). A binary BCH code that can correct up to t errors is defined by (Sweeney, 2002):

$$n \rightarrow 2^m - 1 \tag{5.58}$$

$$(n - k) \rightarrow m\, t \tag{5.59}$$

$$d_{\min} \geq 2t + 1 \tag{5.60}$$

where m is an integer ($m \geq 3$).

Although originally devised as binary codes, BCH codes were subsequently extended to non-binary (multilevel) codes.

Finite Fields

BCH codes are based on the abstract arithmetic of *finite fields*, also known as Galois fields. Galois fields are an abstract concept describing a field (algebraic structure) that contains a finite number of elements, together with two mathematical operators, addition and multiplication (and their inverses), that act on these elements (Sweeney, 2002). Galois fields have the key property that, for any element in the field, addition, multiplication and their inverses also result in elements within the same field. For any prime number p there is a Galois field with p elements, denoted by GF(p). For any positive integer m there is also a so-called *extended field* with p^m elements, denoted by GF(p^m).

We shall not delve any deeper into the mathematics of finite fields here, except to quote a relatively familiar example for binary codes: the Galois field with just two elements, denoted by GF(2). GF(2) comprises two elements $\{0, 1\}$ plus addition and multiplication operators, where the addition operator in GF(2) is the exclusive-OR (*XOR*) logic operation, while the multiplication operator is the logic *AND* operation.

5.3.2.5 Reed–Solomon Codes

A particularly important subclass of non-binary BCH codes are Reed–Solomon (RS) codes, which were discovered by Reed and Solomon (1960). Reed–Solomon codes achieve the largest possible code minimum distance for any linear code. Although discovered independently, RS codes were subsequently shown to be special cases of non-binary BCH codes.

Reed and Solomon codes are generally specified as RS(n, k), with m-bit symbols. RS codes are often used with $m = 8$, that is, using the extended field GF(2^8). As cyclic codewords from GF(q) have length $n \rightarrow (q - 1)$, a Reed – Solomon symbol size of 8 bits forces the longest codeword length to be 255 symbols (when using the generator polynomial approach).

The ability to correct multibit symbols means that RS and other non-binary BCH codes are especially useful for correcting errors that arrive in *bursts*. Furthermore, the ability to correct $(n - k)/2$ symbols *per block* means that it can correct bursts of errors that are spread across multiple symbols. RS codes are still widely deployed in satellite communications, most usually in conjunction with other codes in the form of a serial concatenated code (see later in this section).

5.3.3 Convolutional Codes

We now consider the second main class of FEC codes – convolutional codes. Convolutional codes are so named because the convolutional encoder performs a *convolution* of the input data stream with the encoder's impulse response.

Convolutional Encoding

The nature of convolutional codes is best illustrated by considering their encoding. A hardware convolutional encoder comprises a K stage shift register, where K is known as the contraint length, as it determines the maximum span of output symbols that are affected by a given input bit. Input data bits are shifted one stage at a time at the data rate, while the encoder output comprises modulo-2 sums of the shift register stages, read out at the encoded bit rate which is r times the input data rate. Longer constraint lengths produce more effective codes, but the decoding complexity increases exponentially with constraint length.

Two example $K = 4$ convolutional encoders are shown in Figure 5.15. There are two forms of convolutional encoder: *non-systematic* (top) and *systematic-recursive* (bottom). In the systematic-recursive case, the output contains the input data sequence. The performance of non-systematic convolutional codes on their own is generally better (i.e. lower error rate) than that of equivalent recursive systematic codes with the same constraint length (Berrou, Glavieux and Thitimajshima, 1993); however, the latter is preferred as a constituent encoder for use in turbo codes (see later in this chapter). In the non-systematic encoder (top), the encoder shift register bits are reset to zero prior to the first user bit. The input data bits are then clocked into the shift register on the left-hand side at rate R_b. Modulo-2 sums are formed from different stages of the shift register and time-multiplexed to form the output at rate R_c (in this case at twice the input bit rate).

Viterbi Decoding

Optimum decoding of convolutional codes is usually achieved using the Viterbi algorithm (Viterbi, 1967). The Viterbi algorithm is an example of a maximum likelihood algorithm. Maximum likelihood estimators find the parameters that make the known likelihood distribution a maximum (i.e. maximize the probability of correct data).

The starting point is for the encoder and decoder to be reset (all zeros). For each possible state change, given the input bits, the Hamming distance (error counts) are determined and stored. The decoding process thus forms a trellis diagram, describing the state transition paths, and an

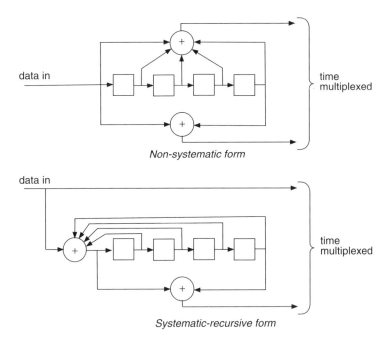

Non-systematic form

Systematic-recursive form

Figure 5.15 1/2 rate, $K = 4$ non-recursive (top) and systematic-recursive (bottom) convolution encoders (after Berrou, Glavieux and Thitimajshima, 1993).

accumulator keeps track of the total errors for each path. Whenever two paths arrive at the same state, only the path with the lowest total number of errors is kept – the other is discarded. At each successive stage, the four best path errors are kept, and four non-productive paths are discarded. What if two paths merge with equal accumulated error metrics? The answer is that, as there is no additional information available that can assist in choosing between the two paths, one of the paths is selected arbitrarily. There is, however, a distinct possibility that this path is not optimum and will be discarded at some later stage anyway (Sweeney, 2002).

At the end of the trellis, the path with the lowest cumulative error is selected as the most likely case, and the number of errors is obtained from the accumulated error total. By tracing the path back along the trellis, these errors may be located. Viterbi decoding of the trellis may be used with a fixed-length sliding window, which results in a fixed decoding latency.

The performance of Viterbi decoding with soft-decision decoding is generally better than that for hard-decision decoding. The process for Viterbi decoding using soft-decision metrics is essentially the same as that for hard-decision metrics, except that, instead of using the Hamming distance between the decoded bit and the possible bit, the Euclidean distance between the symbol probabilities is used.

Punctured Convolutional Codes

Generating codes with code rate r higher than 1/2 can be problematical, and, as the code rate tends to unity, the computational effort in the decoder increases significantly (Proakis, 2001). In order to overcome this problem, high-rate codes are typically implemented using low-rate codes (such as a 1/2 rate convolutional code) in which some of the coded bits are periodically deleted from the output of the encoder. This process, known as *puncturing*, also results in more efficient decoding.

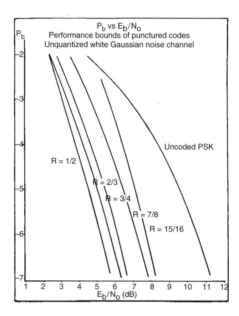

Figure 5.16 Bit error performance bounds for the memory 6, rate 1/2 original code, and the punctured rates 2/3, 3/4, 7/8 and 15/16 derived from it (Wu *et al.*, 1987). Reproduced by permission of © 1987 IEEE.

Bits are usually deleted according to a *puncturing matrix*. The performance of $K = 6$ convolutional codes at different code rates is illustrated in Figure 5.16.

5.3.4   Interleaving and Code Concatenation

Convolutional codes tend to generate a burst of uncorrected errors at their output when the code is unable to correct all the transmisson errors.

5.3.4.1   Interleaving

Where errors occur in bursts, the use of *interleaving* after the encoder and subsequent de-interleaving before the decoder permits a reduction in the number of errors per codeword by dispersing the errors across multiple codewords. The performance of convolutional codes, in particular, may be improved by the use of an interleaver and de-interleaver, although such interleaving increases latency.

The choice of interleaving method varies with particular encoding method and channel characteristic. A particularly simple example of interleaving involves the temporary storage of bits in a matrix with data read in row-wise and read out columnwise but numerous other, more sophisticated, interleaving methods suited to different channel conditions exist.

5.3.4.2   Serial Concatenated Codes

It was stated previously that non-binary BCH codes, notably Reed–Solomon codes, have the attractive property that they can correct errors in bursts. It is therefore common to use concatenated

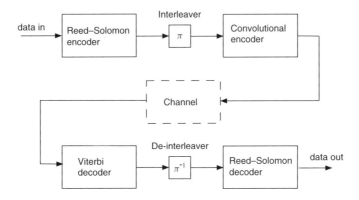

Figure 5.17   System employing concatenated RS/convolutional code.

coding formed from the combination of convolutional codes (inner code) with RS codes (outer code), as shown in Figure 5.17 – typically in conjunction with an interleaver (indicated by π) and a de-interleaver (denoted by π^{-1}). The data are first encoded using an RS code, and the output of the RS encoder is then interleaved and encoded a second time using a convolutional code. At the decoder, the process is reversed, with the (outer) RS decoder used to correct any error bursts from the output of the Viterbi decoder for the (inner) convolutional code. Such a scheme is referred to as a serial concatenated code. The improvement in BER performance with a serial concatenated code is illustrated in Figure 5.18, which compares the performance of uncoded, convolutional coded (only) and serial concatenated coded data.

5.3.5   Turbo and LDPC Codes

For many decades, up until the mid-1990s, the ultimate in practical error control code performance was represented by the use of serial concatenated codes. Yet, the performance of these codes fell significantly short of the theoretical Shannon channel capacity limit. Then, in 1993, Berrou and co-workers announced their work on parallel concatenated convolutional codes with iterative decoding, which they called *turbo-codes* (by analogy to the use of turbo-chargers in motor vehicle engines).

5.3.5.1   Turbo Codes

Turbo-Code Encoding
Turbo codes (Berrou, Glavieux and Thitimajshima, 1993) utilize two (or more) simple constituent encoders, together with an interleaver, such that the two parallel encoders use the same input bit sequence (but in different order) to produce different permutations of the same data stream. An example of a generic turbo-code encoder is illustrated in Figure 5.19 (top). The figure shows two constituent encoders (usually systematic-recursive convolutional encoders), with the input data to the second encoder passing through an interleaver. The outputs of the two constituent encoders are time multiplexed prior to transmission.

Good turbo codes are produced by short-constraint-length convolutional codes (Andrews *et al.*, 2007). Berrou's original turbo-codes incorporated two 1/2 rate systematic-recursive convolutional encoders with a constraint length of 4.

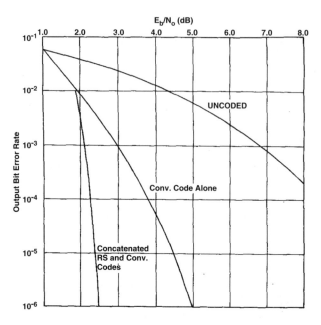

Figure 5.18 Theoretical performance of RS (255, 233) with convolutional $R = 1/2$, $K = 7$ code (Wu *et al.*, 1987). Reproduced by permission of © 1987 IEEE.

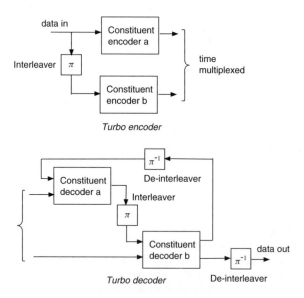

Figure 5.19 Generic turbo-code encoder (top) and decoder (bottom), after Andrews *et al.* (2007).

Turbo-Code Decoding

A generic turbo-code decoder is illustrated in Figure 5.19 (bottom). The decoding of turbo codes employs two elementary decoders arranged in a form of a feedback loop with associated interleaving and de-interleaving. A Viterbi decoder may be used for one of the constituent decoders, while the other uses a soft-input, soft-output, 'forward-backward' BCJR algorithm (named after its inventors: Bahl, Cocke, Jelinek and Raviv (Bahl *et al.*, 1974)), with the error rate decreasing with each iteration. Typically, 3–10 iterations are performed, depending on the signal-to-noise ratio and the delay constraints. Figure 5.20 illustrates the improvement in BER with iteration for Berrous' original turbo code.

Turbo codes were the first error-correcting codes to approach the Shannon capacity limit. Berrous' original code achieved within 0.7 dB of the Shannon limit, while Boutros *et al.* (2002) described a 1/3 rate, infinite-length turbo code that performed within 0.03 dB of the Shannon limit. The performance of some practical commercial modems employing turbo codes is given in Table 5.2.

The principal issues with the use of turbo codes are the relatively high decoding complexity and relatively high latency (owing to the use of interleavers and de-interleavers and the need to iterate in the decoder).

5.3.5.2   LDPC Codes

Turbo codes were the first practical codes closely to approach the Shannon limit, and their intro-duction restimulated research into high-performance codes. With the advent of iterative decoding methods in the late 1990s, MacKay and Neal (1996) and MacKay (1999) rediscovered Low-Density Parity Check (LDPC) codes, which had been introduced decades earlier by Gallagher (1963) but at

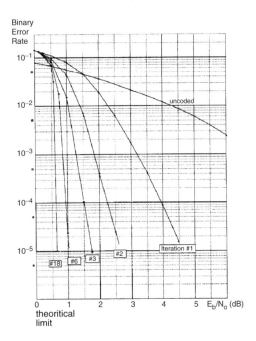

Figure 5.20   Example turbo-code error performance versus iteration number (Berrou, Glavieux and Thitimajshima, 1993). Reproduced by permission of © 1993 IEEE.

Table 5.2 Performance of some commercial FEC implementations for QPSK at BER $= 10^{-6}$ (adapted from COMTECH, 2010)

Modulation	Code type	Code rate	Block size	$\frac{E_b}{N_0}$
QPSK	None	1	N/A	13 dB
QPSK	Convolutional/Viterbi decoding	$1/2$	N/A	5.5 dB
QPSK	Turbo	$7/8$	16 kb	4.0 dB
QPSK	Turbo	$3/4$	4 kb	3.3 dB
QPSK	Turbo	$1/2$	4 kb	2.6 dB
QPSK	LDPC	$3/4$	16 kb	2.7 dB
QPSK	LDPC	$1/2$	16 kb	1.7 dB

the time were considered impractical owing to the computational effort required. MacKay demonstrated that, under certain conditions, LDPC codes can outperform turbo codes.

Low-density parity check codes are linear block codes that are characterized by a *sparse* parity check matrix **H** (i.e. with few non-zero elements). Gallagher's original LDPC codes are now known as *regular* LDPC codes. Regular LDPCs are those in which the number of ones in each column/row of the parity check matrix is a constant (determined by a particular set of constraints). Subsequently, however, Richardson, Shokrollahi, and Urbanke (2001) generalized Gallagher's concept, introducing *irregular* LDPC codes in which the number of ones is not the same in each row/column. Typically, irregular LDPC codes outperform regular LDPCs and also turbo codes for longer block lengths.

LPDC Encoding

Generating good sparse parity check matrices is relatively easy – indeed, it seems that completely randomly chosen code matrices perform quite well. The challenge is that the encoding complexity of such codes is usually high, as the parity check matrix **H** is large, and, although **H** is sparse, the generator matrix **G** is generally not.

LPDC Decoding

LDPC codes also employ iterative decoding, based on a message-passing algorithm, known as the *belief propagation* algorithm. With reference to Figure 5.21, which shows a Tanner graph (a graphical representation of the parity check matrix), messages are passed in both directions along the edges (connections) between the *variable nodes* (received codeword bits) and *check nodes* (equivalent to the rows in the parity check matrix). These messages are probabilities, or *beliefs*.

A message passed from a variable node to a check node is the probability that the received bit has a certain value – initially, the observed value of that received codeword bit. Each check node receives beliefs from *all* of the variable nodes to which it is connected (i.e. from both the dataword and parity bits). A message is then passed from each check node to the relevant variable nodes, indicating the probability that the check node has a certain value given all the messages passed to it in the previous round. The variable nodes then update their messages to the check nodes, incorporating this information. This iterative process repeats until the algorithm terminates.

Chung *et al.* (2001) published a design for a $1/2$ rate, irregular LDPC code with a block length of 10^7 bits for a bit error rate of 10^{-6}, which is theoretically capable of operating within just 0.0045 dB of the Shannon limit. Again, the performance of some practical commercial modems employing LDPC codes is given in Table 5.2.

Today, both turbo codes and LDPC codes have been adopted by a number of systems and standards. LDPC codes are used in DVB-S2, while turbo codes are used in DVB-RCS. Meanwhile,

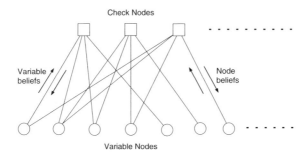

Figure 5.21   Tanner graph and belief propagation.

their development continues, and it is now possible to construct turbo and LDPC error-correcting codes that require extremely low E_b/N_o by increasing the block size. An advantage cited for LDPC codes is that they have more degrees of freedom than turbo codes (Andrews *et al.*, 2007).

5.3.6   Lower Bound on Code Performance

Better performance with turbo codes and LDPC codes is generally achieved at the expense of increased block size, which impacts on latency. The dependence of the minimum energy per bit to one-sided noise spectral density ratio on code block size is therefore of interest.

Sphere-Packing Bound

Shannon (1959) calculated the lower bound on the performance of error-correcting codes and its dependence on code block size and code rate. Shannon's sphere-packing bound provides a lower limit for the error rate achievable by an arbitrary code of a given block size and code rate r for the case of equal-energy signals under the assumption of AWGN. Figure 5.22 shows the dependence of the Shannon sphere-packing bound for $P_e = 10^{-4}$ (Dolinar, Divsalar and Pollara, 1998).

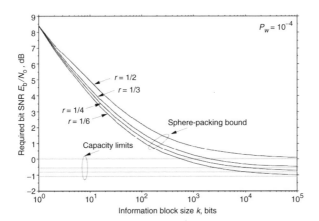

Figure 5.22   Shannon sphere-packing bound limit on code E_b/N_o with block size for $P_e = 10^{-4}$ (Dolinar, Divsalar and Pollara, 1998). Reproduced from Caltech/JPL/NASA.

Latency

In order to approach within $<1\,$dB of the Shannon limit requires code block sizes of $>1\,$kb. As the processing delay due to channel coding is at least equal to the block length (in bits) divided by the bit rate, it is apparent that, the larger the block size, the longer is the coding delay. The impact of this delay will clearly depend on whether the data being sent are delay sensitive – such as, for example, speech or videoconferencing.

Revision Questions

1. What is a matched filter? Give the maximum output signal-to-noise ratio per symbol. Describe two methods by which one can implement such a filter.
2. What is the significance of the Nyquist criterion for the throughput of data?
3. Given a baseband channel of width 1 MHz, what is the maximum symbol rate that avoids ISI? What is the equivalent symbol rate for a 1 MHz passband channel?
4. Estimate the occupied bandwidth for 1 Mb/s binary raised-cosine pulses with a roll-off factor of 0.33.
5. What are the minimum distances for binary ASK, minimum FSK and PSK?
6. Give the number of bits per symbol and maximum spectral efficiencies for BPSK, QPSK, 8-PSK and 64-QAM.
7. In a binary scheme, if the probabilty of a 'one' is twice that of a 'zero', what would the optimum threshold be? Calculate the probability of a bit error.
8. Estimate the BER for QPSK and 16-PSK for $E_b/N_o = 10\,$dB.
9. By approximately how many dB is the performance of QPSK and 64-QAM worse than the Shannon limit?
10. How does the sensitivity of 64-PSK compare with that of 64-QAM for $E_b/N_o = 10\,$dB? Which modulation scheme would you use for maximum sensitivity? Which for use with a non-linear channel, and why?
11. What are the advantages and disadvantages of GMSK? Find out how you would demodulate GMSK.
12. Under what circumstances would you consider using OFDM or COFDM?
13. Explain the differences between a block code and a convolutional code.
14. What is a syndrome and how does it help with error correction?
15. How many errors can a Hamming code correct, and why? Demonstrate this limit on error correction using the {7, 4} Hamming code in the text, employing example codewords.
16. What is a cyclic linear block code and what are its advantages? How are datawords encoded? How are errors detected?
17. Why is an interleaver often used with convolutional encoding? What are the implications for latency?
18. Why is a Reed–Solomon code often used as the outer code in a serial concatenated code? Why is an interleaver typically placed between the two encoders?
19. Using the example data of Table 5.2, estimate the coding gain (in dB) achieved by this modem at BER $= 10^{-6}$ for each code type.
20. What are the pros and cons of turbo codes versus LDPC codes? What is the difference between regular and irregular LDPCs?

References

Andrews, K.S., Divsalar, D., Dolinar, S, Hamkins, J., Jones, C.R. and Pollara, F. (2007) The development of turbo and LDPC codes for deep space applications. *Proceedings of the IEEE*, **95**(11), 2142–2156.

Bahl, L., Cocke, J., Jelinek, F. and Raviv, J. Optimal decoding of linear codes for minimizing symbol error rate. *IEEE Trans. on Information Theory*, **IT-20**(2), 284–287.

Berrou, C., Glavieux, A. and Thitimajshima, P. (1993) Near Shannon Limit Error – correcting coding and decoding: Turbo Codes. *IEEE International Conference on Communications*, Geneva, Switzerland. Technical Programme, Conference Record, Vol. 2, pp. 1064–1070.

Bose, R.C. and Ray-Chaudhuri, D.K. (1960) On a class of error correcting binary group codes. *Information and Control*, **3**, 279–290.

Boutros, J., Caire, G., Viterbo, E., Sawaya, H. and Vialle, S. (2002) Turbo Code at 0.03 dB from capacity limit, Proceedings of IEEE International Symposium on *Information Theory*, June–July, Lausanne, Switzerland, p. 56.

Chung, S.-Y., Forney, G.D., Jr., Richardson, T.J. and Urbanke, R. (2001) On the design of low-density parity-check codes within 0.0045 dB of the Shannon limit. *IEEE Communications Letters*, **5**(2), 58–60, 2001.

COMTECH (2010) www.comtechefdata.com/articles_papers/LDPC%20and%208-QAM.pdf [accessed February 2010].

Dolinar, S., Divsalar, D. and Pollara, F. Turbo codes and space communications. *Proceedings of Space Operations Conference SpaceOps '98*, Tokyo, Japan, 1–5 June.

Gallagher, R.G. (1963) *Low Density Parity Check Codes*. Monograph, MIT Press. Available: http://www.inference.phy.cam.ac.uk/mackay/gallagher/papers/ldpc.pdf [accessed February 2010].

Goldsmith, A. (2005) *Wireless Communications*, Cambridge University Press, Cambridge, UK.

Hamming, R.W. Error detecting and error correcting codes. *Bell System Technical Journal*, 1950, **24**(2).

Hartley, R.V.L. (1928) Transmission of information. *Bell System Technical Journal*, July.

Hocquenghem, A. (1959) Codes correcteurs d'erreurs. *Chiffres*, **2**, 147–156.

MacKay, D. (1999) Good error correcting codes based on very sparse matrices. *IEEE Trans. Information Theory*, March, 399–431.

MacKay, D.J.C. and Neal, R.M. (1996) Near Shannon limit performance of low density parity check codes. *Electronics Letters*, July.

Murota, K. (1985) Spectrum efficiency of GMSK land mobile radio. *IEEE Trans. on Vehicular Technology*, **34**(2), 69–75.

Nemer, E. (2005) Physical layer impairments in DVB-S2 receivers. 2nd IEEE Consumer Communications and Networking Conference, 3–6 January, Las Vegas, NV, pp. 487–492.

Ngo, N.H., Barbulescu, S.A. and Pietrobon, S.S. (2005) Proceedings of 6th Australian Communications Theory Workshop, 2–4 February, pp. 79–83.

Nyquist, H. (1924) Certain factors affecting telegraph speed. *Bell System Technical Journal*, **3**, 324–346.

Pahlavan, K. and Levesque, A.H. (1995) *Wireless Information Networks*. John Wiley & Sons, Inc., New York, NY.

Proakis, J.G. (2001) *Digital Communications*. McGraw-Hill, New York, NY.

Reed, I.S. and Solomon, G. (1960) Polynomial codes over certain finite fields. *SIAM Journal of Applied Mathematics*, **8**, 300–304.

Richardson, T.J., Shokrollahi, M.A. and Urbanke, R.L. (2001) Design of capacity-approaching irregular low-density parity-check codes. *IEEE Trans. on Information Theory*, **47**(2), 619–637.

Saleh, A.A.M. (1981) Frequency-indepenent and frequency-dependent nonlinear models of TWT amplifiers. *IEEE Trans. Communications*, **COM-29**(11), 1715–1720.

Shannon, C.E. (1948) A mathematical theory of communication. *Bell System Technical Journal*, **27**(July/October), 379–423 and 623–656.

Shannon, C.E, (1949) Communication in the Presence of Noise. *Proceedings of the IRE*, **37**, 10–21 (reprinted as Shannon, C. E. (1984) *Proceedings of the IEE*, **72**(9), 1192–1201).

Shannon, C.E, (1959) Probability of error for optimal codes in a Gaussian channel. *Bell System Technical Journal*, **38**, 611–656.

Sweeney, P. (2002) *Error Control Coding*. John Wiley & Sons, Ltd, Chichester, UK.

Viterbi, A.J. (1967) Error bounds for convolutional codes and an asymptotically optimum decoding algorithm. *IEEE Trans. on Information Theory*, **IT-13**, 260–269.

Wu, W.W., Haccoun, D. Peile, R. and Hirate, Y. (1987) Coding for satellite communication. *IEEE Journal on Selected Areas in Communications*, **SACJ**(4).

6

Satellite Access and Networking

6.1 Introduction

The previous chapter introduced a number of concepts related to digital modulation and coding. Armed with this understanding, we are ready to consider in more detail the power budget, occupied bandwidth and *capacity* of practical satellite links. Calculation of the required RF carrier power is usually known as a *link budget*.

For mass-market applications, the *total* capacity of a satellite service is important, as it determines how many users, on average, the service can support at any given time (from which an estimate of the probability of unavailability of service to a user may be estimated). Moreover, as the number of potential users may well exceed the total capacity (although typically not all users require service at the same time), *multiple-access* techniques – by which the available capacity is shared – are clearly critical to the provision of cost-effective services. Furthermore, many such access schemes provide access *on demand* for maximum efficiency and flexibility. In this chapter, we compare and contrast the three dominant methods of implementing multiple-access schemes – frequency, time and code division – and their hybrids; in addition, a fourth technique – space division – can substantially increase overall capacity through the introduction of frequency *reuse* in non-overlapping coverage regions.

Having introduced and quantified these multiple-access techniques, we go on to consider how satellites may be utilized in *networks*. Specifically, we discuss generic architectural, payload and transport concepts commonly used in satellite networks.

6.2 Satellite Access

6.2.1 Single Access

Let us first consider the performance of a single carrier per (physical) channel. In addition to being a valid method of accessing satellite capacity, this will also serve as a useful baseline for our subsequent comparison of multiple-access techniques. Specifically, let us consider the link power budget of such a system.

6.2.1.1 Minimum Signal-to-Noise Spectral Density

We considered the estimation of the received signal-to-noise (density). In Chapter 5 we then considered the estimation of the minimum normalized signal-to-noise ratio $(E_b/N_o)_{min}$ needed to achieve

Satellite Systems for Personal Applications: Concepts and Technology   Madhavendra Richharia and Leslie David Westbrook
© 2010 John Wiley & Sons, Ltd

a specified Bit Error Rate (BER). Clearly, we require the received signal-to-noise ratio to equal or exceed the minimum to achieve the desired BER.

As AWGN noise power scales linearly with receiver bandwidth, and as, for given modulation and coding schemes, the noise bandwidth must be increased in proportion to the bit rate, it is useful to continue to normalize the required received signal-to-noise ratio to the noise bandwidth. The minimum mean received signal (carrier power)-to-noise spectral density $\left(\frac{C}{N_o}\right)_{min}$ is related to the bit rate R_b via

$$\left(\frac{C}{N_o}\right)_{min} \geq R_b \left(\frac{E_b}{N_o}\right)_{min} \tag{6.1}$$

Equation (6.1) thus gives the minimum acceptable value of C/N_o for data rate R_b.

6.2.1.2 Link Margin

The link (power) margin may be defined as the excess power (given as a ratio) over and above the minimum needed for the link to function as required at the specified BER for a given bit rate. Given the available signal-to-one-sided noise spectral density, the link margin may be expressed in terms of the actual value of C/N_o as follows:[1]

$$margin \equiv \frac{\frac{C}{N_o}}{R_b \left(\frac{E_b}{N_o}\right)_{min}} \tag{6.2}$$

6.2.1.3 Capacity for Single Access

Power-Limited Capacity

From equations (6.1) it is apparent that the power-limited information capacity (in b/s) of a single channel can be expressed as

$$(R_b)_{max} \leq \frac{\frac{C}{N_o}}{\left(\frac{E_b}{N_o}\right)_{min}} \tag{6.3}$$

Comparison of equation (6.2) with equation (6.3) suggests that an alternative interpretation of the link margin would be the ratio of the power-limited channel capacity to the actual bit rate.

It can sometimes be instructive to consider the information capacity of a satellite link *per square metre of effective aperture* of the user's terminal antenna, because both EIRP and G/T of the user's terminal increase with aperture area. For the downlink (which is often the limiting case), it is straightforward to show that the capacity per square metre of effective aperture is equal to the satellite power flux density (PFD) incident on the user's antenna, divided by the required (minimum) energy per bit. If we may assume that the receiver noise spectral density (and hence the minimum energy per bit) does not vary significantly with the aperture area,[2] then for a given PFD, the link capacity will be proportional to the size (effective area) of the user's antenna, as expected.

---

[1] Given in dB, the link margin is

$$(margin)_{dB} = \left(\frac{C}{N_o}\right)_{dB} - 10 \log (R_b) - \left(\frac{E_b}{N_o}\right)_{min\ dB}$$

[2] In practice, sky noise may vary with antenna aperture.

Table 6.1 Example uplink power budget

Uplink parameter	Forward link	Return link
Frequency (GHz)	14.3	14.3
Satellite altitude (km)	35 786	35 786
Terminal altitude (km)	0	0
Terminal elevation angle (deg)	35.9	38.1
Satellite nadir angle (deg)	7.04	6.84
Earth coverage angle (deg)	47.1	45.1
Range (km)	38 106.3	37 928.7
Propagation delay (ms)	127	126
Terminal max. EIRP (dBW)	63.5	44.8
Terminal OBO (+ve dB)	0	0
Terminal pointing loss (+ve dB)	0	0
Terminal EIRP (dBW)	63.5	44.8
FSL (+ve dB)	207.2	207.2
Atmospheric Losses (+ve dB)	11	11
Other losses (+ve dB)	0	0
PFD at satellite (dBW/m^2)	−110.1	−128.7
Satellite max. G/T (dB/K)	5	5
Satellite pointing loss (+ve dB)	0	1
Satellite G/T (dB/K)	5	4
C/N_o for uplink (dBHz)	78.9	60.3

Bandwidth-Limited Capacity for Single Access

The spectrum occupied by a power limited single access increases with the symbol rate. Ultimately, however, the maximum symbol rate is limited by the available channel bandwidth. For ASK, PSK and their derivatives with spectral efficiency for a single-channel access given by ζ_{ch}, the channel bandwidth-limited capacity is

$$(R_b)_{max} \leq \zeta_{ch} \, B_{ch} \tag{6.4}$$

where B_{ch} is the single-access channel bandwidth.

6.2.1.4 Example Link Budget

We have now covered all of the topics needed to perform a useful link budget. By way of example, Tables 6.1 to 6.3 give a link budget calculation for a duplex link between two ground terminals (a large hub station and a user's remote terminal).[3] Details for the uplink (i.e. ground terminals to satellite) are given in Table 6.1, while those for the downlink (i.e. satellite-to-ground terminals) are given in Table 6.2. Table 6.3 provides analysis of the total C/N_o for up (Earth-to-satellite) and down (satellite-to-Earth) links and gives the theoretical channel capacity, together with the power margin for the particular system parameters.

---

[3] Link budget adapted from Nera (2002).

Table 6.2   Example downlink power budget

Downlink parameter	Forward link	Return link
Frequency (GHz)	11	11
Satellite altitude (km)	35 786	35 786
Terminal altitude (km)	0	0
Terminal Elevation angle (deg)	38.1	35.9
Satellite nadir angle (deg)	6.84	7.03
Earth coverage angle (deg)	45.1	47.1
Range (km)	37 361.6	37 194.9
Propagation delay (ms)	125	124
Satellite max. Boresight EIRP (dBW)	50	50
Satellite back-off/PDF (+ve dB)	9	30.7
Satellite pointing loss (+ve dB)	2	0
Satellite EIRP (dBW)	39	19.3
Downlink FSL (+ve dB)	204.7	204.7
Atmospheric losses (+ve dB)	11	11
Other losses (+ve dB)	0	0
PFD at terminal (dBW/m^2)	−134.4	−154.1
Terminal boresight G/T (dB/K)	18	32
Terminal pointing loss (+ve dB)	0	0
Terminal G/T (dB/K)	18	32
C/N_0 for downlink (dBHz)	69.9	64.2

Table 6.3   Maximum data rate and power margin calculation

Parameter	Forward link	Return link
C/N_0 for uplink (dBHz)	78.9	60.3
C/N_0 for downlink (dBHz)	69.9	64.2
C/N_0 at terminal (dBHz)	69.4	58.8
Modulation type	QPSK	QPSK
Bits/symbol	2	2
Code rate	0.75	0.75
Required E_b/N_0 (dB)	5.5	5.5
Power limited data rate		
Max. data rate (kb/s)	2679.2	727.6
Power margin – defined system		
System symbol rate (kS/s)	1333.3	128
System channel rate (kb/s)	2666.7	256
System uncoded bit rate (kb/s)	2000	192
Required C/N_0 (dBHz)	68.6	58.4
Power margin (dB)	0.8	0.3

6.2.2   Multiple-Access Methods

Communications satellites are expensive assets, and each satellite typically provides only a finite capacity. It is therefore rarely practicable or desirable to allocate a whole satellite transponder to a single satellite service, let alone to a single user. Instead, the capacity of each satellite transponder will be shared by multiple services, and typically each service may, in turn, be shared by a large number of users.

The method by which capacity is shared and user access facilitated is referred to as *multiple access*. Multiple-access methods are particularly important for personal satellite systems where the number of potential users significantly exceeds the available capacity but not all users wish to use the service at the same time, and the costs must be shared by a large user base.

Multiplexing versus Multiple Access

Where multiple data streams (to multiple users or user applications) share a *single* transmission, the transmission is said to be *multiplexed*. This might, for example, be the case for the forward-link transmission from a single central hub to multiple user terminals (see the Networking section later in this chapter). However, where multiple data streams have *multiple* transmissions that must share access to the same spectrum (and satellite transponder), they typically employ a *multiple-access* technique. This might be for the return-link transmissions from multiple users to a central hub. These techniques are closely related, with the distinction between multiplexing and multiple access essentially depending on whether it is necessary to accommodate multiple independent transmitters. In general, multiplexing will be more efficient, as it does not need to allow for transmitter timing (and Doppler frequency) variability. Given their similarities, we shall consider multiplexing and multiple-access techniques together.

There are five basic multiplexing/multiple-access methods (together with various combinations thereof):

- *Random access.* Users transmit data packets as and when required (in an apparently random fashion) and accept that some transmissions will suffer collisions with those of other users. The user's equipment will usually rely on an acknowledgement of successful transmission from the receiver to determine whether or not a collision took place and it is necessary to retransmit the packet – whereupon the whole process starts again.
- *Frequency Division Multiplexing (FDM) and Frequency Division Multiple Access (FDMA).* The allotted spectrum is divided up (sliced) into smaller-bandwidth FDM/FDMA channels – one per user. FDM/FDMA may be combined with other multiplexing and multiple-access techniques.
- *Space Division Multiplexing (SDMA).* The allotted spectrum is *reused* across multiple beams that do not overlap significantly (and hence do not result in interference). SDMA is almost always used together with other multiple-access techniques, particularly FDMA.
- *Time Division Multiplex (TDM) and Time Division Multiple Access (TDMA).* The channel is divided up into sequential time intervals (time 'slots') – one per user. TDMA may be used together with FDMA and SDMA.
- *Code Division Multiplex (CDM) and Code Division Multiple Access (CDMA).* The channel is accessed by multiple users using CDMA spread-spectrum, with a different CDMA spreading code per user. CDMA may be used together with FDMA and SDMA.

FDMA, TDMA and CDMA may usefully be visualized as dividing up the available resources along different dimensions, as shown in Figure 6.1 (note that we could equally have chosen one of the axes in Figure 6.1 to be 'Space' – for SDMA).

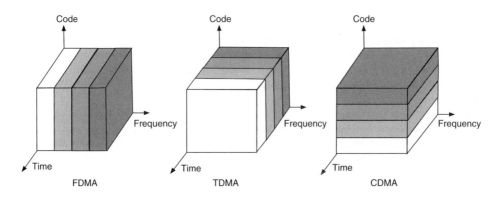

Figure 6.1 Conceptualization of the main multipexing/multiple-access methods: FDM/FDMA, TDM/TDMA and CDM/CDMA.

6.2.3 Random Access

6.2.3.1 Overview

In random access, user terminals transmit packets as and when needed, with little or no regard for the potential for interference with the transmissions of other user terminals. In general it is not practical to sense whether the spectrum is in use before transmitting in a satellite service, because, in a geostationary system, the round-trip time is around 250 ms, and a large amount of data can be sent in this time. It is therefore usual for the receiving end to send an acknowledgement for each transmission to indicate it was successfully received. If no acknowledgement is received within a given time, the data are retransmitted, and so on. The acknowledgement itself may also have to contend with other transmissions.

The most basic random-access contention protocol is the *ALOHA* protocol. In ALOHA, if no acknowledgement is received, terminals assume a packet 'collision' and retransmit the packet after a random delay. An obvious weakness of the ALOHA protocol is that the potential exists for two ALOHA packets to collide for just a brief overlap period at the end of one burst and the start of the next; whereupon, the access protocol process is disrupted for up to two burst lengths. A simple modification to the basis ALOHA protocol is to require that the access requests occur within particular time slots – so-called *slotted ALOHA* – which results in efficiency of twice that of ALOHA.

6.2.3.2 Efficiency of Slotted ALOHA

For a truly random process, the probability of arrival of a random ALOHA packet at the receiver is governed by Poisson statistics. If the total number of packets transmitted per time slot (transmission period) is G, then it can be shown that the mean number of packets S successfully received (without collisions) per time slot is given by

$$S = \begin{cases} G \exp(-2G) & \text{for ALOHA} \\ G \exp(-G) & \text{for Slotted ALOHA} \end{cases} \tag{6.5}$$

It turns out that the throughput of ALOHA peaks at just over 18% and then tails off because, at high traffic loads, almost all packets are lost in collisions. The maximum throughput efficiency of slotted ALOHA is twice as high, at approximately 36%. For loading above these values, both access methods become unstable due to multiple retransmissions.

6.2.3.3   Use as a Reservation Mechanism

The modest throughput efficiency of random access using slotted ALOHA, means that it is not generally attractive except for the transmission of short messages, or perhaps when propagation is intermittent (such as in a helicopter). However, it commonly plays a part within the signalling of most multiple-access schemes which allow for access *on demand*, where it is used as part of a capacity *reservation* mechanism. Although the sharing of satellite capacity may be on a permanent basis – sometimes referred to as Permanently Assigned Multiple Access (PAMA) – in general, the number of potential users will exceed the available capacity, but not all users will need to use the service at the same time. In such cases, communications capacity will be allocated *on demand* on the basis of requests issued from user terminals to some central network control system. When the user terminal no longer requires the service, the capacity is normally returned to a central pool for reuse by other users. This is generally referred to as Demand Assigned Multiple Access (DAMA). At one level of sophistication a DAMA system may allocate a fixed-data-rate 'circuit', such as a voice circuit which exists for the duration of the call, while at another level of sophistication the user connection is nominally *always on*, with satellite capacity allocated on a packet-by-packet basis.

How does a terminal obtain access on demand? One method would be to poll each terminal in turn and interrogate it as to whether it needs to send data. There are obvious drawbacks to this approach:

- *Latency.* If the number of terminals is large, a given terminal may have to wait a significant time before it is polled.
- *Wasted capacity.* Polling of large numbers of terminals wastes resources that could be used for useful data traffic.

The answer is to allow terminals to request access on a contended basis. Although ALOHA and slotted ALOHA may not be appropriate for use in multiple access on their own, they are commonly used as a contention protocol for reserving systems capacity. A terminal wanting access transmits a signalling burst indicating its requirements and collisions of such bursts can usually be tolerated.

6.2.4   FDM and FDMA

6.2.4.1   Overview

In FDM and FDMA, the channel is effectively 'sliced' in frequency to form multiple channels, with each carrier having a different centre frequency, and with channels spaced so as to minimize adjacent channel interference, as illustrated in Figure 6.2. Where FDM/FDMA channels are preassigned, this scheme is sometimes referred to as Single Channel Per Carrier (SCPC).

It is generally necessary to incorporate so-called guard bands between adjacent FDM/FDMA channels, so as to reduce the potential for interference between adjacent signals, and these guard bands inevitably reduce the bandwidth available to useful signals. The principal difference between FDM and FDMA is that FDMA channel spacing must also take into account transmitter frequency accuracy and Doppler shift – resulting in wider spacings.

The advantages and disadvantages of the FDMA scheme are listed below:

- Advantages: simple to implement; no network timing signals required.
- Disadvantages: can be relatively inflexible (particularly in respect of adapting to varying throughput requirements, and therefore typically used with circuit-switched services with preassigned fixed rate channels; inefficient use of power if implemented in a single satellite

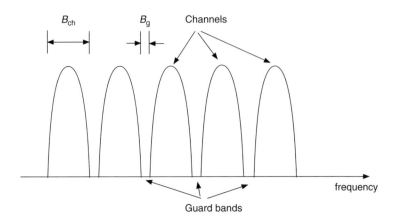

Figure 6.2 FDM/FDMA.

transponder (output power must be reduced in order to avoid excessive intermodulation 'noise' – consequently, FDMA cannot exploit the maximum transponder output power).

6.2.4.2 Bandwidth Limited Capacity of FDMA

The efficiency of multiple access schemes is important. If the (average) useful channel bandwidth of each FDM/FDMA signal is B_{ch} and the (average) guard bandwidth is B_g, then FDMA bandwidth efficiency η_B is the ratio of the total useful channel bandwidth to the total bandwidth (including guard bands)

$$\eta_B \, [\text{FDMA}] = \frac{\Sigma B_{ch}}{\Sigma B_{ch} + \Sigma B_g} \tag{6.6}$$

Expressed in terms of the number of channels, for the case of M equiwidth channels occupying a total allocated bandwidth B_a

$$\eta_B \, [\text{FDMA}] \simeq 1 - M \frac{B_g}{B_a} \tag{6.7}$$

Hence, to a first approximation, the bandwidth efficiency of FDMA decreases linearly with the number of channels. The overall spectral efficiency of FDM/FDMA is then equal to the product of the FDM/FDMA bandwidth efficiency and the average spectral efficiency of each channel. It is no surprise that the bandwidth-limited capacity of an FDMA system is somewhat less than that for a single channel, and according to equation (6.7) decreases with the number of accesses owing to the unused spectrum in the guard bands.

Iridium example. Consider, by way of example, the Iridium satellite service FDMA scheme (Iridium uses a combination of FDMA, SDMA and TDMA). Key parameters for the Iridium communications system are given in Table 6.4 (ICAO, 2006; Hubbel, 1997). In this scheme, 240 FDMA traffic channels are used; the channel bandwidth is 31.5 kHz, on a 41.67 kHz channel spacing, yielding an FDMA bandwidth efficiency of 75.6%.

6.2.4.3 Intermodulation Noise

A significant limiting factor in FDM/FDMA systems that use a common electronic channel is the impact of nonlinear elements in the system. In particular, electronic amplifiers become more

Table 6.4 Some parameters used in estimating the capacity of the Iridium and Globalstar communication systems (ICAO, 2006; Dietrich, Metzen and Monte, 1998; Mazella *et al.*, 1992; Hubbel, 1997; Chang and de Weck, 2005)

Parameter	Iridium	Globalstar
Allocated bandwidth	10 MHz	16.5 MHz
Active satellites	66	48
Beams per satellite	48	16
Active beams	2150	(all)
SDMA cluster size	12	1
FDMA channels	240	13
Channel bandwidth	31.5 kHz	1.23 MHz
Channel spacing	41.67 kHz	1.23 MHz
TDMA burst rate	50 kb/s	–
TDMA frame length	90 ms	–
TDMA burst length	8.28 ms	–
TDMA bursts per frame	8	–
Duplex method	TDD	FDD
Nominal user bit rate	2.4 kb/s	4.8 kb/s
Modulation	QPSK	QPSK
Code rate	$3/4$	$1/2$
Required E_b/N_o	6.1 dB	5.7 dB (U/L)
Target Link Margin	15.5 dB	6 dB

nonlinear as the signal level is increased towards the point at which their output saturates (i.e. no more output power is available), and hence nonlinear effects are generally of greatest significance in the output of high-power amplifiers (HPAs). A lesser contribution can occur from sensitive low-noise preamplifiers, the inputs of which are overloaded, and from passive structures – such as transmit antennas – which experience very high powers. This is the so-called 'rusty bolt' effect, which typically results from rectification at the junction between dissimilar metals.

Nonlinearities result in intermodulation (mixing) between any signals in the same nonlinear element, and in an FDMA satellite system the satellite transponder is most likely to experience the greatest number of simultaneous carriers.

A typical characteristic of a high-power satellite travelling-wave tube amplifier (TWTA) under sinusoidal excitation is illustrated in Figure 6.3. The characteristic nonlinear shape of the measured input versus total output power curve depends on the number of input sinusoids present.

From a power efficiency perspective, we would like to operate the transponder (or other HPA) as close as possible to its saturated output level. However, distortion caused by the nonlinear characteristic increases nonlinearly with total input power below the saturation point.

Given input FDM/FDMA channel signals with carrier frequencies f_1, f_2, f_3, etc., it may be shown that the centre frequency of any intermodulation (mixing) product f_{IM} is determined from

$$f_{IM} = m_1 f_1 + m_2 f_2 + m_3 f_3 + \cdots \tag{6.8}$$

where m_i are integers and the order of any intermodulation product is given by $\Sigma_i |m_i|$.

As it is usual to incorporate a bandpass filter at the output of such amplifiers (in order to limit interference to other systems), we are primarily concerned with those intermodulation products whose frequencies are *in band*, close to the input carrier frequencies. Even so, prediction of intermodulation spectra under realistic carrier-loading scenarios is computationally demanding, and

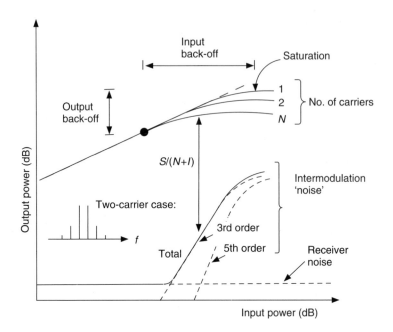

Figure 6.3   Intermodulation in a nonlinear device.

therefore we restrict our discussion to simple examples. Consider initially the simple case of two input carriers, at frequencies f_1 and f_2. In addition to the amplified input carriers at f_1 and f_2, the output of the nonlinear device will contain third-order intermodulation products at frequencies $(2f_1 - f_2)$ and $(2f_2 - f_1)$, fifth-order products at $(3f_1 - 2f_2)$ and $(3f_2 - 2f_1)$, seventh-order products at $(4f_1 - 3f_2)$ and $(4f_2 - 3f_1)$, and so on.

In general, third-order products (which are the products of three terms) are larger than fifth-order products, which in turn are larger (more significant) than seventh-order products, and so on; thus, third-order products tend to dominate. In the case of multiple carriers, third-order products exist with frequencies $(2f_1 - f_2)$ and $(f_1 + f_2 - f_s)$, and, if we have N input carriers, it can be shown that the number of third-order products of the type $(2f_1 - f_2)$ is $N(N-1)$, while the number of third-order products of the type $(f_1 + f_2 - f_3)$ is $N(N-1)(N-2)/2$ (Bond and Meyer, 1970). Therefore, if we have, say, 100 input carriers, the number of third-order products alone will be of the order of half a million (although a significant number of these will be degenerate).

The spectral widths of intermodulation products is determined by a convolution of the contributing signal spectra (mixing terms), and consequently spectra of intermodulation products are typically broader than those of the input signals, with fifth-order products broader than third-order products, and so on. It is thus not hard to see that, for a large number of modulated carriers, the intermoulation products overlap in such a fashion as to appear to form an increased noise floor – hence the term *intermodulation noise*.

Output Back-Off

This level of intermodulation noise increases nonlinearly with the total input carrier power. Consequently, for a given number of carriers, there will be a particular total output power level at which the signal-to-total noise is a maximum. By way of illustration, Figure 6.3 presents the dependence of carrier-to-intermodulation noise ratio with output power back-off ratio. This figure shows the effect of varying the output power level on the signal-to-noise and signal-to-intermodulation noise ratio for a multichannel FDMA system in a nonlinear amplifier. The signal-to-thermal noise ratio increases with increasing power, while the signal-to-intermodulation noise ratio decreases. The signal-to-total noise therefore exhibits a maximum somewhat below the saturation point. It is usually convenient to measure this optimum drive level (operating point) relative to the saturated output power level, and it therefore occurs at some specified number of decibels below saturation. This difference is known as the *output back-off*. The corresponding level below the input power corresponding to output saturation is known as the *input back-off*.

6.2.4.4   Power-Limited Capacity of FDMA

The back-off in output power needed must therefore be included in the satellite EIRP figure when considering the power link power budget for FDM/FDMA – with the result that there is consequent reduction in signal-to-noise ratio, as compared with single access. The amount of back-off needed will depend on the particular HPA characteristic and the number and type of carriers; however, an output back-off of 6 dB is not uncommon, which effectively means that only a *quarter* of the available transponder power is being used with this FDM/FDMA scenario. The power-limited capacity of FDM/FDMA is therefore potentially significantly less than for single carrier access. If the output back-off is OBO (expressed as a linear ratio) then the FDMA power efficiency η_P (the ratio of useful power/capacity to that for a single access) for a shared transponder is

$$\eta_P \, [\text{FDMA}] = \frac{1}{\text{OBO}} \tag{6.9}$$

Relative throughput efficiencies of FDMA, TDMA and CDMA for the case of a common transponder without frequency reuse are illustrated in Figure 6.9.

6.2.4.5   OFDM and OFDMA

Orthogonal FDM (OFDM) and Orthogonal FDMA (OFDMA), mentioned in Chapter 5, are special cases of FDM and FDMA respectively. Here, multiplexing is used to improve the performance of modulation schemes in a multipath environment. In essence, the total channel (bandwidth) is subdivided into a relatively large number of subchannels (usually of equal width). Channel data bits are divided among these smaller subchannels, with each subchannel operating at a lower data rate (reduced according to the number of channels). Hence, if the number of subchannels is N, we effectively have N parallel datalinks, each carrying $1/N$ times the overall data rate, and, provided the bandwidth of the subchannels is less than the channel coherence bandwidth (the inverse RMS delay spread due to multipath), then these subchannels will experience flat fading (as opposed to frequency-selective fading for the whole channel). No symbol distortion occurs, and we just have to cope with amplitude fading. Guard intervals are usually introduced in each low rate subchannel to counter any transmitter timing error or multipath spread. In OFDMA, multiple access is achieved by assigning subsets of OFDM subcarriers.

If this frequency division multiplexed arrangement were to be implemented using conventional FDM/FDMA, using a large number of independent datalinks, a large number of modulators would be required and guard bands needed to avoid adjacent channel interference, with resultant poor spectral efficiency and high implementation cost. In OFDM/OFDMA, the subchannels are formed using a single I and Q modulator and are made orthogonal with minimum spacing, thereby maximizing spectral efficiency. As illustrated in Figure 6.4, the subchannel signal spectra are generally given by sinc functions (for rectangular input pulses).

It is common to employ QPSK for the modulation on each OFDM/OFDMA subchannel. At the modulator input, the input serial bit stream is converted to N parallel data streams. Each is then

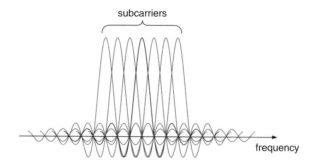

Figure 6.4   Amplitude spectrum of OFDM subcarriers.

applied to a constellation mapping function that outputs the complex coefficients that describe the symbol on the I and Q constellation diagram. The constellation coefficients are applied to an Inverse Fast Fourier Transform (IFFT), the output of which is converted to analogue I and Q signals. These are then used to modulate cosine and sine signals in the usual way. At the demodulator, the I and Q signals are digitized and input to a Fast Fourier Transform (FFT) to produce N constellation coefficient outputs. These are then detected according to their proximity to the nearest symbol, and the parallel data are converted back into a serial bit stream.

Although OFDM/OFDMA avoids frequency-selective fading, the performance of individual sub-channels is still affected by the flat fading, and OFDM/OFDMA is therefore normally always used together with an appropriate error-correcting coding technique. One particular combination of OFDM and convolutional error coding with interleaving, used for digital audio broadcasting (DAB), is known as Coded OFDM (COFDM).

6.2.5   SDMA (Frequency Reuse)

6.2.5.1   Overview

An increasingly common and very effective method of increasing capacity is Spatial Division Multiple Access (SDMA) – more commonly known as *frequency reuse*. Frequency reuse using a cellular-type beam pattern is illustrated in Figure 6.5. Such schemes commonly divide the available spectrum among a small cluster of adjacent cells (beams), and the cluster pattern is replicated multiple times (although it can be more efficient to divide the spectrum unequally if the cell loading distribution is very uneven).

The capacity for a given bandwidth allocation may be increased substantially by reusing the same spectrum again and again in multiple non-overlapping beams – either on multiple beams on the same satellite or non-overlapping beams on multiple satellites in a constellation of satellites.

6.2.5.2   Capacity Enhancement with S-FDMA

In such schemes, the cluster size K defines the reuse factor (the number of beams having different spectrum allocations). Cluster sizes of 4, 7 and 12 are typical. Suppose that we have N_B beams in the constellation sharing the same spectrum (i.e. across all satellites). The average bandwidth efficiency of SDMA is

$$\eta_B\,[\text{SDMA}] = \frac{\kappa\,N_B}{K} \tag{6.10}$$

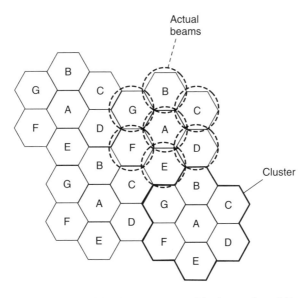

Figure 6.5 Cellular beam pattern with cluster size of 7.

where the factor κ (with $\kappa \leq 1$) is introduced to account for any overlap in the coverage of beams from different satellites. The SDMA bandwidth efficiency (enhancement) is equal to the ratio of the bandwidth-limited capacity of SDMA to that for a single-channel system. The efficiency of frequency reuse in an FDMA system (S-FDMA) is thus found by multiplying the relative efficiencies of FDMA and SDMA.

Provided the number of non-overlapping beams is large, the capacity can be increased significantly compared with no frequency reuse. Such frequency reuse is commonly employed in conjunction with FDMA, but may also be employed with TDMA and CDMA or combinations of all three.

Iridium example. Continuing with our Iridium example, there are 66 satellites, and each satellite has 48 beams. This makes 3168 beams in all. However, owing to significant beam overlap near the poles in this near-polar orbital scheme, at any one time only 2150 beams are actually active (ICAO, 2006). The beam overlap factor is thus $k \rightarrow 0.68$. The beam cluster size is $K \rightarrow 12$, yielding a substantial SDMA efficiency (frequency reuse factor) across the constellation of 179.5.

6.2.6 TDM and TDMA

6.2.6.1 Overview

We turn our attention next to Time Division Multiplexing (TDM) and Time Division Multiple Access (TDMA). Here, the channel is sliced in time, with successive slices being allocated to different data streams.

TDM

Figure 6.6 illustrates the operation of a generic TDM multiplexer. Multiple data streams arrive at the inputs to the multiplexer (left-hand side of the figure) at some data rate. The output of the multiplexer comprises successive time slices (samples) of the inputs (one after the other). The

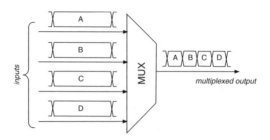

Figure 6.6 Time division multiplexing.

multiplexed data are hence transmitted at a higher data rate (approximately the sum of the input data rates). At the demultiplexer, the reverse process occurs. Although some overhead bits are typically introduced when multiplexing the data – in order to allow more efficient and flexible demultiplexing – in general the efficiency of TDM is close to unity. TDM is thus the most efficient mutiplexing technique.

TDMA

In TDMA, as with TDM, the channel is 'sliced' in the time domain. However, TDMA must accommodate transmissions from successive independent terminals occurring in non-overlapping *bursts*, as illustrated in Figure 6.7. To avoid the potential for overlap of successive bursts, guard intervals are introduced between successive bursts. Furthermore, some time must be allocated at the start of each burst to the recovery of carrier and bit timing, and for data synchronization.

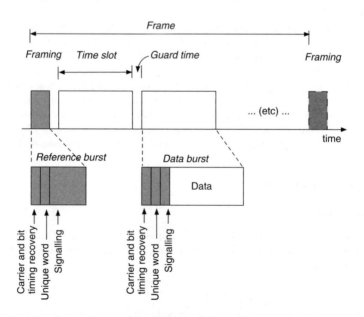

Figure 6.7 TDMA frame structure.

TDMA bursts generally occur within fixed-length *frames*, and therefore each terminal must buffer a frame's worth of data (both input and output) so as to accommodate the burstiness of TDMA transmission. Within each frame, timing reference and access signalling is provided by a reference burst, usually at the start of each frame.

Advantages of TDMA include:

- *Flexibility.* A TDMA timeframe may be divided up into arbitrary combinations of bit rates.
- *Power efficiency.* In a pure TDMA scheme, as only one signal occupies the transponder at any given time, intermodulation is not a serious issue, and the transponder may therefore be operated in saturated mode for maximum power output.

Disadvantages of TDMA include:

- *Synchronization.* TDMA generally requires network-wide synchronization of timing in order to prevent bursts from overlapping. As such, TDMA networks can be vulnerable to loss of network timing signals.
- *User transmit power.* As each terminal must buffer its (nominally continuous) data and compress these into bursts, each burst requires higher transmit power to achieve the same bit error ratio.
- *Latency.* The TDMA framing structure increases the latency over the satellite link.

TDMA Link Budget

Given a particular burst rate R_B, the link budget calculation for TDMA follows the same procedure as for single-channel access, except that the channel is accessed at the burst rate rather than the coded bit rate R_c.

TDMA requires that a frame's worth of user data be stored in a buffer, with the buffer filled/emptied during the appropriate burst. The coded bit rate R_c (i.e. including error coding) is therefore related to the burst rate R_B via

$$\frac{R_c}{R_B} = \frac{T_B}{T_F} \tag{6.11}$$

where T_B is the average period of transmission of useful data in the user's burst. It follows that the required terminal C/N_o and transmitter EIRP are higher than for single access by a factor $\frac{R_c}{R_B}$.

6.2.6.2 Efficiency of TDMA

The bandwidth-limited efficiency and power-limited efficiency of TDMA are the same, and are determined by the fraction of the TDMA frame period T_F available for useful data transmission (i.e. excluding preambles, etc.):

$$\eta_B\,[\text{TDMA}] = \eta_P\,[\text{TDMA}] = \left(\frac{\Sigma\,(T_B)}{T_F}\right) \tag{6.12}$$

where ΣT_B is the total useful transmission time.

Again, expressing the efficiency in terms of the number of TDMA bursts, if we can assume M equilength bursts, then

$$\eta[\text{TDMA}] \simeq 1 - M\frac{T_w}{T_F} \tag{6.13}$$

where T_w is the average time per burst that is not available for useful data transmissions (including guard time, preamble, signalling, etc). Hence, to a first approximation, the efficiency of TDMA decreases linearly with the number of simultaneous users. The relative throughput efficiencies of FDMA, TDMA and CDMA for the case of a common transponder without frequency reuse are illustrated in Figure 6.9.

Iridium example. Let us complete our Iridium example. Each Iridium FDMA channel comprises eight simplex (one-way) TDMA bursts (permitting four duplex calls per FDMA channel). The TDMA frame length is 90 ms and each TDMA burst lasts 8.28 ms, yielding a TDMA frame efficiency of 73.6%. We can now estimate the bandwidth-limited capacity of the entire Iridium system (i.e. ignoring power limitations). Iridium employs QPSK modulation (2 bits per symbol) at a burst rate of 50 kb/s and a code rate of 3/4. Hence, the modulation spectral efficiency for a single 31.5 kHz channel is 1.19 (b/s)/Hz. Multiplying this by the FDMA efficiency (0.756), the SDMA efficiency (179.5) and the TDMA efficiency (0.736), we obtain a total spectral efficiency of 118.91 (b/s)/Hz. The allocated bandwidth is 10 MHz, giving 1.19 Gb/s of bandwidth-limited capacity for this system. Taking into account the burst rate, code rate and fraction of each frame occupied for a single burst, the maximum simplex data rate is 3.45 kb/s. Each duplex (two-way) circuit requires two TDMA bursts (i.e. 7.9 kb/s in total), implying a maximum of $\sim$172 000 duplex circuits across the constellation.

6.2.6.3 MF-TDMA

TDM is ideal for the forward link between a large central hub (e.g. Internet service provider) and multiple users having bursty traffic, as the allocation of burst may be adapted according to the bursty nature of the demand. A TDM (or potentially TDMA) link may be operated at maximum transponder power (and hence efficiency) using a large hub terminal, with high EIRP.

For the return link, however, the much smaller user terminals are typically unable to produce the high EIRP needed to achieve maximum transponder power efficiency. In such cases the combination of FDMA (which requires lower EIRP) and TDMA (for flexibility) is very attractive. Such a scheme is known as MultiFrequency TDMA (MF-TDMA). By way of example, MF-TDMA is the multiple-access method used in DVB-Return Link, over Satellite (DVB-RCS), with the forward link being provided by DVB-S (TDM).

6.2.7 CDM and CDMA

6.2.7.1 Overview

In Code Division Multiplexing (CDM)/Code Division Multiple Access (CDMA), multiple signals share the same spectrum simultaneously but each access is distinguished by a unique spread spectrum code.

As before, we need to distinguish between the use of code division for multiplexing and multiple access.

CDM/Synchronous CDMA

Code Division Multiplexing or *synchronous CDMA* codes are appropriate where multiple data streams share a single transmitter. As these codes can be synchronized at the (common) transmitter, they can be made orthogonal. Consequently, there is no mutual interference between codes. The requirement to synchronize all codes is not a burden for multiplexed signals (CDM) transmitted from a shared hub (e.g. hub to mobile). Examples of CDM/synchronous CDMA spreading codes include Walsh–Hadamard and Orthogonal Variable Spreading Factor (OSVF) codes – the latter being used in UMTS to support different spreading factors (and therefore data rates).

Asynchronous CDMA

Asynchronous CDMA codes are those that are not synchronized – as might be used by multiple independent transmitters sharing the same spectrum. Such codes are not strictly orthogonal, with the consequence that mutual interference occurs between signals. These codes are typically used for the user–hub link, where the hub receives multiple unsynchronized transmissions. Examples of codes for asynchronous CDMA are general PseudoRandom Noise (PRN) codes and *gold* codes (which are quasi-orthogonal).

Advantages of CDMA include:

- *No channelization required*. In principle, all users share the same spectrum.
- *Resilience to interference*. Enables more efficient frequency re-use than other schemes.
- *Reduced timing synchronization*. Unlike TDMA, precise network timing is not generally required.
- *Mitigation against multipath*. Provides mitigation of frequency-selective fading multipath (for example, with a Rake receiver).
- Reduced sensitivity to Doppler spread compared with FDMA.

Disadvantages of CDMA include:

- *Mutual interference.* For non-orthogonal codes, the level of interference increases with the number of simultaneous users, and at some point the required signal-to-noise plus interference can no longer be achieved.
- *Power control.* The control of mutual interference requires careful (automatic) power control.
- *Increased complexity.* Acquisition of a CDMA signal involves code acquisition (alignment of the de-spreading code at the receiver with the spreading code). CDMA places additional demands on code acquisition and tracking (these topics are covered in Chapter 7).

6.2.7.2   Spread Spectrum

CDM and CDMA are a form of Spread Spectrum (SS). Spread spectrum is characterized by the modulated signal being spread over a frequency range much greater than the bit rate (and therefore over a much greater frequency range than it would normally occupy using conventional modulation). Potentially, this spreading factor can be several orders of magnitude. Further, spread spectrum signals are characterized by an apparent 'randomness' – thereby minimizing their spectral density and hence interference to other spectrum users, and ensuring demodulation only by the appropriate receiver. Strictly speaking, spread spectrum is not a modulation scheme; rather, it is a method of spreading the bandwidth of a particular modulation method.

Spread spectrum on its own is a very inefficient use of the spectrum for a single access. However, if multiple users share the same enlarged spectrum but are able to recover their data from other signals by utilizing different spreading *codes*, the overall efficiency can be recovered and we have the notion of code division multiplexing and code division multiple access (i.e. multiplexing/multiple access using only different codes).

Spectral spreading occurs at the transmitter, while the reverse process – *despreading* – occurs at the receiver. Referring to Figure 6.8 (which relates to the Direct-Sequence Spread Spectrum (DS-SS)), the user's data signal is spread using an appropriate code sequence having a much broader spectral shape. A consequence of spreading the signal power over a much broader spectral range is that the power spectral density (power per Hz) is reduced (in inverse relation to the increase in

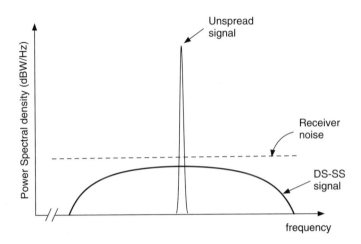

Figure 6.8   Direct-sequence spread-spectrum modulation.

spectral width). The low average power spectral density of the spread spectrum means that it can coexist with other SS signals, even if they are not orthogonal.

On reception, the same code sequence is used to despread the input signal, with the result that the wanted signal is returned to its original narrowband form (recall that one implementation of the matched filter is to multiply the de-spread input signal by the expected signal). Although noise and wideband interference are not affected by the despreading process, any narrowband interference is spread, resulting in significant resilience to such interference. The increased immunity to interference and improved covertness from detection of the spread spectrum are of significant importance to military users.

In summary the potential advantages of spread spectrum are:

- *Increased immunity to narrowband interference.* As a consequence of the despreading process at the SS receiver, the transmitted signal is less affected by both unintentional and intentional narrowband interference (narrowband compared with the spread spectrum signal). This feature makes SS particularly suitable for use where robust and secure communications are vital (such as strategic and tactical military communications), where deliberate jamming may be anticipated. Civilian applications include its use in multiple-access schemes, where interference occurs from other system users sharing the same spectrum.
- *Reduced transmitted power spectral density.* This feature makes SS attractive for use in covert communications where messages may more easily be hidden in background noise (i.e. the spectral density at a receiver is less than that of the receiver's background noise), where there is a requirement of low probability of detection and eavesdropping. This aspect can also be utilized to minimize interference to other users of the spectrum (e.g. on other satellites) when the angular discrimination produced by the antenna is not sufficient (for example, in mobile VSAT terminals).
- *Increased immunity to multipath.* Multipath may be considered to be a form of self-interference.

Processing Gain
Spread-spectrum systems are typically characterized by their *processing gain*, usually quoted in dB. Although processing gain may be defined in different ways, it is often defined as the ratio of

the spread-spectrum bandwidth B_{SS} to the unspread bandwidth B needed to transmit the data

$$G_p \equiv \frac{B_{SS}}{B} \tag{6.14}$$

$$\sim \frac{B_{SS}}{R_b} \tag{6.15}$$

The significance of the processing gain is apparent when one considers the ability to spread the spectrum to extract signals from narrowband interference. The improvement in signal to interferer levels is generally equal to the processing gain.

6.2.7.3 Implementations of the Spread Spectrum

There are two basic approaches to implementing spread spectrum:

- *Direct-Sequence Spread Spectrum (DSSS).* In DSSS, the low-bit-rate data sequence is multiplied by a high-bit-rate spreading code sequence prior to modulation.[4] DSSS is the predominant approach to spread spectrum used in commercial systems.
- *Frequency-Hopping Spread Spectrum (FHSS).* In FHSS, the carrier frequency of a modulated signal is periodically 'hopped' from one frequency to another (achieved by synthesizing the hopping carrier frequency). The IF signal comprising the modulated data is mixed with (multiplied by) a hopping carrier wave, the sequence of carrier frequencies being determined by a pseudorandom code. Depending on the method used, the hop rate may either be comparable with or slower than the information symbol rate (known as slow frequency hopping) or be much faster than the symbol rate (fast frequency hopping). FHSS can potentially take advantage of greater spectral ranges than DSSS, and consequently greater processing gain can be achieved by this method. For this reason (and others), FHSS is usually the method of choice for robust military communications. However, the practical implementation of FHSS is significantly more complex than DSSS.

6.2.7.4 Processing Gain for Direct-Sequence Spread Spectrum

We shall focus here on DSSS, as this is the principal spread-spectrum technique employed in personal satellite applications. For DSSS CDMA, the processing gain may be more conveniently defined in terms of the ratio of chip (symbol) rate R_s to uncoded bit rate R_b

$$G_p \rightarrow \frac{R_s}{R_b} \tag{6.16}$$

For example, in the case of the GPS C/A code, the chip rate is 1.023 MHz and the bit rate (navigation message) is 50 b/s. The processing gain in this case is approximately 43 dB.

6.2.7.5 Interference-Limited Capacity of CDMA

For non-synchronous CDMA, the codes are non-orthogonal codes, and interference results between users' signals. Ultimately, if too many users are transmitting simultaneously, this interference may prevent successful reception. This limiting capacity of CDMA using non-orthogonal codes is sometimes referred to as the 'soft capacity' (as opposed to the 'hard capacity' of FDMA and TDMA).

For convenience, let us assume M identical accesses and perfect power control, so that all CDMA signals have the same power spectral density at the receiver (e.g. at a central hub).

---

[4] Note that modulo-2 addition of two data sequences is equivalent to multiplication of their NRZ sequences.

The effective noise spectral density N_o' – the noise density N_o plus interference density I_o, including the noise-like interference of the other $(M - 1)$ users – may be expressed as (Viterbi, 1985; Gilhousen et al., 1990)

$$N_\mathrm{o}' = N_\mathrm{o} + I_\mathrm{o} = N_\mathrm{o} + \alpha\,(M - 1)\,\frac{R_\mathrm{b}E_\mathrm{b}}{B_\mathrm{ss}} \tag{6.17}$$

where R_b is the bit rate, E_b the energy per bit and B_SS is the spread-spectrum bandwidth. The noise density contribution from the other $(M - 1)$ users is essentially their signal power $R_\mathrm{b}E_\mathrm{b}$ divided by the 'noise' bandwidth. The factor α allows for consideration of other factors, such as voice activity, which potentially reduces the *average* interference power, and adjacent cell interference (in SDMA systems), which increases it. These factors will affect the average interference power.

In the limiting case, we may assume that the energy-to-one-sided noise plus interference ratio is equal to the system minimum value $\left(\frac{E_\mathrm{b}}{N_\mathrm{o}}\right)_\mathrm{min}$, and that the actual $E_\mathrm{b}/N_\mathrm{o}$ is equal to this value times the power margin. From equation (6.17) we can determine the maximum number of users

$$(M)_\mathrm{max} = 1 + \frac{B_\mathrm{SS}}{\alpha R_\mathrm{b}\left(\frac{E_\mathrm{b}}{N_\mathrm{o}}\right)_\mathrm{min}}\left(1 - \frac{1}{\mathrm{margin}}\right) \tag{6.18}$$

The spectral efficiency of CDMA is the ratio of the total bit rate $(M\,R_\mathrm{b})$ divided by the occupied spectrum B_SS, and may be expressed in terms of the average signal-to one-sided noise spectral density (per user) $\frac{C}{N_\mathrm{o}}$

$$\xi\,[\mathrm{CDMA}] = \frac{1}{\left(\frac{E_\mathrm{b}}{N_\mathrm{o}}\right)_\mathrm{min}}\,\frac{M\left(\frac{C}{N_\mathrm{o}B_\mathrm{SS}}\right)}{\left(1 + \alpha\,(M - 1)\left(\frac{C}{N_\mathrm{o}B_\mathrm{SS}}\right)\right)} \tag{6.19}$$

The interference-limited throughput efficiency for CDMA is the ratio of the spectral efficiency to that of a bandwidth-limited single access. For large numbers of users, the spectral efficiency saturates at the inverse of α times the effective energy per bit to one-sided noise density. Relative throughput efficiencies of FDMA, TDMA and CDMA for the case without frequency reuse are illustrated in Figure 6.9. It is seen that the capacity of CDMA under these conditions is generally lower than for FDMA (which is itself lower than for TDMA). This is a consequence of the interference effect of other unsynchronized CDMA user terminals (Viterbi, 1985).

Effect of Frequency Reuse and Voice Activity

The relative capacities of CDMA versus FDMA depends on a number of factors and remains a somewhat controversial subject, however, the efficiency of CDMA may be enhanced by a number of factors (Gilhousen et al., 1990).

Specifically, CDMA may take advantage of quiet periods in speech calls. It is estimated that, on average, in a telephone call, each user talks only 35% of the time. A CDMA terminal may effectively turn its transmissions off (or at least reduce its power) during quiet periods, thereby reducing interference (Gilhousen et al., 1990). The Globalstar mobile satellite service reduces its data rate according to the user's speech patterns.

The relative performance of CDMA is further enhanced when SDMA (frequency reuse) is utilized. Since CDMA has some built-in immunity against interference, the same channel frequency may be reused in all 'cells' of a frequency reuse scheme (a cluster size of 1), whereas FDMA can only reuse frequency in non-overlapping cells (in the Iridium system example, the cluster size is 12). However, the resultant adjacent channel interference from reusing the same frequency in each cell increases the interference noise by a factor of ~2–3 (Zhang and Zhu 2006).

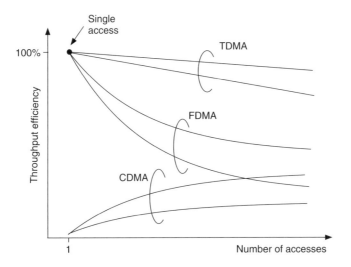

Figure 6.9 Comparison of FDMA, TDMA and CDMA efficiencies (no frequency reuse case).

6.2.7.6 MF-CDMA

Although, in principle, spread spectrum may utilize the entire spectrum allocation for that service, in practice the spread bandwidth of DSSS CDMA is generally limited, and it is typically used together with FDMA as MultiFrequency CDMA (MF-CDMA). Globalstar is an example of an MF-CDMA system.

6.3 Payloads

Before moving on to discuss how multiple access schemes allow the formation of satellite networks, we briefly consider the principal types of satellite communications and broadcasting payload architectures, since these will have an bearing on the type of networks supported.

6.3.1 Unprocessed Payloads

6.3.1.1 Transparent Repeater

The simplest and the most extensively deployed payload configuration, for both communications and broadcasting, is the *transparent repeater* or transponder. Here, the received signal is merely amplified, and its uplink frequency is translated to the downlink frequency before transmitting the signals towards the service area. This proven architecture is simple, reliable and mature and, significantly, presents the flexibility to change the service modulation and multiple access characteristics *after the satellite has been launched*. The transparent repeater is often referred to as a 'bent-pipe' transponder, as the output is merely a replica of the input.

Figure 6.10 shows a simplified block diagram of a transparent repeater payload. Signals are received by an antenna system, bandpass filtered to remove extraneous noise and then amplified in a Low-Noise Amplifier (LNA) before being down-converted to the transmit frequency band and amplified further. As, in almost all cases, the allotted spectrum of satellite services (and particularly those in the C, X, Ku and Ka frequency bands) exceeds by some margin the occupied bandwidth

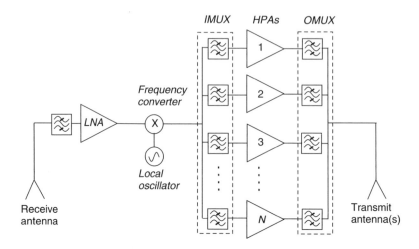

Figure 6.10   Multichannel bent-pipe transponder payload.

for any single carrier, some form of frequency division multiplexing or frequency division multiple access is almost always used (potentially in conjunction with TDM/TDMA or CDM/CDMA). Since the total transponder power required to support a very large number of signals may exceed that achievable in a single TWTA, and since we have already established that the throughput capacity of each satellite communications transponder when using FDM/FDMA will decrease with the number of carrier signals present, the total bandwidth (and power) is generally divided into smaller bandwidth segments at the High-Power Amplifier (HPA) stage, each bandwidth segment having a separate channel amplifier. The number of accesses per channel amplifier is thus reduced, and the total payload power output comprises the *sum* of the power outputs for each channel amplifier.

Typically, the number of channel amplifiers (referred to as transponders) on-board a satellite is quite large. By way of example, current Intelsat satellites have between 24 and 72 transponders per satellite. Broadcast systems may also have between 24 and 32 transponders in C and Ku bands. Typical telecommunication and broadcast transponders are 27–72 MHz wide.

6.3.2   Processed Payloads

Although a majority of current satellite communications and broadcasting payloads comprise transparent transponders, there is a gradual shift towards payload architectures where repeaters incorporate more complex functionality involving space-hardened Digital Signal Processors (DSPs). Figures 6.11 and 6.12 illustrate two different architectures employing such processing payloads – often known as On-Board Processed (OBP) payloads.

6.3.2.1   Switched-IF Processed Payload

In the first payload architecture (Figure 6.11), instead of block converting from the uplink RF frequency to the downlink RF frequency, the received FDM/FDMA signals from each beam are down-converted to an IF frequency, in blocks of contiguous spectrum. Each input beam/spectrum block may then be switched to any output beam/spectrum block via a switch fabric. The output IF signals are then up-converted to the transmission frequency band. Processing is performed at an intermediate frequency without demodulation, where received signals are routed to an appropriate

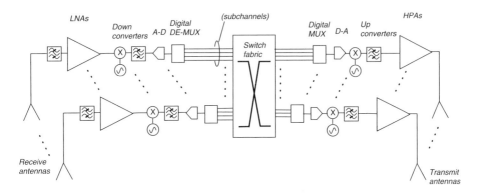

Figure 6.11   Non-regenerative processed payload.

beam either in a static but reconfigurable arrangement or in a time-synchronized manner. Thus, small segments of spectrum may be switched into different spot beams, or, if a phased-array antenna is utilized, phase shifting may also be done in DSP at IF and beam shapes may be altered through ground commands. By way of example, in Inmarsat-4 satellites, 200 kHz blocks of spectrum may be routed into any beam in near real time.

6.3.2.2   Fully Regenerative Payload

In the second type of processing payload shown in Figure 6.12, the received signals in each beam are demodulated to baseband data (e.g. as NRZ), and a baseband data switch – potentially even a space-hardened on-board circuit or packet switch – is used to route signals between beams. The outputs of the on-board switch are then remodulated prior to transmission.

In principle, a regenerative transponder may feature various degrees of processing of the content of the received signal. The processing can be quite simple – a lightly coded TDMA signal transmitted from a large station can be coded more heavily and retransmitted in a TDM format for reception on small personal terminals. The processing can be more involved where the content of the digital signals (packets) are processed and acted upon.

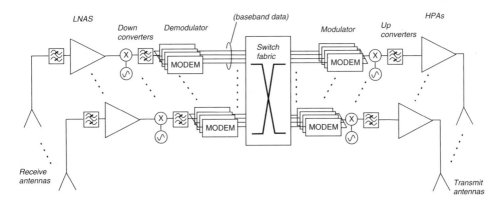

Figure 6.12   Fully regenerative processed payload.

A notable current example of a regenerative payload is that found on Iridium satellites, in which individual calls (circuits) are routed via intersatellite links to the appropriate satellite servicing the end-user. All the conventional ground-station functions can theoretically be performed on-board satellite.

The regenerative packet-switched payload is probably the most complex and challenging of all the communication satellite payload architectures. Hispasat Amazonas satellites include an on-board DVB-S/DVB-RCS regenerative switching payload, while recent development of fully regenerative IP packet-switching payloads has been given a boost by the Cisco/Intelsat/DoD Internet Router in Space (IRIS) programme.

6.4 Networks

We have discussed the methods by which a large number of user terminals may share finite satellite payload power and bandwidth. The user terminals constitute a communications *network*, and in this section we shall discuss the salient features of a generic satellite communications network. Specific network services and technologies are discussed in Chapters 11 and 12.

6.4.1 Network Architectures

Before continuing our discussion of satellite networks, we should mention, in passing, the simplest satellite network architecture of all – the *point-to-point network*, in which the network comprises just two satellite terminals. In general, however, a service will support multiple users. Depending on the application and network requirement, a satellite network is generally arranged in either a *star* or a *mesh* configuration, as illustrated in Figure 6.13.[5] Hybrids of these two forms (e.g. partial mesh) are also sometimes used.

6.4.1.1 Star/Hub–Spoke

In a star network configuration (also called a *hub–spoke* network), user terminals, for example, satellite phones or Very-Small-Aperture Terminals (VSATs), connect with each other and external users through a large satellite Earth station, known as a *hub*. To carry a public service, the hub is connected to the terrestrial public network (telephone network or Internet), whereas in a private network it terminates within the facilities of the private enterprise.

A significant advantage of the star network configuration is that the link budget is eased for small terminals, as they are typically linked to a hub with a large antenna and high power (and hence high EIRP and G/T). Performance is thus dominated by the C/N_o achievable over the user terminal – satellite link. A further advantage of the star configuration is that all the forward links from the hub to multiple users may be synchronized and thus more efficiently multiplexed – for example using TDM. However, for communications between two user terminals in the star, the amount of satellite resources (power and bandwidth) needed is double that for communications between the user terminal and the hub alone.

6.4.1.2 Mesh

In a mesh configuration, all the Earth stations are interconnected or 'meshed', and communications between user terminals does not include a hub. One of the meshed terminals may provide terrestrial connectivity.

---

[5] In terrestrial computer networks, other topologies, such as bus, ring and tree, may also be found.

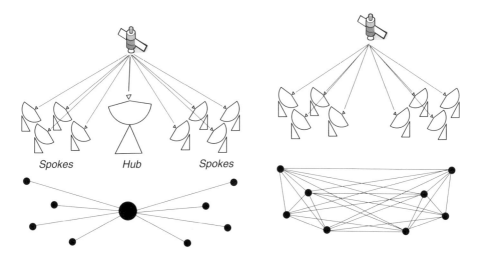

Figure 6.13 Satellite network topologies. Left: star (hub–spoke); right: mesh. The upper figures show the physical configuration, while the lower figures show the logical interconnections.

Significant advantages of the mesh network configuration are reduced latency (delay) and the increased efficiency in the use of satellite resources for the case where the bulk of the communications is between meshed terminals, rather than between user terminals and terrestrial networks (via some hub). However, for a bent-pipe satellite, the meshed link budget is more stringent than for the star configuration owing to the smaller terminal size at both ends of each link. Referring to Equation 4.37, the overall C/N_o may be up to a factor of 2 (3 dB) worse for a mesh satellite link than for a hub-spoke link, for the same user terminal characteristics. Nevertheless, with the advent of high-powered satellites capable of generating hundreds of small spot beams, supplemented with regenerative transponders, mesh connectivity between VSATs and satellite phones is becoming more attractive.

6.4.2 Network Models

6.4.2.1 Switching Models

Communications networks, other than simple point-to-point connections, require a means of effecting connections between two or more users (or user applications) from the many sharing the service – in effect a method of switching between different routes – and these may be broadly categorized into two generic switching models:

- *Circuit-switched networks.* A circuit (end-to-end connection) is established prior to transmission, and this provides a fixed throughput capacity for the duration of the connection – regardless of whether this capacity is fully utilized. This is the familiar connection model in conventional telephony and mobile voice communications (including most mobile satellite communications). Advantages of circuit switching include guaranteed throughput capacity and constant delay. However, circuit-switched networks suffer from relatively low efficiency, as information throughput is rarely constant.
- *Packet-switched networks.* The data are split into a sequence of finite-length data packets. Individual packets traverse the network separately and are reassembled at the receiver. Packet switching

is particularly suitable for bursty traffic, where the required data rate varies significantly with time, and is thus well suited to communications between computers. Packet-switched networks benefit from the use of *statistical multiplexing*, by which packets are time multiplexed, but with buffering of the input packet streams to accommodate surges. Significant advantages of packet switching are its flexibility, scalability and efficiency – particularly with bursty traffic. However the packetizing, statistical multiplexing and routing introduce additional *latency* (delay) and *jitter* (variation in delay). In addition, packet-switched networks can be less effective for the transport of time-critical information (voice and real-time video), than circuit-switched networks. Depending on the implementation, most packet-switched networks also provide few guarantees regarding Quality of Service (QoS).

Both circuit- and packet-switched satellite networks exist. However, rather than have separate networks for voice, video, data, etc., it is apparent that a single network capable of supporting all types of medium is very attractive (and potentially more cost effective). It is thus easy to see why many service providers are migrating towards the use of a single *packet-switched* network.

Packet-switched networks can further be categorized as being:

- *Connection oriented.* A *virtual* end-to-end connection (route) is established and all data packets travel through the network over the same path. As a virtual connection is established, it is possible to provide some guarantees of QoS. Because all packets follow the same path, the switching (routing) of packets may be very efficient and packets need only carry a simple label indicating the virtual circuit – so-called *label switching*. Examples of connection-oriented packet switching include the Asynchronous Transfer Mode (ATM), frame relay and MultiProtocol Label Switching (MPLS).[6]
- *Connectionless.* Each packet is routed individually, as it arrives, with packets potentially taking different routes. Each packet must thus carry the full network address of the intended recipient, and each router/switch must work out the best route for that packet address. Typically, packet delivery is on a best-effort basis. The dominant example of connectionless packet switching is the Internet Protocol (IP).

6.4.2.2   Protocol Models

Networks are typically characterized by dividing their functionality logically into *layers*, in order to structure and simplify network design, and to facilitate inter/intrasystem communication and standardization. Each network layer performs a distinct function in peer–peer communication conducted through a set of rules known as a *protocol*. An analogous comparison would be the hierarchical arrangement in an organization where a manager receives instruction from his boss, gives instructions to his subordinates and discusses common issues with a peer of the same level. A satellite system can integrate seamlessly with a terrestrial system by adapting identical protocols. Mobile satellite systems such as Inmarsat, Thuraya and Iridium offer seamless terrestrial roaming by sharing the upper-layer protocols of the partner terrestrial system while optimizing the lower layer protocols for satellite delivery.

OSI Model

A conceptual reference architecture known as Open Systems Interconnect (OSI), developed by the International Standards Organization and adopted by the ITU, is commonly used for understanding

---

[6] MPLS is a protocol used to provide the benefits of label switching (routing efficiency and traffic engineering) to connectionless packet-switched services – such as IP.

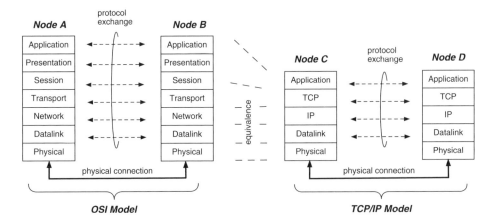

Figure 6.14   OSI (left) and TCP/IP (right) network layer models.

and benchmarking telecommunication network architectures. The OSI seven-layer 'stack' is illustrated in Figure 6.14 (left).

Each layer performs a distinct set of functions:

1. The *physical layer* specifies functionalities related to the transmission medium, for example, the radio links.
2. The *datalink layer* provides error-free data to the higher layers. It includes a sublayer called the Medium Access Control (MAC), used to manage access to the shared physical layer resources (by FDMA, TDMA, etc.).
3. The *network layer* routes information from the source address to the destination address, including functions such as network addressing, congestion control, accounting, coping with heterogeneous network protocol, etc.
4. The *transport layer* provides transparent transfer of information between the two entities and includes a provision for reliable data delivery.
5. The *session layer* connects and controls application processes between local and remote entities and includes synchronization to enable restart in case of failure, remote log-in, etc.
6. The *presentation layer* manages data transformation, formatting and syntax, for example, conversion between different datasets, encryption and data compression.
7. The *application layer* defines the interaction between user and communication system, for example, e-mail, file transfer, etc.

Interlayer communication with levels above and below are conducted vertically through the stack. From a satellite networking perspective, the physical, datalink and network layers are of considerable significance.

TCP/IP Model

The OSI seven-layer model is an abstract one, and, although various communications technologies were developed in line with the OSI concept, not all have strictly adhered to it, for practical reasons. An alternative model, the Transport Control Protocol/Internet Protocol (TCP/IP) model, was developed independently of the OSI model to describe a data communication system for computers, which eventually became the Internet. TCP/IP differs from the OSI model in its structure

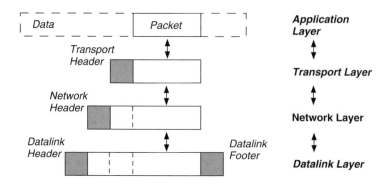

Figure 6.15   Data encapsulation.

and partitioning (IETF, 1981; IETF, 1988). Nevertheless, owing to the runaway success of the Internet, and the drive for convergence, the Internet Protocol (IP) has largely superseded alternative technologies in recent years. Thus, an increasing number of applications are now built over the IP – for instance, Voice over IP (VoIP) and IP television (IPTV).

Figure 6.14 (right) illustrates the TCP/IP protocol model. The five-layer TCP/IP model[7] effectively compresses the upper three layers of OSI into a single application layer (although the mapping is not exact).

Data Encapsulation

In a packet-switched network, protocol instructions for the peer at the far end are encapsulated in *headers* appended to the incoming packet and passed on to the layer below. This process of encapsulation is illustrated in Figure 6.15. A footer is also added at the datalink layer (e.g. redundant bits used to detect transmission errors). At the datalink layer, such encapulated packets are often known as *frames*.

The process of encapsulation results in a reduction in throughput compared with synchronous data transmission owing to the overhead for the headers and footers. The amount of overhead depends on both the header and data length. For example, asynchronous transfer mode packets, which are optimized for voice transmissions, are of fixed length (53 bytes), comprising a 5 byte header and 48 bytes of data (resulting in an overhead of 9.2%). IP packets, on the other hand, are optimized for data, but the header length is much longer. Header overhead is a significant issue for voice over IP (VoIP), as the voice packets are very short. In fact, uncompressed VoIP typically has a overhead in the region of ~50%.

Flow Control

In a two-way data network, in order to indicate whether packets were received successfully (i.e. no unrecoverable errors detected) and reassembled, the standard approach is to number the packets and acknowledge each packet, or group of packets, using some form of ACKnowledgment (ACK) – in the form of a short return packet.

---

[7] In some descriptions, the TCP/IP model has four layers, and, with reference to the OSI model, it does not specify the physical layer – leaving the choice of transmission technology flexible. We use a five-layer description to allow easier comparison with OSI.

In a packet-switched network, the potential exists to *saturate* the receiver with packets at a rate faster than it can process, and some form of *flow control* is needed to ensure efficient use of the network. The main forms of flow control are:

- *Stop and wait.* The transmitter sends a single packet and then stops transmitting and waits for the ACK for that packet. Although simple to implement, this is a somewhat inefficient method of flow control, particularly in satellite networks, as the delay between the end of transmitting the packet and receiving the ACK is at least twice the one-way transmission delay.
- *Sliding window.* The receiver notifies the transmitter of the maximum number of packets it will accept. The transmitter sends this number of packets before waiting for the acknowledgement for the first packet, whereupon it sends another packet, and so on. Each packet is given a sequence number so the receiver can indicate which packets were received. The sequence number is sent with the acknowledgement. The effect is that of a sliding window of packets that may be transmitted, moving as the ACKs are received, as illustrated in Figure 6.16. In addition to sending ACKs, the receiver may restrict the flow of packets at times of congestion by altering the size of the window.

Automatic Repeat Request

A failure to deliver a frame may be indicated by a *timeout*, in which no ACK was received within a specified time, or by a Negative ACK (NAK) from the receiver requesting retransmission. Such requests are known as Automatic Repeat Requests (ARQs). The principal types of ARQ retransmission schemes are:

- *Stop-and-wait ARQ.* In a stop and wait flow control system, rather than waiting for the timeout at the transmitter, the receiver notifies the transmitter that a frame was received but contains errors.
- *Go-back-N ARQ.* The receiver in a sliding-window flow control system indicates that a frame with a given sequence number contains errors (or was received out of sequence). The receiver then discards all subsequent packets until the missing packet and subsequent packets are retransmitted.

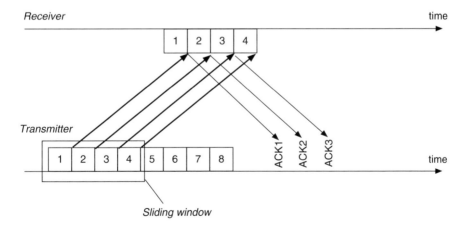

Figure 6.16   Sliding window packet network flow control.

- *Selective repeat ARQ.* The receiver sends an NAK for the missing or corrupted frames and only these are resent (after the transmitter has transmitted its window). Although clearly more efficient than the go-back N scheme, selective repeat ARQ requires the receiver to maintain a buffer, as packets may be received out of sequence. It also requires that each packet be acknowledged separately, whereas go-back N may acknowledge several frames with one ACK.

6.4.2.3   Limitations of Sliding-Window Flow Control in Satellite Networks

Maximum Window Size

The performance of sliding-window-type protocols (such as those used in the ubiquitous TCP/IP protocol of the internet) is constrained by the product of the transfer data rate and the round-trip delay. This *bandwidth–delay product* effectively defines the amount of unacknowledged data that the protocol must handle at any given time in order to keep the data 'pipe' full. Problems arise when the bandwidth–delay product is large, a condition referred to as a 'long fat pipe' (IETF, 1992).

Congestion Control

In addition to error detection and flow control, data network transport layers, in particular the Transport Control Protocol (TCP), also implement *congestion control*. The behaviours of these congestion control algorithms (which typically throttle back throughput) can further compound the problem of the long fat pipe.

6.4.2.4   Performance-Enhancing Proxies

A common approach to accelerating packet data performance when using flow control over long fat pipe satellite links is to utilize Performance-Enhancing Proxies (PEPs) over the satellite portion of a data network – with one PEP at each end of the link, as illustrated in Figure 6.17. In this example, TCP is substituted for another open standard transport layer protocol (XTP) over the satellite link.

PEPs generally utilize two main approaches:

- *Spoofing.* Spoofing involves the transmitting PEP sending fake ACKs (which would normally come from the receiving application). The transmitter PEP maintains a buffer of those packets sent since the last true ACK emanating from the receiver application. The transmitter PEP intercepts the true ACKs to determine when to empty the buffer, and must also prevent these reaching

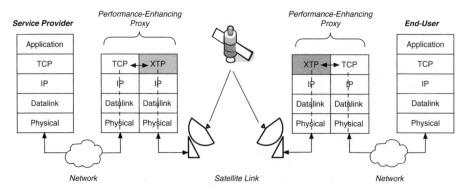

Figure 6.17   PEP satellite gateways utilizing the XTP protocol.

(and confusing) the data source. Should any data be lost in transmission, the transmitter PEP retransmits the data from its buffer. Spoofing thus largely overcomes the problem of link latency by masking it from the application, although the round-trip user application response time is unchanged.

- *Splitting*. Splitting is the separation of the network into the satellite link and the non-satellite parts at either end of the satellite link. A different flow control protocol one specifically optimized for use over satellites – is then used over the satellite link (i.e. between the PEPs). One such example is the Xpress Transport Protocol (XTP) (Caini, 2006), originally created by Silicon Graphics Inc. and now being developed by the XTP forum, and widely used in commercial products.

Revision Questions

1. Compare and contrast the different methods of supporting multiple access.
2. What is the difference between TDM and TDMA? When might TDM be used as part of a multiple-access service? What is a Unique Word used for?
3. What is intermodulation noise and when does it appear? How many third-order intermodulation products might you expect with 10 accesses? What is the principal mitigation technique against intermodulation in FDM/FDMA, and what are the consequences on system capacity?
4. Describe the different types of satellite communications payload architecture and list their advantages and disadvantages.
5. What are the advantages of using OFDM/OFDMA in satellite systems and which types of service would benefit most from its use?
6. Compare and contrast the principal types of satellite network topology. Which topology would most likley be used for a private VSAT system?
7. Describe the sliding-window flow control method for packet data networks. How does this relate to issues with the use of TCP/IP over satellite links. What techniques can be used to mitigate problems with TCP/IP over satellite?
8. Create link budgets for both the Iridium and Globalstar systems. Plot the variation in the capacity of the Globalstar constellation with the voice activation and adjacent cell interference parameter α.

References

Bond, F. and Meyer, H. (1970) Intermodulation effects in limiter amplifier repeaters. *IEEE Trans. communication Technology*, **18**(2), 127–135.

Caini, C., Firrincieli, R., Marchese, M., de Cola, T., Luglio, M., Roseti, C. *et al*. (2006) Transport layer protocols and architectures for satellite networks. *International Journal of Satellite Communications and Networking*, **25**, 1–26.

Chang, D.D. and de Weck, O.L. (2005) Basic capacity calculation methods and benchmarking for MF-TDMA and MF-CDMA communications satellites. *International Journal of Satellite Communications and Networking*, **23**, 153–171.

Dietrich, F.J., Metzen, P. and Monte, P. (1998) The Globalstar System. *IEEE Trans. Antennas and Propagation*, **46**(6).

Gilhousen, K.S., Jacobs, I.M., Padovani, R. and Weaver, L.A. (1990) Increased capacity using CDMA for mobile satellite communications. *IEEE Journal on Selected Areas in Communications*, **8**(4), 503–514.

Hubbel, Y.C.A. (1997) A comparison of the Iridium and AMPS systems. *IEEE Network*, March/April, 52–59.

ICAO (2006) *Technical Manual for Iridium Aeronautical Mobile Satellite (Remote) Service*, 19 May.

Mazella, M., Cohen, M., Roufetty, D., Louie, M. and Gilhousen, K.S. (1992) Multiple accesss techniques and spectrum utilization of the Globalstar mobile satellite system. 18th International Symposium on *Space Technology and Science*, 18th, Kagoshima, Japan, 17–22 May, Vols 1 and 2, A95-82299, pp. 1537–1544.

Nera (2002) *Digital Video Broadcasting, Return Channel via Satellite (DVB-RCS) Background Book*. Available: http://www.dvb.org/documents/white-papers/RCS-backgroundbook.pdf [accessed February 2010].

IETF (1981) RFC793: Transmission control protocol DARPA Internet program protocol specification.

IETF (1988) RFC1072: TCP extensions for long-delay paths.

Viterbi, A. (1985) When not to spread spectrum – a sequel. *IEEE Communications Magazine*, **23**(April), 12–17.

Zhang, Y. and Zhu, X. (2006) Capacity estimation for an MF-CDMA cellular model based on LEO satellite constellation. International Conference on *Wireless Communications, Networking and Mobile Computing, WiCOM 2006*, 22–24 September, Wuhan, China, pp. 1–4.

7

Doppler and Pseudorange (Navigation)

7.1 Introduction

Previous chapters have addressed the transmission of *digital information*. However, radio signals may also be used for navigation or position location (for example, the location of an emergency beacon in a distress system), where the information broadcast is of secondary importance to the user. In this chapter we discuss the Doppler shift due to the relative motion of the satellite and the user terminal, and go on to show how this shift can be used to *locate* a transmitter beacon. Also in this chapter we discuss the concepts behind *ranging* from satellites – that is, measuring the user's range from various satellites. If the positions of the satellites are accurately known, then the users' position may be estimated, and this is the basis of various global satellite navigation systems.

Satellite range is normally estimated using the *pseudorange* – a distance derived from the time difference between the transmission time and the time of reception of a known signal (i.e. the time of flight), assuming a given velocity (essentially, the speed of light). Pseudorange measurements form the basis of measurements in the popular Global Positioning System (GPS).

Accurate pseudorange measurements require signals with rapid transitions and the most common approach is to employ Direct-Sequence Spread Spectrum (DSSS) – usually in the form of Code Division Multiple Access (CDMA) to allow ranging signals from different satellites to share the same spectrum (each signal being distinguished by a unique spreading code). In this chapter we discuss the basic concepts behind code-phase (time) measurements and discuss how multiple pseudorange measurements, together with knowledge of the satellite positions, may be used to establish an estimate of the user's position. Of significant importance in any positioning system is *accuracy*, and those factors that impact on accuracy are explored in this chapter and in Chapter 13.

Accuracy of code-phase pseudorange measurements can be improved by the use of *differencing* methods, where the pseudorange measurements (and their errors) are broadcast from fixed sites (whose positions are accurately determined) to nearby pseudorange receivers, which can then apply these corrections to their own measurements. If the error sources are common to both, then the user position accuracy is enhanced. Finally, the ultimate in positional accuracy is to be obtained using carrier phase measurements as opposed to code-phase measurements. Unfortunately, the measurement of carrier phase is not without problems, and these are briefly outlined.

Satellite Systems for Personal Applications: Concepts and Technology   Madhavendra Richharia and Leslie David Westbrook
© 2010 John Wiley & Sons, Ltd

7.2 Doppler

7.2.1 Doppler Shift

The frequency of the received signal (be it at the satellite or at the user's terminal) will be altered by relative motion between emitter and receiver owing to the well-known Doppler effect. The Doppler-shifted frequency increases when the emitter and receiver are moving towards each other and decreases as they move away. The amount of Doppler frequency shift Δf at the receiver is related to the transmitter signal frequency f via

$$\frac{\Delta f}{f} = \frac{v_{\text{rel}}}{c} \tag{7.1}$$

where v_{rel} is the relative velocity between transmitter and receiver and c is the speed of light. The relative velocity is equal to the rate of change in satellite *range*. Hence, the Doppler shift may be expressed as

$$\Delta f = -\frac{1}{\lambda}\frac{\partial r}{\partial t} \tag{7.2}$$

where r is the satellite range and $\lambda = c/f$ is the wavelength of the transmitted signal.

In vector notation, the range and its rate of change are

$$r = |\mathbf{s} - \mathbf{p}| \tag{7.3}$$

$$\frac{\partial r}{\partial t} = (\dot{\mathbf{s}} - \dot{\mathbf{p}}) \cdot \mathbf{u} \tag{7.4}$$

where $\mathbf{s}$ is the satellite position vector (in some reference frame), $\dot{\mathbf{s}}$ is its velocity vector, $\mathbf{p}$ is the user's position vector, $\dot{\mathbf{p}}$ is the user's velocity vector and $\mathbf{u}$ is the unit vector describing the direction of the satellite from the user

$$\mathbf{u} = \frac{(\mathbf{s} - \mathbf{p})}{|\mathbf{s} - \mathbf{p}|} \tag{7.5}$$

For a geostationary satellite, the dominant Doppler contribution arises from user terminal motion, with a small contribution from any residual satellite motion. For non-geostationary satellites, the satellite motion usually dominates. In the case where a signal is passed between two (different) user terminals via a satellite, there may be Doppler shift on both the uplink (the link to the satellite) and the downlink (the link from the satellite), and therefore the total Doppler experienced at the terminals will be the sum of these contributions.

7.2.2 Position Location Using the Doppler Shift

Shortly after the launch of the very first satellite, Sputnik 1, researchers at John Hopkins University were able to determine its orbit simply by measuring the Doppler shift of its transmissions. It was soon realized that the Doppler effect could be used to determine a user's position if the satellite orbit was accurately known. This discovery evolved into the US Navy's TRANSIT satellite navigation system, which was in active use between 1964 and 1996. Position location using Doppler continues to be used today – for example, by the Argos and COSAT/SARSAT systems – and in this section we discuss the principles behind these systems.

There are two types of Doppler navigation/location system:

- *Doppler navigation systems.* One or more satellites broadcast continuously, and users on the Earth use the Doppler shift (and satellite orbit) to determine their own position. This was the basis of the TRANSIT system which predated GPS.

- *Doppler position location systems.* The user's equipment broadcasts *intermittently* (usually periodically) to one or more satellites, which measure the Doppler shift, and this information is used (at some central location) to determine the user's position. This is the basis of the Argos and COSAT/SARSAT systems.

The principal difference between these systems from a technical perspective is that, in the case of the former, the broadcast phase is continuous and allows optimum processing of the Doppler shift by counting phase transitions, while, in the case of the latter, a position fix must be effected from a sequence of transmission *bursts*, and as a result there is no phase continuity between bursts. We focus here on processing of the intermittent broadcasting system.

Figure 7.1 illustrates the Doppler shift observed for low and medium Earth orbit satellites. The general characteristic is observed to be 'S'-shaped, with the Doppler changing from positive to negative at the point where the satellite is closest to the user.

7.2.2.1 Solution for Periodic Transmissions

Graphical Method
Consider the case of a periodic broadcast between an unknown user location and a satellite orbiting in a near-polar orbit. We observe that the amount of Doppler shift at a given time is directly related to the apex angle of a hypothetical cone projected from the satellite's position and passing through the user's location, as illustrated in Figure 7.2 (left). The cosine of the apex half-angle α is related to the Doppler shift via

$$\cos(\alpha) = \frac{v_{rel}}{v} = \frac{\lambda}{v}\Delta f \tag{7.6}$$

where v is the orbital velocity of the satellite along its track.

The intersection of this hypothetical cone and the Earth's surface (with local surface altitude being obtained from digital elevation data) should yield the user's position, in principle. In fact, a better estimate of the user's location is obtained from the intersection of two such cones, obtained from two widely spaced Doppler measurements, as this reduces the bias caused by any uncertainty in the

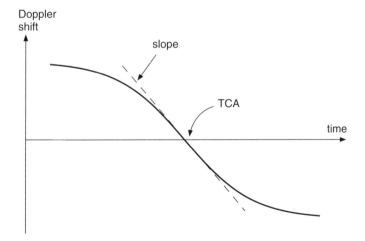

Figure 7.1 Example Doppler shift, illustrating the TCA and slope.

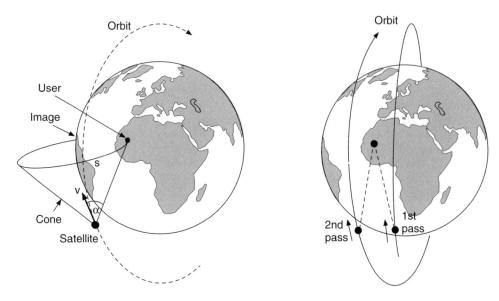

Figure 7.2 Intersection of a hypothetical cone and the Earth's surface (left) and variation in the user's relative position with each pass in a Doppler position location system (Earth-fixed view).

actual transmitted frequency (wavelength) (Argos Online, 2010). It only remains to eliminate the ambiguity as to which side of the ground track the user is located, and solutions from two successive orbits (each with different ground tracks) are generally required to eliminate this ambiguity, as illustrated in Figure 7.2 (right).

Doppler Slope Method

There is considerable redundancy in the Doppler curve data, and the user's position may be determined using a number of different methods. With reference to Figure 7.1, the observed Doppler shift exhibits a point of inflection at the so-called Time of Closest Approach (TCA). The satellite position at this time effectively gives the position of the user as measured in the direction parallel to the satellite track, while the slope of the Doppler curve at this point can be used to give us the minimum range to the user (at TCA) *across* the satellite track.

In the vicinity of TCA it is generally acceptable to analyse a rectilinear approximation, where the Earth is assumed to be flat and the satellite to be moving in a straight line, as illustrated in Figure 7.3 (Chung and Carter, 1987; Levenon and Ben-Zaken, 1985). Under this approximation, the range and its first derivative with respect to time are given by (neglecting user motion)

$$r \simeq \sqrt{x_o^2 + y^2 + h_s^2} \tag{7.7}$$

$$\frac{\partial r}{\partial t} \simeq \frac{vy}{r} \tag{7.8}$$

where x_o is the user position offset at TCA across the satellite track, and h_s is the satellite altitude at TCA.

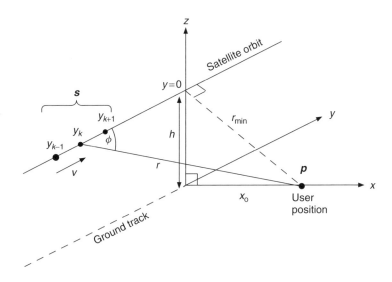

Figure 7.3   Doppler geometry near to TCA (rectilinear approximation).

It is straightforward to show that the minimum range r_{min} is proportional to the slope of the Doppler curve at TCA (i.e. at $y \to 0$)

$$r_{min} \to \ -\frac{v^2}{\lambda \left[\frac{\partial(\Delta f)}{\partial t} \right]_{y=0}} \tag{7.9}$$

from which the offset in user position x_o may be determined using equation (7.7).

General Solution

As the user's position may be determined from all parts of the Doppler curve, it stands to reason that, the more measurement data used, the greater is the final positional accuracy. Because we may not know the exact transmitted frequency, we also need to solve for this as well. It turns out that, by accurately modelling the rotation of the Earth in a single satellite orbit, it is possible to remove the positional ambiguity discussed earlier, and a first position fix may be effected in a single orbital pass (with subsequent passes improving the accuracy and reliability of the first fix).

Consider a generic system where the frequency is measured in equispaced bursts over the entire period of visibility of the satellite (i.e. over the whole Doppler curve). The measured frequency for the kth burst measurement may be expressed as

$$f_k = \frac{1}{\lambda} \left(c - \frac{\partial r}{\partial t} \right) + \varepsilon_k \tag{7.10}$$

where a term ε_k has been introduced to describe any unmodelled errors that affect the position accuarcy.

In the general case, we have four unknowns: the user's position coordinates (in an appropriate reference frame – or some parametric representation thereof (Levanon and Ben-Zaken, 1985)) plus the unknown transmitter wavelength λ. If we may continue to assume that the user is located on the Earth's surface, then the number of unknowns is reduced to three. As we have more measurements than unknowns, the method of nonlinear least squares, described in Section 7.3.2.1, is typically used to obtain the user's position.

7.3 Pseudoranging

The Doppler-based TRANSIT system was ultimately replaced by the more accurate Global Positioning System (GPS) which relies on pseudoranging and a modified version of trilateralization (position estimation using distances from three known points). The user's position is estimated by first estimating the range (distance) to a number of satellites whose instantaneous position is known to a high accuracy (as illustrated in Figure 7.4).

7.3.1 Pseudorange

As it is not possible to measure directly the satellite range, it is usual to measure the pseudorange, defined by

$$\text{Pseudorange} = \text{time difference} \times \text{speed of light}$$

where the time difference in question is that between the time a radio signal was received at the user's position and the time it was transmitted from a satellite with known position (i.e. the time of flight).

Consider the situation where the pseudorange is determined for a number of satellites, each broadcasting a unique signal. The pseudorange ρ_k for the kth navigation satellite is the difference between the time t_R at which a navigation signal was received by the user's receiver and the time $t_{T,k}$ it was transmitted *times* the speed of light

$$\rho_k \equiv c\left(t_R - t_{T,k}\right) \tag{7.11}$$

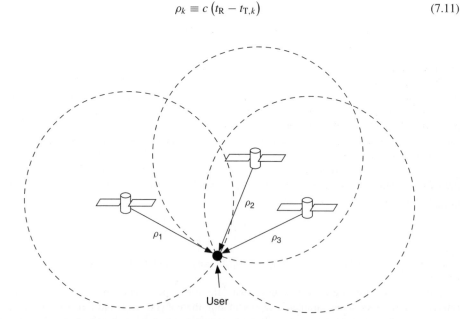

Figure 7.4 General principle of pseudoranging trilateralization.

In principle, if both the receiver and transmitter times were known very precisely, then the user's position could be determined using just three range measurements. However, the accuracy of the user's receiver clock is generally inadequate for an accurate position fix, and consequently, we must also determine the receiver clock offset from true system time. This requires at least four pseudorange measurements (from four different satellites).

7.3.2 Position Determination Using Ranging

7.3.2.1 Estimation of Receiver Position and Clock Offset

Let us consider how the user's position is determined given pseudorange measurements from at least four satellites.

Let the user's true position in Earth-Centred Earth-Fixed (ECEF) coordinates be given by the vector $\mathbf{p} = \{x_p, y_p, z_p\}$, and the position of the kth navigation satellite (in the same coordinate system) be $\mathbf{s}_k$. It is important to note that the satellite position is the position at the time the ranging signal was transmitted, because, by the time the signal is received, the satellite may have moved by several tens of metres.

The *true* range to the user's position from the kth satellite is given (in vector notation) by

$$r_k = |\mathbf{s}_k - \mathbf{p}| \tag{7.12}$$

The measured pseudorange ρ_k is derived from the apparent reception time and hence comprises the true range plus a contribution due to the (unknown) user receiver clock offset τ_R from navigation system time plus a residual measurement error ε_k (due to receiver noise, satellite clock errors, etc.) – assuming the receiver has already incorporated any pseudorange corrections.

$$\rho_k = r_k + c\,\tau_R + \varepsilon_k \tag{7.13}$$

where c is the speed of light.

As the user's position has three vector components and we also have to determine the receiver clock offset, we have four unknowns (i.e. $\{x_p, y_p, z_p\}$ and τ_R). We therefore require at least four simultaneous pseudorange equations (and hence four pseudorange measurements) uniquely to determine all four values. Typically, however, more than four satellites will be visible at a given time, and, intuitively, one expects that the highest accuracy will be achieved by utilizing as many of the visible satellites as possible, using an appropriate method. Indeed, many modern receivers have more than four channels, with 12 channels being fairly common, and typically use the method of nonlinear least squares.

Weighted Nonlinear Least-Squares Estimation Method

In the method of nonlinear least squares we seek to find the set of values for a set of unknowns that minimizes the (weighted) sum of the residual errors squared. For convenience of notation, let $\mathbf{x}$ be a vector containing the unknowns, and let $\mathbf{f}$ be a vector containing the residual (measurement) errors ε_k. The weighted least-squares algorithm seeks to minimize the function

$$F = \frac{1}{2} \sum_{k=1}^{N} w_{kk}\, \varepsilon_k^2 \tag{7.14}$$

$$= \frac{1}{2} \left(\mathbf{f}^{\mathrm{T}} \mathbf{W} \mathbf{f} \right) \tag{7.15}$$

where w_{kk} are positive weights given to the error ε_k for the kth measurement, the allotted weight reflecting the confidence in that measurement. $\mathbf{W}$ is a diagonal matrix[1] comprising the weights w_{kk}, while $\mathbf{f}^T$ denotes the transpose[2] of $\mathbf{f}$.

Expanding the error function $\mathbf{f}$ about $\mathbf{x}$ as a first-order Taylor series gives the error at some vector increment $\Delta\mathbf{x}$

$$\mathbf{f}(\mathbf{x} + \Delta\mathbf{x}) \approx \mathbf{f}(\mathbf{x}) + \mathbf{J}\Delta\mathbf{x} \tag{7.16}$$

where $\mathbf{J}$ is the matrix of first-order partial derivatives of the error vector $\mathbf{f}$ with respect to the vector components of $\mathbf{x}$ (known as the Jacobian matrix).

It can be shown that, given some initial estimate for the vector of unknowns, the offset $\Delta\mathbf{x}$ that results in the least-squared error is given by the expression

$$\Delta\mathbf{x} = -\left[\mathbf{J}^T \mathbf{W} \mathbf{J}\right]^{-1} \left(\mathbf{J}^T \mathbf{W}\right) \mathbf{f} \tag{7.17}$$

If the underlying equations are linear, then a single iteration is sufficient. If, however, the equations are nonlinear (as is the case for pseudorange navigation), then several iterations may be necessary with the Jacobian revaluated at each step, and typically a (weighted) Gauss–Newton method (a variation of Newton–Raphson) is used.

Weighted Least Squares for Pseudorange Measurements

For our pseudorange measurements, we have the following definitions for the vector of unknowns and the error vector

$$\mathbf{x} \rightarrow \begin{bmatrix} x_p \\ y_p \\ z_p \\ -c\,\tau_R \end{bmatrix}; \quad \mathbf{f}(\mathbf{x}) \rightarrow \begin{bmatrix} \varepsilon_1 \\ \varepsilon_2 \\ \cdot \\ \cdot \\ \cdot \\ \varepsilon_N \end{bmatrix} \tag{7.18}$$

In this case, it is interesting to note that the Jacobian $\mathbf{J}$ used to obtain the solution is given by

$$\mathbf{J} = \begin{bmatrix} \partial f_1/\partial x_1 & \partial f_1/\partial x_2 & \ldots & \partial f_1/\partial x_4 \\ \partial f_2/\partial x_1 & \partial f_2/\partial x_2 & \ldots & \partial f_2/\partial x_4 \\ \ldots & \ldots & \ldots & \ldots \\ \partial f_n/\partial x_1 & \partial f_n/\partial x_2 & \ldots & \partial f_n/\partial x_4 \end{bmatrix} \rightarrow \begin{bmatrix} u_{x,1} & u_{y,1} & u_{z,1} & 1 \\ u_{x,2} & u_{y,2} & u_{z,2} & 1 \\ \ldots & \ldots & \ldots & 1 \\ u_{x,N} & u_{y,N} & u_{z,N} & 1 \end{bmatrix} \tag{7.19}$$

where $u_{x,i}$, $u_{y,i}$ and $u_{z,i}$ are just the direction cosines – the elements of the unit (direction) vector $\mathbf{u}_k$ from the user to the kth satellite.

7.3.2.2 PN Code Pseudorange

How does the receiver estimate the time since the signal was transmitted? Typically, each satellite transmits a Pseudo Random Number (PRN) code (a long binary number sequence that has the appearance of being random), which begins at some defined epoch. The particular epoch is transmitted (along with other data) in a so-called *navigation message*.

---

[1] A diagonal matrix is a square matrix containing non-zero elements only on the diagonal (i.e. when row and column index are the same).

[2] The transpose of a vector or matrix is simply formed by the transposition of row elements to column elements, and vice versa.

In the case of GPS (and also Galileo), all satellites transmit on the same frequency, but each transmits a different PRN code (CDMA), and the receiver selects the appropriate signal by correlating the received waveform with the relevant code. The use of Direct-Sequence Spread Spectrum (DSSS) allows multiple satellites to share the same frequency and affords processing gain, which allows the navigation signals to be extracted from noise and narrowband interference. In the case of the Russian GLONASS system, each satellite transmits its PRN code on a unique frequency (Kaplan and Hegarty, 2006).

Autocorrelation

In order to determine the PRN code delay relative to the relevant epoch, the receiver generates a delayed replica of the PRN and forms the correlation of the replica and the received signal (in effect, an autocorrelation). Autocorrelation involves computing the overlap of a delayed version of the signal with itself.

The autocorrelaton function $R(\tau)$ for binary phase shift keyed modulation with rectangular pulses (BPSK-R) of amplitude A and chip period T_c, assuming a perfectly random (infinite) PRN code, is given by

$$R(\tau)\,[\text{BPSK-R}] = \begin{cases} A^2\left(1 - \frac{|\tau|}{T_c}\right) & |\tau| \le T_c \\ 0 & |\tau| > T_c \end{cases} \tag{7.20}$$

where t is the time offset. It is apparent that for correlator delays greater than one chip, the signal and its delayed form are uncorrelated – as the adjacent bit is assumed to be random – and the correlator output is zero. For delays less than one chip, the autocorrelation varies linearly with delay – as the area (integral) of the overlap between the pulses varies (resulting in a equilateral triangle with a total width of two chips). Maximum autocorrelation occurs for zero code offset delay, thereby allowing estimation of the code 'phase' (timing).

Figure 7.5 shows the autocorrelation function for a finite, Maximal-Length Sequence (MLS). The significant differences between this and the autocorrelation function for the ideal random sequence are that there is an autocorrelation maximum every $N_c T_c$ (where N_c is the code length) and the autocorrelation floor is negative valued. Such MLS sequences provide good autocorrelation but have relatively poor cross-correlation characteristics for multiple sequences sharing the same CDMA spectrum. For this reason, the GPS Coarse/Acquisition (C/A) signal uses so-called *Gold* codes, formed from two maximal-length sequences.

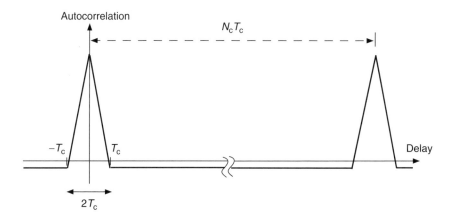

Figure 7.5   Autocorrelation of a finite-length maximal-length PN sequence.

Autocorrelation for More Complex Signals

We have seen the autocorrelation functions for BPSK-R (antipodal) signals. In the pursuit of greater accuracy and flexibility, there is a trend towards the use of more complex modulation for navigation signals, with the future GPS and Galileo systems opting for variants of Binary Offset Carrier (BOC) modulation. BOC is based on applying a square-wave subcarrier modulation to a BPSK-R signal. BOC modulation is defined by two parameters, BOC (n, m), where n is the subcarrier frequency and m is the chip rate (usually both given in multiples of 1.023 MHz).

The result of using BOC is that the peak spectral energy is offset from the carrier frequency – by the subcarrier frequency – and the increase in high-frequency components helps to sharpen the autocorrelation offset resolution. The narrower peaks of the BOC autocorrelation function provide for reduced tracking ambiguity and reduced sensitivity to noise; however, the multiple peaks make code acquisition and tracking somewhat more difficult.

In 2007, The United States and the European Union announced their intention to work towards a common civilian navigation based on Multiplexed BOC (MBOC). The definition of MBOC is in the frequency domain and comprises predominantly BOC(1,1) with a fraction of BOC(6,1) – to increase high (offset) frequency components – according to

$$\text{MBOC}(f) = \frac{10}{11}\text{BOC}(1, 1) + \frac{1}{11}\text{BOC}(6, 1) \tag{7.21}$$

The normalized power spectral densities of MBOC and BOC(1,1) are illustrated in Figure 7.6 (Dovis *et al.*). The spectrum peaks are offset from the carrier centre frequency by the subcarrier modulation frequency. The multiple-peaked autocorrelation function for the Galileo form of MBOC is shown in Figure 7.7 (Dovis *et al.*, 2008). The autocorrelation of BOC(1,1) is also shown for comparison (straight-line function). It will be noted that, while the widths of the two autocorrelation functions are the same at the half-power points, the Galileo MBOC autocorrelation width at the peak is narrower than for BOC (which is itself narrower than for BPSK-R), permitting improved accuracy in code delay estimation.

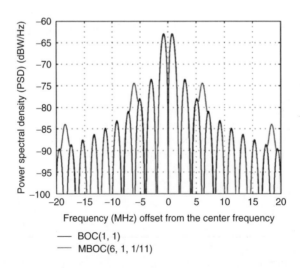

Figure 7.6 Unit power spectral densities comparison of BOC(1,1) and MBOC(6,1,1/11) (Dovis *et al.*, 2008). The MBOC spectra display increased power at around 6 and 18 MHz offset. Reproduced courtesy of *International Journal of Navigation and Observation*.

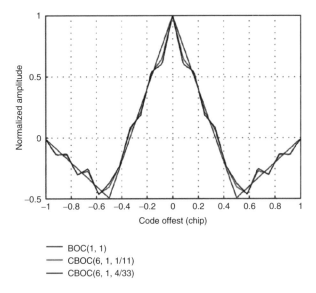

Normalized amplitude

Code offest (chip)

— BOC(1, 1)
— CBOC(6, 1, 1/11)
— CBOC(6, 1, 4/33)

Figure 7.7 Normalized autocorrelation for the Galileo MBOC (CBOC) signal and comparison with BOC (1,1) (Dovis *et al.*, 2008). The CBOC autocorrelation peaks are sharper. Reproduced courtesy of *International Journal of Navigation and Observation*.

Acquisition

During start-up, and when a new satellite becomes visible, the receiver must estimate the amount of Doppler shift caused by the satellite motion prior to obtaining carrier phase lock, and also determine the PRN coarse code delay prior to obtaining code-phase (tracking) lock.

The receiver achieves these goals by searching for the signal carrier frequency and code delay. The combination of one Doppler *bin* and one code bin is known as a *cell*, and of the order of 40 000 cells must typically be searched during signal acquisition, with the highest 'bin' providing the best estimate of Doppler frequency offset and code delay (Krumvieda *et al.*, 2001), after which code and phase/frequency tracking loops are activated. This search is illustrated by the two-dimensional surface in Figure 7.8. The need to search for this pair of parameters results in the familiar delay when a GPS receiver is switched on, and, the longer the PRN code length, the more computationally intensive this search becomes. Speeding up this acquisition phase is an area of intense study and typically involves specialized hardware with multiple parallel correlators.

In general, code acquisition has the highest received power requirement, and therefore ultimately limits the sensitivity (and thereby usefulness) of a pseudoranging navigation receiver. The required carrier-to-noise density C/N_0 for satisfactory acquisition in a satellite navigation receiver will depend on a number of parameters – most notably on the receiver integration time for each search cell, which, in turn, determines the time needed to complete satellite code acquisition (Schmidt, 2004). In particular, without special provisions, the coherent integration of the autocorrelaton signal is generally limited to times shorter than the code period. As a result, basic GPS receivers are typically capable of acquiring satellite signals with an input carrier-to-noise density ratio (C/N_0) of ~34 dBHz (Soliman, Glazko and Agashe, 1999). For those CDMA navigation signals containing data (navigation message data), the coherent integration period is also generally constrained to less than the length of one data bit, in order to avoid a data bit boundary – although modern high-sensitivity navigation receivers use a number of advanced techniques to extend the non-coherent integration time, to achieve receiver sensitivities such as those indicated in Table 7.1 (AMTEL, 2010).

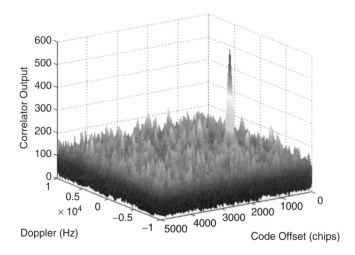

Figure 7.8   Signal acquisition search. Reproduced by permission of Data Fusion.

Table 7.1   Example of high-performance GPS C/A
code pseudorange receiver sensitivities (AMTEL, 2010)

Phase	Power	
	dBm	dBW
Code tracking	−158.5	−188.5
Code acquisition	−142	−172

Code Tracking

Once acquired, a delay locked loop (DLL) is used to track the delay for the PRN code so as
to maintain maximum autocorrelation (the receiver code aligned with the received code). If the
navigation signal contains data (the navigation message data), then a non-coherent tracking loop
is typically used. DLLs utilize *early* and *late* detector samples just before and just after the actual
sample. The difference between these sample amplitudes forms an error signal that is used to speed
up or slow the tracking oscillator (as the error is zero at code optimum alignment). Typically, the
code position can be aligned to about one-hundredth of a chip.

7.3.2.3   Carrier Phase Pseudorange

Pseudorange systems use autocorrelation to determine the code phase in order to estimate the time
difference between transmit and receive times. How accurately can one estimate the code offset
(since this affects pseudorange accuracy)?

It was stated earlier in this section that, as a rule of thumb, modern PRN code pseudoranging receivers can typically achieve about one-hundredth chip accuracy in estimating PRN
code delay (and hence time of flight). In the case of the GPS C/A code, the chip rate is
1.023 MHz, giving a chip period of about 1 ns, or an ultimate accuracy (ignoring all error other
contributions) of about 0.3 m, and this level of accuracy is generally adequate for a wide range of
user applications.

Certain applications – such as surveying – demand higher levels of accuracy, and in these specialized applications the carrier phase is also used to determine pseudorange. Carrier phase measurements can typically be made to about one-hundredth of a cycle. In the case of GPS, the civilian L1 frequency is 1575.42 MHz, corresponding to a wavelength of 190 mm. Therefore, in principle, pseudorange measurements based on carrier phase measurements can be estimated to approximately 2 mm – about 1500 times more accurate than for C/A code pseudorange (assuming one can neglect other error sources).

Unfortunately, there are significant difficulties in using carrier phase tracking. While it is possible to determine the carrier phase with great accuracy, the cyclic nature of the phase means that there is uncertainty over the exact number of cycles (and therefore wavelengths) in the path between satellite and receiver. In terms of the carrier phase pseudorange, this ambiguity may be expressed as an additional error term

$$\epsilon_k = N_k \lambda + \cdots \tag{7.22}$$

where λ is the free-space wavelength and N_k is the integer ambiguity in the phase measurement to the kth satellite. General methods of estimating carrier phase ambiguity are currently an area of significant research.

Nevertheless, if one is interested in the accurate measurement of *relative* position, then one may use the differencing technique (with two receivers – one fixed and one mobile), as one can more easily keep track of the smaller carrier phase ambiguity between two closely spaced receivers. Real-Time Kinematic (RTK) surveying systems perform these corrections in real time.

7.3.3   Accuracy and Dilution of Precision

7.3.3.1   Pseudorange Errors

There are various errors associated with range measurement, and hence the raw measurement is in effect a 'pseudorange' *estimate*. Sources of error in pseudorange measurements can be assigned to one of four types:

- errors in broadcast satellite ephemeris;
- errors in satellite clock timing;
- errors in estimating atmospheric propagation delay (particularly, ionospheric delay in single-channel receivers);
- Multipath effects.

These errors and their mitigation are discussed further in Chapter 13. Significantly, the severity of these errors is magnified according to the particular user–satellite *geometry* – the effect is known as Dilution of Precision (DOP).

User Equivalent Range Error

If we can assume that the pseudorange measurement errors are uncorrelated Gaussian random processes with zero mean and standard deviation[3] σ, then it is known that the optimum least squares weight matrix is proportional to the inverse of the measurement variance co-variance matrix[4] $\mathbf{C}_\rho$. We thus assume the covariance matrix is given by $\mathbf{C}_\rho = \sigma^2 \mathbf{W}^{-1}$ (note: for the case of

---

[3] The statistical variance of a variable is the mean of the squared excursion from the mean. The square root of the variance is called the standard (or RMS) deviation.

[4] Covariance is a measure of how much two statistical variables change together. For vectors of statistical variables the covariance takes the form of a square matrix – the variance covariance matrix.

unweighted least squares, the weight matrix becomes the identity matrix: $\mathbf{W} \rightarrow \mathbf{I}$). The RMS error in pseudo-range σ is usually known as the User Equivalent Range Error (UERE).

Using the law of the propagation of errors, the covariance matrix $\mathbf{C_x}$ for our vector of unknowns $\mathbf{x}$ in ECEF is found to be

$$\mathbf{C_x} \rightarrow \sigma^2 \left(\mathbf{J}^\mathrm{T} \mathbf{W} \mathbf{J} \right)^{-1} \tag{7.23}$$

We are usually more interested in the positioning errors in a local coordinate system, since this makes it easier to separate horizontal and vertical positioning accuracies. We saw in Chapter 2 that we may transform our ECEF coordinates to the local tropocentric horizon coordinates (or similar) by using an appropriate (compound) transform Matrix $\mathbf{M}$ – which we can further augment to allow for the extra receiver time offset variable. Again, applying the law of propagation of errors, the co-variance matrix in the user's local coordinate frame is

$$\mathbf{C}_\mathrm{local} = \mathbf{M} \mathbf{C_x} \mathbf{M}^\mathrm{T} \tag{7.24}$$

$$\equiv \sigma^2 \, \mathbf{D} \tag{7.25}$$

where the resultant matrix $\mathbf{D}$ describes the Dilution Of Precision (DOP) due to satellite geometry, as compared with the UERE.

7.3.3.2 Dilution of Precision

We are typically interested in the horizontal and vertical positioning accuracies. We thus define:

- *Horizontal Dilution Of Precision (HDOP)*. This describes the dilution of precision of the two-dimensional horizontal accuracy.
- *Vertical Dilution Of Precision (VDOP)*. This describes the dilution of precision of the one-dimensional vertical accuracy.
- *Positional Dilution Of Precision (PDOP)*. This describes the overall dilution of precision of the three-dimensional position accuracy. The PDOP is equal to the sum of the squares of the HDOP and VDOP.
- *Time Dilution Of Precision (TDOP)*. This describes the overall dilution of precision of the time accuracy.
- *Geometric Dilution Of Precision (GDOP)*. This describes the overall dilution of precision for the complete solution (three-dimensional position and time accuracy). The GDOP is equal to the sum of the squares of the PDOP and TDOP.

The relevant DOP quantities may be determined from the elements of matrix $\mathbf{D}$:

$$\mathrm{HDOP} = \sqrt{D_{11} + D_{22}} \tag{7.26}$$

$$\mathrm{VDOP} = \sqrt{D_{33}} \tag{7.27}$$

$$\mathrm{TDOP} = \sqrt{D_{44}} \tag{7.28}$$

$$\mathrm{PDOP} = \sqrt{\mathrm{HDOP}^2 + \mathrm{VDOP}^2} \tag{7.29}$$

$$\mathrm{GDOP} = \sqrt{\mathrm{PDOP}^2 + \mathrm{TDOP}^2} \tag{7.30}$$

The significance of the various DOP values may be expressed as follows. A standard deviation of 1 m in pseudorange observation will result in an RMS error of HDOP in the horizontal position, VDOP in the vertical position and 1/c times TDOP in the receiver time offset.

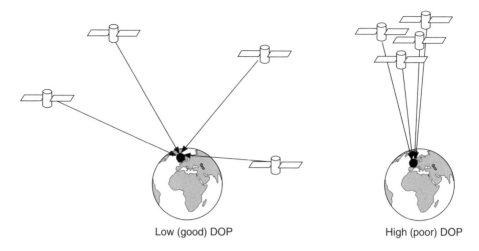

Low (good) DOP   High (poor) DOP

Figure 7.9 Illustration of good and bad geometric dilution of precision.

The concept of the geometric dilution of precision is more clearly illustrated in Figure 7.9. When the visible satellites are widely spread in the sky, pseudorange error is minimized; however, when visible satellites are closely spaced, the uncertainty is increased.

PDOP Mask
Given the effect of dilution of precision on position accuracy, a navigation receiver will typically allow the user to define an PDOP mask – that is, a value of PDOP above which the position is not estimated owing to the high dilution of accuracy.

Elevation Mask
A navigation receiver will also typically allow the user to define an elevation mask – effectively a minimum satellite elevation. Any satellites at lower elevations will not be used in the position estimation, as it is known that multipath effects can degrade the positioning accuracy with satellites whose elevation is less than 15° or so. The choice of elevation mask is a compromise, as setting this value too low will increase errors due to multipath, while setting it too high will restrict the number of 'visible' satellites and hence increase the DOP value.

7.3.4   Differential Positioning

We conclude our discussion of navigation using pseudorange measurements by noting that pseudorange errors may be reduced (and accuracy improved) by combining measurements of the same satellite from two receivers, using *differencing*, as illustrated in Figure 7.10.

Differencing combines the pseudoranges of each satellite measured at two spatially separated receivers, that are in communications range. If the error contributions are sufficiently similar for both receivers, they may be substantially reduced by differencing. This technique may be used to eliminate satellite clock bias, and, depending on the separation of the receivers, can significantly reduce the impact of ionospheric and tropospheric delay errors for single-frequency receivers. Indeed, it is estimated that, using single differencing, ionospheric and tropospheric delay errors are negligible for GPS receivers within 30 km.

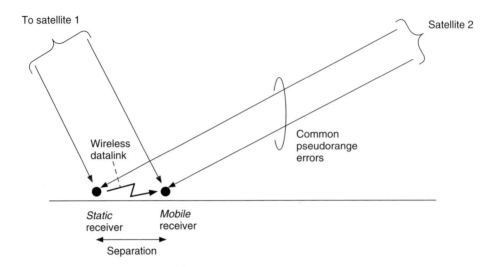

Figure 7.10   Pseudorange differencing.

The use of differencing where the location of one receiver is fixed and very precisely known also allows the elimination of satellite ephemeris errors, and forms the basis of various differential GPS systems, which broadcast corrections to GPS to nearby differential GPS-capable receivers. Prior to 2000, the use of differential GPS was particularly effective in mitigating the deliberate degradation of the satellite clock in GPS – known as Selective Availability (SA).

The errors terms for each pseudorange measurement will comprise a contribution $\Delta t_{s,k}$ due to the satellite clock, plus ionospheric and tropospheric delay, plus satellite ephemeris $\varepsilon_{eph,k}$, etc. Hence, for the fixed site A, the difference between the measured pseudorange and the true range for the kth satellite is thus

$$\varepsilon_k(A) = c\left(\Delta t_{s,k} + \Delta t_{IONO,k}(A) + \Delta t_{TROP,k}(A)\right) + \varepsilon_{eph,k}(A) + \cdots \qquad (7.31)$$

This error is broadcast to the mobile site B and subtracted from the measured pseudorange at that receiver to give

$$\rho_k(B) - \varepsilon_k(A) = r_k(B) + c\,\tau_R(B) + (\varepsilon_k(B) - \varepsilon_k(A))$$
$$\simeq r_k(B) + c\,\tau_R(B) + \cdots \qquad (7.32)$$

As the error term due to the satellite clock is the same in both measurements, the differencing removes this contribution. In addition, if the satellite paths are sufficiently similar, then the error terms due to incorrect satellite ephemeris and imperfect modelling of the ionopheric and tropospheric delays also cancel to a large part (the receiver clock error for the fixed site is assumed to be accurately known).

Revision Questions

1. What are the two main techniques used to determine the position of a user using satellites?
2. Describe how one might determine the position on the surface of the Earth of an intermittent transmitter using Doppler measurements to a near-polar satellite. What is the dominant source of error in such a system?

3. What is the difference between satellite range and pseudorange? In pseudoranging, how is the time of flight determined?
4. Why does signal acquisition in a pseudorange satellite navigation system take so much time? What signal parameters are being determined during acquisition? What limits the sensitivity?
5. What are the advantages and disadvantages of using carrier phase pseudorange over code phase?
6. What is user equivalent range error, and how does this differ from the error in local coordinates? What is meant by dilution of precision?
7. List the principal sources of error in a satellite pseudorange navigation system. Which is usually dominant? Why is it attractive to use two measurement frequencies? Show mathematically how this helps.
8. Describe the advantages of differencing for pseudoranging, and give examples of such systems from the literature.

References

Argos Online (2010) http://www.argos-system.org [accessed February 2010].

AMTEL (2010) *Measuring GPS Sensitivity*. Available: http://www.amtel.com/dyn/resources/prod_documents/doc4933.pdf [accessed February 2010].

Chung, T. and Carter, C.R. (1987) Basic concepts in the processing of SARSAT signals. *IEEE Trans. Aerospace and Electronic Systems*, **AES-23**(2), 175–198.

Dovis, F., Presti, L.L., Fantino, M., Mulassano, P. and Godet, J. (2008) Comparison between Galileo CBOC candidates and BOC(1,1) in terms of detection performance. *International Journal of Navigation and Observation*, 2008, Article ID 793868.

Kaplan, E.D. and Hegarty, C.J. (2006) *Understanding GPS: Principles and Applications*. Artech House, Norwood, MA.

Krumvieda, K., Madhani, P., Cloman, C., Olson, E., Thomas, J., Axelrad, P. and Kober, W. (2001) A complete IF software GPS receiver: a tutorial about the details. Proceedings of the 14th International Technical Meeting of the Satellite Division of the Institute of Navigation (ION GPS 2001), 11–14 September, Salt Palace Convention Centre, Salt Lake City, UT, pp. 789–829.

Langley, R.B. (1999) Dilution of precision, *GPS World*, May.

Levanon, N. and Ben-Zaken, M. (1985) Random error in ARGOS and SARSAT satellite positioing systems. *IEEE Trans. on Aerospace and Electronic Systems*, **AES-21**(6).

Minimum Operational performance standards for global positioning system/wide area augmentation system airborne equipment. Technical Report RCTA/DO-229A, 1998.

Schmid, A. and Nevbaver, A. (2004) Comparison of the sensitivity limits for GPS and Galileo receivers in multipath scenarios. Position Location and Navigation Symposium, 26–29 April, Monterey, CA, pp. 503–509.

Soliman, S., Glazko, S. and Agashe, P. (1999) GPS receiver sensitivity enhancement in wireless applications. IEEE MTT-S Symposium on *Technologies for Wireless Applications*, 21–24 February, Vancouver, Canada, pp. 181–186.

8

Compression, Speech, Audio and Video Encoding

8.1 Introduction

In previous chapters we have explored various concepts related to the transmission of *digital* information between a satellite and a user's terminal.

The Shannon channel capacity theorem relates to the efficient transmission of *information*, and in this chapter we shall expand on the definition of the amount of information in a given *message*. Another aspect of the efficient transmission of information is the removal of redundant information prior to transmission. For digital data this is known as data compression, and the subject of *lossless* versus *lossy* data compression is addressed in this chapter.

In practice, a significant proportion of the source material of interest to humans is *analogue* in nature. In particular, two-way speech communication is very important for humans, while the use of teleconferencing (two-way speech and video communications) is also increasing in popularity. In the field of broadcasting, high-fidelity (hifi) audio (e.g. music) and high-quality video are dominant media forms.

The transmission of such analogue signals clearly involves some form of transducer (for example a microphone or video camera at one end and a loudspeaker or video display at the other). Once the sound or optical signals are converted to electronic signals, they must be sampled at an appropriate rate, prior to transmission, and herein lies a problem. The raw digital bit rates (sample rate times number of quantization bits) associated with speech, audio and video are non-optimum and hence greater than we would wish for maximum efficiency of a satellite communications or broadcast service – as the cost per user of a satellite service increases with the amount of bandwidth and power used.

The answer is to exploit redundancy in the source information and limitations in our human senses in order to allow these analogue signals to be compressed without the signal degradation being noticeable to the end-user. In the case of the encoding of human speech, the answer is to use a vocoder (voice encoder), which uses a model of the human vocal tract to mimic speech electronically at the receiving end, transmitting not the sampled speech but a set of numerical parameters that tell the electronic vocoder at the receiver how to sound at a given instant. In the case of the encoding of hifi audio, the answer is to exploit the limitations of human hearing, and in particular the 'masking' of quiet sounds by louder sounds. In the case of the encoding of video, the answer is to exploit the limitations of human vision and the redundancy (similarity) between successive pictures. Many of these techniques are instances of *lossy compression* – in which some

Satellite Systems for Personal Applications: Concepts and Technology   Madhavendra Richharia and Leslie David Westbrook
© 2010 John Wiley & Sons, Ltd

of the original (sampled) information is lost. However, if done carefully, the end-user is relatively insensitive to this loss.

In this chapter, we aim to acquaint the reader with the generic techniques used for the encoding of speech, hifi audio and video. The encoding of these media types is a very dynamic area, with significant rewards in terms of efficiency for progress. Our aim here is to give a flavour, only, and the interested reader is directed to the literature for details of the latest encoding techniques.

8.1.1 Why Digital?

Why digital? After all, our personal experience of the world is essentially analogue in nature: sight, sound, etc., all of which involve (for all practical purposes) continuous variations in loudness/ intensity, pitch/colour, etc. The answer is that digitizing these analogue signal communications prior to communications (or storage) offers significant advantages, many of which are already familiar to us:

- *Robustness.* All analogue signals are degraded, to some degree, as a result of transmission, switching or storage owing to the inevitable introduction of noise and distortion in the information flow chain. Digital signals, and binary digital signals in particular, generally have substantially higher resilience to noise and distortion, and can, to a significant degree, be transmitted, switched and stored without degradation, that is, with no loss of information compared with the digital original.
- *Ease of regeneration and replication.* Digital signals may be regenerated at will (i.e. signals may be cleaned up before accumulated noise or distortion present becomes severe enough for errors to be introduced). In addition, any number of identical copies can be made, again with no loss of information.[1]
- *Ease of switching.* Use of digital signals permits any information, regardless of its source, to be switched using the same electronic circuits as those used in digital computers, permitting switching between information sources and/or destinations and thereby allowing commonality (and the efficiency that results therefrom) in their transmission and storage. Furthermore, disparate information sources may be readily multiplexed together to form new *multimedia* information streams.
- *Suitability for mathematical manipulation.* As digital signals are, in effect, just streams of numbers, they can be manipulated – using Digital Signal Processors (DSPs) – in order to improve their quality with regard to communications, switching or storage. Coding may be utilized to compress digital signals in order to reduce the time needed to transmit them over a communication system or the amount of resources needed to store them. Coding may also be used to increase their robustness by introducing redundant information that permits error checking (to identify when they have been corrupted) and even error correction (with or without the need for retransmission).

8.2 Lossless Data Compression

Error control coding is vital to the efficient transmission of information bits. However, we have not yet considered the efficiency of the raw data prior to coding, which may contain significant redundancy and thus require higher transmission rates than are strictly necessary (or longer transmission times for a given rate). In this section we look at measuring the average information content of a complete message and methods to eliminate redundant information without loss of fidelity.

---

[1] It is noted that the ease with which it is possible to replicate digital content also poses a significant challenge for the protection of copyright – for example in the distribution of copyrighted digital video and music.

8.2.1   Lossless versus Lossy Compression

Prior to transmission (or storage) of digital messages, compression is commonly used – often without the user's knowledge – to reduce the time taken to transmit them over a fixed bandwidth channel or the amount of storage space needed to archive them.

Such compression may either be *lossless* or *lossy*. Lossy compression is typically associated with encoding of analogue information, such as voice, music, still images and video, where some degree of degradation of the original (which may, for example, contain more information than the human senses can detect) is acceptable. By comparison, lossless compression exploits some redundancy in the information in a file or message and achieves compression by removing or reducing this redundancy in such a way that it can be restored with no loss of accuracy.

In the following subsections, we outline the more common forms of lossless data compression. We go on to discuss lossy compression in later sections.

8.2.2   Entropy Encoding

We require a measure of the *information* contained in a message. In Chapter 5 we cited Hartley's *logarithmic* definition of information. Shannon (1948, 1949), built on Hartley's idea for measuring information, and argued that the amount of information in a message is related to the inverse of the *probability* of that particular message being selected. Indeed, if the message selected is always the same, then there is actually no useful information transmitted. If, on the other hand, a particular message occurs very infrequently, then its value on reception is very high.

Given a message i selected from a finite set of possible messages, and having probability p_i, Shannon gave the amount of information H_i in the message as:

$$H_i = \log_m \left(\frac{1}{P_i} \right) \tag{8.1}$$

where the base m of the logarithm is chosen to be appropriate to the type of message. The most common base for digital information is base-2 (binary) and the resulting information unit is the binary information unit or *bit* – a term first coined by J. Tukey (Shannon, 1948).

Shannon further reasoned that the average information, or *entropy*, contained in a message, which may take one of M forms, is (in bits) (Shannon, 1948)

$$\overline{H} = \sum_{i=1}^{M} P_i \, \log_2 \left(\frac{1}{P_i} \right) \tag{8.2}$$

Shannon chose the word 'entropy' because of similarities with Boltzmann's statistical entropy, used in thermodynamics.

If each message possibility has equal probability (equal to $P_i \rightarrow \frac{1}{M}$), then the average information is $\overline{H} \rightarrow \log_2 (M)$ bits. As a sequence of n binary numbers – perhaps the outputs of n flip-flop circuits – can take $M \rightarrow 2^n$ different values, the average information in such a sequence, where the probability of each outcome is the same, is $\log_2 (2^n)$ or n bits. These days we are quite familiar with the concept of 'bits', and this result may seem obvious, but it is important to realize that it only holds true if the probability of transmission of each symbol is the same. We shall see in the following that, where some symbols are less likely than others, scope exists to encode the message more efficiently.

It is often the case that the probabilities of each outcome are unequal. An example might be when sending or storing English text, where it is well known that certain letters of the alphabet – such as 'e' and 'a' – are used more frequently than others. The optimal codeword length (in bits) of a message is given by equation (8.2). Shannon (1949) outlined a method of entropy encoding that he attributed to Fano of MIT. It was, however, Fano's student, David Huffman, who created an optimal method of entropy encoding.

8.2.2.1 Huffman Encoding

Huffman's encoding method measures the probability of each symbol in the message and, using this information, creates a codebook by building what we would recognize today as a binary tree, according to simple rules that take into account the probability of each symbol or combination of symbols. Each branch of the tree is then allocated a bit code [0 or 1], and the code for a given symbol is obtained by traversing the tree from its root to the leaf node for the wanted symbol.

By way of example, consider Huffman encoding of the phrase: 'every dog has its day'. For each symbol (i.e. character) in the message ('a', 'd', 'e', etc.), a leaf node is created, containing both the symbol and its probability in the message. A binary tree is then built from the bottom up by pairing the nodes with the lowest probabilities to create a new parent node with probability given by the product of its two child node probabilities. Each of the child nodes is then labelled either 0 or 1. This process is repeated until a single binary tree is obtained as shown in Figure 8.1. Once the tree is complete, symbols are allocated codewords, obtained by concatenating the binary labels given to each branch, from the root to the leaf node containing the wanted symbol.

Table 8.1 gives the probabilities for each symbol and resultant Huffman codewords for the example phrase using the binary tree in Figure 8.1. In this example, the symbol 'a' (probability 0.09) is encoded with just 3 bits (100), whereas the symbol 't' (probability 0.04) is encoded with 5 bits (10110).

A Huffman code is an example of a *prefix code* – a code in which no codeword is the prefix of another codeword (preventing ambiguity over which symbol is being sent). Huffman encoding requires evaluation of the frequency of each symbol in the message and assignment of codewords. Decoding is straightforward, although there is a small overhead associated with transmission of the decoding table itself.

8.2.2.2 Arithmetic Coding

Huffman coding is simple to implement and is optimal for the case where each symbol/character has a separate codeword – in the sense that the length of each codeword is minimal to the nearest

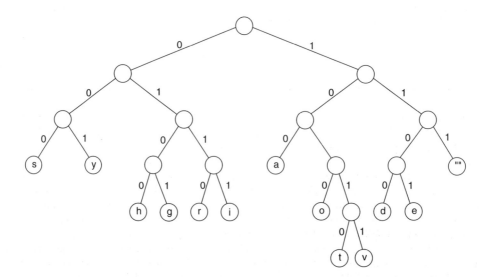

Figure 8.1 Huffman coding binary tree for the phrase *'every dog has its day'*.

Table 8.1 Huffman codes for the phrase
'every dog has its day'

Symbol	Probability	Binary code
	0.19	111
y	0.09	001
s	0.09	000
a	0.09	100
e	0.09	1101
d	0.09	1100
o	0.04	1010
g	0.04	0101
h	0.04	0100
i	0.04	0111
r	0.04	0110
v	0.04	10111
t	0.04	10110

integer bit. Another type of entropy encoding, known as *arithmetic coding*, can achieve a coding efficiency within 1 bit of the entropy of the *total* message (Howard, 1994) and is sometimes used as an alternative to Huffman coding. It does this by generating a single codeword for the complete message, rather than for individual symbols/characters.

In order to illustrate the process involved in arithmetic coding, consider the artificially simple example of encoding the words 'BILL GATES', and further suppose for convenience that these nine symbols are the only valid symbols (Nelson, 1991). The probability of each symbol in the message is 1/10, except for the letter 'L', for which the probability is 2/10. The numerical range 0 to 1 is segmented with an interval allotted to each symbol that reflects its probability in the message, as illustrated in Table 8.2. The first symbol in the message is 'B', and this lies in the range 0.2–0.3. This range is further subdivided according to the same rules. The symbols 'BI' would thus lie in the range 0.25–0.26. Repeating this subdivision results in a value in the interval 0.2572167752–0.2572167756. Thus, any number in this range will uniquely code the original

Table 8.2 Arithmetic code intervals for
the first symbol in the phrase 'BILL
GATES', after Nelson (1991)

Character	Probability	Range
	0.1	0.00–0.10
A	0.1	0.10–0.20
B	0.1	0.20–0.30
E	0.1	0.30–0.40
G	0.1	0.40–0.50
I	0.1	0.50–0.60
L	0.2	0.60–0.80
S	0.1	0.80–0.90
T	0.1	0.90–1.00

message (although, in practice, the encoding is typically implemented using integer rather than floating point maths).

The increase in coding efficiency with arithmetic coding comes at the expense of significant additional complexity and computational effort. Furthermore, a message encoded using arithmetic coding is susceptible to corruption by transmission errors, and its decoding can not begin until the whole message is received – thereby increasing latency.

8.2.3   Dictionary Encoding

Entropy encoding ignores any patterns that result from repeated sequences of symbols. Consequently, it is often used in conjunction with another compression algorithm that takes into account repetition of symbol patterns.

8.2.3.1   Run Length Encoding

One of the simplest methods of lossless compression of patterns results from the use of Run Length Encoding (RLE).

Consider a message comprising the following sequence of symbols:

   A B C C C C C C C C A A A A A A A A A

Run length encoding simply substitutes repeated symbols with a single symbol and a count. Hence, the above sequence might become something like

   A B C8 A9

RLE is particularly efficient where the message contains a long sequence of zeros (or other value) and is used extensively in the encoding of audio and video.

8.2.3.2   Lempel–Ziv LZ77

A more sophisticated method of encoding repeated symbols is obtained by sequences of repeated patterns of (different) symbols. The so-called LZ77 algorithm, named after its inventors Abraham Lempel and Jacob Ziv (Ziv and Lempel, 1977), replaces a repeated pattern sequence with two integer values: the first is the distance (in symbols) to a previous identical pattern sequence, while the second is the number of symbols in the repeated pattern.

Thus, the sequence

   A A A A A A A A A A B B C A A A A A A B B C

might become

   A (1,10) B (1,1) C (9,9)

where (B,L) is a codeword meaning go back B and copy L symbols. Note that, if L is longer than B, the pattern is repeated to obtain L symbols.

In order to keep track of previous symbol patterns, the LZ77 algorithm maintains a buffer, or 'sliding window' (LZ77 is also known as the sliding-window algorithm). The need to keep comparing new patterns with those in the buffer means that LZ77 encoding can be time consuming; however, LZ77 decoding is generally very efficient. LZ77 and Huffman encoding are the main components of the DEFLATE algorithm (LZ77 being applied first), which in turn forms the basis of many lossless file compression applications – including ZIP and GZIP.

8.2.3.3 Lempel–Ziv–Welch

In 1978, Ziv and Lempel published another general-purpose lossless encoding algorithm, generally known as LZ78 (Ziv and Lempel, 1978). Instead of maintaining a sliding-window buffer, LZ78 builds up a dictionary of symbol sequences previously encountered while encoding. The most common form of LZ78 algorithm is an implementation by Welch known as the Lempel–Ziv–Welch (LZW) algorithm (Welch, 1984). Operation of the LZW algorithm can, perhaps, be best understood by reference to the pseudocode set out in Algorithm 8.1 (Welch, 1984).

Algorithm 8.1 LZW encoding

```
w=NIL
WHILE {more symbols}
read next symbol k
IF {sequence w in dictionary}
w=wk
ELSE
output index of w
add wk to dictionary
w=wk
ENDIF
ENDWHILE
```

In the process of encoding, symbols are read in, one by one, and the new symbol is appended onto the longest sequence of immediately preceding symbols already contained in the dictionary. If this appended sequence is not already contained in the dictionary, the dictionary index of the unappended sequence is emitted and the appended sequence is entered into the dictionary. The appended sequence then becomes the immediately preceding sequence for the next input symbol.

As the dictionary might otherwise grow without limit, there is a limit on dictionary length; once full, the dictionary remains static. Note that there is no need to transmit the dictionary, as both encoder and decoder build up the dictionary as coding (decoding) proceeds. LZW coding forms part of the Adobe Portable Document Format (PDF).

8.3 Digitizing Analogue Signals

8.3.1 Sampling

We turn now to the encoding of analogue signals. During encoding of analogue signals, the signal is usually first converted into a raw (uncompressed) digital data sequence in the form of Pulse Code Modulation (PCM).

In PCM, the (continuous) input analogue electrical signal is periodically sampled using an Analogue-to-Digital Converter (ADC), wherein the amplitude of each sample is measured and *quantized* (rounded to the nearest of a defined set of amplitude levels – usually an integer multiple of some minimum value). Each quantized sample value is then binary encoded, and sequential datawords are emitted as a raw digital sequence. At the source decoder, the encoded digital signal is converted back into an output analogue signal – using a Digital-to-Analogue Converter (DAC).

It is well known that the minimum sampling rate needed to permit accurate reconstruction of the original analogue signal is at least twice the highest (wanted) frequency component of the input signal – the Nyquist rate. This is the Nyquist–Shannon sampling theory, and is related to our discussion of the maximum symbol rate for a given bandwidth in Chapter 5 (Shannon, 1949). The intuitive justification for the Nyquist criteria (Shannon, 1948) is that, if a signal is band limited to f_{max}, it cannot change to a substantially new value in a time interval (sample period) less than half of one cycle of the highest frequency component present (i.e. $\frac{1}{2f_{max}}$). PCM sampling must thus be performed at or above twice the highest wanted frequency component contained in the input signal.

Conversely, the input waveform must be free of (unwanted) frequency components above half the sample rate, and, in order to assure this, the signal is usually low-pass filtered before sampling. We can interpret the sampling process as multiplication of the input analogue signal with a periodic impulse (comb) function. As the periodic impulse function also has a periodic frequency spectrum, with frequency components spaced by the sampling frequency, the effect of sampling is to produce a sampled frequency spectrum that periodically replicates the input spectrum, with these spectral *images* spaced by sampling rate, as shown in Figure 8.2. A low-pass reconstruction filter is used in the DAC to remove any frequency components greater than half the sampling rate. Provided the bandwidth of the input spectrum is narrower than half the sampling rate, the final reconstructed output signal at the DAC contains only (and all of) the input signal. If the input signal bandwidth exceeds half the sample rate, however, the output signal will be distorted. The original input spectrum will be truncated, and some of the (truncated) image spectrum will fall within the output filter – a condition known as *aliasing*.

Quantization

The process of quantization approximates the true sample amplitude to the nearest PCM quantization level, and this inevitably introduces amplitude errors in the analogue output signal in the decoder owing to the difference between the true signal amplitude and the quantized signal level at each sample.

Referring to Figure 8.3, assuming linear quantization, the *quantization error* generated by the PCM sampling process is approximately uniformly distributed between ±0.5 of the amplitude of

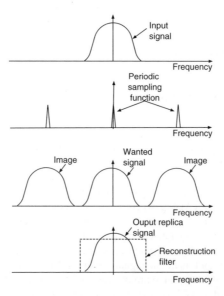

Figure 8.2 Frequency domain representation of sampling where the input bandwidth is less than half the sample rate.

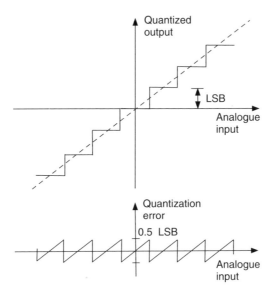

Figure 8.3   Quantization noise.

the ADC's Least Significant Bit (LSB). It is easily shown that, for a sinusoidal input signal[2] whose amplitude is equal to the maximum input range of an ideal n-bit ADC, the resulting signal-to-quantization noise ratio (SQNR) is given by[3]

$$\text{SQNR} = \frac{3}{2}\left(2^{2n}\right) \tag{8.3}$$

The PCM signal-to-quantization noise ratio doubles (increases by 3 dB) for every additional bit of ADC resolution. Clearly, the number of ADC/DAC bits must be chosen so that this so-called quantization 'noise' is negligible within the context of the intended application. Similarly, the linearity (accuracy) of both the ADC and DAC must also be sufficient for the output analogue signal not to be significantly distorted compared with the input.

PCM Data Rate

What data rate is needed to transmit this raw PCM signal? The minimum transmitted bit rate, R_{PCM} (in bits per second), needed to accommodate the resultant (raw) digital code stream is at least equal to the product of the sampling rate and the number of bits per sample (ignoring any overhead for synchronization, etc.). For an n-bit ADC, the raw data rate will be

$$R_{\text{PCM}} \geq 2n f_{\text{max}} \tag{8.4}$$

As the ADC resolution $n \gg 1$, it is apparent that the PCM data bit rate is substantially higher than the original analogue bandwidth of the original source. Consequently, significant additional source encoding is generally desirable in order to reduce the digital bandwidth required for analogue sources. This additional encoding and its inverse decoding typically involve some form of lossy

---

[2] Other input waveforms produce slightly different signal-to-quantization noise ratios.
[3] Expressed in dB, the signal to quantised noise for a sinusoid is

$$(\text{SQNR})_{\text{dB}} = 1.76 + 6.02n \tag{8.5}$$

compression and must be tailored to the particular analogue source type, as efficient encoding tends to exploit known characteristics of the source (or indeed the human recipient) in order to encode the input information with greatest efficiency. As a result, the source encoding signal processing stages are quite different for voice, high-fidelity (hi-fi) audio, still imagery and video signals, and we shall address these separately.

8.4   Speech Encoding

To begin our discussion of lossy source encoding methods, let us focus first on the encoding of human speech. The majority of the spectral energy (i.e. frequency content) in human speech is contained between approximately 200 Hz and 2.8 kHz, and an analogue bandwidth of approximately 4 kHz is generally found to be sufficient to transmit speech in analogue form with acceptable quality. A Nyquist sampling rate of 8 kSamples/s is therefore appropriate for encoding speech into a raw PCM digital signal.

Typically, speech is initially digitized at 16-bit (linear) resolution in order to ensure acceptable dynamic range, and at 16 bits linear resolution the minimum data bit rate for raw digitized speech would be

$$16 \times 8000 = 128\,000 \text{ (b/s)}$$

With binary digital modulation, this is clearly a substantial increase in bandwidth requirement over the original analogue signal.

8.4.1   Waveform Encoding

In terrestrial, landline, voice communications, waveform encoders are used to reduce the PCM transmission data rate while retaining toll quality. Waveform encoders/decoders attempt to reproduce the original waveform as faithfully as possible without prior knowledge of the type of audio signal, and a common form of waveform encoder used widely in telecommunications is based on logarithmic compression and expansion.

Typically, in such encoders, the initial 16-bit samples are reduced to 8-bit codewords by *companding* – a combination of compression (at the source) and expansion (at the destination) – using a non-linear numerical mapping process between input and output bits. Two forms of logarithmic companding are widely used worldwide, both of which are defined in International Telecommunication Union (ITU) recommendation G.711: μ-Law encoding is used in the United States and Japan, while A-Law is used in Europe and elsewhere. The 8-bit encoding reduces the bit rate to 64 kb/s (ITU-T G.711, 1988).

While a 64 kb/s waveform encoded PCM may be acceptable for wire line telephony – such as Integrated Switched Digital Network (ISDN) telephony – wireless and satellite communications resources (predominantly radio-frequency power and spectrum) are more limited and significantly more expensive by comparison. Quantization methods that exploit the similarity of successive samples – such as Differential PCM (DPCM) – can reduce this bit rate somewhat. Nevertheless, there is significant motivation *substantially* to reduce the bit rate for the transmission of speech in these systems, with the objective of achieving acceptable speech quality in the smallest transmitted bandwidth.

8.4.2   Vocoders

There are many forms of speech encoding, and the interested reader is directed to the comprehensive review by Spanias (1994). However, for low bit rates, the dominant compression technique is the voice encoder or *vocoder*. Vocoders can achieve substantial reductions in the data rates needed for human speech – down to a few kb/s. Vocoders are based on speech production models designed to produce human-speech-like sounds – often mimicking (in an electronic form) the properties of the human vocal tract. With vocoders, the information passed between the source encoder and decoder is not the sampled input speech signal but rather instructions (numerical parameters) to the decoder that tell it how to recreate (mimic) the sounds received at the input. This information is typically transmitted every 10–20 ms (50–100 Hz).

There are essentially two types of sound produced in human speech: *voiced* and *unvoiced*. Voiced speech is produced when air from the lungs forces the vocal chords to vibrate and hence produces quasi-periodic sounds that resonate at particular frequencies, or *pitch*. Examples of voiced speech include the vowels – such as 'e' and 'a'. In contrast, unvoiced speech is produced by forcing air through an opening without causing the vocal chords to vibrate, and hence unvoiced sounds are essentially noise-like. Examples of unvoiced speech include 's', 'f' and 'sh' sounds. In both voiced and unvoiced speech the sounds are modified by the spectral filtering produced by the shape of the vocal tract – the throat (pharynx) and mouth (and also the nose for some sounds). Resonant modes of the vocal tract are known as *formants*. The amount of air forced through the larynx determines the speech *volume*.

Vocoders employ an *analysis-synthesis* approach, pioneered by Dudley in the late 1930s (Spanias, 1994). This involves analysing the original speech at the encoder and then synthesizing it at the decoder. Most vocoders employ a similar physical model of the way human speech is generated. It is useful, when attempting to understand vocoder operation, to focus on the operation of the decoder part illustrated in Figure 8.4. The role of the encoder is to estimate the parameters needed by the decoder and encode that information efficiently prior to transmission. As the characteristics of the vocal tract are found to change on a timescale of 10–100 ms, this usually determines the update rate of the vocoder parameters (with 20 ms being common).

Because bandwidth is such a valuable commodity in wireless communications, vocoder technology continues to evolve rapidly, with each new generation of enhancements striving to improve the quality of transmitted speech for a given bit rate, or reduce the bit rate for a specified speech quality. As a result, there is no dominant vocoder standard, and in this book we focus on three generic vocoder types that illustrate the main concepts of vocoding: linear predictive coding, codebook excited linear prediction and multiband excitation.

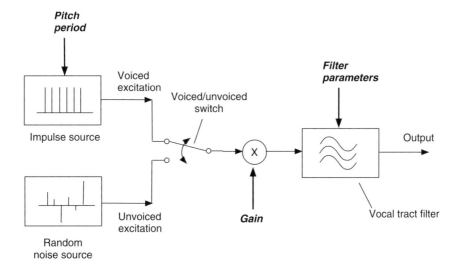

Figure 8.4 Basic speech synthesis model.

8.4.2.1 Linear Predictive Coding (LPC)

A popular class of vocoder is the Linear Predictive Coder (LPC) – being relatively simple to implement. It also forms the basis of the vocal tract filter model for most other vocoder types. In the basic LPC decoder, the excitation for voiced speech is a (periodic) impulse generator, while that for unvoiced speech is a noise source, as illustrated in Figure 8.4. These excitation sources are then digitally filtered – imitating the effect of the vocal tract. The LPC vocal tract model uses a *linear predictive* filter. Linear prediction is a general technique for predicting the future values of a discrete time signal by using linear combinations of previous samples.

In LPC-based vocoders, each speech sample is assumed to be a linear combination of the excitation and weighted previous samples. The nth voice sample, s_n, is thus

$$s_n = g\, u_n + \sum_{i=1}^{M} a_i\, s_{n-i} \tag{8.6}$$

where u_n is the excitation, g is the gain and a_i are the LPC coefficients. The LPC filter is illustrated in Figure 8.5, where $A(z)$ refers to its z-domain discrete time filter representation.

The LPC encoder samples and analyses the input speech and estimates the appropriate excitation type (voiced or unvoiced), its pitch, amplitude (gain) and the necessary filter coefficients. These parameters are compressed and transmitted to the decoder, which mimicks the original speech. Typically between 8 and 14 LPC coefficients are required for satisfactory voice quality, depending on the application. Filter coefficients are found by minimizing the energy of the innovation (excitation) signal required, using a sliding window (Spanias, 1994).

LPC example. By way of example, US Federal Standard FS1015 defines an LPC vocoder that operates at 2.4 kb/s using 10 filter coefficients, known as LPC-10. The transmitted parameters thus comprise these 10 coefficients plus the gain and the voice/unvoiced decision and pitch period combined into a single parameter (making 12 in all). After quantization, these parameters occupy 48 bits, which are sent every 20 ms (thus requiring a bit rate of 2.4 kb/s).

Ultimately, the LPC vocoder is limited by its use of only two simple excitation sources. As a result, at low bit rates, LPC speech tends to sound synthetic and unnatural.

8.4.2.2 Codebook Excited Linear Prediction (CELP)

Almost all advanced vocoders employ the same vocal tract model (filter) as that used in LPC encoders, but with a more sophisticated excitation process. These are typically based on *analysis by synthesis*, a closed-loop approach whereby the encoder constantly compares the synthesized waveform to the original speech in order to determine the best excitation signal to use at the decoder. A generic analysis by synthesis encoder is illustrated in Figure 8.6 (Spanias, 1994). Note that in the synthesis part of the encoder (replicated in the decoder) the voiced and unvoiced excitations are replaced by a vector (list) of more complex excitations, with the most appropriate

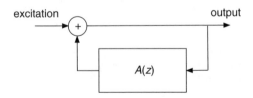

Figure 8.5 LPC vocal tract filter.

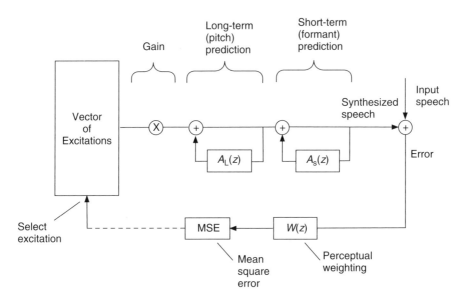

Figure 8.6 Generic analysis-by-synthesis encoder (adapted from Spanias, 1994).

being selected so as to minimize the weighted mean square error between the synthesized speech samples and the actual speech samples. There are both long- and short-term prediction filters, the former determining the pitch period, and the latter the vocal tract filtering.

A significant example of the analysis-by-synthesis approach is the Codebook Excited Linear Prediction (CELP) vocoder, and its derivatives. In a CELP vocoder, rather than use simple impulse and noise excitations for voiced and unvoiced speech respectively, a 'codebook' vector of different excitation samples is used instead, in order to allow a better match to the input speech. Searching the codebook for the optimum excitation is time consuming, and various methods have been developed to speed up this search. In some CELP vocoder variants, such as US Federal Standard 1016 4.8 kb/s CELP, the excitation is obtained using a small, structured codebook (called the fixed codebook) and a so-called 'adaptive' codebook containing previous excitation sequences (with the adpative codebook searched first). In the Vector CELP (VCELP) vocoder, the optimum excitation is found from a combination of two structured codebooks plus an adaptive codebook.

A 16 kb/s CELP vocoder provides voice quality that is indistinguishable from the ITU 64 kb/s logarithmic companded PCM speech standard, and CELP vocoders can produce intelligible speech at 4.8 kb/s. At still lower data rates, CELP vocoder speech tends to exhibit a buzzy quality when the input contains both voiced and unvoiced contributions, or when voiced speech is accompanied with background noise.

CELP vocoders and their derivatives are widely used in terrestrial mobile communications. VCELP vocoders are used in the half-rate Global System for Mobile communications (GSM) service, while another variant, the Algebraic Code Excited Linear Prediction (ACELP) vocoder, is used in the enhanced full-rate GSM service.

8.4.2.3 MultiBand Excitation (MBE)

A limitation of both LPC and CELP vocoders is the reliance on a single excitation source. In voice quality tests, another form of vocoder, the MultiBand Excitation (MBE) vocoder, and its derivatives, has been shown to perform particularly well at lower bit rates (around 2.4 kb/s), where LPC and CELP struggle.

MBE vocoders split the speech into different frequency bands and continually classify different parts of the input frequency spectrum as either voiced or unvoiced. Input speech is usually analysed in the frequency domain, and a binary decision is made as to whether each frequency band in the spectrum is voiced or unvoiced. In the MBE decoder, unvoiced bands are excited using filtered white noise, while voiced bands are excited using harmonic oscillators. Some MBE variants utilize a weighted combination of voiced and unvoiced excitations in each band.

As a result of their ability to provide acceptable voice quality at low data rates, MBE-type vocoders and their derivatives are widely used today in personal satellite communications. The Improved MBE (IMBE) vocoder is used in the Inmarsat M service. Another MBE derivative, the Advanced MBE (AMBE) vocoder, is used in the Inmarsat Mini-M, Iridium, Thuraya and AceS satellite voice services.

8.5 Audio Encoding

8.5.1 Audio Source Encoding

We turn our attention next to efficient encoding of general audio waveforms, and in particular high-fidelity audio – including stereo music and multichannel surround-sound content. Whereas efficient speech encoding is generally modelled on the human *vocal tract*, source encoding for high-fidelity audio is tailored towards the characteristics of human *hearing*.

The human ear can distinguish frequencies between *approximately* 20 Hz and 20 kHz – depending on the listener's age. Compact Disc (CD) quality stereo audio is therefore sampled at a Nyquist rate of 44.1 kSamples/s, at 16-bit resolution.[4]

The minimum raw binary data bit rate for CD quality stereo is thus

$$16 \times 44.1 \times 2 = 1411.2 \, \text{kb/s}$$

In fact, the actual data rate for a 16-bit resolution CD, including synchronization, framing and error correction overheads, is 4.32 Mb/s.

Although a slight reduction in the input frequency range (e.g. to 16 kHz) can be utilized with minimal impact on quality, effective audio encoding is primarily based around (lossy) *perceptual encoding*, together with conventional lossless data compression techniques. Perceptual encoding relies principally on two properties of human hearing whereby a weaker sound is masked by a stronger one: *simultaneous masking* and *temporal masking* (Noll, 1997; Pan, 1996). Such techniques are examples of lossy compression, in the sense that deleted information cannot be recovered.

8.5.2 Psychoacoustic Encoding

Simultaneous Masking
Referring to the spectrum plot of Figure 8.7, simultaneous masking is a frequency domain phenomenon that causes masking of weak sounds in the vicinity of the frequency of a masker (a stronger sound). In the figure, the *threshold of hearing* – the amplitude at which a sound cannot be detected – is seen to vary with frequency. The effect of the masker is to cause the threshold of hearing to be increased in the vicinity of the masker frequency. The effect of this masking decreases with frequency separation of the strong and weak sounds and depends also on the absolute frequency

---

[4] Note that newer audio formats – such as Super Audio CD (SACD) and DVD Audio – now support up to 24-bit resolution.

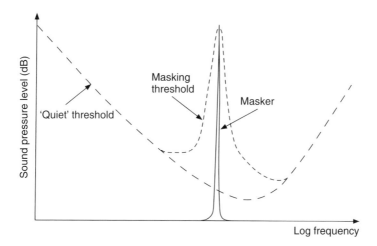

Figure 8.7   Simultaneous audio masking (frequency domain).

and amplitude of the masker. Furthermore, there may be several simultaneous maskers, each with its own masking threshold.

Temporal Masking

Referring now to Figure 8.8, temporal masking occurs when a weak sound is masked by a nearly coincident (in time) stronger sound. Masking occurs both after the stronger sound has ended (*post-masking*) and, perhaps somewhat surprisingly, just *before* it begins (*pre-masking*). It should be recognized that the human perception of sound results from a *combination* of the physical response of the ear and subsequent processing within the brain. Depending on the strength of the masker, post-masking occurs for up to 50–200 ms after the end of the masker, and pre-masking for up to 5–20 ms before it begins (Noll, 1997).

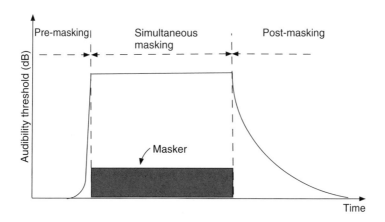

Figure 8.8   Temporal audio masking (time domain).

Critical Frequency Bands

A further property of the human auditory system is the existence of about 24 so-called critical bands – frequency ranges within which we are unable to distinguish signals with different frequencies. The auditory system effectively blurs sounds within each of these critical bands, which are approximately logarithmically spaced in centre frequency.

8.5.2.1 Perceptual Coding Methods

In essence, perceptual coding discards masked audio information – information that will not be audible to the majority of users. The remaining audio information is then encoded using standard lossless compression techniques – run length encoding and Huffman encoding. In the encoder, the input signal is first transformed into the frequency domain and analysed with regard to a psychoacoustic model. This analysis is used to determine the masking thresholds for each frequency range.

MPEG example. In MPEG Audio Layer II (MP2) audio encoders, the audio spectrum is divided up into 32 frequency bands using a polyphase quadrature digital filter[5] (Rothweiler, 1983). These 32 sub-bands crudely equate to the 24 critical frequency bands, although the equal width of these digital filters does not match the approximately logarithmically spaced critical bands. Polyphase quadrature filters are used in MP2 because they are computationally less demanding than other transforms. However, a drawback of this type of filter is that the combination of the polyphase filter transform (at the encoder) and its inverse (at the decoder) is not lossless and results in some (albeit inaudible) audio information being lost. By contrast, in MPEG 2 Audio Layer III (MP3), each band is further divided into 18 sub-bands using a Modified Discrete Cosine Transform (MDCT) (Pan, 1996). Once transformed into the frequency domain, component frequencies can be allocated bits according to how audible they are, using the masking levels in each filter. MP3 also exploits any interchannel redundancies – for example, where the same information is transmitted on both stereo channels.

8.6 Video Encoding

We turn next to the encoding of images, and in particular video.

8.6.1 Image Encoding

The need for compression of digital images is apparent. Techniques used by both still and moving image compression have many common aspects (although the resolution of the latter is generally lower), and it is therefore relevant to introduce the encoding of video images by looking first at the encoding of still images. One of the most widely used still picture compression standards is that produced by the Joint Photographic Experts Group (JPEG). JPEG has two basic compression methods: a Discrete Cosine Transform (DCT) method for lossy compression and a predictive method for lossless compression. Both of these techniques are also employed in video coding.

JPEG provides for the following encoding modes (Wallace, 1992):

- *Sequential.* In this mode the image is encoded in a single pass, from left to right and top to bottom.

---

[5] Polyphase quadrature filters use a sliding transform – similar to the Discrete Cosine Transform (DCT – a type of real-valued Fourier transform) – to split a signal into a specified number (usually a power of 2) of equidistant sub-bands.

- *Progressive.* In this mode, designed for applications where image transmission speed is limited and the recipient wishes to see the image build up, the image is encoded in multiple scans with progressive increases in resolution – from coarse to fine.
- *Lossless.* In this mode, the image is encoded in a lossless fashion (with lower compression compared with lossy modes).
- *Hierarchical.* In this mode, the image is encoded at multiple resolutions, allowing lower-resolution versions to be viewed without decompressing the high-resolution versions.

By far the most common method of JPEG image compression is the 'baseline method' which uses a subset of the sequential DCT (lossy) encoding method, together with Huffman entropy encoding.

8.6.2   Lossy DCT Image Encoding

We now focus on the two-dimensional Discrete Cosine Transform (DCT) method of encoding. The original image is divided into blocks of pixels – typically 8 × 8 pixels per block. Each colour component block (e.g. one each for red, green and blue) is then transformed to the *spatial frequency* domain, using a two-dimesnional DCT. The output of the DCT is a set of numerical coefficients that describe the relative amplitudes of the DCT basis functions (spatial patterns) within the pixel block.

The 64 monochrome basis function patterns (spatial frequencies) of the 8 × 8 DCT are shown in Figure 8.9. The encoded picture data define the amount of each basis function pattern that goes to make up the original pixel block. In the figure, low spatial frequencies are towards the top left, and high spatial frequencies towards the bottom right, with the top left-hand basis function having no variation (known as the DC component – by analogy with electric Direct Current).

The essence of DCT encoding is that the spectral content of most blocks is generally dominated by the DC and low (spatial) frequency components. Higher spatial frequency components can therefore usually be discarded with minimal impact on the picture quality for most applications.)

After DCT, the frequency domain coefficients for each block are quantized – this quantization is the principal source of loss of information in DCT encoding. The resulting matrix of quantized

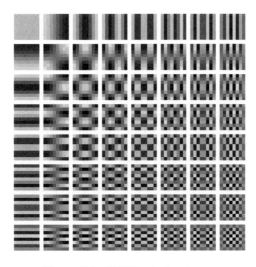

Figure 8.9   DCT basis functions.

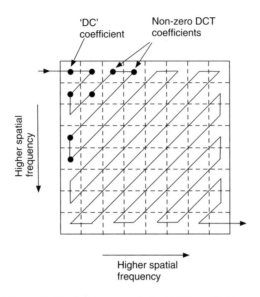

Figure 8.10   Read-out path for DCT coefficients.

spatial frequency coefficients is read out in a zig-zag fashion as shown in Figure 8.10, with the result that the lowest spatial frequency DCT coefficients (in both directions) are read out first.

The combined effect of quantization and the zig-zag read-out method is that the majority of the latter coefficients in the output sequence are zero, and the subsequent use of run length encoding results in significant data compression. Furthermore, as the likelihood is that the DC (constant) spatial components of adjacent blocks will be similar, only the *difference* between the current DC coefficient and that for the previous block need be encoded. However, because the eye is most sensitive to low spatial frequency changes, this coefficient must be encoded at higher (typically 10-bit) resolution. Finally, Huffman (or arithmetic) entropy encoding is used to ensure that the most frequently occurring codes (lists of run-length-encoded coefficients) are the shortest ones.

The process at the decoder is the reverse of encoding. The DCT coefficients for each block are decoded and applied to an inverse DCT (IDCT) to obtain the individual pixel values.

8.6.3 Encoding Video Signals

The encoding of digitized video and television (TV) pictures uses the same concepts as those developed for still images, but adds the use of motion compensation for additional bit rate compression.

ITU Recommendation BT.601-6 (ITU BT.601-6, 2007) defines the pictures in a standard-definition (SD) digital television frame as comprising 720×576 pixels for Europe (720×486 pixels for the United States and Japan), with frames transmitted at an update rate of 25 Hz (30 Hz). In order to avoid flicker, these frames consist of two successive *interlaced* fields comprising alternate horizontal lines of pixels, so as to produce the effect of updating at 50 Hz (60 Hz in United States and Japan).

If each frame were transmitted as a sequence of red, green and blue values (RGB) using 8 bits per colour per pixel, the raw bit rate with no overhead would be (for Europe)

$$25 \times 720 \times 576 \times 3 \times 8 = 248\,832\,000 \, \text{b/s} = 237.3 \, \text{Mb/s}$$

Clearly, without further compression, transmission of video would be prohibitive, except for the most demanding (and profitable) applications. Just as audio encoding exploits the limitations of human hearing, so video encoding exploits limitations of human vision to reduce this data rate.

Colour Model

Standard-definition TV defined in ITU BT.601-6 (2007) uses a different colour model to the more familiar red, green and blue (RGB) model. Instead, the red, green and blue intensities are combined into luminance (i.e. grey level) plus two colour difference values. This is partly historical (as black and white TV required only a luminance value) and partly due to the different resolving power of the human eye to intensity and colour. For colour displays, the luminance Y is defined as the sum of the individual colour components, weighted according to the response of the human eye to each colour.

Historically, these signal values are adjusted to compensate for the nonlinear characteristics of video display electronics. The original video display, the Cathode Ray Tube (CRT), has a nonlinear light intensity to input current characteristic. For a monochrome display, the actual luminance Y is related to the input luminance signal Y' via the power law

$$Y = (Y')^\gamma \tag{8.7}$$

where the parameter γ for a CRT is in the region of 2, being primarily associated with the physics of the electron gun.

As a result of this potential nonlinearity in the display, luminance information Y' is actually encoded as the *gamma-corrected* value, so that the combined effect of this encoding and the nonlinearity of the display will be to produce an intensity proportional to that in the original scene. The two colour difference values – known as *chroma* – are given by the differences between each gamma-corrected blue and red colour and luma values:

$$C_b = B' - Y' \tag{8.8}$$

$$C_r = R' - Y' \tag{8.9}$$

This colour model is known as $Y'C_bC_r$ – meaning luma plus the difference of the gamma-corrected blue value and the luma and the difference of the gamma-corrected red value and the luma (note that a green difference chroma value is not needed, as this can be determined from the other two values plus the luma). ITU BT.601-6 (2007) defines standard formulae to allow conversion between uncorrected RGB and the gamma-corrected $Y'C_bC_r$ model.

Chroma Subsampling

As the human eye is less sensitive to spatial variations in colour than those in intensity, it has long been common to exploit these different resolutions through the use of chroma subsampling. Here, colour is sampled at lower spatial sampling rates than is used for intensity, with little detectable impact on image quality. This is another example of lossy compression, as it is irreversible.

A typical chroma subsampling scheme is 4:2:0 ($Y'C_bC_r$), these three digits being:

1. The relative horizontal sampling rate for the luma (relative to the basic sampling rate).
2. The relative horizontal sampling rates for the blue and red chroma (relative to the luma signal).
3. The relative vertical sampling rates for the blue and red chroma, or 0 if a 2:1 vertical subsampling ratio is used for both C_b and C_r, compared to the luma.

Use of 4:2:0 chroma subsampling (the most common) implies that the two chroma signals are spatially sampled (both vertically and horizontally) at half the rate of the luma signal, resulting in a bandwidth reduction of 75% compared with 4:4:4 (RGB). On average, therefore, for a block of four 4:2:0 pixels we will have four luma values but only one red and one blue chroma value each (making a total of six values). At 8 bits per value, this gives a bit rate for the European standard of

$$720 \times 576 \times 25 \times 8 + 360 \times 288 \times 25 \times (8 + 8) = 124\,416\,000 = 118.6\,\text{Mb/s}$$

while using 4:2:2 chroma subsampling and the US standard gives the following bit rate:

$$720 \times 486 \times 30 \times 8 + 360 \times 486 \times 30 \times (8 + 8) = 167\,961\,600 = 160.1\,\text{Mb/s}$$

(note that 1 Mb/s = 1024 kb/s and 1 kb/s = 1024 b/s).

Video coders/decoders exploit both temporal and spatial redundancy in the video signal, and the limitations of human vision. Temporal redundancy occurs when a part of the picture in one frame also exists in the next frame, and this includes objects that have moved between frames. Spatial redundancy occurs where the luma and chroma values are the same (or similar) in several adjacent pixels.

Currently, the dominant video encoding standards are those of the Motion Picture Expert Group (MPEG) – notably MPEG-2 for standard definition video and, increasingly, MPEG-4 for high-definition video. In fact, MPEG video decoders contain a 'toolbox' of compression algorithms grouped into a number of subsets called 'profiles'. Furthermore, profiles support a number of levels (combinations of image size, frame rate, etc.), and decoders may implement a subset of profiles.

8.6.4 Motion Estimation

Rather than transmit the complete luma and chroma information for each frame, video encoding algorithms aim to transmit differences between frames where possible and use a DCT to encode the difference information into the frequency domain, discarding insignificant high spatial frequency information that would not normally be noticeable.

Referring to Figure 8.11, a video sequence comprises a Group Of Pictures (GOP) – a sequence of *frames*. Each frame or picture is divided into *slices* along the raster direction, and each slice is further divided into *macroblocks*, for use when estimating interframe motion. In MPEG-2, each slice comprises a string of macroblocks of 16×16 pixels, and the number of macroblocks in a slice may vary. Each macroblock further comprises *blocks* of pixels, used in the DCT. In MPEG-2, each block contains 8×8 pixels.

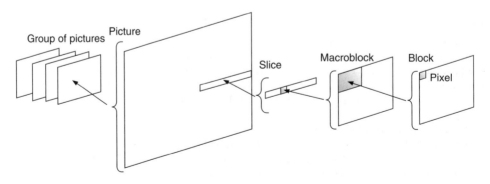

Figure 8.11 Video entities.

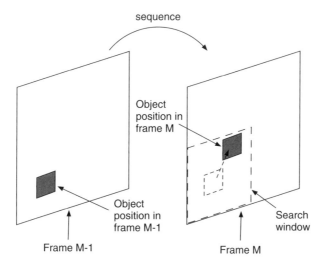

Figure 8.12   MPEG macroblock motion estimation.

During encoding, each macroblock is compared with the current reference frame (potentially the last complete frame). Any similar macroblocks present in both the current and reference frames, within the limits of a specified search window, are detected, and relative motion (if any) is estimated, as illustrated in Figure 8.12 (Sikora, 1997). This information is transmitted as *motion vector* information, and the relevant macroblock luma and chroma values are then *subtracted* from the current image frame. The size of the macroblocks represents a compromise between accuracy of the motion matching process and the number of separate motion vectors that are generated per frame. After motion estimation and subtraction, the remaining 'difference' image blocks (both luminance and chrominance) are then transformed into the spatial frequency domain using a DCT.

In the case of MPEG video encoding, three different types of encoded picture are used with regard to motion estimation:

- *Intra-Pictures (I-Pictures).* These pictures are coded using only information from the current frame. Coding employs the DCT transform but not motion estimation. I-Pictures provide key references for other types of picture, and also provide random access points into the encoded video stream.
- *Predicted-Pictures (P-Pictures).* These pictures are coded with motion compensation with respect to the nearest previous I- or P-Pictures. Coding uses both motion compensation and DCT. This allows greater compression than I-Pictures.
- *Bidirectional Pictures (B-pictures).* These pictures are coded with motion compensation using both past and future pictures as references. This allows for greater compression than P-Pictures – at the expense of greater computation effort and latency.

Figure 8.13 indicates the reference frames used by a P-Picture and B-Picture. In the case of MPEG-2, typically every 15th frame is an I-Picture (Sikora, 1997).

Figure 8.14 summarizes the processing flow in a generic video encoder. The motion prediction is carried out first, and any common blocks are subtracted from the original pictures. The residual images are transformed using a DCT, after which the DCT coefficients are quantized nonlinearly. These quantized coefficients are then entropy encoded for maximum transmission efficiency. The

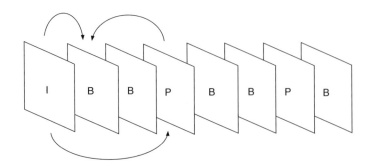

Figure 8.13   MPEG motion estimation reference frames.

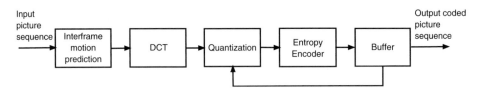

Figure 8.14   Schematic of a generic video encoder.

quantization levels may be adjusted dynamically using feedback from the state of the output buffer, so that the peak data rate does not exceed the specified maximum rate.

Revision Questions

1. What is meant by information entropy? Explain the differences between Huffman and arithmetic entropy coding.
2. Describe three types of dictionary encoding and use two to encode a line from this book.
3. Explain the distinction between lossless and lossy encoding techniques, and give examples of each.
4. Determine the numerical range for arithmetic encoding of the message 'STEVE JOBS'.
5. Explain the differences between a waveform encoder and a vocoder. Describe the vocoder speech production model.
6. What is analysis by synthesis and how is it used to improve the performance of vocoders?
7. What are the critical bands. Describe how masking makes it possible to compress audio signals.
8. A full High-Definition TV (HDTV) picture comprises 1920 × 1080 pixels. Assuming a European standard refresh rate of 25 Hz and 8-bit 4:2:0 chroma subsampling, what is the raw video bit rate in Mb/s.
9. Using a diagram, illustrate the complete encoding process for video frames.

References

Howard, P.G. and Vitter, J.S. (1994) Arithmetic coding for data compression. *Proceedings of the IEEE*, **82**(6), 857–865.

ITU BT. 601-6 (2007) Recommendation BT. 601-6, Studio encoding parameters of digital television for standard 4:3 and wide screen 16:9 aspect ratios.

ITU-T G.711 (1988) Recommendation G.711, Pulse code modulation of voice frequencies.

Nelson, M. (1991) Arithmetic coding + statistical modeling = data compression. *Dr Dobb's Journal*, February.

Noll, P. (1997) MPEG digital audio coding. *IEEE Signal Processing Magazine*, September, 59–81.

Pan, D. (1996) A Tutorial on MPEG audio compression. *IEE Multimedia*, **2**, 60–74.

Rothweiler, J.H. (1983) Polyphase quadrature filters – a new sub-band coding technique. Proceedings of the International Conference IEEE ASSP, Boston, MA, pp. 1280–1283.

Shannon, C.E. (1948) A mathematical theory of communication. *Bell System Technical Journal*. **27**(July/October), 379–423.

Shannon, C.E. (1949) Communication in the presence of noise. *Proceedings of the IRE*, **37**, 10–21 (reprinted in Shannon, C. E. (1984) *Proceedings of the IRE*, **72**(9), 1192–1201).

Sikora, T. (1997) MPEG digital video coding standards. *IEEE Signal Processing Magazine*, September, 82–100.

Spanias, A.S. (1994) Speech coding: a tutorial review. *Proceedings of the IEEE*, **82**(10), 1541–1582.

Wallace, G.K. (1992) The JPEG still picture compression standard. *IEEE Trans. Consumer electronics*, **38**(1), xviii–xxxiv.

Welch, T.A. (1984) A technique for high-performance data compression computer, **17**, 8–19.

Ziv, J. and Lempel, A. (1977) Universal algorithm for sequential data compression. *IEEE Trans. Information Theory*, **23**(3), 337–343.

Ziv, J. and Lempel, A. (1978) Compression of individual sequences via variable-rate coding. *IEEE Trans. Information Theory*, **24**(5), 530–536.

Part II

Techniques and Systems

9

Digital Broadcasting Techniques and Architectures

9.1 Introduction

According to a recent Satellite Industry Association's report, over 130 million direct satellite television subscribers are spread around the world – a clear indication of the success of satellite broadcast systems. The key enablers attributed to this success include modern satellites capable of powerful transmissions over vast expanses, efficient transmission technology, highly integrated, low-cost receiver systems and a vast variety of rich content at an affordable price.

This chapter introduces the concepts of satellite broadcasting systems applied in the context of personal reception. The chapter covers three broad areas – the first part builds on the concepts in Chapter 8 related to audio and video encoding and introduces their transport; the second part explains the essentials of direct-to-home (DTH) transmission systems; and the third part addresses direct-to-individual broadcast system architecture.

Current audio and video compression techniques dwell specifically on the Motion Pictures Expert Group (MPEG) multimedia standards which constitute the industry's standards for compression and transport. A majority of direct broadcast satellite systems beaming Standard-Definition TV (SDTV) either use or can support the MPEG-2 standard, while High-Definition TV (HDTV) and multimedia broadcast systems rely firmly on the standard's upgraded version, MPEG-4.

To encourage mass-scale acceptance, regulatory bodies promote standardization of the satellite broadcast service. The International Telecommunication Union (ITU) recognizes at least six DTH broadcast systems that encompass television, data and multimedia. We introduce these ITU recommendations and compare their technical merits, with an emphasis on the Digital Video Broadcast–Second Generation (DVB-S2) standard, because of its highly efficient and flexible delivery architecture achieved through recent advances in transmission technology.

Modern satellite and handheld receiver technologies enable television and radio broadcasts to handheld and portable personal devices. Numerous commercial systems that offer sound and television services to the individual and groups have emerged in the last decade. The final part of the chapter introduces and compares the direct-to-individual broadcast systems, discussing in detail system E and Digital Video Broadcast–Satellite Handheld (DVB-SH) standards – both of which support commercial mobile television services.

Satellite Systems for Personal Applications: Concepts and Technology   Madhavendra Richharia and Leslie David Westbrook
© 2010 John Wiley & Sons, Ltd

9.2 MPEG Multimedia Standards

At the time of writing, the encoding of both high-fidelity (non-speech) audio signals and video signals is dominated by the standards defined by the International Standards Organization (ISO) Motion Pictures Expert Group (MPEG), subsequently adopted by both the ISO/IEC and the ITU. The first standard, MPEG-1, 'Coding of moving pictures and associated audio for digital storage media at up to about 1.5 Mbit/s', issued in 1992, and its successor MPEG-2, 'Generic coding of moving pictures and associated audio information', issued in 1994, define generic coding for both moving pictures (i.e. video) and associated audio information. Work on an MPEG-3 standard was discontinued. More recently, MPEG-4 primarily focuses on new functionality, but also includes improved compression algorithms, and for the first time includes speech compression algorithms – such as CELP. Each standard comprises a number of parts – for example, MPEG-4 has 23 parts – each of which focuses on a particular aspect of the encoding (video, audio, data encapsulation, etc). Significantly, MPEG standards only define in detail the source decoder, providing a toolbox of algorithms, and standardized bit stream formats. A detailed encoder architecture is typically not defined (although example implementations are given), thereby allowing equipment manufacturers to differentiate their products while ensuring compatibility of user's equipment.

9.2.1 Audio Broadcasting

MPEG-2 Audio Layer II (MP2) Encoding

MPEG-1 provided for three different types of audio encoding with sampling rates of 32 kHz (kSamples/s), 44.1 kHz and 48 kHz for monophonic (mono), dual mono and stereophonic (stereo) channels. MPEG-2 added support for up to five audio channels and sampling frequencies down to 16 kHz. Subsequently, MPEG-2 Advanced Audio Encoding (AAC) added sampling from 8 to 96 kHz with up to 48 audio channels (and 15 embedded data streams). MPEG-2 audio layer I is the simplest encoding scheme and is suited for encoded bit rates above 128 kb/s per channel (Pan, 1995). Audio layer II, known as MP2, has an intermediate complexity and is targeted at bit rates around 128 kb/s. MP2, also known as Musicam, forms the basis of the Digital Audio Broadcasting (DAB) system and is incorporated into the Digital Video Broadcasting (DVB) standard.

MPEG-2 Audio Layer III (MP3) Encoding

MPEG-2 audio layer III, known as MP3, provides increased compression (or, alternatively, improved quality at the same data rate) compared with MP2 at the expense of slightly increased complexity and computational effort. Today, MP3 is widely used for the storage and distribution of music on personal computers via the Internet and for digital satellite broadcasting by 1worldspace (see Chapter 10). The MP3 psychoacoustic model uses a finer-frequency resolution than the MP2 polyphase quadrature filter band provides (Pan, 1995). MP3 divides the audio spectrum into 576 frequency bands and processes each band separately. It does this in two stages. First, the spectrum is divided into the normal 32 bands, using a polyphase quadrature filter, in order to ensure compatibility with audio layers I and II. In MP3, however, each band is further divided into 18 sub-bands using a Modified Discrete Cosine Transform (MDCT). The MP3 MDCT is a variant of the discrete cosine transform that reuses a fraction of the data from one sample to the next. We have learnt in Chapter 8, that following the transformation into the frequency domain, component frequencies can be allocated bits according to their audibility using the masking levels in each filter. We also noted that MP3 exploits inter-channel redundancies, for example in situations when the same information is transmitted on both stereo channels. Typically, MP3 permits compression of CD-quality sound by a factor of ~12.

9.2.2   Video Broadcasting

As with audio encoding, the dominant video encoding standards are currently the Motion Picture Expert Group (MPEG) standards, notably MPEG-2 for standard-definition video and, increasingly, MPEG-4 for high-definition video. MPEG-2 video decoders contain a 'toolbox' of compression algorithms grouped into a number of subsets called 'profiles'. Furthermore, profiles support a number of levels (combinations of image size, frame rate, etc.), and decoders may implement a subset of profiles.

MPEG-2 Video Encoding

MPEG video encoding algorithms aim to transmit differences between frames where possible and use a DCT (a form of Fourier transform) to encode the difference information into the frequency domain, discarding insignificant high spatial frequency information that would not normally be noticeable. For this reason, they are sometimes referred to as hybrid block interframe Differential PCM (DPCM)/Discrete Cosine Transform (DCT) algorithms. A video buffer is used to ensure a constant bit stream on the user side.

Typical bit rates for MPEG-2 encoded standard-definition video are in the region 3–15 Mb/s, that is around 10–50:1 compression is achieved over the raw PCM bit rate.

High-Definition TV and MPEG-4

Even greater compression ratios are required for the transmission of high-definition TV video. HDTV frame resolutions of up to 1920×1080 pixels are in use (at 25 Hz) – with up to 5 times as many pixels per frame compared with Standard-Definition TV (SDTV). While MPEG-2 High Level can support resolutions of up to 1920×1080 (sometimes referred to as full-HD), HDTV generally requires the increased compression available in MPEG-4 to be viable. MPEG-4 Advanced Video Coding (MPEG-4 part 10 AVC), jointly developed with the ITU Video Coding Experts Group (as H.264), employs additional techniques to achieve compression ratios greater than those for MPEG-2, and is used by several satellite services for broadcasting HDTV. MPEG-4 AVC is utilized in BluRay video discs. Specifically, MPEG-4 AVC utilizes a number of features to achieve higher compression than MPEG-2. Some of the more significant features of MPEG-4 AVC are:

- Up to 32 reference pictures may be used for motion compensation, rather than just 1 (I-Pictures) or 2 (B-Pictures).
- The macroblock size used for motion compensation may be varied from 16×16 to 4×4 pixels with subpixel precision.
- New 4×4 and 16×16 pixel block transforms.
- Improved non-linear quantization size control.
- Improved entropy encoding.

9.2.3   Multiplexing and Transporting

The systems part of the MPEG-2 specification (ISO/IEC 13818-1:2000(E)) specifies syntactical and semantic rules to combine MPEG-2 video and audio elementary streams (output of an encoder), including other types of data content, into a single or multiple stream to enable storage or transmission.

Figure 9.1 (ISO/IEC 13818-1:2000(E), 2000) illustrates packetization of encoded video and audio elementary streams to produce a 'packetized elementary stream' (PES). PES packets are next combined with system-level information to form transport streams (TS) or programme streams (PS).

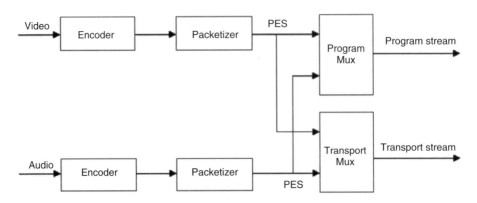

Figure 9.1 MPEG programme and transport stream formation from video and audio PES. This figure, taken from ISO/IEC 13818-1:2000 Information Technology – Generic coding of moving pictures and associated audio information systems, is reproduced with the permission of the International Organization for Standardization (ISO). This standard can be obtained from any ISO member and from the web site of the ISO Central Secretariat at the following address: www.iso.org. Copyright remains with the ISO.

The *programme stream* consists of one or more streams of PES packets of common time base into a single stream. The stream is useful for operation in relatively error-free environments such as interactive multimedia applications.

The *transport stream* consists of one or more independent programmes into a single stream. This type of stream is useful in error-prone environments as satellite broadcasts. Packets are 188 bytes in length. Transport stream rates may be fixed or variable. Values and locations of Programme Clock Reference (PCR) fields define the stream rate. The transport stream design is such as to facilitate:

- retrieval and decoding of a single programme within the transport stream figure (9.2);
- extraction of only one programme from a transport stream and production of a different transport scheme consisting of the extracted programme;
- extraction of one or more programmes from one or more transport streams and production as output of a different transport stream;
- extraction of contents of one programme from the transport stream and production as output of a corresponding programme stream;
- conversion of a programme stream into a transport stream to carry it over a lossy environment, and its recovery at the receiver.

Figure 9.2 illustrates the concept of demultiplexing and decoding of a single programme from a received transport scheme containing one or more programme streams. The input stream to the demultiplexer/decoder includes a system layer wrapped about a compression layer. The system layer facilitates the operation of the demultiplexer block, and the compression layer assists in video and audio decoding. Note that, although audio and video decoders are illustrated, these may not always be necessary when the transport stream carries other classes of data.

9.3 Direct-to-Home Broadcast System

Recognizing the advantages of digital television transmissions, and their potential to transport HDTV efficiently, considerable effort was directed in the 1980s and 1990s (notably in the United States, Europe and Japan) towards the development of digital transmission systems, resulting in the

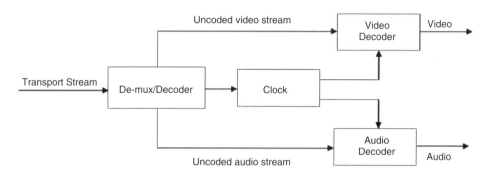

Figure 9.2 Demultiplexing and decoding of an MPEG transport stream. Reproduced by permission of © ITU.

design of several terrestrial and satellite systems. The convergence of computing, telecommunications and broadcast disciplines led the developers to adapt a generic architecture that would offer a variety of enabling services in addition to SDTV and HDTV (Dulac and Godwin, 2006).

The key enablers to promote user acceptability of a DTH broadcast system are:

- easy interoperability of the DTH receiver and terrestrial television sets;
- reliable, low-cost and small outdoor installation, controllable easily from indoors;
- low-cost set-top box at par in all aspects with other consumer products;
- cinema-quality television pictures on sleek television sets;
- availability in remote areas – rich and interactive content equalling, if not bettering, the terrestrial television offerings;
- high-quality commercial practices.

Enabling technology and services include a digital video recorder to facilitate recording of programme directly in a digital format, interactivity (e.g. multichannel display or multiple camera angle displays), receiver-enabled home networking, reception on mobile vehicles (KVH Industries Online, 2010), as shown in Figure 9.3, and aircraft (JetBlue Airways Online, 2010), etc.

ITU's Broadcast Satellite Service (BSS) radio frequency plan provides a useful framework with guaranteed availability of spectrum in Ku band to each member country, allowing high-power radio transmissions amenable for reception at home via small non-obtrusive antenna on low-cost receivers.

In the remainder of this section we present a generic architecture of a direct broadcast system and an Integrated Receiver Decoder (IRD). Next, we highlight the salient characteristics of prevalent transmission standards recognized and recommended by the ITU. We highlight the technology of the Digital Video Broadcast-Second Generation (DVB-S2) satellite system, as it incorporates a wide range of recent technology advances to provide a highly flexible and efficient medium, and moreover it is uniquely identified by the ITU as a broadcast system for digital satellite broadcasting system with flexible configuration.

Chapter 10 presents examples of operational broadcast systems.

9.3.1 Architecture

Transmission System Architecture
A typical direct-to-home system architecture is represented in Figure 9.4 (after Dulac and Godwin, 2006).

(a)   (b)

(c)

Figure 9.3 (a) A low-profile phased-array outdoor unit feeds signals to a DTH receiver for back-seat viewing, (b) back-seat viewers and (c) A DTH antenna and receiver system in a luxury yacht. Reproduced from © KVH Industries.

The system comprises incoming signals to the uplink facility from one or more sources – a studio, a local terrestrial broadcast signal feed, pre-recorded material, etc. Occasionally, additional material such as advertisements may be added locally at pre-agreed points of the incoming programme. The incoming signals are monitored, routed within the facility and if necessary, readjusted and synchronized. The prerecorded material is checked for quality, edited, when necessary, and read into video file servers to be played at the broadcast time.

Commercial systems incorporate a facility known as conditional access to facilitate reception solely by the authorized users. Other functionalities include additions of Service Information (SI) and Electronic Programme Guide (EPG), analogue-to-digital conversion, compression, multiplexing to create a suitable transport stream, error control and modulation. The SI and EPG equipment creates signals for displaying programme related information, for example programme title, start/end time, etc. The compression equipment is typically MPEG-2, although migration to MPEG-4 is

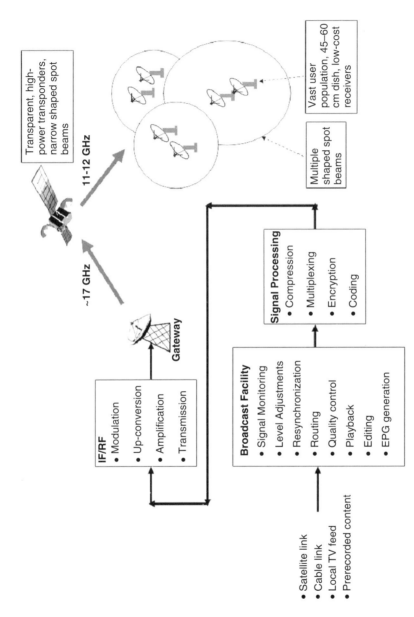

Figure 9.4   A typical direct-to-home system architecture (after Dulac and Godwin, 2006).

endemic because of its tighter encoding. Several channels are multiplexed to provide a single high-rate channel for transmission. The stream is forward error corrected, modulated and transmitted.

The ITU BSS plan recommends uplink transmissions in the 17.5 GHz band and downlink transmissions to the user community in the 12.5 GHz band. Some commercial systems operate outside the ITU plan in the Fixed Satellite Service (FSS) part of the Ku band, but their transmission powers are restricted by the FSS regulations.

BSS satellites are placed in geostationary orbits. The satellites use transparent transponders capable of transmitting very high powers through spot, and often, shaped beams to be able to provide the high signal quality necessary for reception on small DTH receivers. As per the BSS plan, incoming frequencies are in the 17.5 GHz band, and the transmission frequencies are in the 12.5 GHz band. Operators often collocate satellites, typically separated by ~0.1°, to maximize available power per orbital slot.

The DTH receivers typically use a 45–60 cm dish at Ku band, depending on satellite EIRP. The relatively high gain of the receive antenna in conjunction with high-power satellite transmissions generally provides sufficient link margins to counter rain fades, which are common in this band. The indoor unit housing the electronics is known as the Integrated Decoder Receiver (IRD).

Generic Reference IRD Model

The core functions of all the direct-to-home television systems are nominally identical, and hence a generic model of the IRD is feasible. Operators may tailor the remaining functions around these core functional entities. The ITU proposes reference architecture of the IRD on this premise. The model provides a structured definition of functionalities to facilitate a generic receiver design.

Consider, first, a protocol stack of a typical broadcast transmission system. Table 9.1 is an interpretation of the stack of a transmission system proposed by the ITU (ITU BO.1516, 2001). The reference IRD model is developed from the stack by separating the core elements from others:

- *The physical and link layers* are responsible for the physical transport of radio transmissions, including the functions of modulation/demodulation, coding/decoding, interleaving/deinterleaving and application/removal of energy dispersal.
- *The transport layer* addresses the multiplexing/demultiplexing of the programmes and associated components, and packetization/depacketization of the content.
- *Conditional access* controls the operation of the encryption/decryption of programmes and the associated functions.

Table 9.1 A generic protocol stack model of a digital broadcast system

Layer	Transmit	Receive
Application	Customer services	Services presentation on TV, PC, etc.
Presentation	Introduce presentation features	Interface to users (e.g. remote control)
Network	Introduce EPG services information, etc.	Obtain EPG, service information, etc.
Session	Conditional access: control and encryption, programme organization	Conditional access: control and decryption programme selection
Transport	Packetization: audio, video, data	Depacketization: audio, video, data
Link	Add energy dispersal	Remove energy dispersal, de-interleave, decode
Physical	Processing for physical transport	Processing to retrieve raw data

- *Network services* perform video and audio encoding/decoding, Electronic Programme Guide (EPG) functions, service information management and optionally, data decoding.
- *The presentation layer* is responsible for the user interface and operation of the remote control.
- *Customer services* deal with the different user applications such as audio and video programme display.

Figure 9.5 derived from the ITU (ITU BO.1516, 2001) illustrates a generic conceptual architecture of an Integrated Receiver Decoder (IRD) to illustrate various functions, demonstrate the organization of the common elements within the IRD and identify elements that may differ between IRDs.

As observed in the preceding section, the *core functions* relate generically to a transmission system. The additional essential functions relate to the service provision, operation of the system and additional or complementary features, which may differ depending on implementation. These functions and units include: a satellite tuner, output interfaces, an operating system and applications, EPG, Service/system Information (SI), Conditional Access (CA), a display, a remote control, Read Only Memory (ROM), Random Access Memory (RAM), FLASH memory, an interactive module, a microcontroller and units to support teletext, subtitling, etc.

A typical DTH receiver comprises a 45–60 cm dish usually with an offset feed configuration, mounted unobtrusively in a location of clear visibility to the satellite. A phase-array configuration is useful for special applications – typically a mobile television receiver to receive BSS transmissions. The low-noise amplifier and block down-converter (LNB) are mounted on the feed to minimize front-end noise, and the L-band IF connected by a cable to the IRD inside the house. An example

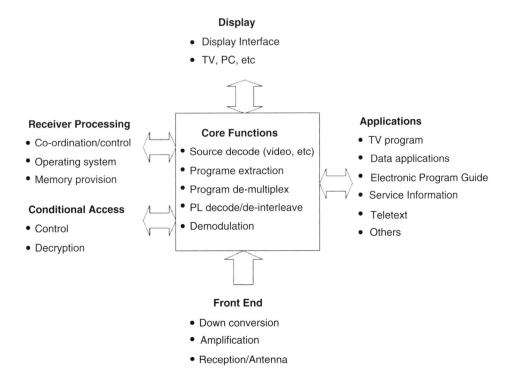

Figure 9.5 A generic model of an IRD illustrating the main core functions shared by each type of IRD and associated functions. Adapted from (ITU BO.1516, 2001).

equipment is shown in Chapter 10. The cable also feeds DC power to the low-noise block. A tuner in the IRD filters the frequency range of the receiving programme from the L-band intermediate-frequency spectrum.

The link layer demodulates, decodes and de-interleaves the signals and removes energy dispersal signals. Energy dispersal signals such as a low-frequency triangular wave avoid spectral spikes in the transmission spectra, thereby reducing the risks of interference with other systems. The packets are next demultiplexed and separated as video, audio and data streams in the transport layer. Under the control of the conditional access, the signals are de-encrypted. The network services layer has the role of decoding the video, audio and data packets and providing network services including EPG and service information. The user interacts with the presentation and customer services layers through a remote control or switches installed on the television set. The chosen programme is converted into analogue signals and fed into the television display unit.

Software-reconfigurable receivers are common as they simplify an upgrade to the receiver. The upgrade may, for example, become necessary to repair a software anomaly, add a new functionality or reconfigure receiver subsystems when a new satellite transponder is deployed. Modern receivers typically include a digital video recorder permitting users to record a programme directly in a digital format, or to store content, ready for an 'instant' video-on-demand display, thereby avoiding the interactive delay. Often L-band IF, signals from a single dish/LNB flow to one or more receivers in the customer's home. Home networking systems permit interworking between receivers, thereby permitting programmes recorded on one set to be viewed by other sets elsewhere in the house, and, in addition, support features such as security and quality of service management.

9.3.2   Transmission Standards

The development of digital television has evolved independently around the world, and hence several types of transmission system are in use. The majority operate in the 11/12 GHz downlink band. Being digital, the systems can support numerous applications and services efficiently – be it television, multimedia, data services or audio.

To assist in the selection of an appropriate system, ITU recommends four systems – system A, system B, system C and system D. Some interesting features of these systems are summarized below, and their main attributes are compared in Table 9.2 (condensed from ITU BO.1516, 2001). System E is treated in some detail subsequently.

System A
- Utilizes the MPEG-2 video and sound coding algorithms.
- The baseband transport multiplex is based on MPEG-2.
- Robust radio-frequency (RF) performance is achieved by a concatenated Forward Error Correction (FEC) scheme using Reed–Solomon and convolution coding, with soft-decision Viterbi decoding.
- Five coding rate steps in the range 1/2 to 7/8 provide spectrum and power efficiency trade-offs.
- Transmission symbol rates are not specified, allowing operators to optimize the satellite transponder power bandwidth.

System B
- Utilizes the MPEG-2 video coding algorithm and MPEG-1 layer II audio syntax and transport.
- Does not use MPEG-2 transport.
- The coding scheme is similar to that of system A.
- Three coding rate steps in the range 1/2 to 6/7 provide spectrum and power efficiency trade-offs.
- The transmission symbol rate is fixed at 20 Mbaud.

Table 9.2 A comparison of a significant parameters of ITU-recommended direct broadcast systems (condensed from ITU BO.1516, 2001) (courtesy of ITU)

Parameter	System A	System B	System C	System D
Services	SDTV and HDTV	SDTV and HDTV	SDTV and HDTV	SDTV and HDTV
Input signal format	MPEG-TS	Modified MPEG-TS	MPEG-TS	MPEG-TS
Rain fade survivability	Determined by transmitter power and inner code rate	Determined by transmitter power and inner code rate	Determined by transmitter power and inner code rate	Hierarchical transmission is available in addition to the transmitter power and inner code rate
Common receiver design with other receiver systems	Systems A, B, C and D are possible	Systems A, B, C and D are possible	Systems A, B, C and D are possible	Systems A, B, C and D are possible
Commonality with other media (i.e. terrestrial, cable, etc.)	MPEG-TS basis	MPEG-ES (elementary stream) basis	MPEG-TS basis	MPEG-TS basis
ITU-R Recommendations	Rec. ITU-R BO.1121 and Rec. ITU-R BO.1294	Rec. ITU-R BO.1294	Rec. ITU-R BO.1294	Rec. ITU-R BO.1408
Net data rate (without parity)	Symbol rate (R_s) is not fixed. The following net data rates result from an example R_s of 27.776 Mbaud: Rate 1/2: 23.754 Mb/s; rate 7/8: 41.570 Mb/s	Rate 1/2: 17.69 Mb/s Rate 6/7: 30.32 Mb/s	19.5 Mbaud, 29.3 Mbaud 1/2: 18.0 Mb/s, 27.0 Mb/s 7/8: 31.5 Mb/s, 47.2 Mb/s	Up to 52.2 Mb/s (at a symbol rate of 28.86 Mbaud)
Modulation scheme	QPSK	QPSK	QPSK	TC8-PSK/QPSK/BPSK
− 3 dB bandwidth	Not specified	24 MHz	19.5 and 29.3 MHz	Not specified (e.g. 28.86 MHz)
Reed–Solomon outer code	(204, 188, $T = 8$)	(146, 130, $T = 8$)	(204, 188, $T = 8$)	(204, 188, $T = 8$)
Interleaving	Convolution, $I = 12$, $M = 17$ (Forney)	Convolution, $N_1 = 13$, $N_2 = 146$ (Ramsey II)	Convolution, $I = 12$, $M = 19$ (Forney)	Block (depth = 8)
Inner coding	Convolution	Convolution	Convolution	Convolution, Trellis (8-PSK: TCM 2/3)
Inner coding rate	1/2, 2/3, 3/4, 5/6, 7/8	1/2, 2/3, 6/7	1/2, 2/3, 3/4, 3/5, 4/5, 5/6, 5/11, 7/8	1/2, 3/4, 2/3, 5/6, 7/8
Packet size	188 bytes	130 bytes	188 bytes	188 bytes
Transport layer	MPEG-2	Non-MPEG	MPEG-2	MPEG-2

(continued overleaf)

Table 9.2   (*continued*)

Parameter	System A	System B	System C	System D
Satellite downlink frequency range	Originally designed for 11/12 GHz, not excluding other satellite frequency ranges	Originally designed for 11/12 GHz, not excluding other satellite frequency ranges	Originally designed for 11/12 GHz and 4 GHz satellite frequency ranges	Originally designed for 11/12 GHz, not excluding other satellite frequency ranges
Video coding syntax	MPEG-2	MPEG-2	MPEG-2	MPEG-2
Video coding levels	At least main level	At least main level	At least main level	From low level to high level
Video coding profiles	At least main profile	At least main profile	At least main profile	Main profile
Aspect ratios	4:3, 16:9 (2.12:1 optionally)	4:3, 16:9	4:3, 16:9	4:3, 16:9
Audio source decoding	MPEG-2, layers I and II	MPEG-1, layer II; ATSC A/53 (AC3)	ATSC A/53 or MPEG-2, layers I and II	MPEG-2 AAC
Teletext	Supported	Not specified	Not specified	User selectable
Closed caption	Not specified	Yes	Yes	Supported

System C
- Supports the MPEG-2 video coding algorithm and ATSC A/53 or MPEG-2 layers I and II audio coding.
- The baseband transport multiplex is based on MPEG-2.
- Supports multiple digital television (and radio) services in Time Division Multiplexed (TDM) format.
- The coding architecture is similar to that of system A.
- Eight coding rate steps in the range 1/2 to 7/8 provide spectrum and power efficiency trade-offs.

System D
- Supports broadcast of multimedia services.
- Integrates various kinds of digital content, for example, Low Definition TV (LDTV), High Definition TV (HDTV), multiprogramme audio, graphics, texts, etc.
- The baseband transport multiplex is based on MPEG-2.
- A series of selectable modulation and/or error protection schemes are available to allow spectrum and power efficiency trade-offs

9.3.2.1   Digital Video Broadcast-Second Generation (DVB-S2) Standard

The Digital Video Broadcast (DVB) project, initiated in Europe but subsequently extended worldwide, defines digital broadcast standards by consensus among the participants representing the manufacturing industry, broadcasters, programme providers, network operators, satellite operators and regulatory bodies (DVB Project Online, 2010). The DVB standards embrace broadcast transmission technology across all media – cable, terrestrial and satellite. The standards are widely used in Europe, North and South America, Asia, Africa and Australia.

The DVB-S specifications were standardized in 1993 (as EN 300 421). A second-generation specification, DVB-S2, was produced in 2003 in response to a growing demand for more capacity and services, taking advantage of the advances in technology. The DVB-S2 system was standardized by the European Telecommunication Standard Institute (ETSI) as EN 302 307 (ETSI Online). In August 2006 the ITU's study group on satellite delivery issued a recommendation (BO.1784) that DVB-S2 be the preferred option for a 'Digital Satellite Broadcasting System with Flexible Configuration (Television, Sound and Data)', entitled system E.

The DVB-S standards support up to 12 categories of transmission medium encompassing a plethora of media contents – among others, standard- and high-definition television, radio and data with or without user interactivity. The standard includes specifications for Internet Protocol (IP) data, software downloads and transmissions to handheld devices. The standards, namely DVB-Satellite (DVB-S), DVB-Return Channel Satellite (DVB-RCS) and DVB-Second Generation (DVB-S2) apply to fixed user terminals. The DVB-Satellite Handheld (DVB-SH) standard, discussed later in the chapter, applies to handheld terminals.

Many parts of the DVB-S specifications are shared between various transmission media. The source coding of video and audio and formation of the transport stream comprises MPEG-2 tailored for satellite systems (MPEG specifications are otherwise too generic). DVB-S also supports H.264/AVC video and MPEG-4 high-efficiency AAC audio, and, additionally, audio formats such as Dolby AC-3 and DTS coded audio. Guidelines are also available for transporting IP content. The specifications support teletext and other data transmitted during the vertical blanking period, subtitles, graphical elements, service information, etc.

DVB-S's enhanced version, DVB-S2, is based on three key concepts (Morello and Mignone, 2004):

- best transmission performance;
- total flexibility;
- reasonable receiver complexity.

The specifications enable delivery of services that could never have been delivered by DVB-S. According to the developers, the DVB-S2 standard is not intended to replace DVB-S in the short term for conventional TV broadcasting applications but is rather aimed at new applications such as the delivery of HDTV and IP-based services, fly-away small DSNG stations, low-cost Internet access to rural areas and in developing countries, etc. The DVB-S2 specifications, in conjunction with recent advances in video compression, have enabled the widespread commercial launch of HDTV services. The supported applications are:

- standard- and high-definition television broadcasts;
- interactive services, including Internet access for consumer applications;
- professional applications: Digital TV Contribution (DTVC) facilitating point-to-multipoint transmissions, and Digital Satellite News Gathering (DSNG);
- content distribution;
- Internet trunking.

In addition to MPEG-2 video and audio coding, DVB-S2 is designed to handle a variety of advanced audio and video formats which the DVB Project is currently defining. In fact, DVB-S2 accommodates most common digital input stream format, including single or multiple MPEG transport streams, continuous bit streams, IP as well as ATM packets.

It is 30% spectrally more efficient than its predecessor, employing an adaptable modulation scheme consisting of QPSK and 8-PSK for broadcast applications and 16-APSK and 32-APSK for professional applications such as news gathering and interactive services.

The modulation and coding schemes may be dynamically adapted to variable channel condition on a frame-by-frame basis. The coding arrangement consists of a Bose–Chaudhuri–Hocquenghem (BCH) outer code and a Low-Density Parity Check (LDPC) inner code. The communication performance lies within 0.7 dB of the theoretical limit. The flexibility offered by variable coding and modulation provides different levels of protection to services as needed.

The specifications support operation on any type of satellite transponder characteristics with a large variety of spectrum efficiencies and associated C/N requirements. The DVB-S2 broadcast services comprise a backward compatible mode and a more optimized version, which is not backward compatible.

The system allows interactive services by return channels established either via satellite or another medium incorporating the added flexibility of adaptive coding and modulation (ACM) to the forward channel through feedback.

Structured as a 'toolkit', DVB-S2 attempts to optimize transmission efficiency and flexibility, keeping receiver costs at an acceptable level.

DVB-S2 Architecture
Figure 9.6 shows a block schematic of functional blocks of a DVB-S2 transmission system.

There are two levels of framing structures – one at baseband (BBFRAME) and the other at physical layer (PLFRAME). BBFRAME includes signalling to configure the receiver for the given specification and service. The PLFRAME comprises a regular sequence of frames, each coded and modulated for the given channel condition and application, containing a few signalling bits for synchronization and physical layer signalling.

Referring to BBFRAME (in Figure 9.6), the mode and stream adaptation block interfaces with the incoming stream, provides synchronization and supports the adaptive coding- modulation schemes. It merges multiple streams and slices them into blocks, each of which are modulated and coded homogeneously. A header of 80 bits containing transmission characteristics is next appended to the baseband data block to assist reception. The information within the header, for example, informs the receiver as to whether the transmission stream is single or multiple, the type of coding modulation schemes, signal format, etc. When the volume of data is insufficient to fill the frame, data padding is introduced, and finally the frame is scrambled and passed over for coding.

The FEC coding-modulation schemes are instrumental in high transmission efficiency. The FEC code comprises a Low-Density Parity Check (LDPC) code that exhibits a low distance from the Shannon limit for the specified decoder complexity (equating to $14\,mm^2$ of silicon $0.13\,\mu m$ technology). LDPC codes are suitable for iterative decoding at reasonable complexity because of their easily parallelizable decoding algorithm which can use simple operations resulting in reduced complexity and storage. To avoid error floors at low error rates, a BCH outer code of the same block

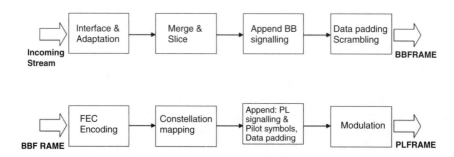

Figure 9.6 A DVB-S2 transmission system.

length as the LPDC code is concatenated. The total block comprises 64 800 bits for applications not sensitive to delays and 16 200 bits for delay-sensitive applications. Code rates vary from 1/4 to 9/10, depending on modulation scheme, radio link conditions and application. The bit rate was targeted as $\leq 10^{-7}$, equivalent to an error of 1 bit per hour for a 5 Mb/s incoming stream.

The coded payload modulates the carrier with a QPSK, 8-PSK, 16-APSK or 32-APSK scheme, as required by the given application data rate and link conditions. The corresponding spectral efficiency ranges from 0.5 to 4.5 b/symbol. Typically, QPSK and 8-PSK schemes are applied for broadcast applications through nonlinear transponders, whereas 16 and 32-APSK schemes are better suited for professional and other applications operating through transponders driven in the quasi-linear region. DVB-S2 provides backward compatibility with DVB-S through a hierarchical modulation scheme.

The physical layer (PL) signals are composed of a regular sequence of frames, within each of which the modulation and coding schemes remain unchanged, that is a homogeneous operation. A header of 90 bits preceding the payload assists in synchronization at the receiver, detection of modulation and coding parameters, etc. This header must be made particularly robust, as the LPDC/BCH scheme is not applied here and the packet will not be detectable unless the header is detected correctly. An interleaved first-order Reed–Muller block code in tandem with the $\pi/2$-BPSK modulation scheme was found to be suitable. The PL framing scheme also introduces dummy frames in the absence of data, and pilot symbols to aid synchronization at the receiver when the signal-to-noise ratio (SNR) is low. Finally, the signal is scrambled to avoid energy spikes in transmissions using a 'signature', which additionally assists in identification of the uplink facility in the case of system malfunction in the uplink.

With regard to the receiver, some of the main issues are synchronization, demodulation and decoding. Synchronization poses a challenging problem in the presence of such a wide range of dynamically changing modulation and coding schemes in low SNR. Various techniques may be used – for example, Gardner's algorithm for clock recovery, differential correlation for physical layer frame synchronization, a pilot-aided scheme for frequency recovery and variants of pilot-aided techniques for phase recovery and digital automatic gain control to assist accurate soft information at the decoder. The LDPC decoding is achievable through an iterative exchange of information between bit and check nodes.

The system can operate at Energy per Symbol to Noise power density ratio (Es/No) from −2.4 dB (QPSK, code rate 1/4) to 16 dB (32-APSK, code rate 9/10) (Figure 9.7). Figure 9.7 (Morello and Mignone, 2006) compares the spectral efficiencies of DVB-S and DVB-S2 system performance in white noise. Note that DVB-S2 offers around 20–35% increase in capacity or 2–2.5 dB advantage over DVB-S and DVB-DSNG performance.

Figure 9.8 presents a block schematic of a DVB-S2 interactive unicast system demonstrating Adaptive Coding and Modulation (ACM). The coding and modulation schemes in the forward direction can change frame by frame to match the link conditions reported back by the user terminal. It has been shown that, compared with the Constant Coding and Modulation scheme (CCM), ACM typically offers a capacity gain factor of 2–3 – a figure likely to improve when propagation conditions worsen.

9.4 Direct Broadcasts to Individuals and Mobile Users

Direct-to-individual satellite broadcast systems are a recent addition to the catalogue of popular personal electronic systems, such as multifunctional mobile phones and music systems, GPS-enabled navigation aids, etc. They provide sound, multimedia, television and similar services on small portable or mobile user terminals. Other applications of such systems include: customized audio or video content for mobiles (e.g. virtual TV channels, podcasts); data delivery (e.g. for ring tones); video-on-demand services; information (e.g. news); location-based services (e.g. traffic); interactive services via an external communications channel (e.g. UMTS), etc. (Kelley and Rigal, 2007).

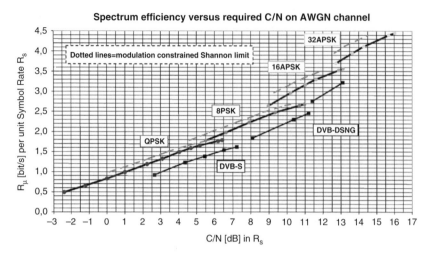

Figure 9.7 A comparison of DVB and DVB-S2 performance (Morello and Mignone, 2006). Reproduced by permission of © IEEE.

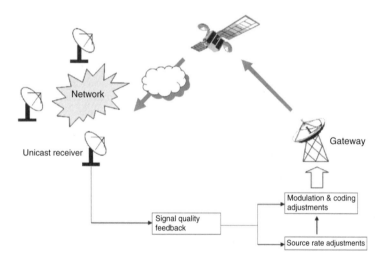

Figure 9.8 A block schematic of a DVB-S2 unicast system illustrating dynamic control of link parameters.

As remarked previously, television receivers mountable on cars, ships, and aircraft for reception of conventional DTH transmissions are now routine. These types of application use receivers that install agile tracking antennas and, where necessary, modification to the receivers to compensate for the Doppler (e.g. KVH Industries Online, 2010; JetBlue Airways Online, 2010). The European Space Agency demonstrated the feasibility of supporting mobile broadcast systems on vehicles via ageing satellites (i.e., those with insufficient fuel to maintain sufficiently low orbit inclination) with the compounded aim of inculcating a fresh lease to such satellites. This class of receivers belongs to the DTH category and hence is not considered in this section. Similarly, the 'suitcase' DTH receivers, which allow easy portability for reassembly elsewhere, are also excluded.

One of the fundamental problems in providing satellite services to handheld and mobile units is due to signal shadowing, compounded by the limited sensitivity of the receivers because of their compact antennas. In spite of high-power transmissions of modern satellites, the link margin available at the receiver is generally insufficient to offset the shadowing and multipath, necessitating robust countermeasures. As a further consideration, it becomes necessary to produce the receivers costeffectively and ergonomically to be commercially attractive.

One technique to solve the shadowing problem is through repeat transmissions from terrestrial repeaters which fill the coverage gaps. Additionally, the transmission format is designed to resist light shadowing and rapid deep fades by incorporating appropriate interleaving and coding schemes.

The ITU recommends five types of S-band satellite systems to promote broadcasts to individual and vehicle units in Recommendation ITU-R BO.1130-4 (2001). ETSI has standardized the Digital Video Broadcast-Satellite Handheld (DVB-SH) system for delivery of video, audio and data services to handheld devices at S-band. The following paragraphs introduce salient characteristics of these standards, with an emphasis on digital system E (ITU-R Recommendation BO.1130-4) and DVB-SH which incorporate some of the latest advances in technology.

Table 9.3 (condensed from ITU BO.1130-4, 2001) compares a few salient features of ITU systems. Chapter 10 presents examples of two operational systems.

The main features of each of these systems are highlighted below. System E is described in detail later in the section.

Digital System A (DAB)

- The system is also known as the Eureka 147 DAB system.
- It applies to satellite and terrestrial broadcasting applications for vehicular, portable and fixed reception.
- It includes complementary terrestrial repeaters.
- It enables production of a low-cost dual-mode receiver.
- It offers various levels of sound quality up to CD quality.
- It supports data services, conditional access and the dynamic rearrangement of services.
- It delivers services in the 1400–2700 MHz band.

Digital System B

- The system was proposed by Voice of America/Jet Propulsion Laboratory.
- It applies to digital audio and data broadcasts.
- It is targeted for vehicular, portable and fixed (indoor, outdoor) receivers.
- It includes complementary terrestrial repeaters.
- It enables the production of a low-cost receiver.
- It uses adaptive equalizers for multipath rejection.
- It delivers services in the 1400–2700 MHz band.

Digital System D_S

- The system is also known as the WorldSpace system.
- It applies to digital audio and data broadcasting.
- It enables the production of an inexpensive fixed and portable receiver.
- Its services are best delivered in the 1452–1492 MHz band.

Digital System D_H

- The system is also known as the hybrid satellite/terrestrial WorldSpace system.
- It provides digital audio and data broadcasting to vehicular, fixed and portable low-cost receivers.

Table 9.3 A comparison of ITU recommended digital broadcast systems to small receivers (condensed from ITU-R BO.2001-4, 2001) (courtesy of ITU)

Characteristics	Digital system A	Digital system B	Digital system D_S	Digital system D_H	Digital system E
Target installation	Vehicular, portable and fixed	Vehicular, portable and fixed (indoor, outdoor)	Portable and fixed	Vehicular, portable and fixed	Vehicular, portable and fixed
Service	Digital audio and data broadcasts	Digital audio and data broadcasts	Digital audio and data broadcasts	Digital audio and data broadcasts	High-quality digital audio, multimedia, data and medium-quality video broadcasts (15 frames/s)
Frequency band (MHz)	1400–2700	1400–2700	1400–2700	1400–2700	2630–2655
System type: mixed (satellite and terrestrial in same band); hybrid (use of on-channel terrestrial repeaters[1] to reinforce satellite signals); satellite delivery alone	Mixed and hybrid	Mixed and hybrid	Satellite delivery	Hybrid	Mixed and hybrid
Layered architecture	OSI compliant, with one deviation	Capable of OSI compliance (not tested)	OSI compliant	OSI compliant	MPEG-2 compliant
Audio range	8–384 kb/s in increments of 8 kb/s	16–320 kb/s per audio channel in increments of 16 kbit/s	16–128 kb/s per audio channel in increments of 6 kbit/s	16–128 kb/s per audio channel in increments of 16 kb/s. Each 16 kb/s increment can be split into two 8 kb/s services	16–320 kb/s per audio channel in any increment size

Audio codec	MPEG-2 layer II audio decoder typically operating at 192 kb/s	Perceptual Audio Codec (PAC) source encoder at 160 kb/s was used for most field tests	MPEG-2 and MPEG-2.5 layer III audio coding	MPEG 2 and MPEG-2.5 layer III audio coding	MPEG-2 AAC audio coding
Modulation	COFDM	QPSK with concatenated block and convolution error-correcting coding	QPSK modulation with concatenated block and convolution error-correcting coding	QPSK modulation with concatenated block and convolution error-correcting coding	CDMA based on QPSK modulation with concatenated block and convolution error-correcting coding
Performance in multipath and shadowing[2]	On-channel repeaters are used to service shadowed areas. Power from multipath echoes arising from on-channel repeaters, and falling within a given time interval, is summed to improve performance	Link margin is improved through time diversity, reception diversity and transmission diversity. On-channel repeaters are used to service shadowed areas. Multipath echoes caused by terrestrial repeaters are cancelled through equalizer	The system is designed primarily for direct reception with link margin for light shadowing	Time and space diversity is possible in satellite reception. Supports terrestrially retransmitted on-channel MultiChannel Modulation (MCM) for service in shadowed areas. Allows power summation of multipath echoes falling within a given time interval, including those from on-channel repeaters	Power summation of multipath is used with a Rake receiver. Shadowing countered through on-channel terrestrial transmitters and interleaving
System status	Terrestrial systems operational in several countries, including gap fillers; field tested over satellite at 1.5 GHz	Prototype stage; field tested over satellite for vehicular reception	Field tested; fully operational via 1worldspace's AfriStar and AsiaStar satellites	Field tested terrestrially and over satellite (Afristar)	Fully operational in Japan and Korea since 2004

[1]Known as the Auxiliary Terrestrial Component (ATC) or Complementary Ground Component (CGC).

[2]Multipath here refers to replica signals received as echos owing to multiple propagation paths; echoes falling on the main signals are typically countered by FEC coding in each system; if they are distinguishable, they could be summed to advantage or else cancelled.

- It includes complementary terrestrial repeaters.
- Satellite delivery is an enhanced version of System D_S to improve line-of-sight reception in areas partially shadowed by trees.
- It includes complementary terrestrial repeaters, based on MultiCarrier Modulation (MCM), which improves upon the techniques that are common in systems such as digital system A.
- It delivers services in the 1452–1492 MHz band.

Digital System E

- The system is also known as the ARIB (Association of Radio Industries and Businesses) system.
- It provides services for high-quality audio and multimedia data (including television) to vehicular, portable and fixed receivers.
- It is based on complementary terrestrial repeaters.
- It optimizes performance for both satellite and terrestrial on channel repeater service delivery in the 2630–2655 MHz band.

9.4.1   Architecture and Standards

The OSI reference model (ITU-T X.200, 1994) is a convenient tool for describing the digital broadcasting systems recommended by the ITU, as it enables a logical partitioning of functionality. Table 9.4 presents an interpretation of the OSI model for digital broadcasting as applied to system-A (ITU-R BO.1130-4, 2001).

Under this interpretation, the *physical layer* modulates, codes, interleaves and transmits the information on radio waves. It also introduces energy dispersal to each stream to minimize spectral spikes in transmitted waveforms. The *datalink layer* arranges the frame structure of transmission to facilitate receiver synchronization. The *network layer* concerns the identification of groups of data as programmes. The *session layer* is concerned with the selection and accessing of broadcast information and provision for conditional access. The *transport layer* deals with the identification of data as programme services and its multiplexing. The *presentation layer* deals with conversion of broadcast information and its presentation. This may involve audio source coding and decoding, presentation as monophonic or stereophonic, linking programmes simultaneously to a number of

Table 9.4   An interpolation of the OSI model for system A (ITU-R BO.1130)

Name of layer	Function	Example features (specific to system A)
Application layer	Deals with use of the system, that is, the application	System facilities; audio quality; transmission modes
Presentation layer	Readies the incoming stream for presentation	Audio encoding and decoding; audio presentation; service information
Session layer	Deals with data selection	Programme selection; conditional access
Transport layer	Groups data	Programme services; main service multiplex; ancillary data association
Network layer	Deals with the logical channel	Audio frames; programme-associated data
Datalink layer	Formats the transmitted signal	Transmission frames; synchronization
Physical layer	Deals with physical (radio) transmission	Energy dispersal; convolution encoding; time interleaving; frequency interleaving; modulation radio transmission

languages and producing programme information such as programme names. The *application layer* services deal directly with the user at the application level. For example, it considers the facilities (e.g. audio programme) and audio quality on offer.

System E

System E broadcasts digital audio, medium-quality video, multimedia and data via geostationary satellites augmented with on-channel terrestrial transmissions (Hirakawa *et al.*, 2006). Although optimized for 2630–2655 MHz, the concept can be applied to other frequency bands. The broadcast content is sent to the feeder Earth station which uplinks it to the satellite at 14 GHz. The satellite retransmits the signals after down conversion to 2.6 and 12 GHz bands. The 2.6 GHz signals are transmitted at a high EIRP for direct reception. The 12 GHz transmissions are feeds for gap-filler terrestrial transmitters.

The physical layer consists of a code division multiplexer (CDM) transmitted over a QPSK modulation scheme and concatenated Reed–Solomon convolution codes. The same scheme applies to both transmission components, that is, satellite and terrestrial. The data are interleaved at the transmitter to spread out the contiguous bits. The interleaving depth is such that the small blockage of signals lasting up to ~1 s that is typically encountered in vehicles owing to small obstructions such as signposts, bridges, a cluster of trees, etc., can be recovered.

Figure 9.9 illustrates a block diagram of the modulation process up to the transmitter section of 64 programme streams. The pilot channel introduces the pilot symbols. This figure illustrates the creation of a single channel of the CDM. Each stream can carry data up to a maximum rate of 236 kb/s; more than one stream is combined to achieve a higher data rate.

The serial data stream of each incoming channel is converted into *I* and *Q* data sequences, each of which are spread by the a unique 64 bit Walsh code used to identify the channel at the receiver (64 bits) and a truncated *M* sequence (2048 bits). The resulting stream is QPSK modulated, filtered (with a 0.22 roll-off) and multiplexed with up to 64 streams. Each channel can be separated at the receiver using the standard Rake receiver techniques when the Walsh code of a channel is known. While the payload is transmitted through the broadcast channel, pilot symbols are transmitted separately. The pilot transmissions aid receiver synchronization, simplify the path search in the Rake receiver and send system control data to assist in receiver functioning. The Rake receiver has

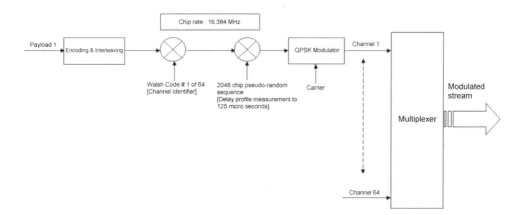

Figure 9.9 Transmitter section up to up-converter stage of a transmitter.

the ability to distinguish each echo and coherently add them to the main signal to give diversity advantage.

The basic bandwidth of the system is 25 MHz. The wave polarization is respectively circular and vertical in the satellite and terrestrial transmissions.

In a heavily shadowed environment, where direct transmissions cannot be received reliably, terrestrial transmitters known as gap fillers are used. The direct amplifying gap-filler arrangement retransmits the S-band received signal after amplification; because of the tendency of such a system to oscillate by constructive feedback, the retransmitted signal levels are low, and hence this type of gap filler is suited for small areas – for example, a narrow tunnel-like street. The frequency translation gap filler receives satellite signals at 12 GHz and retransmits them at a relatively high power after down conversion to the S-band. This arrangement can service an area of ∼3 km radius.

The MPEG-2 system is used for *service multiplexing*, thereby allowing transport of all types of data and permitting interoperability with other broadcast systems. Multimedia and data broadcasting are based on an ITU recommendation (ITU-R BT.1699) that is applied in various digital broadcasting systems such as DVB-S. The MPEG-2 AAC scheme is adopted for audio source coding, although lower bit rates such as 32 kb/s may be used. The MPEG-4 visual and AVC are used for multimedia and data broadcasting. The bit rate for video stream is 370 kb/s at 320×240 pixels spatial resolution and 15 frames/s time resolution.

The receiver architecture for mobile video reception used in the Mobile Broadcasting Corporation's (MBCO) broadcasting service in Japan employs both *antenna diversity* (reception on multiple antennas) and a Rake receiver, which enables useful reception of multipath echoes in a multipath environment. The front end comprises two antennas (for space diversity), followed by Rake CDM demodulation, decoding/de-interleaving and diversity combination. Conditional access deals with user verification of the chosen service; the demultiplexer recreates the desired service stream, followed by audio and video decoding, the outputs being displayed to the user.

9.4.1.1   Digital Video Broadcast–Satellite Handheld (DVB-SH)

In addition to the five ITU-recommended systems mentioned above, the Digital Video Broadcasting–Satellite services to Handheld (DVB-SH) standard is under active consideration in the European region and deployed in the United States. The DVB-SH is an ETSI S-band (<3 GHz) transmission standard (TS 102 585 and EN 302 583) designed to provide video, audio, data and IP-enabled multimedia services to a genre of modern handheld personal devices, such as mobile telephones and Personal Digital Assistants (PDAs) (DVB-SH Online, 2010). This is a hybrid system that combines the large area coverage of satellite systems with gap-filler terrestrial systems – known in Europe as Complementary Ground Components (CGCs) – which service the satellite coverage gaps caused by shadowing of satellite signals (as discussed in the preceding section).

Figure 9.10 illustrates the DVB-SH network concept.

The broadcast content is distributed to terrestrial and satellite transmitters for simultaneous broadcast. Three types of terrestrial transmitter are proposed: terrestrial broadcast infrastructure transmitters possibly located on existing cellular base stations, personal local gap fillers retransmitting on the same or translated frequency in a small area, for example, within a building, and mobile broadcast infrastructure transmitters located on moving trains, etc. Some of the terrestrial transmitters derive the transmission content directly from satellite signals. Both time division multiplex and Orthogonal Frequency Division Multiplex (OFDM) may be used in the terrestrial component.

The variant DVB-SH-A uses OFDM in both links, and DVB-SH-B uses TDM on satellite links and OFDM for terrestrial transmissions. The DVB-SH-A system must operate by backing off the high-power amplifier to a quasi-linear region of transponder, whereas DVB-SH-B can operate near full power. A forward error correction (FEC) scheme (3GPP2 turbo code), in conjunction with a channel interleaver, provides robust countermeasures against shadowing. The channel interleaver

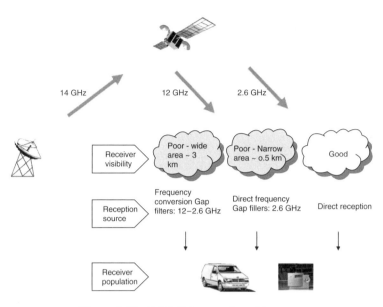

Figure 9.10   DVB-SH network architecture.

offers time diversity from about one hundred milliseconds to several seconds, depending on the targeted service level and the memory size available in the given terminal class.

At present, DVB-SH does not define protocols beyond the physical layer, although work is in progress to address this aspect to facilitate DVB-SH deployment.

The European Commission has confirmed availability of the S-band spectrum for mobile satellite services shared with complementary ground components of a hybrid system. DVB-SH technical pilots were successfully completed in Europe, with plans of an operational service through a joint consortium of Intelsat and SES-ASTRA, and DVB-SH-enabled services should be available commercially in the United States in 2010.

9.5   Military

Although military systems tend to be proprietary, the advantages of standards-based commercial technology by way of lower costs, wide interoperability, MPEG encapsulation support, asymmetric transmission capability allowing a different return medium, etc., have instigated military system planners to consider commercial broadcast technology as a viable transport platform for non-critical information.

NATO has demonstrated a DVB-enabled Satellite Broadcast System (SBS) aiming to highlight the split-path asymmetric and scalable architecture that can support secure high-speed IP services tailored for broadcast applications over military and civilian satellites using commercially available technology (Segura, 2002). Transmit and receive satellite gateways, which encompass DVB, IP encryption and standard routing devices, were attached to terrestrial wide-area network access nodes to create a satellite broadband overlay. Unicast and multicast IP flows were encapsulated into an MPEG-2 transport stream and transmitted. The transmissions could be received by small and light receive dish terminals for point-to-point and point-to-multipoint broadband data transfers.

Full-duplex communication was achieved over forward and return channels differing in bandwidth, latency, encryption, transmission medium and technology. Secure narrowband communications channels over any alternative medium could be used to carry return traffic from consumers. The asymmetric networking techniques used Internet Engineering Task Force standards so as to make the link transparent to the upper protocol layers, thereby enabling services over the Internet Protocol Suite (TCP/IP), reliable data dissemination, real-time video streaming and channel subscription using IP multicast. By diverting the most bandwidth-intensive streams over the satellite path, terrestrial networks could potentially be relieved of congestion. In this configuration, multicasting streams could be tunnelled and routed from dispersed sources to an uplink injection point and selectively encrypted if necessary. It was also possible to route the MPEG-2 streams via the terrestrial DVB-T system to demonstrate satellite-terrestrial broadcast interoperability and an efficient, hybrid transport architecture. DVB-T transmissions over UHF channels to relay the satellite broadcast to mobile receive suites were demonstrated. Return links into the injection point were established over Inmarsat and other personal mobile communications systems.

DVB-S technology is also used in the US Global Broadcast Service (GBS), a joint-service programme. The GBS system provides rapid, asymmetric, high-volume data such as imagery to small, mobile receivers over commercial direct broadcast satellite technology either as multicast or unicast on demand. GBS comprises information sources, uplinks, broadcast satellites, receiver terminals and management processes for requesting and coordinating the distribution of information products.

Revision Questions

1. What are the key enablers for wide user acceptability of a direct-to-home broadcast system?
2. With the help of a diagram explain the main characteristics of a typical direct broadcast system.
3. State the advantages of a generic reference model pertaining to an Integrated Receiver Decoder (IRD).
4. Illustrate a typical protocol stack of an IRD and identify the functions that can be shared between the receivers of various ITU-recommended systems.
5. Outline the differences between DVB-S2 and DVB-SH standards.
6. What are the main technical considerations for provision of sound and multimedia broadcast to handheld receivers?
7. With the help of a block diagram, describe the main features of system E. Compare its characteristics with those of system D, highlighting the main differences and their effects on system performance.
8. State reasons for preference of Ku band for the DTH broadcast service and S band for direct-to-individual broadcasts? (Hint: consider propagation and antenna.)

Project. Survey and compare the technical and commercial features of commercial DTH systems available in your region (e.g. Europe, United States, Asia, etc.). These features include but are not limited to: ITU system identification, operational frequency band, whether BSS or FSS band is in use, characteristics of the feeder station, characteristics and coverage of the satellite (e.g. coverage contour, EIRP), receiver characteristics, services (HDTV, SDTV, Internet, video on demand, etc.), transponder and uplink ownership, number of free and subscription channels, the cost of a typical commercial package, etc.

On basis of the information gathered:

(a) Compare their characteristics and write a critique.
(b) Compare features of any one DTH system with a terrestrial broadcast and a cable system available in your area. Include total cost to a consumer for year 1.
(c) Estimate an approximate yearly cost of running a DTH system for any one system of your choice.

(d) Estimate revenue and profit loss as a function of time – in a 7 year timeframe – using data estimated in (c), an assumed capital amortization and a typical subscription/user. Consider slow/medium/high linear growth of subscribers, assuming a reasonable number of start-up subscribers in year 1. State clearly the rationale for each assumption.

(e) For the model developed in (d), determine the period to break even for each growth model.

(f) Develop a link budget for the DTH downlink at the centre and the edge of coverage, stating all the assumptions.

References

Dulac, S.P. and Godwin, J.P. (2006) Satellite direct-to-home. *Proceedings of the IEEE*, **94**(1), 158–172.

DVB Project Online (2010) http://www.dvb.org [accessed February 2010].

DVB-SH Online (2010) http://www.dvb.org/technology/fact_sheets/DVB-SH-Fact-Sheet.0709.pdf [accessed February 2010].

ETSI Online (2010) http://www.etsi.org/deliver/etsi_en/302300_302399/302307/01.02.01_40/en_30237v0102010.pdf [accessed February 2010].

Hirakawa, S., Sato, N. and Kikuchi, H. (2006) Broadcasting satellite services for mobile reception. *Proceedings of the IEEE*, **94**(1), 327–332.

ISO/IEC 13818-1:2000(E) (2000) Information technology – generic coding of movin pictures and associated audio information systems, 2nd edition.

ITU-R BO.1516 (2001) Recommendation BO. 1516, Digital multiprogramme television systems for use by satellites operating in 11/12 GHz frequency range.

ITU-R BO.1130-4 (2001) Recommendation BO.1130-4, Systems for digital sound broadcasting to vehicular, portable and fixed receivers for broadcasting satellite service (sound) bands in the frequency range 1400–2700 MHz.

ITU-T X.200 (1994) Recommendation X.200, Information technology-open systems interconnection-basic reference model: the basic model. Available: http://www.itu.int/rec/T-REC-X.200-199407-I/en [accessed February 2010].

JetBlue Airways Online (2010) DirecTV Inflight Service. Available: http://www.jetblue.com/whyyoulllike/directv/guide/TribuneTVFront [accessed February 2010].

Kelley, P. and Rigal, C. (2007) DVB-SH – Mobile digital TV in S band. EBU Technical Review, July, 9pp.

KVH Industries Online *Mobile Satellite TV*. Available: http://www.kvh.com/tracvision_kvh/ [accessed February 2010].

Morello, A. and Mignone, V. (2004) DVB-S2 ready to lift off. *EBU Technical Review*, October.

Morello, A. and Mignone, V. (2006) DVB-S2: the second-generation standard for satellite broadband services. *Proceedings of the IEEE*, **94**(1), 210–227.

Pan, D. (1995) A tutorial on MPEG/audio compression. *IEEE Multimedia*, **2**, 60–74.

Segura, R. (2002) Commercial broadcast system extends military reach. *SIGNAL magazine*, February. Available: http://www.afcea.org/signal/articles/anmviewer.asp?a=440&print=no [accessed February 2010].

10

Broadcast Systems

10.1 Introduction

Owing to their capability of unbiased service over vast expanse, satellite systems offer a unique broadcast platform to service large regions economically and effectively. In an accomplished satellite broadcast system, the huge capital outlay is offset by millions of subscribers, thereby making the system economically viable. Broadcasters often exercise the option to lease the space segment from a specialist satellite operator to minimize the associated investment costs and risks. Estimates place the number of subscribers at well over 100 million worldwide, promoting DTH receivers as one of the most successful consumer products of the satellite industry to date. As an adjunct, the criticism regarding the propagation delay inherent in geostationary satellite systems does not apply to non-interactive broadcasts.

Chapter 9 described the underpinning techniques of satellite broadcast systems. Here, we introduce representative digital satellite broadcast systems to demonstrate their applications. We selected the examples purely on the basis of their technical merits and diversity, intending to expose the reader to a variety of concepts. The commercial nature of these undertakings may necessitate changes to their affiliations, service offerings, configuration and architecture. Hence, we recommend that the interested reader consult the company's online literature to keep abreast.

The first part of the chapter deals with satellite radio and multimedia broadcasts to small portable and mobile receivers. The first-generation systems primarily targeted sound broadcasts; the second-generation systems, facilitated by evolution in spacecraft and handset technology, offer a wider portfolio of service, including multimedia.

The second part of the chapter examines practical issues essential in the implementation of Direct-To-Home (DTH) receivers, gives a general global review of DTH system status; and concludes with examples of prevalent European and US systems. Finally, we present a case study of a military broadcast system addressing the US Global Broadcast Service (GBS) used by the US military in various theatres.

10.2 Satellite Radio Systems

Satellite radio broadcasts targeting portable, mobile and handheld receivers are called variously satellite radio, Satellite Digital Audio Radio Service (SDARS) or Satellite-Digital Audio Broadcast (S-DAB). The concept of direct radio broadcasts by satellites was proposed by Arthur Clarke in 1945 (Clarke, 1945), but its realization had to wait decades. Considerable effort was directed towards developing the technology in the 1990s, leading to its fruition near the turn of the

Satellite Systems for Personal Applications: Concepts and Technology Madhavendra Richharia and Leslie David Westbrook
© 2010 John Wiley & Sons, Ltd

century (1999–2001). Commercial services are now available in the United States, Africa, Asia, Europe, the Middle East and the Asia-Pacific region. Main concerns when planning commercial service were whether the general audience, used to a free terrestrial service, would accept subscription services, without a local feel as satellite broadcasts target a widely dispersed audience, and whether receivers suitable for a consumer market could be produced. It was believed that the large service area and uninterrupted CD-quality digital satellite radio channels bundled with value-added services unaffected by terrain, interference, noise and signal variability would be an attractive alternative to the vehicular audience compared with the traditional local FM channels which suffered regular fade-outs and required regular retuning during a long travel. It can now be said that the inhibitions were exaggerated, as evidenced by the millions of satellite radio listeners worldwide, particularly in Canada and the United States. The technology has evolved rapidly since – invariably, all the recent systems either include mobile television and multimedia service or intend to do so.

Both geostationary and non-geostationary orbits are in use, augmented with terrestrial repeaters. Operators who service high-latitude regions favour the highly elliptical orbit where geostationary satellites appear at low elevation angles and hence receivers experience unacceptable signal blockage. In addition to the open transmission standards, proprietary schemes are in use.

In this section we will present the technologies of three commercial satellite radio systems – XM Satellite Radio Inc., Sirius Satellite Radio Inc. and 1worldspace (formerly, WorldSpace). The former two companies have merged into a single entity known as Sirius-XM Radio Inc., but here we will consider each separately because of the differences in their system architecture. 1worldspace applies a different technical solution and is therefore interesting for our purposes. Table 10.1 compares their technical attributes.

10.2.1 XM Satellite Radio Inc.

XM Radio, founded in 1989 as the American Mobile Satellite Corporation, changed its name to XM Satellite Radio in 1997 and began commercial service in 2001 after surmounting numerous financial, regulatory and commercial hurdles. In July 2008 the company merged with its rival Sirius Satellite Radio Inc. to form Sirius XM Radio Inc. Its customer base grew from ~6 million in 2006 to ~7.5 million in 2007. Table 10.1 summarizes the main features of the system, comparing them with Sirius and 1worldspace systems.

The company provides digital radio and datacast services in S band to mobile, portable and fixed receivers located in North America and parts of Canada. Its optimized system architecture comprises two geostationary satellites located at 85° W and 115° W and a network of terrestrial repeaters. The company targets road travellers who tend to lean heavily on radio entertainment on the premise that terrestrial radio stations are unable to match the uninterrupted CD-quality broadcasts on offer. XM Satellite Radio's business plan assumed that the vast coverage area of satellite would provide a sufficient number of subscription-paying customers.

The space segment consists of two powerful satellites, each capable of emitting an Effective Isotropic Radiated Power (EIRP) of >68 dBW from a 5 m aperture antenna. The Continental United States (CONUS) – shaped coverage is also shaped in the z-axis to direct emissions towards the population centres and the beam edges. The antennas are fed through a parallel combination of 16 high-power Travelling Wave Tube (TWT) of 216 W each. The high-efficiency solar array is designed to produce 15.5 kW at the end of life. The orbital lifetime of the satellite is estimated as 15 years, aided with xenon-ion thrusters.

To ensure high-quality, uninterrupted service to mobile listeners, the signal quality is enhanced by various techniques – interleaving, powerful concatenated coding, a robust modulation scheme, coupled with space/time diversity, and terrestrial retransmissions. Furthermore, the orbital positions of the two geostationary satellites are chosen such that they appear at relatively high elevation (typically ~45°) over the intended service area in order to reduce signal shadowing.

Table 10.1 Main features of satellite radio systems (compiled from various sources – Layer, 2001; Sirius Online, 2008; XM Radio Online, 2008; 1worldspace Online, 2008; Sachdev, 1997; ITU-R BO. 1134-4, 2001)

	1worldspace	XM Satellite. Radio Inc.	Sirius Satellite. Radio Inc.
Orbit/number of satellites	Geostationary/2	Geostationary/2	24 H, highly elliptical (Tundra)/3
Frequency band (MHz) uplink/downlink	7025-7075/1452-1492	7050–7075/ 2320–2324 and 2328.5–2332.5	7060–7072.5/ 2332.5–2336.5 and 2341–2345
Target service areas	Africa, Asia, Middle East, Far East, and Europe (planned)	CONUS and Canada	CONUS and Canada
Location	21° E and 105° E	85° W and 115° W	Time variant, moving around 100° W (see Figure 2.7)
Service	Radio and data broadcasts to fixed and portable receivers. Not optimised for mobile reception	Radio and datacast to fixed, portable and primarily mobile receivers	Radio and datacast to fixed, portable and primarily mobile receivers
Modulation/ Access: Satellite [terrestrial]	QPSK/TDM [No repeaters]	QPSK/TDM [COFDM]	QPSK/TDM [COFDM]
Channel Coding	Reed–Solomon – outer; rate half-convolutional – inner	Reed–Solomon – outer; rate half-convolutional – inner	Reed–Solomon – outer; rate half-convolutional – inner
Source coding	MPEG Layer III	Propriety	Propriety
Transmission rate (Mb/s) before channel coding	1.536 per TDM; 2 TDM per spot beam (three spot beams per satellite)	4.0	4.4
Number of CD quality (64 kb/s) channels	48 per spot beam	50	50
Receiver	Portable and fixed	Plug and Play, vehicular, desktop, personal portable and cockpit mounted	Plug and play, integrated with vehicle, desktop, personal portable (add-on: Wi-Fi connectivity for streaming, storage)
Approximate number of terrestrial repeaters	Not used (planned)	1500 in 60 markets	150 in 45 markets
Number of RF channels (bandwidth per RF channel)	6 (2.3-3 MHz)	6 (2 MHz)	3 (4 MHz)

A Reed–Solomon code concatenated with convolutionally coded FEC is applied to the time division multiplexed programme stream and a QPSK modulation scheme. Up to 50 CD-quality programmes can be transmitted at a transmission rate of 4 Mb/s (Table 10.1).

Time, space and frequency diversity are introduced by simultaneous but time-offset transmissions of programmes from the widely separated satellites at different frequencies. Time diversity is obtained by introducing a time delay typically of 4–5 s between the respective transmissions, including those from the corresponding terrestrial repeater (discussed next). The delay overcomes the effects of long fades by reconstructing the content from the available parts of the signals. Both received signal components are processed simultaneously; if one of the signals is shadowed, the receiver continues to provide signals through the other. The wide separation between the satellites reduces the probability of simultaneous fading. In addition, the signals are intereleaved to overcome short glitches. The separation in transmission frequencies provides a countermeasure in conditions where a fading event is frequency dependent.

The fade countermeasures are less reliable when signals are repeatedly shadowed over long periods or suffer deep fades, as, for example, in dense urban areas. These environments are serviced by terrestrial retransmissions – the so-called 'gap fillers'. Satellite transmissions are received in S band by directional antennas and retransmitted in S band at a slightly different frequency. The directivity of the directional rebroadcast antenna is chosen such that the S-band retransmissions do not overload the frontend of the satellite receiver. At the planning stage, about 1500 repeaters in 60 markets were considered essential; however, about 1000 repeaters are in use at the time of writing – a typical US/Canadian city requires about 20 such transmitters. In practice the number of repeaters is altered periodically to optimize signal quality and number of repeaters.

The Coded Orthogonal Frequency Division Multiplex (COFDM) scheme is used in terrestrial transmitters. This scheme outperforms TDM/QPSK in a terrestrial multipath environment.

Various types of receiver are in use, depending on the target clientele and the manufacturer. Figure 10.1 shows a sample of an XM Radio receiver.

10.2.2 Sirius Satellite Radio

The Sirius Satellite Radio system targets the same types of audience as XM Radio. The company was founded as Satellite CD Radio, Inc.; its name was altered in 1999 to Sirius, a bright star, also

Figure 10.1 An early model of an XM Satellite Radio receiver. Courtesy Delphi Automotive LLP (Note: now discontinued by Delphi).

known as the Dog Star – hence its logo depicts a dog and a star. The broadcast services began in July 2002. The number of subscribers has escalated rapidly in recent years – from ~3 million in 2006 to ~6 million in a year. The company merged with XM Radio in July 2008 to form a new venture called Sirius XM Radio Inc. The company broadcasts music, news, talk, sports, proprietary channels, ethnic channels, entertainment, non-stop commercial-free music for restaurants, hotels, etc. (see Table 10.1 for technical attributes).

Sirius's approach to minimizing receiver shadowing is to deploy satellites in a highly elliptical geosynchronous orbit that favours visibility over the United States. The apogee of the orbit is chosen to reside at a high latitude over the North American continent, while the satellite moves around a nominal longitude of $100°$ W to achieve an average elevation angle of approximately $60°$ at a dwell time of about 8 h over the United States. By phasing three satellites appropriately, it is possible to arrange a satellite to rise just when another sets. The broadcast is handed to the new satellite during the transition. The orbital arrangement is depicted in Figure 2.7 (Layer, 2001).

In spite of the movement of satellites relative to receivers, the Doppler frequency shift does not cause problems to the receiver design – rather, the Doppler spread caused by fast-moving vehicles poses a more demanding receiver adjustment. The receivers use non-directional and non-tracking antennas.

The high elevation angle, however, does not guarantee direct visibility at all times. Signals may fade in dense urban streets, tunnels, inside buildings, etc. These areas are serviced by terrestrial repeaters installed where there is sufficient traffic demand for commercial viability. Each repeater, instead of receiving the programmes in S band directly, receives them in Ku band on a fixed antenna receiver. A directive tracking antenna would be essential at S band, as otherwise local transmitters would feed back into the repeater, causing oscillations. Realizing that there would be close to 150 repeaters (in 45 markets), Sirius engineers decided to retransmit the programmes in Ku band via a transponder leased from a satellite located in close orbital vicinity. The Ku-band transmissions are received via static directional antenna (~1 m dish). Note that the number of repeaters is considerably lower than the XM system because of the advantage accrued by higher elevation angle visibility.

The system uses proprietary audio compression (Lucent perceptual audio coder) and conditional access systems. The receiver comprises a low-gain antenna for reception of both satellite and terrestrial components. A tuner down-converts the signal to an Intermediate Frequency (IF) of about 75 MHz. The main receiver digitizes, demodulates, decodes, de-interleaves and decrypts the programmes. The tuner and the receiver can each be realized in an Application-Specific Integrated Circuit (ASIC). The baseband processor stores up to 4 s of undelayed signal, and combines it with the delayed version in a maximal ratio combiner. The digital signals are next decoded, amplified and played to the listeners. Descriptive text, such as the title of the music, is displayed on the receiver's display unit. Figure 10.2 shows a Sirius receiver.

10.2.3   1worldspace

1worldspace was founded in 1990, 'with a mission to provide digital satellite audio, data and multimedia services primarily to the emerging markets of Africa and Asia' (1worldspace Online, 2008). The company, renamed from worldspace to 1worldspace in 2008, launched the Afristar satellite in 1998 and the Asiastar satellite in 2000. Table 10.1 summarizes the main features of the system, comparing them with those of the Sirius and XM systems.

In addition to subscription radio, the system supports data services including IP multicast to facilitate a vast range of applications – disaster warning, alert delivery, miscellaneous data delivery, digital weather, distance learning, navigation overlay, content and webcasting. Broadcast channels offer news, sports, music, brand name content and educational programming.

Figure 10.2   A Sirius Satellite Radio receiver. Reproduced by permission of © Sirius.

The broadcast services are provided within the 1452–1492 MHz WARC 92 allocation band. Uplink stations operate in the 7025–7075 MHz band. A satellite broadcasts programmes in three spot beams, each spot beam illuminating approximately 14 million km^2.

Figure 10.3 presents the service areas originally envisioned by 1worldspace. Afristar 1, located at 21° E longitude, illuminates Africa, the Mediterranean basin countries, the Middle East and parts of Europe. The Asiastar satellite, located at 105° E longitude, covers India, China, the southern part of Russia and South-east Asia. 1worldspace intends to provide coverage in Europe in accordance with the ETSI Satellite Digital Radio open standard (ETSI TR 102 525, 2006). The footprints

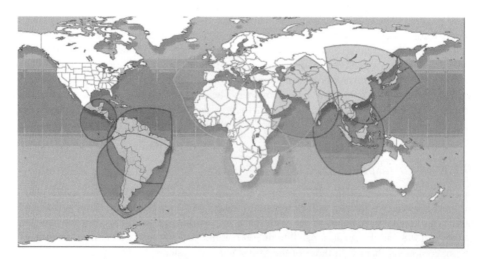

Figure 10.3   Coverage map of 1worldspace as envisioned initially. The Asiastar covers the Middle East, India, China and the FarEast. The Afristar satellite covers Africa, major parts of Europe and the Mediterranean region. The Americastar satellite has yet to be launched. Reproduced by permission of © European Broadcasting Union.

of the 1worldspace satellite constellation encompass 130 countries. Satellite orbital maneuvering lifetime is 15 years.

The service is based on the ITU D_S transmission standard (refer to Chapter 9). This class of system can operate reliably on direct or lightly shadowed paths. The system design relies on the premise that, in the underdeveloped and developing regions, this mode of broadcast would offer a more cost-effective solution rather than a system optimized for vehicular or urban environs. Figure 10.4 shows a block diagram demonstrating the key functionalities of the system for each region (Sachdev, 1997).

A reasonably high EIRP of 53 dBW, assisted by powerful coding consisting of concatenated Reed–Solomon and convolution codes and high-efficiency MPEG layer 3 code in conjunction with high-performance receiver architecture, enables reception on small portable receivers (see Chapter 1). The 1worldspace satellite can provide up to 50 high-quality (64 kb/s) channels per beam. The satellite incorporates two classes of transponders – transparent and regenerative. The power is shared equally between the two transmission modes. There are three processing and three non-processing transponders per satellite. The transparent transponder, designed to support large hubs (up to three large feeder link hubs), translates, amplifies and retransmits the received signal in L band. The processing transponder – entrusted to serve V-SAT (fixed or mobile) feeder stations – converts the incoming signals to baseband and multiplexes and transmits them as a single TDM stream. The hub approach is suited to broadcasters who prefer a dispatch to a central hub, whereas the V-SAT solution suits broadcasters who wish to transmit programmes directly. Although up to 288 V-SAT feeder stations are feasible, between 50 and 100 feeders per region are more likely in practice (1worldspace Online, 2008). A transmission unit of 16 kbps is called a

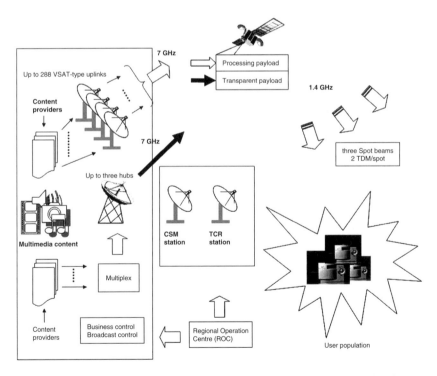

Figure 10.4 End–end functional block schematic (Key: CSM = Communication System Monitoring; TCR = Telemetry, Command and Ranging).

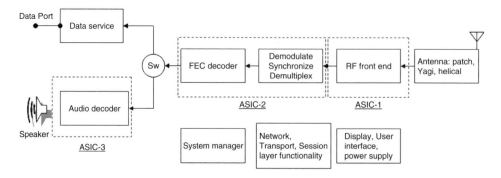

Figure 10.5   Block schematic of a 1worldspace receiver.

primary rate channel. Up to eight primary rate channels may be combined to constitute a broadcast channel of 128 kbps.

1worldspace radio receivers are equipped with a data port that transforms the receivers into a wireless receive-only modem able to download data to personal computers at rates of up to 128 kb/s to receive multimedia content. The receiver can select up to 192 channels of AM quality (16 kb/s MPEG-1 layer 3 encoding). In practice, the choice may be of the order of 50 channels of high quality.

The business and broadcast control facilities coordinate and oversee the region's respective activities (Figure 10.4). The Communication System Monitoring (CSM) monitors communication performance of the system, and the telemetry, command and ranging stations provide the requisite support to a satellite control centre for maintaining the spacecraft.

The receivers must be tuned to one of the two frequencies transmitted in the beam – one for the processing and the other for the transparent transponder. The receivers may use different types of antenna – a simple patch, helix or a short or long Yagi. Typical antenna gain/noise temperature (G/T) of the receiver is -13 dB/K.

Figure 10.5 illustrates a block schematic of the receiver. The high performance, reliability and affordability in this representation is achieved through three ASIC chips (known as the StarMan chipset) marked in the figure (Sachdev, 1997). The receiver has a capability to support up to 128 kb/s data applications such as Power Point presentation. A processing unit performs various network, transport and session layer functions. The performance of the system is controlled and synchronized by a system manager.

10.3   Direct Multimedia Broadcast

Direct multimedia broadcast to portable and vehicular user sets is a recent introduction to the satellite-enabled consumer portfolio. As true of any new technology, frontrunners are always fraught with risks. The technology will not be commercially viable if the masses do not accept it in sufficiently large numbers. This would appear to be the way that the satellite mobile TV systems are headed in Asia (Holmes, 2008). Notwithstanding the reported financial difficulties of the Mobile Broadcasting Corporation (MBCO) and TU multimedia, we present their examples, as these systems are the frontrunners in audio and mobile television broadcasting.[1]

---

[1] The reason for the said financial difficulty is attributed to a competition with free-to-air terrestrial mobile television; at the time of writing, a synergistic commercial arrangement appears to be under consideration.

10.3.1   MBCO and TU Multimedia

Satellite-Digital Multimedia Broadcasting (S-DMB) is a satellite mobile television service supplemented with audio and data services, recognized as system E by the ITU (section 9.4). An S-DMB service provided by the MBCO has been available commercially in Japan since October 2004 – mainly targeting the vehicular sector. Its satellite digital multimedia broadcasting service, called the Mobile Broadcasting Service (brand name 'MobaHO!') provided 30 audio channels, eight video channels and 60 data-services titles in 2008 (MBCO Online, 2008). There were about 100 000 subscribers by mid-2008, although the company is reported to be in financial difficulties.

In Korea, TU Multimedia has targeted an identical service to the cellular phone sector since May 2005. The company offers 40 channels in total – 22 video and 18 audio channels on cellular phones. There were some 1.5 million subscribers in mid-2008.

The MBSAT satellite (also called, Hanbyul), launched in March 2004, is shared by both service providers (Figure 10.6, right). A block schematic of the system of the MBCO transmitter system was illustrated in Chapter 9 (Figure 9.9), and the significant features are listed in Table 10.2 (Hirakawa, Sato and Kikuch, 2006).

The feeder links operate in the 14 GHz band. The satellite is located at 144° E, transmitting 67 dBW in the S band through a 12 m aperture antenna, and 54 dBW in the Ku band.

The impact of shadowing is overcome through a combination of interleaving, coding and antenna diversity for the short fades ($\sim$1 s) typical of highway travel. Two types of terrestrial

Table 10.2   Features of the MBCO S-band multimedia broadcast system

Parameter	Specification
Service	High-quality audio, medium-quality video, multimedia and data
Feeder link	14 GHz band
Service link	2630–2655 MHz
Repeater feeds	12 GHz band and service link
Shadowing countermeasures	Coding/interleaving/antenna diversity (fades <1 s) and terrestrial repeaters (heavily shadowed service areas)
Access/modulation schemes	CDM/QPSK
Coding	Concatenated Reed–Solomon and convolutional, including interleaving
Receiver antenna	Low directive gain (e.g. 2.5 dBi)
Receiving technique	Rake
CDM chip frequency	16.348 MHz
CDM channel identifier	Walsh code
Spreading	Pseudorandom sequence 2048 bits
Number of CDM channels	>30 traffic; 1 pilot
Capacity/channel	236 kbps
Service multiplexing	MPEG-2
Multimedia and data broadcasting	Based on ITU-R BT.1699
Audio coding	MPEG-2 AAC (ISO/IEC 13818-7)
Video coding	MPEG-4 Visual (ISO/IEC 14496-1) and AVC (ISO/IEC 144960)
Video stream spatial resolution	370 kb/s
Video stream time resolution	15 frame/s
Receiver type	Portable (nomadic and handheld), vehicular and fixed

gap filler provide coverage to the heavily shadowed regions. Direct transmission gap fillers cover small, localized areas, for example, a dense city street ($<\sim500$ m) and frequency-conversion-type transmitters cover larger areas, for example, a small town of a few kilometres (~3 km). S-band transmissions are amplified and retransmitted by the direct transmission gap fillers. The Ku-band transmissions are translated to the S band and retransmitted at high power by frequency conversion repeaters. The Ku-band system avoids the problem of oscillations, which tends to occur in direct reception systems at high power owing to feedback.

Three types of receiver are available to subscribers – handheld receivers, as illustrated in Figure 10.6 (left), vehicular receivers and receivers interfaced to PCs through the Personal Computer Memory Card International Association (PCMCIA) interface. The first-generation receivers comprised five LSI signal-processing chips. The new generation of chipsets has reduced the size of the handheld units to that of the cellular phone.

10.3.2   European Initiatives

Numerous research initiatives have been conducted in Europe and elsewhere in recent years to optimize critical aspects of direct broadcast to portable and mobile units (Evans and Thompson, 2007). Recognizing the opportunity, the European Telecommunications Standards Institute (ETSI) has developed an open standard for transporting digital multimedia (ETSI TR 102 525, 2006; see Chapter 9) to exceed the ITU recommendations (ITU-R BO.1130-4). Similarly, the Digital Video Broadcasting (DVB) project's DVB-Satellite Handheld (DVB-SH) standard covers multimedia broadcasts including television and sound.

Commercial services have yet to establish themselves in Europe and remain uncertain at the time of writing. The continent comprises a large number of countries hosting a variety of culture, interests and regulatory regimes. The region has numerous high-quality terrestrial radio stations. Thus, the satellite environment has to contend with terrestrial competition and a difficult regulatory regime.

1worldspace intends to enhance its service reliability and coverage by installing terrestrial repeaters. Europa-Max and ONADA have chosen a three-satellite HEO constellation on the premise that the HEO architecture is better suited for Europe where a significant part of the market lies at high latitudes. Associated with an increase in the number of terrestrial retransmission sites for a

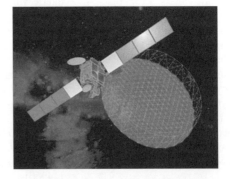

Figure 10.6   Left: Handheld receiver [now discontinued] in the MBCO multimedia broadcast system. Courtesy Toshiba. Right: MBSAT satellite. Reproduced by permission of © Space Systems Loral.

GEO system is the task of obtaining regulatory approval for the terrestrial repeaters from dozens of regulators and the logistics of determining appropriate transmitter sites. In the United States, where a majority of locations are more southerly (south of London) and hence better suited for geostationary coverage, the Sirius HEO system uses a much lower number of terrestrial repeaters than the XM GEO system. However, associated with the HEO system is an unfavourable ITU radio regulation (RR22.2) that assigns a higher priority to a GEO system in the frequency band of interest. Europa-Max planned to commence service by early 2010, with 25–30 radio channels and 10 video channels – increasing to around 50–90 audio channels and 20–30 video services in spot-beam overlap regions. Each satellite will deploy seven spot beams targeting major markets. The contents are said to be thematic audio, proprietary channels like the BBC and CNN, and video services consisting of music, drama, sports, news, etc., for back-seat passengers. Interactivity options include return link via a cellular service or satellite.

ONDAS (ONDAS Online, 2008) is a Spain-registered company proposing a pan-European L-band system aiming to provide sound and multimedia broadcasts, telematics and data service to automobiles, trucks, homes and mobile and portable devices via a HEO constellation of three satellites (elevation more than ~70 deg). Planned services include: vehicle functionality and information, driver and passenger information, emergency calls, in-car video service/integrated entertainment system, enhanced data applications, etc. The satellites will provide 14 kW DC power and will use a 15 m deployable antenna in L (1452–1492 MHz) or S band (2170–2200 MHz). Terrestrial repeaters will be deployed in difficult situations such as inside tunnels. There are plans to sell portable radios that can retransmit signals locally to minimize the number and roll-out of terrestrial repeaters.

10.4 Direct-to-Home Television

In Chapter 9 we introduced the concepts of DTH systems, including the transmission standards recommended by the ITU. Here we will address general issues that relate to the implementation of the DTH systems. We will then review example systems in various parts of the world, emphasizing the European and US regions, which have the most mature markets.

10.4.1 Implementation Issues

A major decision that an operator makes at the outset is the selection of an appropriate downlink frequency band. The ITU has formulated a Broadcast Satellite Service (BSS) plan known as World Administrative Radio Conference 77 (WARC 77) for ITU regions 1 and 3 (countries other than America), and Regional Administrative Radio Conference 83 (RARC 83) for ITU region 3. The plans assign to each signatory, depending on its needs, a number of channels in the 11.7–12.7 GHz band together with the main system parameters including orbital location, spacecraft EIRP and interference threshold. Some DTH operators transmit in the FSS band, ensuring that all the regulations applicable to the chosen band are enforced. The advantage of the BSS plan is that a system can be introduced without protracted regulatory negotiations. The WARC plans are available in Appendix S30 and S30A of the radio regulations (ITU Online). Table 10.3 lists a few key parameters of the BSS plan.

The BSS plans were based on technologies of 1970s and 1980s and hence the assumptions are now outdated, leading to transmission inefficiencies in today's terms. For example, the receiver noise has reduced by an order of magnitude – from ~1000 K to below 100 K; the spectrum efficiency achievable from today's advanced transmission method techniques is several times better and the simple broadcast-only paradigm has evolved to include features such as user interactivity, video on demand, etc. In the BSS plan, satellite locations were also set to the west of the service region to minimize satellite load during eclipse; modern satellite technology has mitigated this constraint, thereby exposing additional orbital slots.

Table 10.3 Parameters of the WARC 77 BSS plan

Feature	Parameter	Comments
Downlink frequency band	11.7–12.7 GHz	
Number of channels	Country dependent	
Typical spacecraft EIRP	63 dBW	
Channels/satellite	5 or multiples	
Polarization	LHCP or RHCP	
Beam shape	Elliptical	Shaped antennas offer better performance
Typical receiver dish	0.9 m	45–60 cm are in use
Receiver temperature	1000 K	75–100 K available now
Eclipse operation	Limited	Pessimistic assumption regarding eclipse power capability reduces flexibility of satellite orbital location as they are always kept to the west of the service region

The plan was first introduced in Japan in the 1980s, and subsequently in the United States and Europe. Nevertheless, significant parts of the spectrum remain unutilized. Analysts believe that the plans need a revision in light of technology and service advancements (Elbert, 2004).

Some operators implement the system in the FSS band via medium-power satellites ($\sim$50 dBW EIRP). Such transponders can be leased on existing satellites to reduce the space segment costs. User interactivity is an emerging requirement owing to a heightening awareness of features like video on demand and bundling of broadcast and Internet services. The FSS band permits a return link to support interactivity, a feature not available in the BSS plan. The antenna of the FSS user terminal would be marginally bigger, but the lower service cost and interactivity outweigh the antenna size issue. The FSS band does not guarantee interference protection from adjacent satellites, unless procedurally coordinated with other operators. The requirement necessitates considerable technical and management effort, and often involves complex technical changes to the system design. For example, a change to the spacecraft antenna transmit pattern may be essential to manage intersystem interference. Table 10.4 presents the frequency bands commonly used for broadcast.

While some broadcasters own the space segment either directly or indirectly, others lease satellite capacity from a specialist operator in order to minimize cost and technical risks. The UK broadcaster BSkyB leases capacity on SES-ASTRA satellites, whereas DirecTV in the United States owns the majority of its fleet. Similarly, operators may either own or lease the uplink facility. The BSS and FSS feeder links share the frequencies. Notably, the Ku-band systems located in equatorial regions must include sufficient link margins because of signal fading incurred during a rainfall. To minimize the penalty attributed to rain fading, some operators in equatorial regions prefer the C band.

Table 10.4 Frequency bands in common use for broadcasts. Use of Ka band is also growing in popularity

Satellite service	Downlink frequency band (GHz)	Comments
Broadcast	2.52–2.67	Community reception
Fixed	3.4–4.2	
Fixed	10.7–12.2	
Broadcast	11.7–12.7	WARC BSS plan

Broadcasters have tended to use proprietary transmission technology, but, with the efficiencies achievable by the MPEG standards, higher-order modulation schemes and economies of scale, broadcasters are migrating towards open standards. These technologies that continue to remain proprietary include conditional access and scrambling. These technologies comply with mandatory privacy and copyright requirements. DVB-S reliant on MPEG-2 is the most common transmission technology. There is a gradual migration towards DVB-S2 and MPEG-4 technologies in new installations as the evolution to HDTV and IP technology takes precedence. One of the impediments in the MPEG-4 migration path is backward compatibility to support the legacy systems (e.g. MPEG-2 source coding).

A wide variety of receiver models are available and variations exist in the following aspects:

- reception from one or more satellites;
- support to one or more home receivers;
- antenna steering;
- capability to vary polarization angle;
- integrated digital recorder;
- home networking;
- HDTV carriage;
- interface with existing systems.

The size of receiver antenna dish ranges from 45 cm to over 1 m, depending on the location of the installation relative to a satellite's orbital location and the frequency band. Some receivers can receive transmissions from multiple satellites in multiple bands. Others were developed with portability and mobility as the prime driver – illustrated in Figure 10.7.

Ku band is commonly used in regions with relatively low rain intensity whereas C-band continues to be in use in regions which are prone to heavy rains.

There are significant variations in local regulations with regard to satellite broadcasting which may impede the growth of the industry. For instance, the growth of the DTH system was stifled in India by unfavourable regulatory constraints until the environment was eased; following which there was a massive surge in growth. It has been observed that affordability of the target population and hence the number of subscribers is central to the success of satellite television operators. In

Figure 10.7 The illustration shows a portable TV (left) and a family viewing at a camp site (right). Reproduced by permission of © DIRECTV.

regions where the DTH services compete with terrestrial operators, the subscribers expect a similar if not better level of service and content.

10.4.2 DTH Services

A vital ingredient to the success of a DTH system is diversity in the programme content such that individuals may select a package of choice at a competitive price. Commercial incentives include a good customer relationship, a simple billing scheme, features to entice the user from competitors, zero-maintenance outdoor and IRD electronics, aesthetics and an easy user interface. Typically, DTH service providers may include some or all of the following features:

- a variety of subscription packages at a cost proportional to the number and uniqueness of the programmes;
- a bundle of free-to-view channels – for example local terrestrial TV channels, news channels;
- pay-per-view to allow a user selectively to choose a programme;
- provision of HDTV channels – this mode of transmission is eventually expected to replace SDTV;
- an IRD with facilities directly to record channels, including functions to pause a channel while viewing, home networking interfaces to enable the recorded material to be viewed from an installation in another room, etc.;
- interactivity to provide video on demand, shop or participate in a live programme;
- technology to enable users to receive programmes in vehicles, aircrafts and ships, and easily assembled kits for out-of-station family trips;
- value-added services such as voice over IP and broadband Internet at discounted price – may or may not be carried over the same satellite;
- other types of broadcast service to diversify clientele, for example, business television and digital signage for the enterprise market.

10.4.3 Representative DTH Systems

General Review

DTH systems enjoy widespread acceptability where other types of medium are either not available or their costs are prohibitive. In the mature market, DTH systems are in direct competition with terrestrial and cable operators. In some countries, DTH services are controlled by the government. Because of the satellites' wide area coverage (in spite of beam shaping and narrow spot beams), cross-border reception remains contentious. Sovereign states do not always desire their population to be influenced by foreign culture, practices or propaganda.

We will briefly outline the DTH status in various parts of the world before addressing the systems prevalent in the European and US regions in detail. Owing to the commercial nature of DTH systems, the market is in a continual state of flux, resulting in mergers, bankruptcies, new entrants, service and technology enhancements or alterations, etc. The reader is encouraged to view online references to get the most recent status (e.g., Wikipedia, 2008).

Africa

DTH services are popular in the African countries. Free-to-view channels introduced in recent years have made DTH widely available to the public. The population centres are widely dispersed, and hence cable systems tend to be more expensive and difficult to introduce and maintain. Terrestrial pay television and some Multichannel Multipoint Distribution Services (MMDS) are also available. MMDS are 2–3 GHz microwave wireless direct-to-home terrestrial distribution services for broadband and television. Transmissions are available in English, French, Portuguese, German and African languages. Countries where DTH services are available include Congo, Kenya, Nigeria, Sudan, Tanzania, Uganda and Zimbabwe. The region is experiencing a rapid evolution of the technology.

Asia

Japan has always been at the forefront of DTH technology. It was the first nation to introduce a system based on the WARC BSS plan. There are two prevalent systems servicing a large number of subscribers.

The Indian Space Research Organization (ISRO) and NASA conducted a highly successful joint direct broadcast demonstration in 1975 via NASA's ATS-6 satellites and ISRO's ground segment consisting of two ground stations and some 5000 receivers. Transmissions in the UHF band beamed education and social welfare programmes to 5000 remote villages. The experiment, known as the Satellite Instruction Television Experiment (SITE), was unique, with ISRO engineers, social scientists and local governments collaborating seamlessly. The demonstrations charted the path to India's domestic satellite system – INSAT. Commercial DTH was launched after the regulatory environment was softened. A number of service providers serve millions of subscribers in the subcontinent, and the service continues to grow.

Opportunity is enormous in Asia, particularly in the burgeoning regions of China and India. In countries where the population is widespread in small islands, such as Indonesia, the DTH service continues to be popular. DTH is also available in Malaysia, Pakistan, the Philippines and Thailand.

Americas

Numerous operators provide services in the United States, but, in terms of number of subscribers, the DTH market is dominated by two operators – DirecTV and Dish Network.

Canada is served mainly by two DTH service providers. Cable television has a stronghold, and MMDS systems are also used in relatively smaller numbers. Systems A and C (Chapter 9) are used in Ku-band FSS allocations typically servicing 45–60 cm dish receivers.

In Latin America the service is popular in Brazil (system A, FSS Ku-band transmissions, 60 cm dish), Mexico (system A, FSS Ku-band transmissions, 60 cm dish) and Argentina (system B, FSS Ku-band transmissions, 60 cm dish) where the service is available through 2–3 operators.

Australia and New Zealand

DTH is well accepted in Australia owing to the sparse population and large distances between cities, available through several DTH service providers. The DTH services in New Zealand, provided by only a few operators, include HDTV transmissions and free-to-view channels.

Europe

A majority of continental services are provided through leased transponders of SES-ASTRA and Eutelsat Hotbird satellites. National systems exist in Spain, Norway, Turkey, Israel and Russia.

In the next two sections we highlight features of European and US systems.

10.4.4   European Region

Two satellite operators dominate the European DTH space segment – SES-ASTRA and Eutelsat. The former owns and operates the ASTRA satellite series, and the latter the HOT BIRD™ satellite series. These satellites carry several thousand channels for pay-TV companies, public broadcasters and international broadcasters to millions of households. Broadcasts channels include news, sports, music, ethnic channels, expatriate channels, interactive services such as shopping and a variety of free-to-air channels.

Broadcasters include BskyB (United Kingdom), SKY Italia (Italy), Canal Digital (Netherlands) and UPC (Central Europe). The DBS segment penetration is said to have reached 21% of the total TV installations in 2005. The market penetration varies depending on local infrastructure, content popularity and service cost. In general, small and more densely populated countries such as the Netherlands possess lower satellite installations owing to heavy cable penetration.

10.4.4.1 ASTRA

ASTRA is part of an international group of satellite operators and network providers known as Société Européenne des Satellites (SES) that provides satellites for numerous services (SES Online, 2008). In the European region, including sub-Saharan Africa, it distributes DTH services via 16 satellites and five orbital locations: 5° E, 19.2° E, 23.5° E, 28.2° E and 31.5° E. The system transmits over 290 channels from 19.2° E satellites and more than 110 channels from 28.2° E. In all, SES-ASTRA distributed close to 1 800 TV and radio channels, in analogue and digital formats, to 40 million homes in Europe and North Africa in the year 2007. SES SIRIUS, an affiliate, distributes more than 140 TV and radio channels to 16.4 million households in the Nordic region. Figure 10.8 (ASTRA Online, 2008) shows the growth in the number of homes and the number of channels served by ASTRA since 1997 over a 10 year period. In the years 2005 to 2007, the growth increased from 29.2 million to 39.9 million, that is, an average of over 18% per year (Figure 10.8).

Figure 10.9 shows the European coverage from an ASTRA satellite, also indicating the size of receiver dish on the coverage contours. The ASTRA consortium began DTH broadcasting in the United Kingdom and Europe for BskyB on medium-power satellite in the FSS band. When the capacity demand increased, ASTRA started to collocate satellites – this approach avoided the need for an elaborate regulatory procedure for obtaining an orbital slot and permitted a seamless, transparent addition of satellites and a hot back-up capability. The receiver dish size ranges from 60 to 120 cm, depending on the location of the receiver, as shown in Figure 10.9.

The company assists the broadcasters by injecting innovative technologies and solutions. For instance: the Duo LNB technique allows simultaneous reception from two ASTRA orbital positions

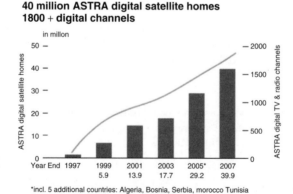

40 million ASTRA digital satellite homes
1800 + digital channels

*incl. 5 additional countries: Algeria, Bosnia, Serbia, morocco Tunisia
Source: SES ASTRA European Satellite Monitor

Figure 10.8   Growth in number of programme channels and subscribers in the past decade. Courtesy © SES ASTRA.

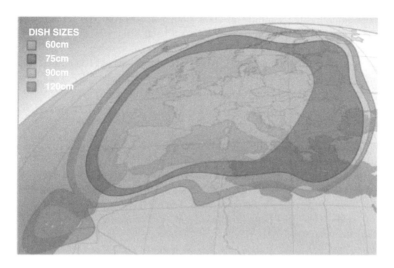

Figure 10.9   European broadcast footprint of an ASTRA satellite. Courtesy © SES Astra.

(19.2° and 23.5° E); Astra2Connect (Chapter 12) provides two-way broadband via satellite and a provision to distribute satellite programme delivery to multiple users; and the Blucom technology provides connectivity between television and the mobile phone.

There is a steady growth in HDTV satellite transmissions, aided by ready availability of space segment capacity. ASTRA authorities expect the number of HDTV channels to reach 100 by 2010.

ASTRA beams are tailored for regional, European and continental coverage. For instance, a dedicated spot beam serves the Nordic and Baltic countries. A wide beam provides for pay TV packages, free TV channels and international broadcasters. An African beam covers the sub-Saharan markets.

Sky Television, introduced in 1989, became the first commercial DBS service in the United Kingdom. A judicious combination of cost-effective technology – medium-power ASTRA satellite transmissions, a video encryption system on PAL broadcast format – led to a successful debut. Sky's conventional technology, constituting a low-cost dish and LNB, proved to be a more viable choice than its rival, British Satellite Broadcast (BSB). To nullify the losses being incurred by both Sky television and BSB, the rivals merged in 1990, forming a new company called British Sky Broadcasting (BSkyB). Gradually, BSB's more expensive D-MAC/EuroCypher technology was replaced by Sky's VideoCrypt system.

Beginning from five channels in 1990, the number of BSkyB channels had increased to around 60 by 1999, necessitating the addition of several satellites. Subsequently, BSkyB launched the first subscription-based digital television platform in the United Kingdom, offering a range of 300 channels broadcast from the 28.2° E location under the brand name Sky Digital.

10.4.4.2   Eutelsat

DTH services are also carried by the Eutelsat consortium on the HOT BIRD™ series of satellites collocated at 13° E, which cover not only Europe but also North America and the Middle East. In addition to numerous free channels, packages broadcasted include American Forces Network Viacom, Eurosport, Sky Italia, NOVA Greece, NOVA Cyprus, etc. It also provides interactive and multimedia services for professional applications such as videoconferencing, LAN interconnection, e-learning, telemedicine, media and corporate file delivery, game networks, etc.

There were three collocated satellites in 2007, anticipated to rise to 6 by 2009. HOT BIRD 6, launched in 2002, comprises 32 transponders (28 Ku-band, four Ka-band) and eight SKYPLEX units for on-board multiplexing. HOT BIRD 8, launched in 2006, offers 64 Ku-band transponders for television and radio broadcasting. HOT BIRD™ employs shaped (contoured) beams, and the coverage regions extend across Europe into the Middle East and beyond.

10.4.5 United States

In the 1970s, enthusiasts began to install dishes in their backyards to intercept signals delivered for cable distribution head ends; although this mode of reception was not permissible, the law was difficult to enforce, leading to delicensing of TV Receive Only (TVRO) systems by the FCC in 1979. Soon, an industry was built and flourished around this type of reception, resulting in millions of 2–3 m dishes being sold in the period up to 1995. The DTH service began in 1994 on the FSS and BSS frequency bands. The DTH home-market penetration continued to grow, reaching over 22% of households in the year up to 2004 (Dulac and Godwin, 2006). Digital systems began to replace analogue systems by 1990. Interestingly, the digital systems were able to comply with the WARC 83 plan which was developed for analogue transmissions for reception on ∼0.9 m dish receivers.

At present, both the FSS and BSS Ku-band are in use. Transport formats include systems A, B and C for reception on 45–60 cm diameter dish receivers. DirecTV (system B) and Dish Network (system A) hold the largest market share; both operate in the BSS band. DirecTV operates via its own satellite fleet, whereas the Dish Network Utilizes Echostar satellites owned by Echostar Technologies LLC.

10.4.5.1 DirecTV

DirecTV (2008) service areas include the United States, the Caribbean and parts of Latin America to provide a bouquet of standard- and high-definition television programmes, interactive services, digital video recorder support, local channel transmissions and pay-per-view services to individuals, businesses, vehicles, ships and aircraft (see Chapter 9).

Founded in 1994, the company had a subscriber base of about 17 million in March 2008. The DirecTV fleet operates in Ku and Ka band from a number of locations (72.5° W, 95° W, 99° W, 101° W, 103° W, 110° W and 109° W). The company owns a majority of its satellite fleet, collocating satellites in many locations to maximize capacity where necessary. The system B transmission format is transported in Ku BSS, Ku FSS and Ka FSS bands, whereas HDTV services use the MPEG-4 and DVB-S2 transmission standards to leverage the latter's high efficiency. DirecTV plans to phase out MPEG-2 transmissions in due course. Recent receivers use slightly elliptical ∼45–50 cm dishes capable of simultaneous view of up to five orbital locations.

10.4.5.2 Dish Network

Echostar created the Dish brand name in 1996 to market satellite television in the United States. Echostar itself was formed in 1980 for the provision of C-band television to individuals. More recently, Echostar separated into two businesses – the Dish Network Corporation is responsible for the satellite television business under the same brand, and Echostar Technologies LLC is responsible for the satellite and technology infrastructure (Wikipedia, 2010).

Dish Network provides SDTV and HDTV (including interactive TV), audio services, pay-per-view subscription, digital video recorder facility and local retransmissions in the United States using a fleet of satellite located at 119° W, 148° W and 61.5° W, etc. The services are carried over the Ku-band FSS and BSS allocations. Dish Network uses the ITU system B transmission system and MPEG-2, but is gradually migrating to MPEG-4 in order to carry HDTV and optimize the space segment capacity. A variety of subscriber equipment is used in order to operate through a satellite fleet with diverse transmission frequencies, EIRP, polarization and orbital location. Beginning from a single LNB 45 cm dish, the most recent dishes can cover a large orbital arc to support services from up to five orbital locations simultaneously. Dish sizes range from 45 to 60 cm, depending on user location and satellite EIRP. Over 14 million subscribers were reported to be in service in December 2009 (Dish Network Online, 2010).

10.5   Military Multimedia Broadcasts

We conclude this chapter with an example of a highly successful implementation of a military multimedia broadcast.

10.5.1   US Global Broadcast Service (GBS)

Origins

In the aftermath of the first Iraq War (1991), the Pentagon identified a shortfall in its ability effectively to provide high-bandwidth data dissemination during that conflict, in particular high data rates to forward and typically mobile users (sometimes referred to as the military equivalent of the 'last mile' in telecommunications). This led to a formal requirement for a high-speed, one-way broadcast of large information products to deployed, on-the-move and garrisoned troops. This requirement led to the introduction of the US Global Broadcast Service (GBS). A key driver for GBS has been the introduction of information products from unmanned aerial vehicles (UAVs) such as Predator and Global Hawk, which typically stream surveillance video.

The GBS programme was formally established in 1996, and the US government fielded a first incarnation of GBS, known as the Joint Broadcast Service (JBS), in Bosnia-Herzegovina, in support of UN peacekeeping activities.

Information carried over GBS is a mix of classified and unclassified data, and the content varies from day to day, according to operational priorities. Classified information sources are encrypted using high-grade encryption. GBS is known to carry such data products as video, audio, files, web pages and situation awareness/common operating picture (GBS Joint Programme, 2010).

Other data products are likely to include surveillance video and images, tasking, intelligence, weather forecasts, database replication, training information, imagery, logistics information, servicing manuals, 24 h news broadcasts, welfare TV, etc.

Although called a broadcast service, strictly speaking GBS is a multicast service, with much of the content being multicast to different user groups in theatre. The current GBS system is based around the concept of smart push, together with a facility for user pull (information requests, which come into information management centres via available backhaul communications). These requests are prioritized, and the requested information is scheduled for subsequent transmission.

GBS Receiver Technology

A key factor in the success of GBS has been the leveraging of low-cost, mass-market direct-to-home satellite TV and satellite Internet technology to provide a compact, affordable solution, thereby allowing large numbers of terminals to be procured. The primary changes needed are in response to the need to make the equipment more rugged and secure. JBS used commercial Ku-band satellite capacity and was based on proprietary Direct Satellite Service (DSS) satellite TV technology from Hughes.

JBS became known as GBS Phase I, and Phase II introduced two key changes: a move to open standard MPEG-2 and DVB-S technology with IP encapsulation, and the use of Ka-band payloads.

Standard GBS terminals comprise both 0.6 and 1 m antennas known as the Next-Generation Receive Terminal (NGRT) and includes all the RF aspects of the GBS terminal – with an L-band

RBM   Man Pack Terminal Antenna

Figure 10.10 GBS Suitcase Portable Receive Suite (GBS SPRS), comprising Receive Broadcast Manager (RBM) and laptop (left) and manpack terminal antenna (right). Reproduced by permission of Raytheon.

feed to associated receiver and user terminal apparatus known as the Receive Broadcast Manager (RBM). More recently, Suitcase Portable Receive Suites (SPRSs) have been introduced, comprising a lightweight antenna and laptop RBM (Figure 10.10). The RBM comprises one or more standard DVB-S Integrated Receiver Decoders (IRDs), an IP crypto unit, an IP router and a ruggedized laptop PC running the RBM software, all housed in a ruggedized enclosure.

In addition to these transportable and man-portable receivers, GBS receivers are deployed on US warships and will in future be deployed on certain larger military aircraft FAB-T.

GBS uses the DVB-S waveform. Although DVB-S was originally designed for video transmission, and uses fixed-length (188 byte) MPEG-2 TS packets; other packet formats can be accommodated within these MPEG-2 packets (with a slight loss in efficiency), and GBS uses DVB MultiProtocol Encapsulation (MPE) to encapsulate IP packets to facilitate IP multicast (diFrancisco, Stephenson and Ellis, 2000).

GBS Broadcast Source

GBS broadcasts originate from one of three static primary injection points (PIPs) located at Norfolk, Virginia, Wahiawa, Hawaii, and Sigonolla, Italy. Broadcasts originate from Satellite Broadcast Managers (SBMs) located at Norfolk and Wahiawa, which merge the various input streams and encrypt it for broadcast (Raytheon GBS Online, 2010). In addition, transportable theatre injection points allow the generation of additional broadcasts in theatre (rather than transmitting the data back to the United States).

GBS broadcast manager software exploits the Kencast forward error correction product called FAZZT (Kencast Online, 2010) to improve reliability of one-way transmissions (where the use of retransmissions to counter errors is not possible), using a proprietary algorithm. FAZZT is also used to compress files to decrease needed bandwidth and maintains an internal database of content, along with metadata information about the content, which a receiver's applications may browse. Remote users can choose which categories of files to receive using user-defined filters, or select files from a 'carousel'. Hot folders can be configured for immediate transmission when files are dropped in them. FAZZT can mirror FTP sites, websites, or data banks.

FAZZT provides transmission queue management, using scheduling and queuing to control transmission priority. Actual transmission bandwidth is controlled by packet shaping (a form of rate limiting/buffering) the output IP streams. The resultant DVB-S stream is then encrypted using a commercial Conditional Access System (CAS) to deter reception by 'enthusiastic amateurs'.

GBS Satellites

At the time of writing, GBS is supported by two UFO block II satellites and several commercial Ku-band transponders, although, with the launch of the first Wideband Global (originally called Gapfiller) Satellite (WGS) in 2008, GBS has also commenced broadcasting over WGS, and Ka-band GBS services will gradually transition completely to further WGSs as they are launched.

The WGS satellites are intended to provide wideband two-way military satellite communications as well as GBS, and Figure 10.11 illustrates the X- and Ka-band processing payload. A digital channelizer with 2.6 MHz quantisation allows cross-banding between X band and Ka band, thereby allowing an in-theatre military X-band terminal to be used to inject the GBS for transmission at Ka band, and, in principle, broadcast of GBS on X band. The diagram shows that phased-array antennas are used for both receive and transmit at X band, while mechanically steered reflectors are used at Ka band.

Characteristics of the WGS satellite Ka-band payload are given in Table 10.5.

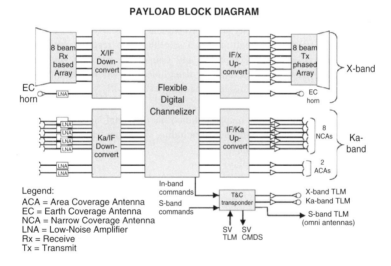

Figure 10.11 WGS payload schematic indicating channelizer and cross-banding. Reproduced by permission of © Boeing.

Table 10.5 WGS Ka-band characteristics

Beam size (°)	G/T (dB/K)	EIRP (dBW)
1.5	8	55–58
4.5	−1	42

Revision Questions

1. Signal shadowing is a major impairment in a satellite radio and multimedia broadcast system. Interruptions, loss in signal quality and repeated disruption cause user annoyance.
   - Outline the technique(s) used to mitigate shadowing impairments, with examples.
   - Explain the trade-offs and rationale applied by operators as cited in the text in selection of fade countermeasures.
2. Compare the salient technical characteristics of all the radio and multimedia systems described in the chapter. Explain the differences.
3. Outline the major system factors and trade-offs considered by a DTH operator during planning.
4. Using the material presented in the text as the basis, compare the main technical characteristics of the BSkyB, DirecTV and Dish Network satellite systems. The reader is encouraged to expand the details (Hint: Consider, at least: frequency of operation, satellite characteristics/ownership, satellite location/coverage, transport mechanism/conditional access, service type, receiver characteristics, financial status, number of subscribers, growth trends).

Project. The chapter introduced features of direct audio/multimedia broadcast systems under active consideration in the European region. The aim of this project is to select two of these systems, and for each:

- determine the most recent technical, regulatory and commercial details;
- determine the status with regard to its implementation – the problems being experienced (e.g. licensing, financial, regulatory).

This part of the project is for a technically advanced student at a final-year undergraduate level. The aim of the work is to develop specifications for a satellite radio receiver and propose compliant handheld or vehicular receiver architecture for any one of the existing satellite radio systems described in the text. An accompanying report will include a survey of the approaches commonly used for this class of receiver. As a minimum, the architecture will:

- identify the main functional blocks, mapping functions to digital signal processors/ASIC;
- propose appropriate input–output interfaces, including their characteristics;
- propose one or more antenna solutions;
- provide alternative solutions for display;
- estimate the processing power;
- estimate prototype cost.

Following the completion of the receiver architecture, the student will embark on a detailed design to specify each part of the architecture.

The project should conclude with construction and testing of a prototype, possibly with a real satellite radio signal, provided that the satellite radio system is available in the region. Instructors can enhance the scope of the project to a master's level by including additional tasks such as computer-aided optimization, field tests, receiver characterization, etc.

References

ASTRA Online (2008) http://www.ses-astra.com/consumer/uk/hdtv/index.php [accessed December 2008].

Clarke, A.C. (1945) Extra-terrestrial relays. *Wireless World*, October, 305–308.

DirecTV (2008) http://www.directv.com [accessed February 2010].

diFrancisco, M., Stephenson, J. and Ellis, C. (2000) Global Broadcast Serivice (GBS) end-to-end services: protocols and encapsulation. MILCOM 2000 21st Century Military Communications Conference Proceedings. Vol. 2, pp. 704–709.

Dish Network Online (2010) http://www.dishnetwork.com [accessed February 2010].

Dulac, S.P. and Godwin, J.P. (2006) Satellite direct-to-home. Proceedings of the IEEE, **94**(1), 158–172.

Elbert, B.R. (2004) *The Satellite Communication Application Handbook*, 2nd edition. Artech House, Norwood, MA.

ETSI TR 102 525 (2006) ETSI TR 102 525 v.1.1.1 (21-09-2006) Satellite Earth Stations and Systems (SES); Satellite Digital Radio (SDR) service; functionalities, architecture and technologies.

Evans, B.G. and Thompson, P.T. (2007) Aspects of Satellite Delivered Mobile TV (SDMB). Mobile and Wireless Communications Summit, 16th IST, 1–5 July, pp. 1–5.

GBS Joint Programme (2010) Global Broadcast Service (GBS) Joint Programme (fact sheet). Available: http://losangeles.af.mil/library/factsheets/factsheet.asp?id=7853 [accessed February 2010].

Hirakawa, S., Sato, N., and Kikuchi, H. (2006) Broadcasting satellite services for mobile reception. *Proceedings of the IEEE*, **94**(1), 327–332.

Holmes, M. (2008) Mobile TV Via Satellite struggling in Asia. *Via Satellite*, December, 56–58.

ITU Online (2008) http://www.itu.int [accessed December 2008].

ITU-R BO.1130-4 (2001) Recommendation BO.1130-4 Systems for digital satellite broadcasting to vehicular, portable and fixed receivers in the bands allocated to bss (sound) in the frequency range 1400–2700 MHz.

Kencast Online (2010) http://www.kencast.com/index.php?option=com_content&view=article&id=12&Itemid=5 [accessed February 2010].

Layer, D.H. (2001) Digital radio takes to the road. *IEEE Spectrum*, July, 40–46.

MBCO Online (2008) http://www.mbco.co.jp/english/03_news/news_archive/050727_02 [accessed in 2008] (Note: MBCO is said to have ceased operation in March 2009. This link does not exist now, although the link http//www.mbco.co.jp is functional in Japanese).

ONDAS Online (2008) http://www.ondasmedia.com/news/ondas-10.htm [accessed February 2010].

1worldspace Online (2008) http://en.wikipedia.org/wiki/WorldSpace [accessed December 2008].

Raytheon GBS Online (2010) http://www.raytheon.com/capabilities/products/gbs/ [accessed February 2010].

Sachdev, D.K. (1997) The Worldspace system architecture, plans and technologies. Annual Broadcast Engineering Conference, South African Digital Broadcasting Association, 11 pp. Available: http://www.sadiba.co.za/PDFfiles/worldspace.pdf [accessed December 2008].

SES Online (2008) http://www.ses.com/ses/ [accessed December 2008].

Sirius Radio Online (2008) http://www.sirius.com/ [accessed December 2008].

Wikipedia (2008) *Satellite Television*. Available: http://en.wikipedia.org/wiki/Satellite_television [accessed December 2008].

Wikipedia (2010) http://en.wikipedia.org/wiki/Echostar_Communications_Corporation [accessed December 2008].

XM Radio Online (2008) www.xmradio.com [accessed February 2008].

11

Communications Architectures

11.1 Introduction

A satellite communications network transports telecommunication services reliably from a source to its destination (or destinations, in the case of multicast networks). By their inherent capability of wide area coverage, satellites can extend the services to regions that would be difficult, uneconomic or impossible to provide using terrestrial fixed or wireless systems. They may operate in various configurations, interface with other networks to constitute a part of the public network or serve as a stand alone private network.

In the first part of the chapter we discuss essential network and service fundamentals, outlining the different roles of satellite networks, categories of telecommunication services, transport protocols and quality of service.

An operational network performs a variety of tasks, in addition to transporting the core service traffic. These include monitoring and managing satellite performance, ensuring that the network performance is healthy and provision of commercial aspects such as billing. We briefly review various interfaces and entities of a satellite network and their functions, followed by an introduction to a network management system.

In the final part of the chapter we explore the topologies, advantages, limitations and the state of the art of High-Altitude Platform (HAP) and hybrid satellite-HAP wireless communications systems.

11.2 Role

The configuration and role of a satellite network depends on the applications at hand. For instance, a private satellite network operates in a stand-alone mode to transport services to a close user group. When part of a public network, the satellite system is only a component of a wider network that may comprise a variety of transmission media – copper wires, coaxial cables, optical fibres, microwave links, cellular wireless systems, etc. This section introduces various *roles* of a satellite system in a network and the manner in which the services are distributed.

Satellite networks are configured in various ways, depending on the desired function and requirements. An *access* network allows a user to communicate with the satellite directly. In a mobile satellite system, users access the serving satellite directly for a connection to a calling party. In a typical Very-Small-Aperture Terminal (VSAT) network, the user terminal, installed on customer premises, provides a direct satellite service such as remote connectivity with headquarters, bank

Satellite Systems for Personal Applications: Concepts and Technology   Madhavendra Richharia and Leslie David Westbrook
© 2010 John Wiley & Sons, Ltd

teller machine support, etc. In a *transit* network, satellites bridge large traffic nodes – for instance, interconnection of regional gateways or Internet service provider nodes. In the transit role, as a satellite network must interwork with other parts of a larger framework, the arrangement requires incorporation of suitable interfaces in conjunction with commercial agreements. In a *broadcast* network, satellites provide point-to-multipoint services such as digital radio, direct-to-home television or multicast.

In the access and transit roles, the satellite system is said to operate in a *point-to-point* configuration because communication takes place between two designated Earth stations. In a broadcast role, where a single location serves several users, the network is configured in a *point-to-multipoint* configuration.

11.2.1 Service Definition

ITU-R defines a satellite service according to its delivery media, mode and capability:

- In a Fixed Satellite Service (FSS), two fixed sites are linked to each other via a satellite.
- A Mobile Satellite Service (MSS) deals with telecommunication services to and from mobile terminals.
- A Broadcast Satellite Service (BSS) addresses television, radio and data broadcasts.

ITU-T defines services generically as *interactive* and *distribution*. Three categories of interactive services are defined:

- *Conversational.* Communication is established in real time, as in telephony.
- *Messaging.* The services are made available through access to a storage device such as a mail box, hence some delay is tolerated.
- *Retrieval.* The services offer retrieval of information from a designated information centre at a time chosen by the user – an example would be a service for retrieving a film and music.

Each category has a specific bandwidth and quality-of-service requirement.

Distribution services refer to broadcast and video-on-demand services that distribute the content to users from a central repertoire. Two categories of distribution services are defined:

- one where the user has no control over the content, such as in live DTH broadcasts;
- the other where the user can receive information on demand from a centre repertory that broadcasts the content cyclically, such as in a video-on-demand service offered by DTH news broadcast channels.

11.3 Circuit-Switched Services

11.3.1 Quality of Service

A telecommunication network must deliver services with a quality compliant to user expectations. The Quality of Service (QoS) measure for a telephonic conversation clearly differs from those of television, file transfer or an e-mail. We also note that the QoS expected of a component *network* differs to that of the *user* who perceives the cumulative degradations of the end–end link. Thus, it is essential to apportion the QoS sensibly when sizing a network's performance budget.

Table 11.1 Target QoS for various applications recommended by ITU-T

Application	Category	Tolerance (Yes/No)	Delay (s)
Conversational voice, video	Interactive	Yes	$\ll 1$
Command/control (e.g. computer games)	Interactive	No	$\ll 1$
Voice and video messaging	Responsive	Yes	~ 2
Transactions (e.g. web browsing)	Responsive	No	~ 2
Video and audio streaming	Timely	Yes	~ 10
Downloads, messaging	Timely	No	~ 10
Facsimile	Non-critical	Yes	> 10
Background services	Non-critical	No	> 10

Table 11.1 lists the user QoS recommended by the International Telecommunication Union for various QoS categories and applications (ITU-T-G1010). These user QoS must translate into network-centric measurable parameters in order to optimize the network performance without the need to involve the end-user. These parameters are typically the signal (or, more usually, carrier)-to-noise ratio, the transmission delay and delay variations, the bit error rate, the packet error rate, the packet delay, packet delay variations, the net throughput, etc.

There is clearly a trade-off between the QoS and economics – high quality is obtainable by allocating larger network resources, thereby increasing the end-user cost; the user may be willing to forego a loss in quality to lower the end-user cost – for instance, users tend to forgo voice quality when using Voice over IP (VoIP) in exchange for the attractive tariff of the service.

11.3.1.1 Grade of Service

A vital aspect of network planning deals with sizing the network capacity to carry the anticipated traffic. The capacity of a circuit mode system is measured in erlangs, E:

$$E = C_{av} H_v \tag{11.1}$$

where C_{av} is the average calling rate (e.g., in calls per hour) and H_{av} is the average call holding time (e.g., in calls per hour).

Noting that the users can tolerate a degree of congestion at the busiest hour, a capacity measure known as Grade Of Service (GOS) is defined as the number of blocked channels (usually at the busiest time) divided by the number of calls attempted.

For sizing the network capacity, the GOS and Erlang traffic are mapped suitably to provide the capacity, measured in terms of number of channels. Two models are in common use – Erlang and Poisson. The Erlang B model, recommended by the ITU-T, is used in Europe. This model assumes that calls arrive at random and blocked calls are lost. The capacity can be fixed assigned to all the participating Earth stations but more often are demand assigned. The Erlang B formula for the GOS for N channels is given by

$$\text{GOS} = \frac{\frac{E^N}{N!}}{\Sigma_{i=0}^{N} \frac{E^i}{i!}} \tag{11.2}$$

It is usual to specify the GOS as a percentage. As this formula becomes difficult to evaluate for large N, GOS values are usually obtained from precalculated tables.

11.4   Packet-Switched and Converged Services

11.4.1   Internet Protocol Suite

The Internet protocol suite has emerged as the dominant packet-switching scheme. It was developed to INTERconnect different computer NETworks and hence works best for computer networks (IETF 1122, 1989; IETF 1123, 1989). The widespread adoption of the protocol in the public domain – led by numerous effective and useful applications – has made the Internet indispensable today. Owing to its pervasiveness, many systems are being developed on the IP model – although, as we shall discover, satellite systems pose some difficulties in transporting IP.

11.4.1.1   Application Layer Protocols

The Transmission Control Protocol/Internet Protocol (TCP/IP) protocol suite comprises a relatively large and growing set of protocols. The largest collection of TCP/IP protocols exists at the application layer and supports the diverse user applications available on the Internet. The pervasiveness of Internet applications has stretched the capabilities of the protocol well beyond its original purpose. We cite only a few key examples here.

The Hypertext Transfer Protocol (HTTP) is the protocol used by the World Wide Web (WWW), which constitutes the system of interlinked files via hyperlinks that enables people to access information and assists in a variety of services and applications. The (older) File Transfer Protocol (FTP) provides for basic transfer and manipulation of files between a local and a remote computer.

The Real-time Transport Protocol (RTP) is an important protocol for use in real-time multimedia applications, such as voice over IP and voice and video teleconferencing. RTP is an application layer protocol used to provide support for end-to-end real-time multimedia transfer. Specifically, it provides facilities for the correct sequencing of data packets and provides some control over jitter (throughput delay variation).

11.4.1.2   Transport Layer Protocols

The transport layer of the protocol suite supports two alternative protocols. The Transmission Control Protocol (TCP) is used for reliable delivery of data, and requires a return path. It adds features to enable detection of errors or lost packets at the receiver, and to request retransmission. TCP attempts to establish reliable communication of variable duration and data size between the client and the host without any knowledge of the application or the network. It employs flow control and various congestion control mechanisms to enable equitable sharing of network resources for the prevailing network conditions. The protocol probes the network for congestion by starting slowly and increasing the transmission speed on the basis of success acknowledgement from the receiver.

By contrast, the simpler, lightweight User Datagram Protocol (UDP) does not enable error detection and is hence less reliable but quicker (with lower overhead), and therefore better suited for real-time applications and one-way communications and multicasts (content distribution from a single source to multiple users).

11.4.1.3   Internet Protocol

At the network (Internet) layer, the Internet Protocol (IP) itself moves packets over any network technology (including satellite). The most widespread version in use today is IP version 4 (IPv4), which employs a 20 byte header followed by up to 65 515 bytes of data. The header contains both source and destination IP addresses (four bytes each), assigned by the Internet authorities. The IP address consists of three parts – network address class (A to E), network identification (Net-id) and host identification (Host-id).

If the source destination lies within the same subnetwork (say, an Ethernet), a host (PC, etc.) is connected to the destination directly. If, on the other hand, the destination Host-id does not reside within the same subnetwork, then the source host sends the request to a *router*. The router forwards the request to the next router, assisted by a routing table. The packet is passed from router to router until the destination is reached. Routers use a routing table and associated routing protocols for directing packets to other routers of the network – an Interior Gateway Routing Protocol (IGRP) within a domain (essentially an autonomous jurisdiction) and an Exterior Gateway Routing Protocol (EGRP) when communicating with external domains.

More recently, the Internet Engineering Task Force (IETF) has developed IP version 6 (IPv6). This evolved version provides a number of improvements to comply with requirements of the new applications and IPv4 address space limitation of IPv4 (at the cost of a 40 byte header instead of 20). IPv6 incorporates a wide range of new features: support of more host addresses, reduction in routing table size, protocol simplification to increase router processing time, mobility support, better security, incorporation of QoS measures, etc.

11.4.1.4   Datalink Protocols

The transport of IP packets can take place over any transmission medium, for example on LAN, terrestrial or satellite wireless systems. The transmission medium utilizes its own addressing scheme, and therefore the IP address has to be mapped to an address appropriate to the transmission system. This is achieved by link layer protocols such as the Address Resolution Protocol (ARP) and the Reverse Address Resolution Protocol (RARP).

11.4.1.5   IP QoS

IP packets traverse the network as a datagram, that is, without any QoS guarantees at the network level, and can hence offer only a 'best effort' service.

In a packet mode system, packet loss and message delay are useful metrics. Interactive services such as a conversation are less tolerant to delay than non-real time services like e-mail, while voice is more tolerant to packet loss than data files. The TCP protocol incorporates measures to improve the quality at the transport layer. Nevertheless, the QoS is generally inadequate for applications such as interactive services.

In a well-designed packet-switched system, the capacity will be dimensioned to ensure that the messages are delivered within the specified delay and congestion packet loss. In times of heavy demand, higher-priority requests may be assigned the resources first while the lower-priority requests wait in a queue. Packets are dropped when the system resource is insufficient. Ideally, network planners will regularly review the resource utilization to assign additional capacity if the network performance degrades.

ITU-T recommendation Y.1540 defines and specifies parameters to quantify IP QoS, taking into consideration the nature of IP data transport which includes performance of the link (e.g. ATM, wireless protocol, etc.), IP and higher layers (TCP, UDP). The parameters include the packet transfer delay and its mean, packet delay variations, the packet error ratio, and the packet loss ratio; the QoS classes vary from the highest 0 (real time, e.g. VoIP) to the lowest 5 (traditional IP, e.g. web surfing, e-mail).

In an effort to assure guaranteed end–end QoS, the IETF defines two advanced IP architectures: *integrated services* (Intserv) and *differentiated services* (Diffserv). The Intserv framework supports a QoS class on a session basis by reserving resource along a virtual path and by exercising admission control. A Resource Reservation Protocol (RSVP) signals the desired resource and traffic features for the applicable packets. Routers along the path, knowing the resource required and assessing the existing load, set up the session when possible. Intserv provides a *guaranteed* service and also a *controlled load* service that offers a lower quality than the guaranteed service but is better than the best effort service. The Diffserv mechanism, rather than assigning resource on a per session basis, assigns each request a class appropriate for the requested QoS. Thus, for example, a session requiring lower delay (or higher priority) is given a higher class that assures a lower failure probability. The Diffserv model is widely used for QoS for IP over satellite.

11.4.2 IP Over Satellite

Following the trend in terrestrial technology, IP is rapidly penetrating the satellite domain. For instance, the satellite extension of terrestrial Universal Mobile Telecommunication System (S-UMTS) is based on IP transport. Given the moves towards convergence of voice, data and even video over a single common network, it is evident that future satellite system for personal application will favour an IP-based solution.

Figure 11.1 demonstrates a generic IP-based satellite network solution. Routers placed at the edges interface with the satellite network, passing on incoming packets to the satellite network and delivering them to the destination. The IP packet translation adds overheads and thereby reduces the transmission efficiency.

11.4.2.1 Performance of TCP/IP Protocols over Satellite Links

Satellite channels degrade the performance when using TCP in a number of ways. Problems arise, for example, when the bandwidth–delay product is large, a condition referred to as a 'long fat pipe' (IETF 1323, 1992). This occurs for broadband systems in high-delay geostationary satellite systems.

TCP uses a 'sliding-window' packet flow control mechanism. As discussed in Chapter 6, there is an upper limit on throughput for a given window size, which defines the maximum number of data bytes that can be transmitted before an acknowledgement is required. The maximum window size in TCP is constrained by the 16 bit field used to report window sizes in the TCP/IP header. Therefore, the largest window that can be used is 64 kB. This limits the effective TCP/IP bandwidth to 64/RTT kB/s, where RTT is the round-trip time (in seconds). As the end to end round-trip time for a geostationary satellite link is of the order of 560 ms, the maximum effective TCP bandwidth for an unoptimized (without window scaling) TCP link over satellite is ~900 kb/s (around 114 kB/s), regardless of the actual satellite link bit rate.

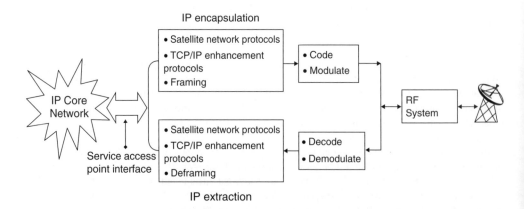

Figure 11.1 A generic configuration of an IP-enabled satellite network.

This situation is exacerbated by TCP's congestion control algorithms, which assume that any packet loss is a result of network congestion. TCP begins a session with the slow start algorithm, the aim of which is to probe the network in order to determine available throughput capacity. Slow start initially sets a congestion window size equal to one data segment (the size of which is negotiated at the start of the session). TCP transmits one segment and waits for the ACK. For every ACK received, the congestion window is increased by one segment (at a rate of approximately one segment per round-trip time). Thus, the number of segments increases from 1 to 2 to 4, etc., resulting in an approximately exponential throughput growth. If an acknowledgement times out, TCP/IP enters its congestion avoidance algorithm. The window size is reset, and a slow start mechanism begins again, however, when the congestion window reaches one-half of the value at which congestion originally occurred, the window size is increased more slowly, and the growth in throughput becomes linear rather than exponential.

The congestion algorithm is not well adapted to the high error rates of wireless systems as compared with wired datalinks. Again, TCP/IP assumes that packet loss because of random errors is due to congestion and throttles back the throughput unnecessarily. As a result, throughput drops rapidly for $BER > 10^{-8}$.

In addition, asymmetry in transmission directions can affect protocol performance, while, in LEO systems, variable round-trip delay and signal drop-outs during handover also degrade TCP performance.

Bandwidth Efficiency of VoIP

The use of Voice over IP (VoIP) is increasingly popular. VoIP offers a low-cost alternative to the traditional telephone network. In addition to the delay introduced by a satellite link, the bandwidth efficiency of VoIP is an issue. Consider the bandwidth efficiency of a common VoIP vocoder standard G.729 (ITU-T G.729). This CELP derivative vocoder has a nominal bandwidth of 8 kb/s. The standard employs 100 ms voice frames and transmits this information in 10 bytes/frame. Typically, two such frames are transported in each IP packet using UDP and RTP. RTP employs a 12+ byte header (plus optional extensions). Thus, each IP packet contains 20 bytes of useful data, while 20 bytes must be added for the IPv4 header (40 bytes for IPv6), 8 bytes for the UDP header and 12 bytes for the RTP header, making a total of 60 bytes. The bandwidth efficiency of this VoIP signal is thus just 33% (Nguyen et al, 2001). Without further compression, the actual required IP bandwidth is thus 24 kb/s (as opposed to 8 kb/s). To this must be added any further overheads in the link layer framing system.

11.4.2.2   Performance-Enhancing Proxy

Alternative TCP/IP implementations exist, some of which incorporate enhancements suited to the use of long fat pipes typically experienced in satellite communications. These include the use of Selective ACKnowledgements (SACK) (IETF 1072, 1988) and window scaling (IETF 1323, 1992). In practice, however, the end-user has little control over which flavour of TCP/IP is used to support his/her application over a network. Slow start enhancement, loss recovery enhancement and interruptive mechanisms are also examples of well-known techniques.

As discussed in Chapter 6, an effective approach to accelerating the performance of the TCP/IP suite of protocols over satellite links is to utilize Performance-Enhancing Proxies (PEPs) over the satellite portion of an IP network – with one PEP at each end of the link. In particular, TCP spoofing places a PEP proxy router near the sender to acknowledge reception before the data are sent over the satellite using a different protocol better suited to the characteristics of the satellite link, thereby increasing the speed of transmission. Both hardware and software PEP solutions exist. Typically in a satellite Internet service, the user equipment PEP may take the form of a software client application (which may even run on the user's PC). At the satellite hub, a powerful hardware-based PEP will typically be used to manage a large number of user links at once. For a more detailed treatment of TCP over satellite enhancement schemes, the interested reader is refered to the literature, for example, Sun (2005).

11.4.2.3   PEP Compression and Prefetch

We have previously discussed the need to use satellite resources efficiently. In terrestrial wired networks the efficiency of the various TCP/IP protocols is less of an issue; however, for certain services, notably VOIP, the low bandwidth efficiency is due to the fact that the total header length represents a significant fraction of the total packet length, and this is a significant issue.

VOIP Header Compression
While voice activity may be exploited to some degree (suppressing packets when the speaker is quiet), the main mitigating approach for satellite TCP/IP networks is to employ a performance-enhancing proxy with header compression. Typically, each 40 byte IP/UDP/RTP header may be compressed to just a few bytes.

Data Compression and HTTP Prefetch
PEPs provide additional features that can enhance the performance of the satellite link when used with typical Internet-type applications. These include on-the-fly data compression for specific file types and HTTP acceleration, which compresses the HTTP, and also by cacheing and prefetching web pages (by predicting subsequent HTTP requests).

11.4.2.4   IP Mobility and Security

Other considerations in relation to IP over satellite systems include mobility management due to both user and satellite movements (in LEO and MEO constellations) and security mechanisms.

   Internet security over satellite channels becomes particularly challenging because of the increased probability of interception. The Internet protocol provides a basic set of security mechanisms at the network level through authentication, privacy and access control (i.e. firewall/password). Security can also be introduced by coding the packets at the IP or TCP/UDP layer, or at the application level. IP security (IPsec) is a security protocol suite implemented at the IP layer, providing confidentiality, authentication and integrity services. Encryption is owned by the data provider rather than by the satellite network operator. The Virtual Private Network (VPN) also comprises firewall protection involving routers at both ends of the network. Security for multicast transmissions can be achieved by distribution of encryption keys to the members of the multicast group.

11.4.3   ATM

Another important packet-switched protocol is the Asynchronous Transfer Mode (ATM) protocol developed to facilitate the integration of broadband telecommunications and computer networks following the standardization of the Broadband Integrated Switched Data Network (B-ISDN). The protocol permits integration of a telephony network of a few tens of kb/s with broadband transmissions of the order of 155 Mb/s and multiples thereof (Jeffery, 1994).

   A key feature of ATM packets (called cells) is their small fixed length. Each cell comprises 53 bytes, eight of which comprise the header and the remaining carry information. The notional efficiency of ATM cells is thus ~85%. The cell size was selected as a compromise between opposing stipulations of the data services and telephone services communities. A larger packet length achieves higher link efficiency for data transmissions, whereas short packets minimize network transmission delay, a desirable property for telephony. As the transport of voice was a driver in the design of ATM, it incorporates sophisticated mechanisms supporting various levels of QoS.

ATM is considered a layer 2 protocol. The header of an ATM cell is too small to carry the source and destination address. Instead, ATM networks use a *label switching* method, in which *virtual circuits* are established and all packets follow the same path, with packets routed along this path using a 'label' in the header.

Because ATM was primarily designed for network operations, the upper network layers, dealing with application and services, are not specified. Hence, few applications were developed specifically for ATM. Nevertheless, the IP datagram can be transported over an ATM network via encapsulation – at ATM Adaptation Layer type 5 (AAL5).

ATM is widely used today in trunk satellite communications (although increasingly supplanted by IP), while the principal use of ATM in personal satellite services is as a layer 2 packet frame for the return link in DVB-Return Channel over Satellite (DVB-RCS).

11.4.4 DVB-RCS

DVB-RCS is an open ETSI air interface standard (EN 301) for an interactive (two-way) Very-Small-Aperture terminal (VSAT) system developed to serve individuals, small businesses and others at data rates in the outbound (forward) direction of up to 20 Mb/s and in the inbound (return) direction of up to at least 5 Mb/s, equivalent to the cable and Asymmetrical Digital Subscriber Line (ADSL) systems. It was developed with a view to assist operators who intended to deploy large-scale wideband VSAT networks (DVB Project Online, 2009).

The standard – introduced in 2000 – is now in widespread use throughout the world, with hundreds of systems deployed serving tens of thousand of users in Ku, Ka, and C bands. Typical DVB-RCS applications include general Internet access (particularly in rural areas), Voice over IP (VoIP) services, telemedicine, tele-education and telegovernment.

In 2007, the system was enhanced to DVB-RCS + M to service mobile user terminals such as those installed in ships or land vehicles with addition of spread spectrum techniques to counter interference to other services. Communications on the move have grown rapidly in recent time, aided by regulatory relaxation in Ku band.

The DVB-RCS network is usually deployed in a star (hub-spoke) configuration (Figure 11.2) consisting of a central hub connected via satellite to a large user base. User terminals may be patched to a local wired or wireless network for provision of Internet or another service.

The forward link comprises a DVB-S or DVB-S2 waveform (see Chapter 9), while the core of the return link is built around a highly efficient MultiFrequency TDMA (MF-TDMA) demand-assigned multiple-access system, allowing efficient transport of different classes of applications such as voice, video streaming, file transfers and web browsing. In the return link, two level 2 packetization schemes are supported, with IP data encapsulated in either ATM or MPEG-2 TS packets. In order to optimize satellite resources dynamically, the access schemes are combined with an adaptive transmission system comprising turbo coding, various burst sizes and efficient IP encapsulation.

DVB-RCS supports four methods of allocating MF-TDMA bursts to user terminals (and combinations thereof):

- *Constant Rate Assignment (CRA).* A constant data rate is negotiated prior to transmission, and capacity is assigned for the duration of the connection. Such a scheme might, for example, be used for the delay-sensitive real-time application such as videoconferencing.
- *Rate-Based Dynamic Capacity (RBDC).* Terminals periodically update their capacity requests – overriding previous requests. A ceiling rate is also indicated, and the scheme endeavours to

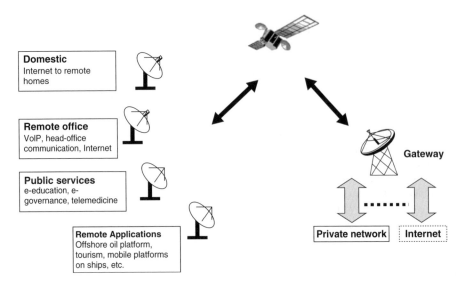

Figure 11.2   A network architecture of DVB-RCS.

guarantee the requested rate up to this threshold. Again, this scheme might be used for real-time applications.

- *Volume-Based Dynamic Capacity (VBDC)*. Terminals dynamically signal the amount of capacity (e.g. in Mbits) needed to empty their internal buffers. Assignments have lower priority than CRA and RBDC, and this is essentially a best effort scheme.
- *Free Capacity Assignment (FCA)*. Any spare capacity is assigned among terminals within the constraints of fairness, and, again, this is a best effort scheme.

11.4.4.1   DVB-RCS Overhead

The use of IP over DVB-S/S2 and DVB-RCS introduces a framing overhead that further reduces bandwidth efficiency for short packets. In DVB-S, the mapping is achieved through a standard known as MultiProtocol Encapsulation (MPE), while DVB-S2 also provides a Generic Stream (GS) encapsulation. Continuing with our example of VoIP over satellite using an 8 kb/s G.729 vocoder (without header compression), the 24 kb/s VoIP IP bandwidth is increased to 32 kb/s when using MPE over MPEG-2 TS packets (i.e. 25% efficiency) (NERA, 2004).

11.5   Satellite Communications Networks

In the previous sections we introduced network technologies applicable to satellite communication systems in the context of personal applications. This section extends those concepts to commercial satellite system networks, taking a mobile satellite service as a representative example.

11.5.1   Topology

In a fixed satellite service, which addresses communication between fixed Earth stations, the ground segment consists of a network of Earth stations interconnected in a mesh or star topology. Typically,

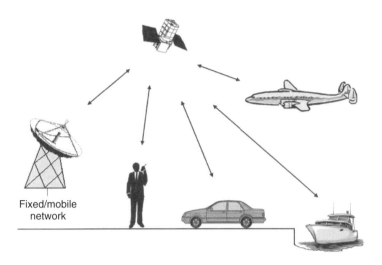

Fixed/mobile
network

Figure 11.3 The main elements of a mobile satellite service network. The systems are generally configured in a star configuration, but some modern satellites permit a mesh topology.

the large stations interwork with a terrestrial traffic node, for instance a telephone exchange or an Internet service provider. The terminals located at the user premises service a small user community or a single user.

Figure 11.3 represents a mobile satellite service network that deals with communication with mobile Earth stations. The ground network consists of one or more gateways and a community of mobile users. The gateways link the mobile users to their destination through a public (or private) network. A majority of the systems are connected in a point-to-point star arrangement.

11.5.2 Network Connectivity, Functions and Interfaces

The satellite networks presented in the preceding paragraphs consist of a number of entities that must interact appropriately for proper functioning. Considering an MSS system as an example, Figure 11.4 illustrates various links used for interconnecting these entities.

A satellite network requires various types of connectivity to support the services. The Service Link (SL) provides the primary radio connectivity of the user with the fixed network; mobile–mobile communication can be established through a Mobile–Mobile Service Link (MMSL); Inter-Satellite Links (ISLs) may be used to interconnect satellites in space: a Feeder Link (FL) connects a gateway to a satellite; a gateway–gateway link (GGL) connects two gateways for exchanging network information and routing calls; intra-gateway links (GLN, for the nth link) connect various functional entities within an earth station (these relate to call and network management); Network Management Links (NMLs) are useful for transporting information related to space and terrestrial network management. The satellite–terrestrial network interface connects the satellite network to the terrestrial private or public network via a terrestrial link (TL).

11.5.2.1 Functions

A satellite network must carry out numerous functions to ensure a satisfactory service. Figure 11.5 presents the functions associated with a mobile satellite service and their inter-relationships

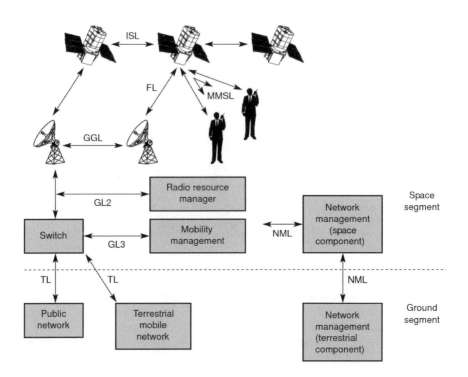

Figure 11.4 Illustration of connectivity links in a mobile satellite service system (Richharia, 2001). Key: GGL = gateway–gateway link, GLN = gateway link N, TL = terrestrial link, ISL = interstation link, FL = feeder link, SL = service link, MMSL = mobile–mobile service link, GNL = gateway network link, NML = network management link. Reproduced by permission of © Pearson Education. All rights reserved.

to demonstrate the main functions of a commercial satellite system and connectivity between entities.

The *core* function of the mobile network is service delivery to mobiles with the desired quality. *Mobile services* represent various services available to the mobile user. A typical portfolio of primary services include voice, data, facsimile, paging, message delivery and emergency calls, and supplementary services would be call transfer, call forwarding, call waiting, call hold, conference calls, etc. The *space segment* consists of one or more satellites that operate in concert. Some network functions (e.g. call routing/switching), usually performed on the ground, can be either wholly or partially executed on-board a satellite when regenerative transponders with on-board processing are used. *Constellation management* (CM) involves standard Telemetry and Telecommand (TT&C) functions related to satellite health monitoring, ephemeris generation, spacecraft orbit raising, orbital adjustment, launch support, initial deployment, replacement of failed spacecraft, etc. The *network planning centre* oversees network management, capacity and business trends and specific events of interest and develops strategies for changes to the network such as expansion of capacity, introduction of new services, redeployment of satellites in the network, etc. The *Network Management* (NM) function deals with real time and off-line management of the network. Real-time functions involve radio resource management, monitoring the radio spectrum to ensure compliance, interference detection, monitoring signal quality and network traffic flow, etc. Off-line functions include collection

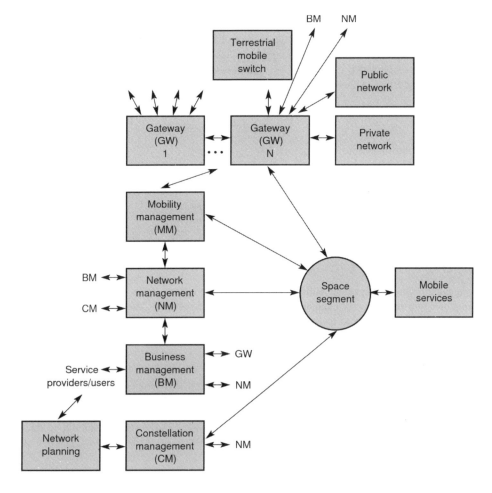

Figure 11.5   Illustration of typical functions performed in a mobile satellite service network (Richharia, 2001).

and dispatch of call data records to the business management system, traffic trend analysis to assist radio resource management and forward planning, fault finding/diagnosis, fraud detection, etc. NM functions provide call data records, receive user profile and other user-related information to the business management system. The *Mobility Management* (MM) system maintains user location in its database and interacts with the NM and gateways (GWs) for call connection and user authentication, user profile, etc. *Gateways* can either belong to the network provider or be owned by individual operators. The *Business Management* (BM) system constitutes part of a company's business centre responsible for customer billing, compiling call records and maintaining user profiles.

A Telecommunication Management Network (TMN) was defined conceptually by the ITU (CCITT M.3010, 1996) to provide a structured approach to facilitate network management. Figure 11.6 (Richharia, 2001) illustrates a derivative of this approach. Functions include administrative management, configuration management, network operations, constellation management, gateway management and switching, security management, maintenance and operational

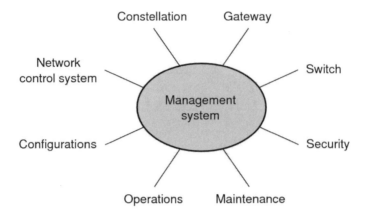

Figure 11.6 A network management system (Richharia, 2001). Reproduced by permission of © Pearson Education. All rights reserved.

management. The management entities communicate with each other on an independent tele-communications network. By maintaining the performance standards, the network operator is able to provide the desired QoS to its customers.

11.5.2.2 Interface

An interface facilitates and formalises connection between different entities by matching their physical and functional characteristics. Figure 11.7 illustrates the main interfaces of a satellite network.

There are two external interfaces: E-1 links the user terminal to the satellite network at the user end, and E-2 links the gateway to the terrestrial network. E-2 depends on the characteristics of the serving terrestrial network, that is, PSTN, Internet, etc.

Internal interfaces for communication include:

- I-1: user–user;
- I-2: user–spacecraft;

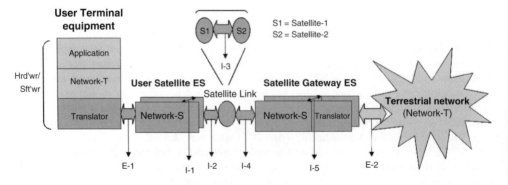

Figure 11.7 Typical interfaces of a satellite network for personal services.

- I-3: spacecraft–spacecraft;
- I-4: spacecraft–gateway;
- I-5: gateway–gateway.

11.6 High-Altitude Platform Systems

In Chapter 2 we introduced the concept of high-altitude platform (HAP) systems, including issues pertinent to their design and development. Here we expand the topic and introduce HAP networking concepts with example applications, and discuss technical issues related to their commercial viability. In Chapter 12 we introduce a few representative HAP research programmes.

11.6.1 Overview

HAP systems offer the potential of delivering broadband fixed and mobile telecommunication services for personal applications. Close proximity of a HAP to the ground allows the use of smaller 'pico' spot beams offering high power flux density. The footprint of a HAP resembles a satellite spot beam. Further segmentation of the beam leads to a granularity at present unachievable by satellite systems. Compare the ∼400 km diameter of the most advanced satellite with the ∼3–10 km diameter spot beam feasible with a HAP system. The finer partitioning enhances the capacity of a HAP system significantly over a limited coverage area. However, this is still very much an emerging technology, and a number of technical, regulatory and commercial issues remain unresolved.

A HAP located at an altitude of 20 km illuminates a circular area of ∼225 km diameter at 10° elevation, thereby covering large areas that would require considerable time and effort to cover terrestrially. Rapid deployment (days to weeks) thus becomes feasible in disaster-struck areas.

The ITU has assigned spectrum for HAP systems in high-frequency bands: 47/48 GHz worldwide (2 × 300 MHz primary allotment), 28/31 GHz (for Broadband Fixed Wireless Access (BFDA) on a secondary basis) in some 40 countries (region 3) and a narrow range around 2 GHz for IMT 2000. Spectrum congestion is minimal at the high frequency end, hence bandwidth has been licensed generously to facilitate broadband communications, for example a BFDA direct-to-home service.

Rain and other hydrometers cause heavy propagation losses for less than ∼1% of time in the 47 and 28 GHz HAP bands. Figure 11.9 illustrates calculated link availability for 28 and 47 GHz links based on ITU recommendation ITU-R P.618 at a point located at sea level and a HAP hoisted 30 km away at an altitude of 17 km for a Mediterranean region (Tozer and Grace, 2009). Terrestrial mobile systems tend to operate in a shadowed region (resulting in Rayleigh fading) and hence suffer considerable propagation loss; whereas, in the slant-path radio links of HAP, propagation conditions can be more favourable, particularly above ∼20° (resulting in Ricean fading).

Table 11.2 illustrates the cumulative distribution of propagation losses estimated at 28 and 47 GHz on the basis of ITU recommendations ITU-R P837, P838 and P839 for a Mediterranean climate. The estimates demonstrate that the lower frequency has a significantly lower fade loss.

Table 11.2 Estimated HAP link margins to guarantee service as a function of percentages of time

Frequency	99.99%	99.9%	99%
28 GHz	32.5 dB	12.3 dB	3.3 dB
47 GHz	64.1 dB	26 dB	7.4 dB

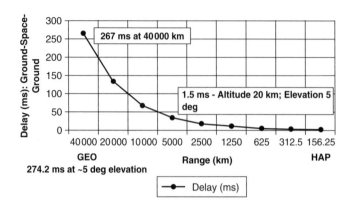

Figure 11.8 Propagation delay versus range in S-band IMT2000 band.

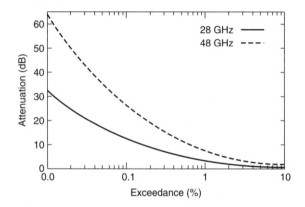

Figure 11.9 Percentage of time that attenuation is exceeded at a point located at sea level and an HAP hoisted 30 km away at an altitude of 17 km in a Mediterranean region. Reproduced by permission of © IET.

Because of the close proximity of HAPs to the Earth, the transmission delay is significantly lower than for satellite systems (see Figure 11.8), thereby minimizing the TCP problem associated with delays of GEO satellite systems. The service area of HAP systems (e.g. ~225 km diameter) compares unfavourably with the wide coverage of satellite systems—nearly one-third of the Earth disc from a GEO (~12 000 km diameter). Hence, HAP systems are particularly suitable for covering narrow regions and in environments such as cities where satellite systems cannot provide the necessary link margin without repeaters. They are unlikely to be commercially attractive to service wide remote areas like oceans, remote mountains and other uninhabited regions of the world.

It is evident that, while HAP solutions are attractive within the constraints, their benefits vis-à-vis satellite systems should not be exaggerated. Rather, an integrated terrestrial–HAP–satellite system is likely to provide a more attractive proposition, where strengths of each medium are harnessed effectively.

11.6.2 HAP Applications

HAP-enabled applications broadly cover the disciplines of telecommunications, broadcasting, surveillance and remote sensing. Telecommunication services include broadband Internet access for fixed personal installation (BWFA), third- to fourth-generation mobile services and telecommunication services for the exploration industry. Example broadcast applications include television/radio broadcast and multicast services in targeted remote areas. Miscellaneous applications include assistance to location-based services, traffic monitoring and reporting, rapid communication support in disaster-struck areas, business television, videoconferencing, remote audio–video medical applications, distance learning, etc.

11.6.3   HAP Network Topology

Various network configurations and their combinations are feasible:

- interconnection of users when they reside either in the same core network or in different core networks of a conventional telecommunication architecture – for example, provision of the Internet to a rural community and wideband wireless access to homes in densely populated areas;
- services to a close user community in a star or mesh configuration – for example, the sale of lottery tickets or intrabank communication;
- broadcast or multicast services.

A basic HAP topology is illustrated in Figure 11.10. The coverage of a single HAP can be extended via inter-HAP radio or optical links or HAP–satellite radio/optical links. Figure 11.11 shows a representation of such an extended network topology.

Consider the use of HAPs for broadband communications to fixed/mobile users. The user community comprises mobile and fixed users, connected to the terrestrial network through a gateway. The radio link consists of the up/down service link to the user community and a high data rate feeder or backhaul links that concentrate the traffic from the user community in the return link and distribute traffic to the user in the forward direction. As discussed previously, in a star network, users interconnect to each other through the gateway, whereas in a mesh network they can directly connect to each other.

11.6.4   Challenges

Aspects that currently require further analysis and developments include:

- the HAP platform;
- the air interface;
- service definition;

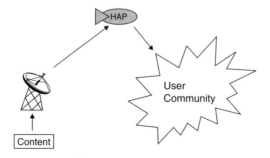

Figure 11.10   A simple HAP telecommunication topology.

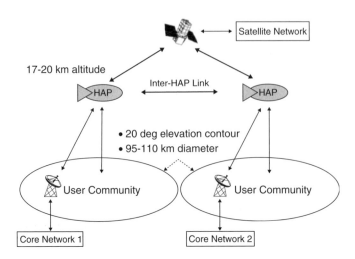

Figure 11.11 A hybrid satellite–HAP constellation.

- user requirements: throughput, size, service charge, connectivity;
- connectivity to the core network.

Detailed discussions of these topics are beyond the scope of this book; the interested reader will find details in references mentioned earlier in this chapter (and also Chapters 2 and 12). Salient aspects are summarized here. The air interface design is crucial. Efficient utilization of resources dictates a trade-off between cost, complexity, antenna spot beam technology and modulation scheme. The schemes must be able to contend with high propagation losses in the Ka and V band, necessitating fade and multipath countermeasures. Robust and novel frequency planning such as a dynamically changeable scheme can potentially enhance the spectrum efficiency. Diversity techniques may be essential for the backhaul – particularly in the tropical regions. Network issues include selection and optimization of handover schemes, roaming arrangements and network architecture and synergy with other systems such as a partner satellite or terrestrial system. Regulatory issues pertain to the licensing of the frequency bands in the service region, obtaining authorization for hosting the HAP in the region. HAP-centric issues include the selection of a reliable technology with the desired platform endurance and stability. Commercial issues include the development of a realistic business plan, obtaining an agreement with potential investors, billing arrangements, etc. Operational aspects following the deployment include the development of robust operational procedures, contingency measures to provide reliable service, maintenance and forward planning.

11.6.5 Technology Status

There are numerous projects in progress across the world that are addressing various aspects of HAP technology. It is believed that, in the short term, the technology will focus on niche applications. Unmanned aerial vehicles and tethered aerostats are already providing services to the military, the surveillance industry, emergency communication, etc. Manned aircraft are under development and in favour for commercial telecommunication applications in a short timeframe, while research is in progress to develop large unmanned solar-powered stratospheric aircraft. The endurance of such aircraft is targeted to be of the order of months in the short term. Stratospheric platforms have yet to be implemented, and hence their *in situ* performance is unproven. Stabilized multispot antennas capable of generating a large number of spot beams – although regularly flown on satellites – have yet to be adapted to suit the weight, motion and dimension constraints of HAPs. Radio and aeronautical regulatory issues, while being addressed, have yet to mature. There are various network issues that require further refinements but are difficult to consolidate in the absence of a commercial commitment, which in turn is hindered by the lack of a plausible business case. Chapter 12 introduces various research and commercial initiatives addressing the present development and challenges in the implementation of HAP systems.

Revision Questions

1. Satellite communication systems are relatively expensive but offer unique solutions. Suggest at least two communication applications for which a satellite solution is attractive and competitive with other transmission media. Suggest a network configuration for each application. State the key factors essential in designing the proposed networks.
2. Outline the problems of IP transmissions over satellite and suggest some methods to alleviate the problem.
3. Describe the internal and external interfaces and functional entities essential in proper functioning of a mobile satellite service.
4. Discuss the advantages of a HAP system. State the main limitations that have hindered their widespread deployment.

References

CCITT M.3010 (1996) Recommendation M.3010, Principles for a telecommunications management network.

DVB Project Online (2009) http://www.dvb.org [accessed July 2009].

IETF 1072 (1988) RFC 1072, TCP extensions for long-delay paths.

IETF 1323 (1992) RFC 1323, TCP extensions for high performance.

IETF 1122 (1989) RFC 1122, Requirements for Internet hosts – communication layers, ed. by Braden, R. October.

IETF 1123 (1989) RFC 1123, Requirements for Internet hosts – application and support, ed. by Braden, R. October.

ITU-T G.729 (2007) Coding of speech at 8 kbit/s using conjugate-structure algebraic-code-excited linear prediction (CS-ACELP).

Jeffery, M. (1994) Asynchronous transfer mode: the ultimate broadband solution? *Electronics and Communications engineering Journal*, June, 143–151.

Nguyen, T., Yegenoglu, F., Sciuto, A., Subbarayan, R. (2001) Voice over IP service and performance in satellite networks. *IEEE Communications Magazine*, **39**(3), 164–171.

Richharia, M. (2001) *Mobile Satellite Communications*. Pearson Education Ltd, London, UK.

Sun, Z. (2005) Next generation Internet (NGI) over satellite, in *Satellite Networking, Principles and Protocols*. John Wiley & Sons, Ltd, Chichester, UK, Ch. 8.

Tozer, T.C. and Grace, D. (2009) *HeliNet – the European Solar-Powered HAP Project*, Available: http://www.elec.york.ac.uk/comms/pdfs/20030506163224.pdf [accessed July 2009].

12

Satellite Communications Systems

12.1 Introduction

This chapter illustrates practical applications of the concepts and techniques introduced in the first part of the book by numerous examples. It highlights features of a few selected communication systems representing mobile, fixed, military and amateur services to illustrate the diverse techniques and the rationale applied.

We categorize the *mobile satellite service* sector by: (i) orbit – GEO and LEO; (ii) throughput – broadband and narrowband. In the GEO broadband category we present two examples that operate on widely separated radio frequency bands. The L-band spectrum allocation is available exclusively to MSS globally, whereas the Ku band allocation is secondary. Inmarsat's L-band Broadband Global Area Network (BGAN) system is representative of the latest mobile broadband technology. In recent years, L band has become increasingly congested because of a limited spectrum allocation in this band shared by a large number of operators. This limitation imposes a severe constraint for broadband services. Ku band provides adequate space segment capacity but does not benefit from the protection status of L band. The availability of economic-space-segment, low-cost mobile VSATs and the insatiable demand for wideband have led to a rapid increase in Ku band broadband mobile systems for routine non-critical applications. We will introduce the state of the Ku band technology, with an emphasis on the aeronautical sector.

The demand for *fixed satellite service* has surged in the personal broadband sector, particularly in the rural and remote area segments. Some operators prefer the Ka band, as there is currently ample spectrum in this band. We introduce a recent Ka band VSAT system called WildBlue to elucidate this trend.

The military always avails itself of satellite technology owing to the strategic advantage of satellites in all theatres. One may debate the pertinence of a military system in the context of personal system. We feel that the military personnel's needs are no less than those of the population at large.

Thousands of radio amateurs thrill in the experience of communicating with each other by means of Orbiting Satellite Carrying Amateur Radio (OSCAR) satellites. The community also contributes to various humanitarian causes – for instance, furthering space technology and satellite-aided societal solutions. Keeping to the theme of the book, a section is devoted to the OSCAR system.

High-altitude platform systems can potentially offer the benefits of a geostationary satellite in synergy with those of a terrestrial system in confined regions. In spite of significant research and development in the past decades, wide-scale implementation of the technology remains elusive. Here, we present the accomplishment of major European research initiatives under the auspices of the European Commission, and introduce some commercial initiatives pioneering the technology.

Satellite Systems for Personal Applications: Concepts and Technology   Madhavendra Richharia and Leslie David Westbrook
© 2010 John Wiley & Sons, Ltd

12.2 Mobile Communications

12.2.1 GEO Broadband Systems

12.2.1.1 L-band: Inmarsat BGAN

The International Mobile Satellite Organization (Inmarsat) is a private company based in London. The company provides global telecommunications services to maritime, land and aeronautical users in L band throughout the world via a constellation of geostationary satellites positioned around the world. Inmarsat services have evolved since 1981 to fulfil user needs in tune with technical advancements. Beginning from the maritime surroundings, the services were extended to aeronautical and land environments by enhancing the air and the terrestrial interfaces as necessary. The size and cost of terminals has decreased continually over the period, while their capability and performance continue to enhance the leveraging of technological advances. Following the launch of Inmarsat's fourth-generation satellites (I-4) in 2005, the company introduced the Broadband Global Area Network (BGAN) service which supports circuit and packet data at a throughput of up to ~0.5 Mb/s on a variety of portable and mobile platforms (Inmarsat Online, 2009).

System Overview

A top-level architecture of the BGAN system is presented in Figure 12.1 (Richharia and Trachtman, 2005). An Inmarsat-4 satellite was shown in Figure 4.9.

User Equipment (UE) consists of Terminal Equipment (TE) such as a PC and a telephone connected through standard interfaces such as a USB port or Bluetooth to a Mobile Terminal (MT). The MT is responsible for radio communications with the BGAN fixed network via an I-4 satellite. A removable UMTS Subscriber Identity Module (USIM) card attached to the MT stores the user's identity and other details as an application. The fixed network consists of a Satellite Access Station (SAS). The system allows SIM-enabled roaming. Authorized subscribers from a partner BGAN or terrestrial mobile networks can insert their SIM to receive BGAN-supportable home services.

The L-band service link operates in right-hand circular polarization in the frequency range 1626.5–1660.5 MHz in the forward link (Earth–space) and 1525–1559 MHz in the return link (space–Earth). The service area is subdivided into three types of zone representing the I-4 antenna

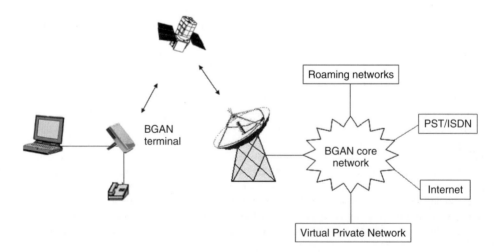

Figure 12.1 BGAN system overview.

beam patterns: narrow, regional and global. Typically, for each satellite the global beam is overlaid by 19 regional beams and over 150 narrow beams during operation. BGAN services operate primarily in narrow beam but are supportable in regional beams. The satellites deploy a transparent architecture comprising C–L band transponders in the forward direction and L – C band transponders in the return direction. The dual polarized feeder link operates in the global beam in the frequency range 6424–6575 MHz in the forward direction and 3550–3700 MHz in the return direction.

BGAN fixed infrastructure consists of three Satellite Access Stations (SASs) – two located in Europe (The Netherlands and Italy) and one in the pacific (Hawaii). The RF system feeds the information received from the UEs to a Radio Network Controller (RNC). The RNC interfaces with the core network which routes and switches calls to the public or private network as necessary and according to the subscriber's subscription profile. The core network interworks with a 3G PP (Third Generation Partnership Programme), release-4 architecture. Mobile Switching Centre (MSC) server nodes support circuit-switched communication (i.e. PSTN and ISDN). A media gateway performs the necessary translation, such as transcoding. The Serving GPRS Support Node (SGSN) and Gateway GPRS Support Node (GGSN) support Internet Protocol (IP) packet-switched communications (Richharia and Trachtman, 2005).

Service

The system provides a suite of packet- and circuit-switched communication services to native BGAN users and authorized roaming customers. Examples include:

- telephony and ISDN calls via the directory telephone number(s) associated with their subscriber identity module (SIM);
- an IP-based Internet/Intranet/virtual private network connection with fixed or dynamically assigned IP addresses on a GPRS platform supported by UMTS streaming, background and interactive quality of service classes;
- SMS messages to and from BGAN users and between BGAN and other networks via circuit and packet modes;
- advanced messaging services such as Multimedia Messaging Services (MMS) as applications supported over the packet-switched GPRS/IP service;
- UMTS location-based services such as offering maps, local travel information, etc.;
- vendor developed content and applications transport.

The voice service includes all the standard enhanced features available in terrestrial fixed-line and mobile networks, for example voicemail, caller line ID, call forwarding, call waiting, conference calling and call barring. The streaming IP data service offers guaranteed bandwidth on demand, enabling live video applications like videoconferencing and video streaming. Non-real-time video can be based on a store-and-forward mechanism that allows optimization of the available resources.

Target users include: media companies, government customers and the military sector, aid agencies and non-governmental organizations, oil and gas companies, construction companies, etc.

Satellite

The satellites are located at 53° W, 64° E and 178° E, corresponding respectively to Inmarsat's Atlantic Ocean Region–West (AOR-W), the Indian Ocean Region (IOR) and the Pacific Ocean Region (POR). Each Inmarsat-4 satellite (I-4) can generate up to 256 beams consisting of a mix of global beam, 19 wide spot beams and more than 200 narrow spot beams reconfigurable through ground commands to provide extra capacity when and where desired. The small size of the beam and the sharp cut-off at the beam edges necessitate an accurate pointing requirement from the spacecraft. This is achieved by using a ground-based beacon and an on-board RF sensing system comprising an amplitude monopulse scheme to ensure correct beam orientation.

The global beam provides a signalling link for all services, and the wide spot beams, in addition to servicing Inmarsat's legacy portfolio, provide a basic BGAN service. In thin route areas where a narrow spot beam has been switched off, the wide spot beams provide a minimum BGAN service prior to the activation of a narrow spot beam. A 9 m space furled reflector antenna fed by 120 helix antennas generates the spot beams. The maximum EIRP deliverable in a narrow spot beam is 67 dBW, and the G/T of narrow spot beams is >10 dB/K. The body dimensions of the satellite are 7 m × 2.9 m × 2.3 m, and the wing spans 45 m, to deliver a DC bus power of 12 kW. Solar cells combine the benefits of silicon technology with advanced gallium arsenide cells. Spacecraft thrusters deploy both chemical and plasma-ion technologies. Payload characteristics are summarized in Table 12.1 (Stirland and Brain, 2006).

User Equipment

Table 12.2 illustrates example characteristics of a set of representative BGAN user terminals (Rivera, Trachtman and Richharia, 2005). The throughput depends on the terminal characteristics (G/T and EIRP) with a maximum rate of ~492 kb/s in the forward direction and ~330 kb/s in the return direction. Note that the table is intended to be representative only; the interested reader should refer to Inmarsat Online to view the company's in-service products and coverage.

User Equipment (UE) is categorised into four broad classes which are further divided into subclasses to distinguish equipment and assist system operation on the basis of UE capabilities and environment.

The land mobile service is accessible through three classes of satellite terminals referred to as class 1 (briefcase size), class 2 (notebook size) and class 3 (pocket size). Although portable, a class 1 terminal is intended for the fixed office environment, while classes 2 and 3 are suitable for people in need of packet- and circuit-switched services on easily portable user terminals (Figure 12.2). Aeronautical and maritime services are available on two terminal variants – a light, cheap, intermediate throughput terminal and a fully fledged variant offering maximum throughput.

Table 12.1 Inmarsat-4 payload characteristics

Spacecraft parameters	Performance
Service link	
Transmit frequency	1525–1559 MHz
Receive frequency	1626.5–1660.5 MHz
Polarization	Right-hand circular
Axial ratio (max)	2 dB
Frequency reuse target	Narrow beams: 20 times (7 colour); Wide beams: 4 times (3 colour)
G/T	
Narrow beam (inner) <6° (viewed from spacecraft)	11 dB/K
Narrow beam (inner) >6° (viewed from spacecraft)	10 dB/K
Narrow beam (outer)	7 dB/K
Wide spot beam	0 dB/K
Global beam	−10 dB/K
EIRP	
Narrow beam (inner)	67 dBW
Wide beam	58 dBW
Global	41 dBW

Table 12.2 Characteristics of a set of proposed BGAN user equipment (UE)

UE class	UE subclass	UE class number	Minimum G/T (dB/K) with satellite elevation of $\geq 5°$	UE EIRP (dBW)
Land portable	Land-A3 class	1	≥ -10.5	20
	Land-A4 class	2	≥ -13.5	15
	Land-A5 class	3	≥ -18.5	10
Aeronautical	Aero high-gain	4	≥ -13	20
	Aero intermediate-gain	5	≥ -19	15
	Maritime high-gain	6	≥ -7	22
Maritime	Maritime low-gain	7	≥ -15.5	14
Land mobile	Land mobile high-gain	8	≥ -10	20.0
	Land mobile low-gain	9	≥ -15	15.0

Figure 12.2 The BGAN system supports a variety of user terminals. The portable land terminals are widely used by people on the move, such as journalists as illustrated above. Courtesy © Nera Satcom.

The latter terminals are evolved versions of those developed for the legacy services attempting to maximize technology reuse – the high-gain terminals target users needing high rate throughput, whereas the low-gain terminals target those who require medium-rate throughput on simpler, lower cost terminals.

Architecture
The BGAN architecture supports seamless interworking with UMTS, leveraging on the fact that UMTS Non-Access Stratum (NAS) protocols support transport over different types of radio access. Accordingly, the BGAN Core Network (CN) architecture is kept the same as that of the terrestrial UMTS, and its air interface is optimized for the best match between terminal characteristics and

the satellite propagation channel. The Radio Network Controller (RNC) uses a standard UMTS interface with the core network (Richharia and Trachtman, 2005).

The BGAN (proprietary) air interface – called Inmarsat Air Interface 2 (IAI2) – protocols ensure that the attributes typical of a geostationary satellite link, that is, high transmission delay, variable error rate, a periodic disruptions, are accounted for. The protocol stack consists of adaptation, bearer connection, bearer control and physical layers.

The communication in between and across remote peers utilizes protocol units and interfaces as mechanisms for control and transfer of information.

The adaptation layer provides functions such as registration management (e.g. spot beam selection, system information handling), GPRS Mobility Management (GMM) handling, mobility management handling and radio bearer control (e.g. handling signalling related to set-up, modification and release of radio bearers).

The bearer connection layer provides functions such as provision of acknowledged mode, unacknowledged mode and transparent mode connections, buffering and flow control of information from the interface to the layer above, QoS policing, segmentation and reassembly of information, Automatic Repeat Request (ARQ) (if required for the particular connection) and ciphering.

The bearer control layer controls the access to the physical layer for each of the established connections. It covers aspects of UE behaviour such as initial timing offset due to UE position, delta timing corrections due to UE or satellite movement, power and/or coding rate adjustments of UE transmissions, frequency corrections due to Doppler variations or UE frequency inaccuracies, etc. It also provides the transport for the BGAN system information from the RNC to the UEs and ensures that an appropriate level of resource is maintained within the RNC.

The proprietary physical layer offers high flexibility and efficiency by adapting modulation, coding rate, transmission power and bandwidth to prevailing channel conditions, UT capabilities and user location. Owing to a wide range of UE antennas and EIRP, a single transmission solution is not feasible. Hence, a judicious combination of modulation, transmission and code rates is used that depends on the situation.

It achieves high efficiency by optimally combining turbo codes, 16-QAM and QPSK modulation schemes, intelligent radio resource management and powerful software-enabled transceivers capable of near-Shannon-limit efficiencies. Channel throughput ranges between $\sim$2 kb/s and $\sim$0.5 Mb/s, depending on UT class, location, received signal quality, the operating beam and transmission direction. Table 12.3 summarizes the core features of the physical layer (Richharia, Trachtman and Fines, 2005).

FDM/TDM access scheme is used in the forward direction and FDM/TDMA scheme in the return direction, assisted by demand-assigned sharing of the radio bearers.

Ku Band – Review

A majority of existing mobile satellite services operate in L and S MSS bands where the technology is mature and low cost and the radio wave propagation conditions are reasonably congenial relative to the Ku or Ka band. MSS allocations in the Ku band have a secondary assignment, which mandates reduced levels of power flux density to protect services of higher status. Nevertheless, bandwidth

Table 12.3   The core features of the IAI2 physical layer

Transmission direction	Access scheme	Modulation schemes	Symbol rates (kSymbols/s)	FEC code	Code rates
Forward	TDM	16-QAM	8.4–151.2	Turbo code	0.33–0.9
Return	TDMA	16-QAM, O-QPSK	4.2–151.2	Turbo code	0.33–0.9

availability in the Ku band is substantially higher. Moving further up in frequency, the Ka band has a significantly larger MSS primary allocation and hence the power flux density constraints are less stringent; however, the technology is relatively new, the numbers of transponders in space worldwide are presently limited and propagation conditions are severer than for lower bands.

Recognizing that the escalating demands for mobile wideband transmission can be offered at relatively low cost in Ku band and the relative abundance of spectrum, the VSAT industry began to introduce telecommunication and broadcast services on ships, aircraft and long-distance trains. However, owing to the regulatory restrictions in the use of the band for mobile communications, the coverage areas are restricted. Recognizing the trend and the ensuing promotion by the industry, such restrictions were lessened by the ITU to encourage mobile operation in parts of the fixed-satellite Ku band.

The key enabling technologies include a low-cost (and preferably low-profile) mobile satellite tracking system (refer to Chapter 4), high-power wide area coverage, efficient air interface methods such as DVB-S2 and robust Doppler compensation algorithms, particularly for airborne systems.

In the remaining part of this subsection, we will briefly review a Ku band aeronautical mobile system, after a few observations about the land and maritime sectors.

In the land mobile sector, National Express, a rail service provider in the United Kingdom, offers Wi-Fi service functioning over a hybrid 3G (terrestrial) 4 Mb/s Ku band VSAT radio link in fast-moving intercity trains at speeds of up to 320 km/h. Similar services are also available in continental Europe; VIA Rail provides satellite broadband services in Canada (Freyer, 2009). Several innovative research and development trials have been conducted in Europe in the past few years (Matarazzo et al., 2006; ESA Online, 2009).

In the maritime sector the technology has grown rapidly in recent years, primarily driven by those for whom mobility and low operational cost are the key drivers rather than global reach or the satellite component of GMDSS communication. Applying an appropriate mix of technology, the cost per megabit in Ku band is lower by up to 6 times than L-band MSS systems such as Inmarsat, although the latter's equipment cost is lower along with global system availability and GMDSS compliance (Freyer, 2009).

Aeronautical System

Consider the 2004 launch of a Ku band aeronautical broadband Internet service by Boeing, known as Connexion. The system was built on the premise that the proliferation of laptops would facilitate a wide acceptance of broadband Internet on long-distance flights in spite of a fee for the service. The business collapsed in 2 years. The technology was flawless, but the equipment was expensive and weighty (around 350 kg), which incurred a fuel overhead. The aircraft downtime, lasting a few weeks during equipment retrofitting, did not find favour from the airlines owing to the ensuing revenue loss; and user demands were insufficient to offset the expenses.

Nevertheless, the belief in the concept prevailed. The notion was reinforced by the recent observation by the airline industry that broadband Internet access and live television were among the most desirable value-added amenities perceived by passengers, and Wi-Fi is embraced by a significant number of air passengers.

Given that the airline industry continues to explore ways to improve passenger experience and operational safety while reducing costs, refined versions of the technology re-emerged. Some of the proposed systems are undergoing trials and others are in service. Examples include the US companies – OnAir's in-flight technology (Inmarsat: L band), Aircell (terrestrial), Viasat/KVH Yonder, ARINC Skylink, Row 44 (Ku band), Intelsat (Ku band global satellite capacity)/Panasonic Avionics (Connexion platform originators) consortium – and Inmarsat (L band). The OnAir system is based on Inmarsat's widely used L-band broadband technology, which is available globally and offers a maximum throughput of nearly 1/2 Mb/s in the forward direction. The Intelsat/Panasonic partnership combines the merits of global coverage over Intelsat's GlobalConnex network, bolstered by the transport efficiency of the DVB-S2 standard and Panasonic's wide experience of in-flight entertainment systems, including proprietary IP router technology and the Connexion experience. The system will utilize an antenna system optimized for aeronautical application. By contrast Aircell offers services over continental US airspace via ground repeaters.

Transmissions of DBS television systems are routinely received on mobile platforms, particularly in the United States. However, the content distribution rights as well as the coverage zones of these services are limited within a specific territory, whereas direct-to-aircraft broadcast services remove these limitations and provide a seamless service throughout a flight by switching between satellites when necessary, bolstered by content distribution arrangements.

There remains some apprehension on the part of airlines and regulatory authorities concerning possible interference of Wi-Fi and cell phone transmissions in the aircraft cabin with the aircraft's operations. User annoyance due to people engaged in long telephone conversations has long been a contentious issue. Regulatory authorities in some countries do not permit pico cell coverage in aircraft because of the potential for RF interference with terrestrial cellular networks. Solutions to resolve such issues may include prolonged flight tests to ensure safety, the voice over IP solution instead of the pico cell approach, blocking of undesirable websites, etc.

Row 44, a California-based US company, has developed a system to support a number of aero-nautical services – broadband Internet, Voice over IP (VoIP), cell phone roaming enabled through on-board pico cells and IP television. The system is compatible with narrow-body aircraft such as the Boeing 737, 757, Airbus A320, A319, etc. (Row 44 Online, 2009). In addition to the passenger services, the system also targets cargo services and flight, cabin-crew and technology operations. Examples of these latter services would be: electronic flight bag, emergency medical support, electronic logbook, cargo bay monitoring, etc. The system weighs less than about 68 kg (150 lb) and is installable in two nights when off air. The hardware comprises a low-profile antenna installed on top of the fuselage, an antenna control unit, a Modem Data Unit (MDU), a server management unit, Wi-Fi and pico cell equipment. The antenna is connected to a base plate that can be customized to fit different airframes. The High-Power Amplifier (HPA), Low-Noise Amplifier (LNA) and up/down converter interface at an intermediate frequency of 950–1450 MHz, with the MDU mounted within the aircraft cabin. The Antenna Control Unit (ACU) is connected to the on-board computer system to collect aircraft position and attitude information, process it and steer the antenna beam towards the satellite. The aircraft–satellite uplink operates in the 14.0–14.50 GHz range. Wi-Fi (802.11b, 802.11g and 802.11n) wireless access units are placed in the airplane cabin with sufficient capacity to service all the passengers; additionally, the system can interconnect with the in-flight entertainment system to grant access to the services on the back-seat entertainment system. Data rates are stated to average 30 Mb/s in the forward direction (i.e. to aircraft) and a maximum of 620 kb/s in the return direction. A second multicast receive-only channel operating at 45 Mb/s is supportable on the same equipment to transport IP television.

12.2.2   GEO Narrowband System

12.2.2.1   ACeS

The Asia Cellular Satellite (ACeS) system is a pioneer in provisioning voice services to cell-phone-sized units. While some of the available satellite voice communication systems rely on a low or medium Earth orbit satellite constellation, ACeS chose to deploy the geostationary orbit. To offset the higher path loss ($\sim$17 dB) relative to a low Earth orbit system (750 km altitude), the ACeS satellite deploys a large antenna reflector capable of generating a vast array of very narrow spot beams in conjunction with powerful transmitters.

The end–end system comprises a fixed infrastructure connected to the terrestrial system, the space segment and the user community. Several types of user terminal – cell-phone size, mobile and fixed dual-mode (terrestrial–satellite) – are supported. Services include high-quality digital voice, 2.4 kb/s duplex data, a high penetration channel for call alerting, several GSM features such as call transfer, seamless roaming (terrestrial–satellite–terrestrial), etc.

The ACeS coverage provided by the Garuda satellite encompasses the Asian market – China and Japan (North), Indonesia (South), Philippines (East) and India and Pakistan (West). ACeS offers these services in collaboration with Inmarsat, a global mobile satellite service provider. (Note: Garuda is a mythical bird mentioned in Hindu and Buddhist mythology.)

The Garuda satellite, based on Lockheed Martin's A2100AX spacecraft bus, is specified for a 12 year life (Nguyen, Buhion and Adiwoso, 1997). Figure 12.3 depicts an artist's impression of the satellite.

ACeS's proprietary air interface uses a TDMA/FDMA scheme. Signalling is based on GSM signalling protocols, after an international standard known as the Geo Mobile Standard 2 (GMR-2) platform developed jointly by the European Telecommunication Standards Institute (ETSI) in Europe and the Telecommunication Industry Association (TIA) in the United States. The forward link operates in C band (6.425–6.725 GHz) towards the satellite, and in L band (1.525–1.559 GHz) towards the users. The return link from users operates in the L-band uplink (1.6265–1.6605 GHz) to satellite and in the C-band downlink (3.4–3.7 GHz) to feeder station. User–user links are established by on-board cross-strapping of the L–L transponder. Gateway-NCC communication for management purposes is carried out via the C–C transponder.

Transmit and receive L-band antenna systems were separated to minimize the noise attributed to passive intermodulation. The deployable reflectors are 12 m in diameter (Figure 12.3) and made of gold-plated molybdenum wire mesh. The satellite's antenna arrangement is increasingly in use for personal systems. A low-level beam-forming network in the transmit chain assigns appropriate amplitude and phase to route signals and feeds them into multiport solid-state power amplifiers (SSPAs). The 'Butler matrix' arrangement allows power sharing between the multiport SSPAs and the feed system sharing in the creation of spot beams. Thus, each SSPA carries signals of different beams, and, similarly, each feed element transmits signals to several spot beams. The arrangement enables efficient use of the SSPAs and feeds and provides flexibility in accommodating load variations across the spot beams. The beam-forming network recreates the spot beams by combining the amplitude and phase of signals from various feeds at the receiver end.

C-band (feeder link) coverage is provided by a single shaped antenna beam formed by a shaped reflector. L-band coverage is provided by 140 spot beams created by two separate 12 m diameter

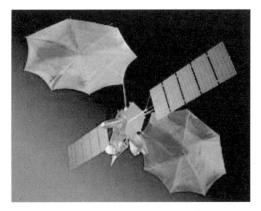

Figure 12.3   An artist's impression of the Garuda-1 satellite. Reproduced by permission of © Lockheed Martin.

antenna systems corresponding to the uplink and the downlink. The configuration permits a frequency reuse factor of 20 with a seven-cell reuse pattern. The G/T of the satellite is 15 dB/K and the total L-band effective radiated power is 73 dBW. These figures are sufficiently high to support up to 11 000 simultaneous handheld voice communication links with a link margin of 10 dB. C-band signals, after being pre-filtered, are down-converted and fed into a digital channelizer, which straps C–L as well as L–L links. Depending on the requirement, the received C-band signals can be routed to the L-band transmit antenna system via the C–L band transponder or to the C-band transmit antenna through the C–C transponder. On the return link, the L-band signals are up-converted and routed to the C-band antenna through the L–C transponder or for mobile–mobile connection to the L-band transmit antenna via the L–L transponder. The routing scheme is configurable through ground commands.

12.2.2.2 Thuraya

Thuraya was founded in the United Arab Emirates by a consortium of leading national telecommunications operators and international investment houses in 1997 (Thuraya Online, 2009). The company owns and operates an L-band geostationary mobile satellite system that provides telecommunication services to small handheld and portable terminals in countries spanning Europe, northern, central and large parts of southern Africa, the Middle East and Central and South Asia.

The service roll-out started in 2001 following the launch in October 2000 of the Thuraya-1 satellite, manufactured by Boeing Satellite Systems (USA). Thuraya-2 was launched in June 2003 to augment network capacity. The satellite is located at 44° E at an inclination of 6.3°. The Thuraya-3 satellite, launched in January 2008, is positioned at 154° W longitude to further expand system capacity and coverage.

The system complements the terrestrial Global System for Mobile Communication (GSM) system, allowing subscribers to switch their dual-mode phone between the satellite and terrestrial networks as required – typically switching into the Thuraya system when outside GSM coverage. The voice quality of the Thuraya telephone service is said to retain the same or similar voice quality as the GSM. Figure 12.4 shows a picture of a Thuraya dual-mode phone in use. The main attributes of the service are: telephony, fax, data, short messaging, location determination (via GPS), emergency services and high-power alerting to penetrate buildings.

Thuraya's service portfolio extends to rural and maritime environments on terminals tailored for these applications. For completeness, we include here Thuraya's service enhancements. Its broadband service delivers shared throughput of 444 kb/s.

The primary gateway of the system, located in Sharjah, UAE, serves the entire Thuraya footprint.

Thuraya satellites can generate 250–300, steerable spot beams from a 12.25 m × 16 m mesh transmit–receive reflector fed by a 128-element dipole L-band feed array. A 1.27 m dual-polarized shaped reflector is used for the C-band feeder communications links. The satellites can support up to 13 750 simultaneous telephone channels.

The Thuraya air interface is constructed on the Geo Mobile Standard 1 (GMR 1) platform developed jointly by the European Telecommunication Standards Institute (ETSI) in Europe and the Telecommunication Industry Association (TIA) in the United States. The GMR standards are derivatives of the GSM standard, developed to facilitate reuse of GSM's rich features, services and hardware/software while enhancing the commonality between the terrestrial and satellite functionality in the dual-mode phone. This would, for example, ease the design of the phone's man–machine interface and the Subscriber Identity Module (SIM), system roaming features, mobile–mobile single-hop connection, etc. GMR-1 shares the upper layer protocols with the GSM architecture to facilitate reuse of the GSM technology such as the visitor location register, the mobile switching centre, etc. The GMR-1 incorporates packet-mode services via GPRS, EGPRS and 3G protocol – a feature utilized in Thuraya's broadband system (Matolak et al., 2002).

Figure 12.4   Thuraya dual-mode phone can switch to a GSM terrestrial system in the absence of a satisfactory satellite signal. Courtesy © Thuraya.

The service link of the system operates in the 1626.5–1660.5 MHz range (Earth–space) and 1525.0–1559.0 MHz (space–Earth). The feeder link frequencies are 6425.0–6725.0 MHz (Earth–space) and 3400.0–3625.0 MHz (space–Earth).

A roaming arrangement between Thuraya and partner terrestrial operators allows users to move in and out of the Thuraya system. Dynamic satellite power control in the service link can support a 10 dB link margin to present a robust performance in the presence of signal fades. However, the link margin of the system is insufficient for indoor communication. Thuraya has resolved the limitation by introducing an interface unit known as the fixed docking unit, which enables access to the system from indoors. Vehicle operation is feasible by using a vehicle docking adapter which also includes a charging facility. Furthermore, a high-power transmission channel can alert indoor users of an incoming call.

The Thuraya narrowband system supports Internet access and data transmission at speeds of up to 60/15 Kb/s (downlink/uplink) using its geo mobile packet radio service. The system offers an integrated solution for control and tracking of vehicle fleets and fixed assets by integrating GPS-based fleet management and satellite packet radio technology.

A GPS-assisted location-based service grants the user the ability to send location details from a handset to designated contacts – family, friends, business associates or emergency services. The GPS locator embedded within a handset locates the user's position. The user can manually send the information by SMS to the operations centre, where an application places the information on a website. The interested party can log on to the website to retrieve the location superimposed on a map, satellite or aerial image available from Google Maps or Windows Virtual Earth.

Thuraya offers maritime communication at a packet data rate of up to 60 kb/s to a variety of marine vessels – fishing boats, supply vessels, merchant fleets, commercial carriers, yachts and leisure craft, etc. An emergency communication facility triggered through a distress button is available to alert preconfigured contacts. Maritime terminals make use of omnidirectional or stabilized directional antenna. Typical applications envisaged include File Transfer Protocol (FTP), Virtual Private Network (VPN), messaging and GPS-based location and position tracking.

The company's rural communication service offers telecommunication access to remote and rural areas. These can be configured as payphones to link remote communities with the mainstream. Thuraya also offers aeronautical services.

The broadband facility operates at shared data rates up to a maximum of 444 kb/s to facilitate Internet services, and at 16–384 kb/s for dedicated streaming IP. Security is maintained by a secure encryption algorithm. TCP acceleration facilities are optimized for the broadband channel.

The broadband satellite modem is made up of a compact, ultra-lightweight A5-sized satellite terminal utilizing a simple plug-and-play mechanism and robust outdoor protection features. Typical broadband applications envisaged by Thuraya are: store-and-forward video, live video and video conferencing, temporary office, etc.

12.2.3   LEO Systems

In contrast to the GEO systems where 3–4 stationary satellites can cover the world, LEO systems utilize dozens of non-stationary satellites. The proponents of LEO systems consider them suitable to service handheld units for a number of reasons:

- lower free-space path loss and propagation delay compared to GEO systems owing to closer proximity;
- low EIRP from user equipment enables easier compliance with radiation limits;
- higher spectral efficiency per unit area;
- true global coverage (when using appropriate orbits).

After a difficult beginnings (some involving bankruptcy), many of these systems have revived through intervention of various economic instruments and incentives. We describe two types of LEO system that have survived the downturn and plan to enhance their network.

The *Iridium* architecture uses regenerative satellites and intersatellite links for network connectivity. The *Globalstar* network uses transparent satellites, where connectivity is achieved via a ground network. Their network architectures are fundamentally different, but the user would hardly be aware!

12.2.3.1   Iridium

Background

Iridium Satellite LLC is a privately held company, based in Bethesda, Maryland. The Iridium satellite system can provide voice and data communication solutions throughout the world (Iridium Online, 2009). The system is named after the element iridium which has 77 electrons in its atom, corresponding to the 77 satellites in the constellation. The size of the constellation was scaled down to 66 for practical reasons, but the original name was held for its commercial appeal (as compared to dysprosium - the element with 66 electrons).

The company targets a clientele in locations where landline or mobile phone connections are 'unavailable, unreliable or overburdened'. The target markets include maritime, aviation, emergency services, oil and gas, forestry, mining, journalism, US government and military, etc.

System Overview

An overview of Iridium system components is portrayed in Figure 12.5. (after Hutcheson and Laurin, 1995). The system consists of a LEO constellation of 66 satellites, and a ground segment encompassing a gateway, a system control segment and the user community. Gateways manage the user community and interconnect with the public network; the system control segment manages

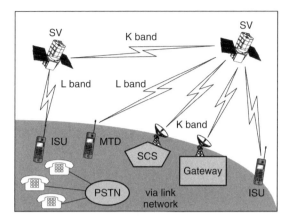

LEGEND:
ISU – IRIDIUM® Subscriber Unit
PSTN – Public Switched Telephone Network
SCS – System Control Segment
MTD – Message Termination Device
SV – Space Vehicle

Figure 12.5   Block schematic of the Iridium system (after Hutcheson and Laurin, 1995).

the constellation, satellite functioning and overall system operation. The user segment comprises a diverse community dispersed across air, land and sea.

The transmission frequency of the service link in both the up and down directions lies in the 1616–1626.5 MHz band. The uplink feeder operates in the 27.5–30.0 GHz range and the downlink in the 18.8–20.2 GHz range, while the intersatellite links operate within the 22.55–23.55 GHz band. Microwave cross-links were chosen because they were simpler in construction and were considered a lower risk than optical cross-links, and the available bandwidth was sufficient for the specified capacity. The air interface deploys a combination of frequency and time division multiplexing/multiple access to maximize the spectrum efficiency. The service link utilizes a variant of time division multiple access known as Time Division Duplex (TDD), where the time slots are shared between the up- and downlinks.

Each call to a terrestrial number is routed to the Iridium gateway located in Arizona, United States. The gateway bridges the satellite network to terrestrial voice and data networks. The radio link design permits direct single-hop links between users (within a satellite's footprint) without the more traditional method of patching the connection through a gateway in a double-hop arrangement.

Space Segment

As illustrated in Figure 12.6, the space segment comprises a constellation of 66 satellites distributed in six orbital planes of 780 km altitude, 86° inclination and 100 min orbital period. The coplanar satellites are equispaced. The design ensures that at least one satellite is always visible at an elevation angle of >8° from any location on the Earth. The footprint of each satellite is ~4500 km in diameter, divided into 48 spot beams. The horizon–horizon visibility period from the ground is of the order of about 10 min. Existing calls of a setting satellite are handed over to a rising one for maintaining real-time communication continuity. As illustrated in Figure 12.6, each satellite is cross-linked in space to four adjacent satellites – two in the same plane and two in the adjacent plane. As a satellite does not have to be in view of a gateway to establish and route a call, the

Figure 12.6   Iridium constellation. Reproduced by permission of © Iridium.

architecture requires a minimum number of gateways. In the limiting case, a single gateway can provide global coverage.

There are 2150 active beams on the surface of the Earth, which enables each frequency to be reused approximately 190 times. Three phased-array antennas are located on side panels of the satellite, each forming 16 spot beams (or cells) to make up the 48-beam pattern. Each cell can support an average of 236 channels.

The satellites converge towards the pole, and hence coverage redundancy increases progressively at higher latitudes, as illustrated in the visibility statistics of the constellation shown in Figure 12.7.

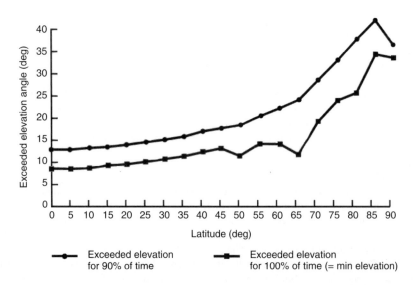

Figure 12.7   Visibility statistics of the Iridium constellation (Krewel and Maral, 1998). Reproduced by permission of © John Wiley & Sons, Ltd.

Notice the bias in the polar region. Spot beams of each satellite are gradually switched off as the satellite advances towards the poles to conserve satellite power, reduce unwarranted coverage and minimize inter/intrasystem interference.

Architecture

The architecture of the system is modelled after the GSM standard for easy integration with terrestrial systems and benefits from advances in the terrestrial standard. Thus, the gateways incorporate GSM functions for call processing and include functionality to manage the Iridium air interface.

The Mobile Switching Centre (MSC) constitutes the main switching element connecting the gateway's Earth Terminal Controller (ETC) to the public network via an international switching centre. Towards the space segment end, the ETC controls three Earth stations: one carries the traffic, the second provides redundancy and rain diversity when separation is adequate and the third acquires the rising satellite to take over a live connection from the setting satellite. The Visitor Location Register (VLR) and Home Location Register (HLR) maintain the location of subscribers in much the same way as in GSM. The Equipment Identification Register (EIR) database keeps the identity of the subscribers' equipment. The Message Origination Controller (MOC) supports Iridium's paging services. Operations, administration and maintenance support is through the Gateway Management System (GMS) which also connects to Iridium's Business Support System (BSS). The BSS has the function of managing the usage and settlement statements with gateway operators.

Each Iridium subscriber has a home gateway that maintains a record of the subscriber in an HLR. An Iridium subscriber could be identified by several types of numbers: the Mobile Subscriber ISDN number (MSISDN), used by a land calling party; a Temporary Mobile Subscriber Identification (TMSI) sent over the radio link while establishing connection, which changes periodically to protect user identity; the Iridium Network Subscriber Identity (INSI), a number stored in the user's phone and sent on the radio path when a valid TMSI is unavailable.

A call is established by querying the location of the user from the HLR and paging the user after establishing an efficient route involving the gateway, visible satellites, intersatellite link and user.

User Segment

The user segment supports an assortment of terminals to service air, land and sea environments. Although the handheld segment remains primary, fixed installations attached to small external antennas are supported. Maritime and aeronautical services were subsumed later in order to extend the user base. A recent innovation is the OpenPort sysem which allows up to 128 kb/s by employing up to 54 channels. Iridium also offers special secure service for military, and a push-to-talk (Netted Iridium) feature.

The handheld design ensures a high link margin of $\sim$16 dB to provide a robust performance even in fading conditions. Each transmission slot is Doppler corrected. A mobile can use up to four 8.28 ms bursts in each 90 ms frame. The peak (burst) transmit power from a handheld terminal is 3.7 W, averaging 0.34 W. The modulation scheme is QPSK at 50 kb/s burst rate, with a carrier spacing of 41.67 kHz. Voice codecs operate at 2.4 kb/s using the AMBE technique. The units resemble a cell phone – typically with a quadrifilar helix antenna of $\sim$1 dBi gain, an EIRP capability of up to 8.5 dBW and a G/T of -23.5 dB/K. The voice quality is said to be sharp, clearly exhibiting a much lower delay or echo than experienced through GEO satellite systems.

Next Generation

Iridium has started preparation for its next-generation system, known as 'NEXT', to replace the existing constellation, which will offer enhanced services leading to the year 2030. The first launch is anticipated in 2013. The transition to NEXT will be seamless, incorporating new capabilities that include higher throughput, end–end IP connectivity and shared payload to support services other than mobile. In addition to increased data rates at L-bands (up to 1.5 Mb/s), NEXT is expected to support wideband LEO services at Ka band. Technical concepts being explored feature secondary payloads including imaging to provide real-time low-resolution images of the Earth.

12.2.3.2 Globalstar

Globalstar is a US company that began personal mobile satellite telecommunication services in 1999 to deliver voice and bearer service up to 9.6 kbps with an external accessory device, bundled with value-added services in more than 120 countries, targeting primarily the mid-latitude region. The applications include Internet and private data network connectivity, position location, short messaging service, call forwarding, etc.

The Globalstar system is formed with the joint experience of Loral Aerospace Corporation in space technology and Qualcomm Inc. in ground and user segments. The system combines numerous technical innovations with the strengths of LEO satellite system technology and cellular CDMA technology (US EIA/71A IS-95 standard). Innovative technologies include spread spectrum–CDMA with efficient power control, high efficiency vocoder with voice activation and satellite diversity using a soft handover technique.

Figure 12.8 illustrates the system concept. The diverse user community is connected via the space segment to the gateways and thence on to the fixed parties.

Space Segment
The Globalstar space segment nominally consists of a 48-satellite 'Walker' constellation. The Walker technique distributes satellites uniformly in the celestial sphere (Walker, 1973). The 390 kg satellites

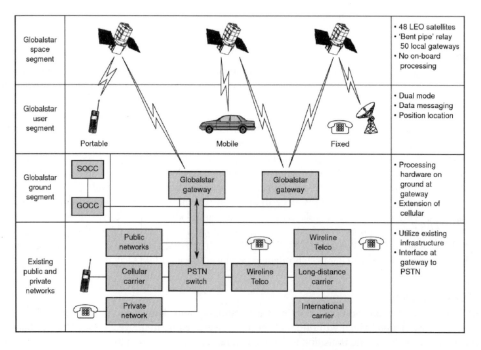

Figure 12.8 Main components of Globalstar's MSS (adapted from Dietrich, 1997).

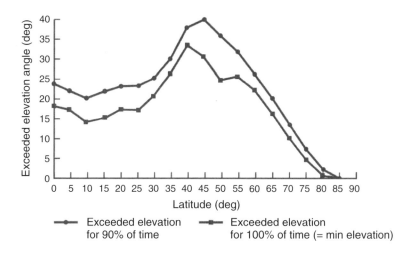

Figure 12.9 Globalstar visibility analysis (Krewel and Maral, 1998). Reproduced by permission of © John Wiley & Sons, Ltd.

are placed in eight 52° inclination, 1414 km altitude, circular orbits with an orbital period of 114 min. Six satellites are placed in each orbital plane. Subsequently, the number of satellites per plane was reduced to 5 making 40 satellites in all. The diameter of a satellite's footprint is about 7600 km, divided into 16 fixed spot beams.

Full coverage with dual-satellite diversity is available up to ±70° latitude, with intermittent coverage up to ±80°, subject to satellite visibility of a gateway. The coverage favours the region within about 30–60° latitudes, where Globalstar perceives its primary market. Figure 12.9 represents satellite visibility statistics as a function of latitude. Notice the high elevation angle visibility at the mid-latitude and compare it with those of the Iridium constellation (Figure 12.7).

Globalstar founders opted for a risk-averse, robust and reliable space segment. Low-risk space-craft technologies include: Sun-tracking solar-cell array and nickel–hydrogen battery cells; on-board GPS receivers for position and attitude control to simplify orbital maintenance and impart on-board timing and frequency reference; intersatellite link exclusion; and transparent transponders. Each three-axis stabilized satellite supports 16 spot beams arranged in two rings around a central beam with transmit and receive beam patterns differing slightly. The beams are generated by phased arrays consisting of 91 elements interfaced to individual high-power amplifiers in the transmit chain and a low-noise amplifier in the receiver. The solar arrays provide a beginning-of-life power of 1.9 kW, backed up by nickel–hydrogen batteries.

Air Interface

The system utilizes frequency division/spread spectrum/CDMA (FD/SS/CDMA) for satellite access in forward and return service links. The forward service link operates in S-band (2.48–2.5 GHz) and the return in L band (1.62–1.63 GHz). The CDMA is a direct-sequence, spread-spectrum scheme in a bandwidth of 1.23 MHz in conjunction with QPSK modulation. The transmission band is spread over thirteen 1.23 MHz channels. The feeder link frequency operates in the 5 GHz band (Earth-space) and 7 GHz band (space–Earth). The carriers are voice activated to increase system capacity. Mobiles transmit at powers of up to 2 W.

The maximum end-to-end delay experienced by users is less than 100 ms, out of which only 18 ms delay is incurred in the satellite link. The accessing scheme allows the combination of signals from multiple satellites and different signal paths (e.g. a reflection), thereby not only mitigating adverse effects of fading but also enhancing signal quality through spatial diversity. The technique also enables a soft handover between satellites. CDMA has the inherent property of rejecting interference, and therefore the system can sustain a tighter frequency reuse between spot beams than would be feasible with a classic FDMA or TDMA.

The radio link provides about 11 dB fade margin in the service link: 1 dB margin for Ricean fading, and up to 10 dB in fading conditions by a power control mechanism that can increment power by up to 10 dB in steps of 0.5 dB. Beyond this the system routes the signal through a better-suited satellite. The CDMA scheme permits graceful degradation of channel quality and at the limit of (nominal) capacity the operator may continue to accept calls at a degraded performance level or redistribute traffic between satellites in view – the provision allows the system to exceed the nominal capacity in peak loading conditions. A variable-rate voice codec that operates between 1.2 and 9.6 kb/s, with 2.4 kb/s average, is used with FEC at 1/3 code rate in the uplink and 1/2 code rate in the downlink. The vocoder rate is changed every 20 ms to suit channel conditions. In the absence of voice, the bit rate drops to 1.2 kb/s to reduce the transmitted power and therefore the intrasystem interference to facilitate optimal use of system capacity.

User Terminal

The system supports three types of user terminal – fixed for residential and rural locations, portable for access solely to the satellite component and portable dual-mode units for accessing both the satellite and terrestrial systems. The terminals utilize three rake receivers at the core to detect signals from multiple signal paths simultaneously and combine the signals coherently for improving the signal quality. Handheld units can transmit at a maximum EIRP of −4 dBW, averaging about −10 dBW. Fixed terminals deploy a small antenna, a 3 W power amplifier and software to manage calls and other supporting functions. Portable units are slightly bigger than conventional cellular telephones, using a specially designed nearly omnidirectional antenna (1–1.5 dBi gain). The antenna protrudes slightly above the user's head to minimize blockage. Battery life is over 2 h for talk and 15 h in standby mode. The car version of the phone incorporates an outdoor antenna attached to a power amplifier. The dual-mode terminal is an integrated Globalstar portable unit and a cellular phone reusing components and features of the cellular set wherever possible. The phone design includes a Subscriber Identification Module (SIM), which facilitates the user to change the handset at will.

Gateway

The system supports as many gateways as required to accommodate different service operators. The gateways incorporate the provision that each operator can exercise an independent control for call routing and management. Thus, a gateway would typically be operated by a local service provider in each region. Each gateway deploys up to four ∼6 m diameter tracking antennas, which ensures that all satellites in view (nominally three) are tracked. If a lower number of satellites are visible at the site, the number of antennas can be reduced in proportion. The gateways interconnect with the PSTN and include an interface with the GSM system.

A call initiated from a mobile is routed through all the satellites in view – up to three satellites in the primary service area. The call is delivered by the satellites to the closest gateway(s), where the signals are combined. As the RF signal traverses three independent paths, the probability of simultaneous signal fade decreases. New satellites are added seamlessly at the receiver to replace satellites that move out of visibility. Note that, in this architecture, failure of a single satellite is unlikely to cause a full outage in a fully populated constellation.

12.3   Fixed Communications

In the following paragraphs we cite examples of a few emerging fixed broadband systems targeting individuals and small businesses.

12.3.1   GEO Broadband Systems

12.3.1.1   HughesNet

Overview

At the time of writing, Hughes Network Services is the World's largest supplier of satellite broadband to consumers and small businesses, and operates its HughesNet services (formerly called DirecWay/DirecPC) to North America, Europe, India and Brazil, having launched the original DirecPC service in 1996. Currently, HughesNet offers forward-link (hub to user terminal) speeds of 1–5 Mb/s and return-link (user terminal to hub) speeds of 128–300 kb/s.

Space Segment

Satellites supporting HughesNet in America include Horizons-1, Galaxy-11 and Intelsat America 8 (IA8). HughesNet is also supported on the Spaceway 3 satellite. Spaceway 3, launched in 2007, uses an on-board processing payload and provides HughesNet services via a special modem.

User Segment

HughesNet terminals conform to the DVB-S and DVB-S2 standards for the forward (hub to user) link and Hughes' IP over Satellite (IPoS) standard for the return (user to hub) link. A typical HughesNet terminal is illustrated in Figure 12.10. C, Ku and Ka frequency band options are available.

Figure 12.10   HughesNet terminal. Reproduced by permission of © Hughes Network Systems, LLC. All rights reserved.

Table 12.4   Ku band parameters for a HughesNet terminal (HughesNet HN700S, 2010)

Parameter	Value
Frequency	11.7–12.7 GHz (downlink); 14–14.5 GHz (uplink)
Antenna diameter	74–180 cm
Terminal EIRP	(not given)
Terminal G/T	(not given)
Polarization	Linear (down/up)
Data rate	up to 121 Mb/s (down); up to 1.6 Mb/s (up)
Modulation	DVB-S/DVB-S2 (QPSK/8-PSK down); O-QPSK (up)
FEC	Concat. RS-convol./LDPC (down); 1/2 to 4/5 turbo (up)
Multiple-access scheme	IPoS (MF-TDMA)
Acceleration	PEP and HTTP acceleration
Encryption	IPSec
Weight	2.2 kg (IDU only)

The DVB-S2 forward link allows for adaptive coding, whereby the FEC coding rate can be changed dynamically on a TDM frame-by-frame basis, and the HughesNet system allows for closed-loop control of the remote terminal's uplink power. Performance parameters for a Ku band terminal (extracted from HughesNet HN7000S, 2010) are indicated in Table 12.4.

12.3.1.2   Astra2Connect–Sat3Play

Overview

Astra2Connect is a European two-way satellite broadband service from SES-ASTRA. It provides always-on connectivity for broadband Internet, as well as Voice over IP (VoIP) telephony – a combination sometimes referred to as two-play (Internet and telephony). These services are primarily targeted at consumers and Small Office/Home Office (SOHO) users who are currently unable to access wired Digital Subscriber Link (DSL) services. At the time of writing, Astra2Connect was offering users four different service levels with downlink (hub to user) speeds ranging from 256 kb/s to 2 Mb/s and uplink (user to hub) speeds ranging from 64 to 128 kb/s.

Astra2Connect currently utilizes Ku band capacity on its ASTRA 1E and 3E satellites, collocated at 23.5° E. In many cases, satellite TV can also be received from adjacent ASTRA satellites with the same antenna through the addition of a second feed and LNB (offset from the first). The central Astra2Connect hub is located at ASTRA's headquarters in Luxembourg.

User Segment

Astra2Connect user equipment comprises Sat3Play user equipment from NewTec (Belgium). The Sat3Play concept evolved from a European Space Agency (ESA) project of the same name. Three-play or *triple-play* is a term used to describe the bundling of broadband Internet, consumer fixed voice (telephony) and TV services from the same service provider. A number of terrestrial DSL and cable network operators already offer such triple-play services. In essence, Sat3Play terminals provide a two-way Internet connection over satellite utilizing low-cost equipment that may be used to provide Internet access, Voice over IP (VoIP) telephony and video and audio content, with facilities to provide additional value-added services such as pay-TV/Video-on-Demand (VoD).

A Sat3Play terminal comprises an OutDoor Unit (ODU) and an InDoor Unit (IDU). A Sat3Play IDU is shown in Figure 12.11. The ODU typically comprises a 79 cm dish (offset reflector antenna)

Figure 12.11   Sat3play terminal. Reproduced by permission of © ESA.

Table 12.5   Sat3Play Ku band terminal characteristics

Parameter	Value
Frequency	10.7–12.75 GHz (down); 13.75–14.0 GHz/14.0–14.5 GHz (up)
Antenna diameter	79 cm or 1.2 m
Terminal EIRP	36 dBW (79 cm antenna)
Terminal G/T	14 dB/K (79 cm antenna in clear weather)
Polarization	Linear (down/up)
Data rate	1–80 Mb/s (down); 160-512 kb/s (up)
Modulation	DVB-S/DVB-S2 (QPSK/8-PSK down); SATMODE (GMSK up)
FEC	Concat. RS-convol./LDPC (down); turbo (up)
Multiple-access scheme	MF-TDMA
Acceleration	TCP/IP-acceleration, HTTP pre-fetching, compression
Encryption	2-way TCP/IP encryption
Weight	0.6 kg (IDU only)

together with an interactive LNB (iLNB) – a combination of a low-noise block down converter and integrated 500 mW transmitter and up-converter. The IDU comprises an L band IP modem providing Ethernet connectivity for user equipment.

The Sat3Play forward link (hub to user terminal) employs DVB-S and DVB-S2 modulation schemes. In order to achieve low consumer equipment costs, the Sat3Play return channel utilizes a low-power transmitter and combines the multiple-access features of DVB-RCS with the modulation scheme of SATMODE, developed under another ESA programme for interactive television (iSatTV) at modest data rates. Further details of a typical Ku band Sat3Play terminal are given in Table 12.5.

12.3.1.3   WildBlue-Surfbeam

Overview
WildBlue is a high-speed satellite Internet service provider in the United States, currently using two Ka band multibeam satellites and low-cost satellite-DOCSIS terminals. At the time of writing, WildBlue is the most significant mass-market exploitation of both cellular multibeam coverage and Ka band satellite technology.

Space Segment
WildBlue has leased the multibeam Ka band capacity on Anik F2 covering the United States (30 of a total of 45 spot beams) since its launch. The launch of the Anik F2 spacecraft by Telesat (Canada) was a ground-breaking development in the deployment of multiple-spot-beam two-way broadband satellite services. The Anik F2 satellite provides 45 Ka band spot beams for subscriber

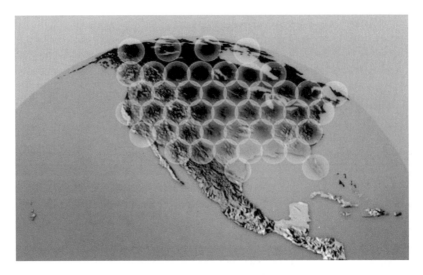

Figure 12.12   WildBlue Ka band coverage on Anik F2. Reproduced by permission of © Boeing.

service laid out across the North American continent, permitting high throughput and significant frequency reuse between coverage areas, resulting in an estimated Ka band payload capacity of about 4–5 Gb/s. In addition, the satellite has six Ka band beams for gateway Earth station links; 6–8 subscriber beams connect to the Internet at each gateway. The cellular beam pattern of Anik F2 is shown in Figure 12.12.

The Anik F2 satellite is particularly noteworthy for its pioneering use of a large number of small-coverage Ka band beams (in a cellular pattern), with each beam yielding higher power densities on the ground and uplink sensitivities, producing substantial cost-per-user advantages and increased total throughput with regard to satellite broadband provision. This is feasible at Ka band owing to the narrower beam widths (and hence coverage areas) achieved for a given antenna size at higher frequencies. Further details of Anik F2 Ka band satellite capability are given in Table 12.6.

Table 12.6   Ka band parameters for ANIK F2

Parameter	Value
Orbit type	Geostationary
Payload type	Bent pipe
Spot beams	45 (Ka band up/down)
Spot beam bandwidth	62.5 MHz
Gateway beams	6 (Ka band up/down)
Gateway beam bandwidth	500 MHz
EIRP	56–59 dBW (Ka band, depending on beam)
G/T	14 dB/K (Ka band)
Polarization	LHCP and RHCP (Ka band down/up)
Predicted life	15 years
Launch mass	5950 kg
DC power	15 kW

Figure 12.13 WildBlue VSAT terminal with Surfbeam Modem. Reproduced by permission of ©
WildBlue Communications, Inc.

In order to compensate for the different propagation conditions at Ka band across the various
climatic regions, 17 of Anik F2's spot beams in the eastern half of North America are designated as
heavy-rain beams – with higher effective isotropic radiated power in order to help combat increased
rain fades in these coverage regions.

In addition to leasing Anik F2's US Ka band beams, WildBlue has also launched its own
dedicated Ka band satellite, known as Wildblue-1 (previously designated 'iSky' or 'KaStar'), which
is collocated with Anik F2, and covers North America with 35 overlapping Ka band spot beams.
WildBlue-1 also has six additional Ka band beams for gateway Earth station connectivity.

User Segment
WildBlue terminals, an example of which is shown in Figure 12.13, incorporate a ViaSat SurfBeam
Satellite Data-Over-Cable Service Interface Specification (S-DOCSIS) modem InDoor Unit (IDU).

The Satellite DOCSIS standard (S-DOCSIS) is an extension of the DOCSIS cable modem networking standard used by millions of terrestrial cable network customers, and is viewed by some as
a lower-cost alternative to DVB-RCS (DVB-Return Channel over Satellite) used in other satellite
broadband systems. Further details of the WildBlue terminal are given in Table 12.7.

12.4   Military Communications

12.4.1   Military Portable Satellite Communications Systems

General Considerations
A discussion on military satellite communications systems could easily fill a book on its own.
In keeping with the main theme of this book, we shall restrict our discussion here to 'personal'
military satellite communications. The term 'personal' is more difficult to define in a military
context; therefore, we use it to mean as being where the end (data)-user is also the SATCOM
equipment operator, and typically this will involve man-portable (manpack) and handheld military
satellite communications equipment.

Table 12.7 Ka band parameters for the WildBlue ViaSat Surfbeam
terminal

Parameter	Value
Frequency	19.7–20.2 GHz (down); 29.5–30.0 GHz (up)
Antenna diameter	68–100 cm
Terminal EIRP	48.3–51.3 dBW
Terminal G/T	15.4 dB/K (typical)
Polarization	LHCP/RHCP (up/down or down/up)
Data rate	0.5–10 Mb/s (down); 128–2560 kb/s (up)
Modulation	QPSK/8-PSK (down); QPSK (up)
FEC	Turbo/concat. RS-convol. (down); turbo/RS (up)
Multiple-access scheme	S-DOCSIS
Acceleration	Embedded PEP and HTTP acceleration

What are the key differences between civilian satellite services and technology and those designed for use employed by the military? Desirable features of military satellite communications include:

- *Rugged.* Military equipment typically needs to be able to operate properly in climatic extremes and cope with transportation and a significant amount of 'rough handling'.
- *Secure.* The information passing over a military system will typically be sensitive, and its interception could put lives at risk.
- *Covert.* In some situations it is clearly desirable to communicate without one's adversary being aware (even if the enemy is unable to decypher the message).
- *Resilient.* It is often vital that communications be protected against enemy jamming (electronic warfare), but also against equipment failure, as military communications are used for passing vital command information.

To the above list should be added portability for a 'personal' military satellite communication system – although, as a general rule, ruggedized portable military equipment is noticeably bulkier and heavier than its civilian equivalent! Security, covertness and resilience are somewhat interrelated, and military users have specific jargon for these areas:

- *INFOrmation SECurity (INFOSEC).* This relates to the protection of all user data (including data contained in computing devices). INFOSEC is normally achieved by data encryption.
- *COMmunication SECurity (COMSEC).* This relates to the protection of all user data sent over communications links. This is also normally achieved by data encryption.
- *TRANSmission SECurity (TRANSEC).* This relates to the use of techniques such as spread spectrum, and in particular Frequency-Hopped Spread Spectrum (FHSS), that are robust against electronic warfare/interference (i.e. antijam) and/or enhance covertness (leading to low probability of detection interception and exploitation); it also refers to the encryption of signalling information.
- *OPEration SECurity (OPSEC).* This relates to the protection of information other than the transmitted message that could potentially reveal useful information to an enemy about a military operation – for example, a user's transmitter location or telephone number (or terminal ID).

Military use of Commercial SATCOM

Before focusing on bespoke military-specific personal satellite communication systems, it is important to recognize that the military are also keen users of commercial services. There are a number of reasons for this:

- Global commercial services are readily available and user equipment is generally highly effective and often more portable than the military equivalent. It is often more convenient, more expedient or simply cheaper to use a commercial service.
- Not all nations have (or can afford) dedicated military communications satellites.
- For those nations with dedicated satellite capability, demand for military communication services typically outstrips supply as a result of the growth in use of video and information technology in military operations (reflecting its wider use in society). Not all military operations demand covert/antijam communications, and hence commercial SATCOM may be used for many non-critical communications (logistics, welfare, etc.).

Use of civilian satellite services is considered acceptable for any operations sanctioned by the United Nations and for peacekeeping operations, although the use of certain civilian satellite communication systems for all-out war fighting is a somewhat grey area that has not yet been tested (as certain satellite services were formed under international treaties that may restrict their use).

Notwithstanding the above, the military use of global personal and mobile communications services – notably Iridium and Inmarsat – and Ku band VSAT services is very widespread.

Ruggedization

Although the military will sometimes use unmodified commercial satellite communications equipment, in many cases it is desirable to use a ruggedized version of a standard commercial terminal in order to protect against the inevitable rough handling. Figure 12.14 shows an example of a ruggedized version of a commercial satellite terminal (Harris, 2010).

Figure 12.14   Ruggedized military BGAN manpack terminal. Reproduced by permission of © Harris Inc.

In the remainder of this subsection, we focus on examples of bespoke military satellite communication systems.

12.4.1.1   NATO UHF Tactical SATCOM

The use of narrowband UHF Tactical SATCOM (TacSAT) is common within the forces of the North Atlantic Treaty Organization (NATO). The popularity of UHF TacSAT with mobile tactical users is due to a number of reasons:[1]

- relative portability of its ground terminals (and in particular its antennas);
- relative lack of criticality in antenna pointing (owing to relatively wide antenna beamwidths);
- relative lack of link degradation due to poor weather conditions;
- ability to penetrate light foliage cover (e.g. in a jungle environment);
- usefulness in urban environments, where line of sight to the satellite may be difficult.

NATO provides dedicated UHF services for use specifically in NATO operations and exercises. Additionally, some NATO countries (notably the United States, the United Kingdom and Italy) have national UHF SATCOM capabilities.

Prior to 2004, NATO owned and operated its own dedicated NATO IV spacecraft. However, in 2004, under the NATO SATCOM Post-2000 (NSP2K) programme, NATO signed a Memorandum of Understanding (MoU) with the defence ministries of Italy, France and the United Kingdom, to provide NATO with UHF and SHF satellite communications services through to 2019. Under the agreement, NATO UHF satellite capacity is provided by Italy and the United Kingdom on SICRAL and Skynet spacecraft respectively, while NATO SHF capacity is provided by France and the United Kingdom on Syracuse and Skynet spacecraft respectively.

Space Segment

NATO UHF military satellite capacity comprises a number of tuneable 25 kHz wide, hard-limited, UHF transponder channels. Historically, 25 kHz channels were first employed in US Navy satellites (5 kHz UHF channels were employed in US Air Force satellites). By convention, UHF satellite channels are hard limited, by which we mean that the output power is maintained constant, regardless of the input power, using a transfer characteristic approximating an ideal limiter. This arrangement affords protection to the channel amplifiers and provides automatic gain control. However, use of hard limiting prevents the sharing of UHF channels using Frequency Division Multiplexing (FDM), even if the signals can be accommodated within the available channel bandwidth. Sharing by Time Division Multiplexing (TDM) is possible, however.

The bulk of NATO UHF capacity is provided by Skynet 5, the United Kingdom's military satellite constellation, and is operated by Paradigm Secure Communications (a UK subsidiary of EADS Astrium) on behalf of the UK Ministry of Defence (MOD) under a Private Finance Initiative (PFI). Under the agreement, the UK MOD has a guaranteed level of service on Skynet 5 (with an option to increase this), but residual spacecraft capacity can be leased to other (friendly) nations and NATO. The Skynet 5 constellation currently consists of three military hardened spacecraft, each of which has 5×25 kHz UHF channels (one channel of which may be re-configured to support 5×5 kHz channels) and 15 SHF (X band) channels. A fourth satellite will be launched in 2013. The SHF payload incorporates a sophisticated phased-array 'nulling' antenna, which is used to mitigate against interference/jamming (Paradigm Skynet 5 Online, 2010). Some additional details on Skynet 5 satellites are given in Table 12.8.

---

[1] As compared with the use of SHF or EHF MILSATCOM.

Table 12.8 Skynet 5 parameters in the public domain

Parameter	UHF	SHF
Orbit type	Geostationary	
Payload type	Bent pipe	Bent pipe
Beams	Earth cover (E/C)	E/C + 4 spots
Transponders	5/9	15
Transponders bandwidth	25/5 kHz	20–40 MHz
EIRP	(not given)	41 dBW (E/C); 56 dBW (spot)
Polarization	RHCP	RHCP
G/T	(not given)	(not given – depends on beam shape)
Design life	15 years (Eurostar-3000 bus)	
Mass	4700 kg	
DC power	5 kW	

Additional NATO UHF capacity is provided by Italy's Sistema Italiano per Comunicazioni Riservate ed Allarmi (SICRAL). SICRAL was designed to provide strategic communications for Italian forces and national authorities. SICRAL 1 was launched in 2001, while SICRAL 1B was launched in 2009. SICRAL 1B was part financed by a consortium led by Alenia Spazio, and residual spacecraft capacity is available to other (friendly) nations. Each satellite provides three UHF, five SHF (X band) and one EHF/Ka band transponders. SICRAL 2 is due for launch in 2010 or 2011, and will replace SICRAL 1.

User Segment

A range of UHF TacSAT terminals are currently in use by NATO forces. One of the more common types is the Falcon II terminal from Harris (also referred to as the AN/PRC117). Figure 12.15 illustrates an AN/PRC117 being used by US forces. The Harris Falcon terminal is actually a multiband V/UHF radio used for both terrestrial and satellite communications. Some parameters for the UHF TacSAT mode are indicated in Table 12.9.

The Harris Falcon II can be used with a number of different antennas, depending on the mode of operation (manpack, communications-on-the move, etc.). Use of low gain antennas comprising either crossed dipoles above a conducting ground plane or quadrifilar helix is common on military vehicles, and no antenna pointing is required. The antenna illustrated in Figure 12.15 is typical of a manpack TacSAT antenna and comprises a Yagi array of crossed dipoles (for circular polarization) and a wire mesh reflector. Typical antenna gain is 8–11 dBiC. The antenna is mounted on a foldable tripod for manual beam pointing. The antenna folds down to fit into a small cylindrical pouch that is carried in the user's rucksack.

UHF TDMA DAMA

It has long been recognized that the number of UHF TacSAT users significantly exceeds the number of available UHF channels, and NATO has initiated a programme to convert some channels for use with a UHF DAMA system. UHF DAMA is a TDMA scheme originally developed by the US military as a way of sharing UHF channels in netted or point-to-point links while supporting diverse data rate combinations. In fact, UHF DAMA is available in two flavours: 25 kHz DAMA (MIL-STD 188-183) and 5 kHz DAMA (MIL-STD 188-182), according to the UHF channel bandwidth.

A number of TDMA frame types are available for both schemes, permitting flexibility in the allocation of channel capacity. For example, a 25 kHz DAMA circuit can be used to support up

Figure 12.15   Harris AN/PRC117 multiband radio and Yagi antenna. Reproduced by permission of U.S. Air Force and Senior Airman, Courtney Richardson.

to five 2.4 kb/s voice or data circuits (time slots). The US later enhanced these standards to create an 'Integrated Waveform' which permits up to thirteen 2.4 kb/s TDMA voice circuits in a single 25 kHz channel. UHF DAMA is effective in increasing the number of low-bit-rate user signals sharing the limited number of UHF TacSAT channels; however, as previously noted, a significant disadvantage associated with the use of TDMA with a low-power terminal with small antennas is that UHF DAMA transmissions occur at the TDMA burst rate rather than the user data rate. Consequently, throughput is governed by the maximum burst rate that the terminal can sustain with its limited power and sensitivity.

Table 12.9   Harris Falcon II (AN/PRC117F) terminal parameters (Harris Falcon II Data Sheet, 2010)

Parameter	Value
Frequency	240–270 MHz (down); 290–320 MHz (up)
Antenna diameter	(antenna separate)
Terminal EIRP	20 dBW (assuming 7 dBiC antenna gain)
Terminal G/T	(receiver sensitivity -120 dBm)
Polarization	RHCP (antenna separate)
Data rate	300 b/s to 56 kb/s
Modulation	MIL-STD-188-181 (shaped BPSK, shaped OQPSK, CPM)
FEC	MIL-STD-188-181 (convolutional coding)
Multiple-access scheme	MIL-STDs 188-181 (SCPC) and 188-182/3 (UHF TDMA DAMA)
Acceleration	None
Encryption	COMSEC and TRANSEC
Weight	4.4 kg (excl. batteries)

12.4.1.2 US MUOS

Overview

The United States is by far the largest user of MILitary SATellite COMmunications (MILSATCOM) and is currently in the process of transforming its legacy MILSATCOM systems into three distinct services:

- *Protected systems.* These provide survivable, resilient (antijam) and covert strategic and tactical communications services at low to medium data rates at EHF, using fully on-board processed Frequency Hopped Spread System (FHSS) payloads and intersatellite links. In future, these services will be provided using Advanced EHF (AEHF) which will replace the existing MILSTAR satellites. Protected communications in polar regions will be provided via the polar-orbiting satellites of the Advanced Polar System (APS). In the longer term, an even more advanced system was planned, to support higher data rates and facilitate on-board IP packet switching, in addition to supporting very-high-data-rate communications using laser (optical) communications. However, the program was cancelled in 2009.
- *Wideband systems.* The Wideband Global System (WGS – previously known as the wideband gapfiller system) will provide high-data-rate, unprotected, fixed communications services at SHF (X and Ka bands).
- *Narrowband systems.* The Mobile User Objective System (MUOS) will provide narrowband (nominally up to 64 kb/s) communications and combat survivor evader location (rescue) services to so-called disadvantaged users – dismounted and/or highly mobile tactical users using low-gain antennas (including handheld). Prior to MUOS, such users would use TacSAT.

MUOS will provide on-demand (with appropriate precedence and pre-emption) narrowband voice, facsimile, low-speed data, alphanumeric short message paging and voice mail services, via an IP core network that leverages terrestrial 3G technology. In addition to point-to-point communications, MUOS will support netted, push-to-talk (PTT) and broadcast services. The maximum data rate will be 64 kb/s, although the potential exists for 384 kb/s.

Space Segment

Four MUOS satellites (plus an in-orbit spare) will replace the US Navy's existing UHF Follow ON (UFO) satellites (each of which currently supports up to seventeen 25 kHz wide and twenty-one 5 kHz wide UHF SATCOM channels). The satellites will be spaced around the Earth to provide MUOS global coverage (Table 12.10 and Figure 12.16). The first MUOS satellite is scheduled for launch in 2010, with full operational capability due in 2014. Each satellite will nominally provide approximately 10 Mb/s of capacity, comprising of the order of 4083 simultaneous MUOS accesses plus 106 legacy (TacSAT) accesses (Huckell and Tirpak, 2007). Digitised Earth feeder links will use Ka band. A key feature of the MUOS satellites will be the use of 16 spot beams formed by a 14 m diameter mesh antenna, in addition to an Earth coverage beam.

User Segment

MUOS leverages 3G terrestrial mobile telephone technology for both the air interface and ground control segment. The air interface will be a modified version of the 3G wideband CDMA multiple-access waveform, known as SA-WCDMA. The most recent public information on the MUOS frequency allocation suggests four 5 MHz WCDMA carriers per beam within 300–320 MHz (down) and 360–380 MHz (up) (Jackson, 2007).

With the emergence of software-defined radios, the United States is aiming to rationalize the number of deployed military radios in future by focusing on a handful of common wideband

Table 12.10 MUOS Satellite parameters in the
public domain

Parameter	Value
Constellation	Four satellites (plus spare)
Orbit type	Geosynchronous
Payload type	Bent pipe
Beams	Earth cover plus 16 spots
Transponders	4 channels
Transponder bandwidth	5 MHz
EIRP	(not given)
Polarization	RHCP
G/T	(not given)
Mass	(not given)
DC power	(not given)

Figure 12.16 Artist's impression of a MUOS satellite, showing the multibeam antenna. Reproduced by permission of © Lockheed Martin.

radios in which different waveforms may be defined in software via the Software Communications Architecture (SCA), thereby permitting improved economies of scale and greater flexibility in implementing new waveforms. The programme is known as the Joint Tactical Radio System (JTRS), and all new US radios are mandated to be JTRS compliant.

As a result, the MUOS air interface will be a JTRS waveform. It is not clear whether all JTRS radios will eventually implement this waveform, and currently the only JTRS radio known to include the MUOS waveform capability is the Handheld, Manpack, and Small Form Factor (HMS) radio (formerly known as JTRS Cluster 5) (JTRS HMS Data Sheet, 2010).

12.4.1.3   UK Talisman SHF Manpack

Overview

Each Skynet 5 satellite has a nominal UHF bandwidth capacity of 125 kHz (Paradigm Skynet 5 Online, 2010), and the total global military UHF satellite spectrum allocation is only a few MHz. By contrast, the total military SHF satellite spectrum allocation at X band is 500 MHz (although not all of this is usable for mobile communications). Furthermore, in principle, the narrower antenna beamwidths achieved at X band permit this spectrum to be reused multiple times. Clearly, therefore, X band can accommodate a greater number of users and at higher data rates. However, the design of manpack terminals at SHF presents some significant challenges. SHF antennas tend to be of the reflector type, which are heavier and bulkier than the collapsible Yagi antennas used for TacSAT. The increased antenna gain at SHF comes with smaller beamwidth, which makes antenna pointing more critical.

Notwithstanding these challenges, SHF manpack terminals are currently in use with the UK and other military forces. We shall focus here on the SHF manpack system developed for the UK MOD, known commercially as Talisman II, which has since been exported to other nations (Janes, 2010). An export version of Talisman has already been supplied to the Turkish Military SATCOM programme, and a derivative of the terminal has been selected by the French Army for use with Syracuse III satellites.

The Talisman II system was originally developed to provide a rapidly deployable secure manpack SATCOM system for UK special forces under the military designation PSC (Portable Satellite Communications)-504 (original commercial name 'Cheetah') (Janes, 2010). This was followed by a version for other regular forces, designated PSC-506.

The Talisman system comprises manpack and transportable headquarters terminals and a base station with associated communications management subsystems. A significant feature is the use of a proprietary Demand-Assigned Multiple-Access (DAMA) waveform to share satellite resources among a large number of users (Janes, 2010).

Space Segment

The space segment for the UK PSC-506 Talisman system comprises the Skynet 5 satellites. Key open-source parameters for Skynet 5 SHF payload were given in Table 12.8.

Ground Control Segment

The Talisman system employs a bespoke DAMA waveform, which shares the allocated SHF spectrum between a very large number of user terminals. Each Talisman DAMA network can support up to 300 terminals, with access control dependent on user priority, and with support for pre-emption by high priority users (Janes, 2010).

The Talisman ground control segment provides the terrestrial connectivity and also the Talisman network management system, which controls user access (via the DAMA system).

User Segment

The Talisman is a triservice (Army, Navy and Air Force) ruggedized SATCOM terminal that allows secure, voice-recognizable speech, messaging and secure data communications between fixed communications at data rates of up to 64 kb/s.

When dissembled, the manpack patrol terminals are designed to fit into the top of a standard Bergen rucksack. Larger, transportable, terminals, with increased power and satellite autotracking, are used in support of mobile headquarters or as transportable tactical terminals in their own right (Janes, 2010). The French Talisman manpack SHF terminal is shown in Figure 12.17.

Figure 12.17 Talisman II SHF manpack terminal in use. Reproduced by permission of ©
Thales SA.

The manpack is designed to be assembled in less than 10 min (even in the dark), and the
assembled terminal comprises two main units:

- *Antenna unit*, which contains the antenna and RF electronics.
- *Data Entry Device (DED) unit*, which contains the handset, keypad and encryption equipment,
  and which can be located up to 100 m away from the antenna unit.

A summary of open-source parameters for the Talsiman manpack SHF terminal is given in
Table 12.11.

Table 12.11 Talisman SHF manpack terminal parameters in the
public domain

Parameter	Value
Frequency	7.25–7.75 GHz (down); 7.9–8.4 GHz (up)
Antenna diameter	60/75 cm
Terminal EIRP	>40 dBW (60 cm antenna)
Terminal G/T	>7 dB/K (60 cm antenna)
Polarization	RHCP
Data rate	up to 64 kb/s
Modulation	(not given)
FEC	(not given)
Multiple-access scheme	(not given)
Acceleration	(not given)
Encryption	COMSEC and TRANSEC
Weight	<10 kg (excl. battery)

12.5 Amateur Communications

12.5.1 Overview

Amateur satellite systems are developed, built and utilized for communication by radio amateur enthusiasts and like-minded communities and organizations who wish to experience satellite communication first hand and promote the technology. The catalogue of amateur satellites and their operational parameters is maintained by the amateur radio operator community, known as 'hams'. The Radio Amateur Satellite Corporation (AMSAT) is an international group of hams to 'foster Amateur Radio's participation in space research and communication' (AMSAT Online, 2009). They 'share an active interest in building, launching and then communicating with each other through non-commercial Amateur Radio satellites' (Baker and Jansson, 1994). To elucidate the prolific activities of AMSAT and similar groups, about 113 amateur satellites were launched in total by mid-2009, of which 32 were operational in the third quarter of 2009, and eight launches were planned (data from AMSAT Online, 2009). The group's satellites have continued to leverage technology advances through volunteered effort, donations and assistance of various governments and commercial agencies in launching the satellites. The community also contributes towards progressing frontiers of the space sciences, education and technology, and assists in humanitarian crises, often as first responders.

Since the launch of the first OSCAR satellite in 1961, a variety of new communication technologies have been pioneered by the group through close cooperation with international space agencies. The satellites have often been built with leftover materials donated by aerospace industries. The space agencies launch satellites at subsidized rates in exchange for useful data gleaned through these satellites. For example, the first voice and store-and-forward transponders were flown on an amateur satellite (Baker and Jansson, 1994). In recent years they have been used in university and school science projects, emergency communications for disaster relief, as technology demonstrators, transmission of earth imagery, etc. (AMSAT Online, 2009).

Because of worldwide interest, and to foster the interest of amateurs, the ITU has allocated spectrum specifically for the amateur satellite service in the following frequency ranges: 29 MHz (wavelength $\sim$10 m), 145 MHz ($\sim$2 m), 435 MHz ($\sim$70 cm), 1270 MHz ($\sim$24 cm) and 2400 MHz ($\sim$13 cm).

We shall introduce features of the Orbiting Satellite Carrying Amateur Radio (OSCAR) satellites and amateur ground equipment. OSCAR represents a series of low-cost, small and simple satellites for the worldwide community of radio amateurs and others to experience satellite communications and participate in space science/technology experiments. AMSAT organizations, its affiliates and like-minded organizations across the world hold the responsibility for the development of each satellite. A satellite is not designated an OSCAR serial number unless the responsible organization requests one; and in doing so, AMSAT assigns a unique sequential number only when the satellite is in orbit. The number is then appended to the nomenclature of the satellite.

12.5.2 OSCAR Satellites

A US, west-coast group of hams conceived the concept of an amateur satellite soon after the launch of the very first satellites – Russia's Sputnik I and the United States' Explorer. A group called Project OSCAR, formed by the pioneering originators, accomplished their goal (Project OSCAR Online, 2009). The first satellite, called OSCAR 1, was launched into a low Earth orbit on the morning of 12 December 1961 as a secondary payload, heralding the dawn of amateur satellite. The satellite, weighing about 4.5 Kg (10 lb), incorporated a small beacon to support measurement of ionospheric propagation and transmit the satellite's internal temperature (Figure 12.18). Although the satellite lasted only for 22 days, burning up on re-entry into the Earth's atmosphere, it was well

Figure 12.18   The OSCAR 1 satellite, weighing about 4.5 kg, was built in makeshift workshops of basements and garages of the Project OSCAR team. Courtesy © AMSAT.

received by the amateurs. Since then, numerous satellites have been launched – constructed on a pragmatic, low-cost approach in procurement, management, development, launch and operational management of the satellites. Subsequent satellites have gradually refined the technology, leading to the introduction of a variety of transponders and improvements in the bus.

On the basis of their capability and orbit, OSCAR satellites can be grouped into distinct phases (Baker and Jansson, 1994). Phase I satellites operated in low Earth orbits, a majority of which used a beacon as the primary source of transmission. OSCARs 1, 2 and 3 and the Russian Iskra 1 and 2 series of spacecraft belong to this category. Phase II satellites operated in low Earth orbit at a higher altitude, where the orbital lifetime would be longer, and hence they were designed for a longer life. OSCARs 6, 7 and 8 and UoSAT OSCARs 9 and 11 belong to this category. The latter two satellites were built by a team of AMSAT members and students at the University of Surrey in England. These satellites were followed by a series of analogue and packet radio satellites, launched by various AMSAT groups spread across several countries. A majority of recent amateur radio satellites, called MICROSATs (or CubeSats), made of 9″ (~23 cm) square cubes to a standard defined by Stanford University and California Polytechnic, carry one or more store-and-forward digital transponders. Phase III satellites were launched into a highly elliptical Molniya-type orbit, exhibiting a longer dwell time in the apogee region and manifesting a footprint 1–2 continents wide from the apogee; further, they deliver higher power and contain a diverse set of communication transponders. Table 12.12 summarizes recent launches of amateur satellites.

Many amateur satellites include a bulletin board that allows connections at up to 9.6 kb/s to personal computers and small transceivers. Packet radio satellites – PACSATs – offer the facility of messaging between any two operators on the Earth in minutes through the store-and-forward technique.

A variety of modulation schemes are in use: Frequency Modulation (FM), Single Side Band (SSB), Audio Frequency Shift Keying (AFSK), Phase Shift Keying (PSK), Binary Phase Shift Keying (BPSK), Gaussian Minimum Shift Keying (GMSK), Frequency Shift Keying (FSK) and Minimum Frequency Shift Keying (MFSK). The Continuous Wave (CW) scheme is also used in the form of Morse code. FM and SSB accessing schemes are in common use. The voice transponders

Table 12.12 A summary of recent amateur satellite launches

Satellite/status	Orbit	Originator	Comments
PARADIGM/ unknown	Altitude: 351 × 343 Km; inclination: 51.64°	University of Texas (Austin)	Launch: 30 July 2009 from Kennedy Space Centre aboard an STS-127; satellite measurements: 5 inch (~12.7 cm) cube
Aggiesat-2/ semi-operational	Altitude: 351 × 343 km, inclination: 51.64°	AggieSat Lab at Texas A&M University	Launch: 30 July 2009 from Kennedy Space Centre aboard an STS-127
Pollux/operational	Altitude: 351 × 343 km, inclination: 51.64°	Naval Research Laboratory	Launch: 30 July 2009 from Kennedy Space Centre aboard an STS-127; satellite measurements: 19 inch (~48 cm) diameter sphere; weight: 63 kg
Castor/operational	Altitude: 351 × 343 km, inclination: 51.64°	Naval Research Laboratory	Launch: 30 July 2009 from Kennedy Space Centre aboard an STS-127; satellite measurements: 19 inch (~48 cm) diameter sphere; weight: 63 kg
STARS/operational	Altitude: 660 × 670 km, inclination: 98.00°	Kagawa University	Launch: 23 January 2009 from Tanegashima Space Centre aboard an H-IIA F15
KKS-1/operational	Altitude: 660 × 670 km, inclination: 98.00°	Tokyo Metropolitan College of Industrial Technology	Launch: 23 January 2009 from Tanegashima Space Centre aboard an H-IIA F15; satellite measurements: 15 cm x 15 cm x 15 cm; weight: 3 kg
PRISM/operational	Altitude: 660 x 670 km, inclination: 98.03°	Intelligent Space Systems Laboratory (ISSL), University of Tokyo	Launch: 23 January 2009 from Tanegashima Space Center aboard a H-IIA F15; satellite measurements: 19 cm × 19 cm × 30 cm; weight: 8 kg

support a variety of operating modes from Morse Code (CW transmission) to SSB, FM and others, as listed above. Some OSCARs incorporate features to transport slow- and fast-scan television pictures to equipped stations.

Many OSCAR satellites incorporate a directive communication antenna. This necessitates an attitude and control system (AOCS) to ensure that the antenna remains pointed towards the earth. OSCAR spacecrafts utilize numerous stabilization techniques – ranging from simple bar magnets to computer-controlled active electromagnet systems as in Phase III spacecraft. These spacecraft correct the stability and spin rate by processing measured Earth and Sun positions to control three sets of on-board electromagnets. Gravity-gradient systems and magnetometry for attitude sensing are also in use. AMSAT's MICROSAT series install bar magnets mounted along the sides of the spacecraft to achieve up-down stability. Some satellites use their antennas to obtain spin stabilization energy directly from the Sun. MICROSATs and other satellites use turnstile-array antennas made from commercially available steel.

To facilitate low-cost launch rides as piggyback on large satellite launches, AMSAT engineers developed the concept of Ariane Structure for Auxiliary Payloads (ASAP) with ESA's support. The structure that fits around the base of Ariane IV's upper stage enables Ariane rockets to launch additional payloads. Six MICROSAT satellites were placed in orbit using this technique in the 1990s. Another approach adapted by AMSAT in securing a launch is to negotiate for non-commercial launches during the developmental phase of a launch vehicle.

Where necessary, the satellites are constructed from military-specified components, and adhering to NASA procedural standards. Judiciously selected commercial-quality components are in regular use. Emphasis is on safe, reliable and proven designs. Back-up and redundancy are built in where essential. Prior to a launch, each satellite undergoes mandatory environmental and thermal tests to assure the launch agency that the satellite is not hazardous to the mission.

The pragmatism of AMSAT engineers is well illustrated by an example. The on-board Internal Housekeeping Unit (IHU) computer for AMSAT's Phase 3-D satellite uses a very simple computer design built around a radiation-hardened 1802 microprocessor chip, which has only 16K of memory – this in preference to a more powerful and modern counterpart. The rationale for the selection was that IHU's tasks did not require more than 16K of memory to perform, 1802 chip's software code was proven and the chip performance was satisfactory on other OSCAR satellites.

Ground Equipment

The OSCAR ground equipment is readily available through numerous outlets. It is relatively inexpensive, and self-assembly kits are widely available. A transceiver typically consists of an antenna system with a rotor, a preamplifier and a modem connected to auxiliary equipment, for instance, a PC or a microphone. As there are numerous satellites of different specifications, off-the-shelf ground equipment generally provides for a subset incorporating expansion flexibility.

Typically, satellites with analogue transmissions support voice and continuous wave (Morse code) signals, and sometimes RTTY (Radio Teletype) and SSTV (Slow Scan Television) signals. Satellites also transmit digital telemetry signals that contain on-board information – satellite EIRP, temperature, solar-cell current, etc. The commercial literature is replete with details to cater for the interested reader (e.g. Ford, 2000).

A transceiver may cover one or more amateur bands. Typically, the frequency tuning increments are 10–20 Hz and the modem supports reception of FM or SSB transmissions and 9.6 kb/s digital signals. Several antenna variants are in use – turnstiles, 'eggbeaters' (crossed magnetic dipoles), helixes, etc. Cross-Yagi with horizontal and vertical elements is commonly used below 1.2 GHz. Typical antenna gains for digital reception are 12–14 dB. The rotor allows easy manipulation of the receive signal for optimum pointing.

Some technically inclined amateurs extend the scope of their activity to receive pictures transmitted from weather satellites, although this is not an amateur satellite service (see Chapter 15). The Meteosat satellites transmit continuously on 1.69 GHz. The transmissions are received via a small satellite dish, a down-converter unit to 137 MHz, a demodulator of 30 kHz bandwidth, an A/D converter and a port to transfer the data to a PC which incorporates software for decoding. Low-orbiting satellites such as NOAA satellites transmit at around 137 MHz, which can be received via a couple of simple, round dipole antennas. Other parts are similar to those of a Meteosat receiver. There is an abundance of tutorials and primers on the subject in the literature for the interested reader (e.g. NOAA, 2009).

Educational Activities

The AMSAT organization supports and pursues educational activities in schools and universities throughout the world. In its University Partnership Programme AMSAT collaborates with universities to obtain assistance at nominal costs in the development of satellites (or projects) in exchange for students' participation. The arrangement provides the students with a rich experience in return for their effort. For instance, a Phase 3-D spacecraft flight model structure was built by a team of students at the Centre for Aerospace Technology at Weber State University in Ogden, Utah.

In another initiative, AMSAT, in partnership with the American Radio Relay League (ARRL) and the National Aeronautics and Space Administration (NASA), is developing new space-qualified amateur radio hardware for NASA's Space Shuttle. The programme, called SAREX (Shuttle Amateur Radio Experiment), has brought schoolchildren of various countries into direct radio contact with the orbiting Shuttle astronauts (see Chapter 1).

12.6   HAP Communications

Although no regular HAP system is yet in regular operation, numerous research activities have been undertaken recently – notably, in Europe, the United States and Japan. Several technical, commercial and regulatory issues remain, however the knowledge has matured to an extent that several pioneering commercial initiatives have emerged in recent years. Table 12.13 summarizes some recent research and commercial initiatives.

We will elaborate upon the Helinet and CAPANINA projects promoted by the European Commission under the auspices of its Framework Programmes (FPs) 5 and 6, and introduce the plans of a few commercial pioneers.

12.6.1   EU Research Programmes

12.6.1.1   Helinet

The Helinet project, funded by the European Framework Programme 5, was a 40 month project commencing in January 2000. The aim of the project was 'to create an integrated infrastructure based on HALE (High-Altitude Long-Endurance) unmanned aerodynamic solar platforms' (Istworld Online, 2009). The infrastructure would be reconfigurable, flexible and capable of quick deployment at minimal environmental and human impact. The project investigated technologies related to (Grace *et al.*, 2005; Tozer and Grace, 2009):

- design of an HAP platform (called Heliplat);
- manufacturing and testing of critical components and realization of a scaled-size demonstrator of the Heliplat and its static tests;
- the definition of the telemetry, tracking and command subsystem of Helipat, including the link budget for the nominal attitude and for the acquisition phase;
- design of a telecommunication network topology, architecture, protocols and a Common Communication Core Interface (CCCI) towards the stated applications (see below);
- an application for vehicle localization based on GPS/GNSS/Galileo and a semi-passive positioning system as a back-up;
- algorithms for on-board processing of environmental data exploiting the particular features of stratospheric platforms and porting them to a suitable optical payload for tests;
- the provision of broadband services.

System Configuration

The system design explored the option to support various broadband services: Internet access, video-on-demand, and videoconferencing, with support for basic telephony. Figure 12.19 illustrates the concept with a possible configuration consisting of 121 spot beams (cells) of $\sim$3 km diameter each to service the broadband users; there is the option of a satellite or/and inter HAP radio or optical link for transporting the backhaul traffic to the terrestrial infrastructure.

Table 12.13 Examples of HAP research and commercial initiatives undertaken in recent years

System	Originator or sponsor	Aim
High-Altitude Platforms for Communications and Other Services (HAPCOS)	COST Action 267: an intergovernmental framework for European cooperation in the field of scientific and technical research – an ongoing initiative since 2005	Increase knowledge of HAP systems for delivery of communication and other services for interested parties such as developers, service providers and regulators
Helinet	Multipartner European Union FP 5 project (2000–2003)	Design a scaled prototype of a solar-powered HAP; develop three types of applications – broadband telecommunications, traffic monitoring and environmental surveillance
CAPANINA	Multipartner European Union 3 year FP 6 project (started 2003); built on outcome of the Helinet project	Develop technology for delivering data rates of up to 120 Mb/s to rural communities, and identify applications and services
SKYNET/National Institute of Information and Communications Technology (NICT), etc.	Japan	Develop a network to provide communication and monitoring services in Japan with about 15 airships
HALOSTAR network	Angel Technologies Corporation/USA	Provide 'ubiquitous service to hundreds of thousands of subscribers residing within a circular footprint of 50 to 75 miles diameter'. Platform: High-Altitude Long-Operation (HALO) piloted, fixed-wing aircraft over each market operating above ~16 km (52 000 ft)
Platform Wireless International	Brazilian company	Markets a tethered airship system (up to 5 km altitude) to support a bundle of communication systems for disaster and emergency events
Global HAPS Network Ltd	UK company	Create a system on a manned stratospheric aircraft to provide security and communications

Platform

The HELIPLAT platform dimensions were determined on the basis of: available solar radiation, mass and efficiency of solar and fuel cells, platform altitude, aerodynamic performance, structural mass and stiffness, electric motor size, propeller efficiency, payload mass and power. Wing spans were anticipated to be up to 80 m. The design methodology considered the energy balance equilibrium between the available solar power and the required power which included the flight.

For practical reasons, the station-keeping tolerance of the Helinet platform was specified to lie in a cylinder of radius 2.5 km and height 0.5 km for 99% of the time, and 4 km radius and 1.5 km height variation for 99.9%. These values are worse than those specified by the ITU (500 m radius), but were considered adequate. The predicted platform motion implies that ground stations would experience angular variations that would require tracking for receivers whose antenna beamwidth was narrower than the variations. The motion would also cause cell wander, which could result in signal drop-outs at the edges. It was observed that the greatest angular variations occurred immediately below the HAP. Containing cell wander would require on-board beam steering together with platform

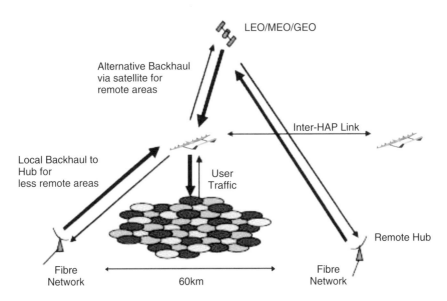

Figure 12.19   A possible system configuration envisaged in Helinet (Tozer and Grace, 2001). Reproduced by permission of © IET.

stabilization. If these measures were to prove inadequate, then cell handover would be necessary to adjust to signal fluctuations, particularly at the beam edges where the antenna radiation pattern rolled out rapidly.

Payload
The Helinet platform is solar powered, and therefore a careful trade-off between power payload weight and payload space is essential. Owing to the restrictions in size of the platform, Heliplat power sources are expected to provide a few kilowatts of power in total, leaving ~1 kW for the three payloads. In the baseline design, 400 W average power was budgeted for broadband communication. The main drainage source would be the power amplifiers (assuming 20% efficiency).

The payload weight and volume were respectively estimated as 100 kg and 2 m^3, and the antenna aperture was budgeted as <2 m^2.

The proposed Helinet antenna system uses a large number of spot beams of uniform size. One consideration in selecting the antenna system was to minimize sidelobe patterns, as they adversely impact upon antenna efficiency and intrasystem carrier-to-interference ratio (C/I); hence, an aperture-type antenna was preferred.

On the basis of the platform constraints, the maximum aperture size was kept as 1 m^2 with a ground coverage diameter of 60 km.

Horn antenna feeds were favoured. The number of horns supportable at 28 GHz was estimated as ~120, which led to a 121-spot-beam configuration, assuming that each horn forms a spot beam. The horns were arranged in five concentric rings. The central region would use a horn of ~17° beam width.

Resource Allocation and Network Protocol
Resource allocation strategies were developed for both packet and circuit modes. The TDMA frame used by the IEEE 802.16 standard operating in conjunction with the Internet Protocol suit was

preferred. In order to improve the quality of service in terms of bit rate, packet loss, packet transmission delay, delay variation, etc., IP QoS architecture models – the integrated services (IntServ)
model and differentiated services (DiffServ) model were evaluated, with consideration of their
applicability to various parts of the Helinet system, and a suitable mapping of QoS parameters
between the IP layer and the MAC sublayer, based on the IEEE 802.16 standard.

12.6.1.2   CAPANINA

The CAPANINA project built upon the results of Helinet with the goal of integrating remote communities into the broadband network from aerial platform networks at low user cost. A Framework
Programme 6 project of the European Commission, which began in November 2003 (Capanina
Online, 2009), investigated, among other things, the delivery mechanism of selected broadband
applications and services. The particular application studied in the project dealt with delivery of
broadband service to individuals travelling in high-speed public vehicles, such as trains, by enabling
access to local wireless LAN access points. Burst data rates of up to 120 Mb/s were considered in
a HAP coverage area of up to 60 km (Grace *et al.*, 2005).

Both optical and radio frequencies were considered for the high-data-rate backhaul links that
could include inter-HAP or HAP–satellite links. Figure 12.20 (Grace et al, 2005) illustrates the
system configuration of the proposed architecture, illustrating interconnectivity and salient components. The potential role of satellite systems was to:

- carry backhaul traffic to the extent feasible from high-traffic areas located remotely;
- provide broadcast content to the HAP network;
- cover areas outside HAP coverage.

It was observed that the estimated backhaul capacity of the order of 14 Gb/s would pose severe
capacity demands; hence, optical backhaul and inter-HAP links to operate in clear sky conditions
in conjunction with the RF bearers for adverse conditions would be beneficial. Inter-HAP links
are able to support optical links readily in the stratosphere because of absence of rain and other
hydrometers at stratospheric altitudes. It was estimated that, in spite of cloud cover extending

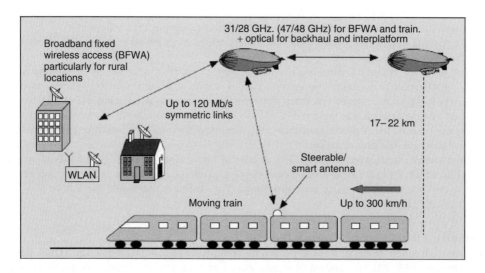

Figure 12.20   CAPANINA network configuration concept (Grace et al, 2005) © 2005, IEEE.

up to 13 km in the mid-latitude regions, optical link distances of between 450 and 680 km could still be served without limitations. The project also planned to develop a reliable optical pointing, acquisition and tracking system.

A proof-of-concept system test bed demonstrated various technologies, applications and services. Three platform technologies were targeted – tethered platform, stratospheric balloon and an *in situ* high-altitude platform. A common communication test strategy facilitated interpretation and comparison. Two trials were held – one in the United Kingdom and the other in Sweden; the UK trial comprised a 300 m tethered circular platform and the Swedish trial consisted of a one-off measurement campaign using a stratospheric balloon. The trials demonstrated the viability of:

- broadband fixed wireless access (BFWA) technology in the 28 GHz band;
- end-to-end network connectivity;
- services such as high-speed Internet and video-on-demand;
- optical communications – high-data-rate backhaul link tests and measurement of atmospheric parameters on the channel during the stratospheric balloon trial.

The investigations on viability of broadband Ka band technology for services to high-speed trains covered standards selection, propagation impairments, radio resource and handover management through to RF, mechatronic design (i.e. synergistic mechanical and electronic design) and adaptive beam-forming techniques for the train and HAP antennas. A selection was made on the basis of the best match for the requirements against a basket of regulatory, technical and commercial criteria involving a wide range of wideband standards, namely the IEEE 802.16 family of standards, IEEE 802.20 (in the early stage of development), HIPERACCESS, DVB-S/S2, DVB-RCS and DVB-T. It was concluded that Since Carrier WiMAX (IEEE 802.16SC including its variations/enhancements – 802.16a and 802.16e) represents the closest match to the predefined requirements, which was also in line with the earlier work of Helinet for fixed users.

12.6.2 Commercial Initiatives

12.6.2.1 Angel Technologies

In the United States, Angel Technologies Corporation (Angel Technologies Online, 2009) and its partners began developing a High-Altitude Long-Endurance (HALO) platform-enabled broadband metropolitan scalable network to provide service on a city-by-city basis throughout the world. The planned HALO™ network would utilize stratospheric, piloted, fixed-winged aircraft with twin turbofan propulsion. The aircraft will fly in a 'distorted torroidal shape' of ~15 km at an altitude of ~16 km in three shifts to provide ubiquitous and dedicated point-to-point multimegabit per second access to hundreds of thousands of users within 80–120 km diameter service areas (5–11 thousand square kilometres). The high-capacity, star-configured, packet-switched network will initially support a total capacity of 16 Gb/s, with plans to extend it to 100 Gb/s (Angel Technologies Online, 2009; Colella, 2009).

The HALO aircraft proof-of-concept model was tested in 1998 by Scaled Composites. It is said to be undergoing flight trials (see Chapter 2). It can operate between about 15.5 and 18.5 km altitude, utilizing two fan-jet engines. The airframe can be tailored for provision of communications services for payloads of up to 1000 kg comprising the repeater, the antenna system and DC power of >20 kW. The antenna will be hoisted below the aircraft in a pod suspended below the fuselage with a capability of generating hundreds of beams (100–1000). For aircraft fixed beams, a beam–beam handover mechanism will be used because of beam movements in the user plane caused by aircraft motion.

The fixed subscriber terminals will use autotracking high-gain antennas in millimetre wave frequencies (~30 cm diameter dish). The high-elevation-angle operation (>20°), in combination with high-gain receiver antennas and high-power transmissions, ensures reliable communication in spite of the adverse RF propagation conditions inherent in millimetre waves. Individual broadband symmetrical bandwidth-on-demand services are planned at 1–5 Mb/s to provide low-cost Internet, video-on-demand, telecommuting, and two-way videoconferencing. Business broadband symmetrical services targeted at 5–12.5 Mb/s would include Internet access, online sales, inventory, Intranets, enterprise networks, extended LANs, offsite training, two-way videoconferencing and access to terrestrial millimetre wave networks. Dedicated services could be offered at up to 155 Mb/s. The initial network capacity is targeted as ~5000 simultaneous two-way DS1-equivalent rate, to be enhanced in the future. Standard protocols such as ATM and SONET will be adopted to interface the system with public networks.

The services will be offered on licensed spectrum allocated for terrestrial broadband. Such arrangements are under discussion with spectrum holders likely to benefit from the proposed network. The Land Mobile Data Service (LMDS) band is believed to be an attractive proposition.

The processing payload of the HALO aircraft can provide user–user connectivity with delays ranging from ~60 to 200 μs. The system is provisioned such that the maximum and minimum data rates are available respectively for >99.7% and >99.9% of the time with outages confined within small areas in the footprint for less than 0.1% of the time.

12.6.2.2 Global HAPs

Global HAPS Network Ltd of the United Kingdom aims to develop a multi-mission platform mounted on a manned stratospheric aircraft to provide a security system and a communications network. The vision is to migrate the operations to an Unmanned Aerial Vehicle (UAV) platform when the technology matures (Global HAPS Network Online, 2009).

The security system (Global HAPS Security System©) will 'monitor, track and help protect strategic assets, vital installations and infrastructure in real-time mode' to provide imagery and data to aid risk management by corporations, homeland security, etc. Beneficiaries would include international communities, governments, insurance underwriters, commercial enterprises and, generally, entities that wish to protect strategic assets and resources.

The broadband communications network (Global HAPS Communications Network©) will provide WiMAX services to fixed and mobile users, assisted by appropriately located backhaul ground terminals over the HAP footprint. The broadband platform will also support a mobile HDTV broadcasting service.

Technologies under investigation include phased-array smart antenna systems and advanced propagation impairment techniques.

The technological and financial risks will be mitigated by deploying systems in response to local needs and demand, augmented with the capability of rapidly replacing or enhancing on-board payload to leverage technological advances and accommodate user- and market-driven changes in a timely manner.

Revision Questions

1. The Inmarsat BGAN system is said to be a satellite component of the UMTS network. Outline the relevant features of the Inmarsat system to justify the statement.
2. Inmarsat's physical layer is different to that of a terrestrial mobile system such as GSM and 3G. Outline the relevant differences in the physical layer characteristics of a terrestrial (such as GSM) and the Inmarsat system.
3. Explain the reasons for differences in the modulation and accessing scheme in the forward and return service links of the BGAN system. Explain the reasons for differences in the forward and return link throughputs.

4. Compare the Iridium and Globalstar systems, considering the orbital features, network topology, service provision and the air interface. Explain the reasons for the different approaches adopted by the system planners.
5. Outline the strengths and weaknesses of the air interfaces of the Iridium and Globalstar systems.
6. Elaborate the statement that HAP systems offer the advantages of both a GEO system and a terrestrial system. List the main issues that you consider need resolution before HAP systems are introduced widely.

Project. The project is directed to those keen to enrich their experience with first-hand knowledge of satellite communication with manageable economic resources. A low-cost rewarding entry point is to utilize an amateur satellite system. The Internet is a rich source of information to progress this work. Obtain a list of amateur satellites that are currently operational (e.g. from the AMSAT website). Determine their orbital parameters, and thence derive the visibility of at least two amateur satellites visible from your location. Familiarization with orbital parameters and the mathematics to estimate the look angle will be a rewarding achievement. Ideally you would develop software to do so. Determine the characteristics of the satellites, such as supported frequency range, satellite transmit power, supported modulation, safe transmission limits, etc., including limitations with regard to the use of the satellite. List the activities you may be able to perform in order to communicate and experiment via the satellite, and shortlist those aspects that you will be most interested in – hardware construction, assembly, tracking software, experimentation, etc. Finally, draw out a detailed plan of activity, cost and schedule, including familiarization with mandatory aspects recommended by satellite operators or AMSAT for safe transmissions. Depending on the resources, proceed with the implementation of the project – if so inclined in a small group of like-minded individuals or as an undergraduate project. More advanced aspects can be added to evolve the project to a postgraduate level.

References

AMSAT Online (2009) http://www.amsat.org/ [accessed August 2009].

Angel Technologies Online (2009) http://www.angeltechnologies.com/ [accessed August 2009].

Baker, K. and Jansson, D. (1994) Space satellites from the world's garage – the story of AMSAT. Presented at the National Aerospace and Electronics Conference, Dayton, OH, 23–27 May.

Capanina Online (2009) http://www.capanina.org/introduction.php [accessed August 2009].

Colella, N.J. (2009) *Broadband Wireless Communications from a Star Topology Network Operating in the Stratosphere*. Available: http://www.angeltechnologies.com/techpaper5.htm [accessed August 2009].

Dietrich, F.J. (1997) The Globalstar satellite communication system design and status. International Mobile Satellite Conference, 17–19 June, Pasadena, CA.

ESA Online (2009) *Broadband to Trains*. Available: http://telecom.esa.int/telecom/www/object/index.cfm? fobjectid=12685 [accessed August 2009].

Ford, S. (2000) An amateur satellite primer. *QST*, April and June. Available: http://www.arrl.org/tis/info/pdf/ 0004036.pdf [accessed August 2009].

Freyer, D. (2009) Commercial satcoms on the move. *Via Satellite*, August, 22–26.

Global HAPS Network Online (2009) http://globalhapsnetwork.co.uk/ [accessed August 2009].

Grace, D., Capstick, M.H., Mohorcic, M., Horwath, J., Pallavicini, M.B. and Fitch, M. (2005) Integrating users into the wider broadband network via high altitude platforms. *Wireless Communications, IEEE*, **12**(5), 98–105.

Harris (2010) Harris Ruggedized Land Portable BGAN Terminal, Data Sheet. Available: http://www.rfcomm .harris.com/bgan/pdf/RF-7800B-DU024_BGAN.pdf [accessed February 2010].

Harris Falcon II Data Sheet (2010) http://www.rfcomm.harris.com/products/tactical-radio-communications/an- prc117f.pdf [accessed February 2010].

Huckell, G. and Tirpak, F. (2007) What the mobile user objective system will bring to UHF SATCOM. MILCOM 2007, 29–31 October, Orlando, FL, pp. 1–4.

HughesNet HN7000 (2010) *Satellite Modem Data Sheet*. Available: http://government.hughesnet.com/ HUGHES/Doc/0/4LJEVCNMMTNK717K16SVNMF6A2/HN7000S_A4_LR_071408.pdf [accessed February 2010].

Hutcheson, J. and Laurin, M. (1995) Network flexibility of the Iridium[R] global mobile satellite system. 4th International Mobile Satellite Conference Ottawa IMSC 1995, cosponsored by Communications Research Centre/Industry Canada and Jet Propulsion Laboratory/NASA, pp. 503–507.

Inmarsat Online (2009) http://www.inmarsat.com [accessed November 2009].

Iridium Online (2009) http://www.iridium.com [accessed August 2009].

Istworld Online (2009) http://www.ist-world.org/ProjectDetails.aspx?ProjectId= 326397e034dd484282e4178e237290f6 [accessed July 2009].

Jackson, M. (2007) Leveraging commercial off-the-shelf solutions for architecting the MUOS ground system. PMW-146-D-07-0055. Available: http://sunset.usc.edu/GSAW/gsaw2007/s2/jackson.pdf [accessed January 2010].

Janes (2010) Talisman II, Jane's C41 Systems. Available: http://www.janes.com/articles/Janes-C41-Systems/Talisman-II-XKu-band-military-satellite-communications-system-UK-PSC-504-United-Kingdom.html [accessed February 2010].

JTRS HMS Data Sheet (2010) http://www.gdc4s.com/content/detail.cfm?item=b8c971d4-9784-41c7-b8e1-1f557f1b2d0d [accessed February 2010].

Krewel, W. and Maral, G. (1998) Single and multiple satellite visibility statistics of first generation non-geo constellation for personal communications. *International Journal of Satellite Communications*, **16**, 105–25.

Matarazzo, G., Karouby, P., Schena, V. and Vincent, P. (2006) IP on the move for aircraft, trains and boats. *Alcatel Telecommunications Review*, 2nd Quarter, 8 pp. Available: http://www.mowgly.org/upload/pub/T0605-IP_Aircraft-EN_tcm172-909771635.pdf [accessed August 2009].

Matolak, D.W., Noerpel, A., Goodings, R., Staay, D.V. and Baldasano, J. (2002) Recent progress in deployment and standardization of geostationary mobile satellite systems. MILCOM 2002 Proceedings, 7–10 October, Anaheim, CA, Vol. 1, pp. 173–177.

Nguyen, N.P., Buhion, P.A. and Adiwoso, A.R. (1997) The Asia cellular satellite system. Proceedings of the 5th International Mobile Satellite Conference, 16–18 June, Pasadena, CA, cosponsored by NASA/JPL and DOC/CRC JPL, Publication 97–11, Jet Propulsion Laboratory, Pasadena, CA, pp. 145–152.

NOAA (2009) *User's Guide for Building and Operating Environmental Satellite Receiving Stations*. Available: http://noaasis.noaa.gov/NOAASIS/pubs/Users_Guide-Building_Receive _Stations_March_2009.pdf. [accessed August 2009].

Paradigm Skynet 5 Online (2010) http://www.paradigmsecure.com/out_services/skynet5 [accessed February 2010].

Project OSCAR Online (2009) http://projectoscar.wordpress.com/ [accessed August 2009].

Richharia, M. and Trachtman, E. (2005) Inmarsat's broadband mobile communications system. GLOBECOM Workshop on *Advances in Satellite Communications: New Services and Systems*, St Louis, MO, 2 December, 6 pp.

Richharia, M. Trachtman, E. and Fines, P. (2005) Broadband global area network air interface evolution. 11th Ka and Broadband Communications Conference and 23rd AIAA International Communications Satellite Systems Conference (ICSSC) 2005, 25–28 September, Rome, Italy, 12 pp.

Rivera, J.J., Trachtman, E. and Richharia, M. (2005) The BGAN extension programme. ESA Bulletin No. 124, p. 62–68.

Stirland, S.J., and Brain, J.R. (2006) Mobile antenna developments in EADS astrium. First European Conference on Antennas and Propagation. EuCAP 2006. Volume, Issue, 6–10 Nov. 2006, pp. 1–5.

Row 44 Online http://www.row44.com/solutions.php [accessed July 2009].

Thuraya Online (2009) http://www.thuraya.com/en/article/technology-1.html [accessed August 2009].

Tozer, T.C. and Grace, D. (2001) High-attitude platforms for wireless communications. *Electronics and Communication Engineering Journal*, June 127–137.

Tozer, T.C. and Grace, D. (2009) *HeliNet – The European Solar-Powered HAP Project*. Available: http://www.elec.york.ac.uk/comms/pdfs/20030506163224.pdf [accessed July 2009].

Walker, J.G. (1973) Continuous whole earth coverage by circular orbit satellites. Presented at International Conference on *Satellite Systems for Mobile Communications and Suveillance*, 13–15 March, London, UK, IEEE conference publication no. 95.

13

Satellite Navigation Techniques

13.1 Introduction

Navigation involves planning a route and ensuring that it is followed throughout a journey. This process requires the navigator to obtain an accurate fix of the present position to steer the course towards the desired route. Mariners used celestial bodies, landmarks, lighthouses and schemes requiring visual sightings until radio systems arrived in the early part of the twentieth century. Early terrestrial radio navigation systems include Loran (Long-range navigation), Decca, Omega, Very-high-frequency Omnidirectional Radio range (VOR) and Distance Measuring Equipment (DME) – and some of them, such as VOR and DME, continue to be in use.

A system called TRANSIT launched in the United States in 1960s heralded the era of satellite navigation systems. The TRANSIT system, developed primarily for military users (initially to obtain submarine fixes) was released to the civil community in 1967. It was rapidly adopted by oceanographers, off-shore oil exploration units and surveyors, and subsequently integrated with civil marine navigation systems.

The Navigation System with Time and Ranging Global Positioning System (Navstar GPS) – commonly called the Global Positioning System (GPS) system – evolved through the experience of TRANSIT in conjunction with those of US programmes 621B (managed by the Air Force Space and Missile Organization) and TIMATION (managed by the Naval Research Laboratory). TIMATION satellites enabled very precise time and time transfer, while Air Force System 621B contributed through its experience with the spread-spectrum ranging technique. The third satellite of the TIMATION satellite became a GPS demonstrator. During the early 1970s, the US Department of Defense joint programme initiative led to the evolution of the GPS system by synergizing the effort of TIMATION and project 621B. A GPS prototype satellite was launched in February 1978, leading to the introduction of an operational system.

A Russian civilian system known as TSIKADA emerged in 1971 – operating on a similar principle to the TRANSIT system. The last TSIKADA satellite was launched in 1995, but the constellation continues to be replenished with a slightly different type of spacecraft. A series of TSIKADA-type satellites were fitted with an auxiliary Cospas-Sarsat rescue beacon locator payload (described later) and were given the name 'Nadezhda (Hope)'. It is believed that TSIKADA is a simplified version of a military satellite system called Parus (or TSIKADA Military) which was developed initially and operated in parallel (Goebel, 2009).

The (former) Soviet Union continued the development, which resulted in the formation of a new system known as GLONASS (Globalnaja Navigatsionnaja Sputnikovaja Sistema; English translation: Global Navigation Satellite System) based on a ranging concept offering a similar order of accuracy to the GPS system. One notable difference of the GLONASS system is the accessing

Satellite Systems for Personal Applications: Concepts and Technology   Madhavendra Richharia and Leslie David Westbrook
© 2010 John Wiley & Sons, Ltd

scheme, which is based on frequency division multiple access rather than the code division multiple access used in GPS. The GLONASS system incorporates a civil mode in addition to the military mode. After a period of decline, the constellation has recently been brought up to a basic operational capability, with plans to replenish it to its full capability in the next few years (targeting 2010); there are plans to upgrade the GLONASS satellites to improve the satellite lifetime to 10 years from the 2–3 years of the first generation.

EutelTRACS and OmniTracks systems are respectively European and US regional navigation systems based on geostationary satellites that provide location-based services respectively in Europe and the United States. Cospas-Sarsat is an internationally approved global distress and safety system based on a low Earth orbit (LEO) constellation. The Argos system is a French LEO satellite system used globally for numerous humanitarian and scientific applications.

The impetus to satellite personal navigation systems arrived with the introduction of the GPS system. Other satellite systems (GLONASS, EutelTRACS, etc.) contributed further to the evolution of the satellite navigation system. Aided by a dramatic reduction in the cost and size of receivers, satellite navigation is now an established feature of personal electronic aids either on its own or embedded in devices such as mobile phones, personal digital assistants, digital cameras, portable PCs, wrist watches and emergency call systems. In-car navigation systems and on-person navigation aids are now routine gear of motorists, hikers, campers, surveyors, geophysicists, etc. The GPS system has by far the largest penetration in the personal domain.

Table 13.1 summarizes the accuracy of various navigation systems and their suitability as personal systems [Source: Schänzer (1995) and research by one of the authors].

A boost to the personal market segment came in the year 2000 when the US government switched off the selective availability mode – which improved the accuracy of fixes for civilian uses to below ten metres ($\sim$7 m, 95%). Since then the accuracy available from GPS receivers has steadily improved. Beginning in 2003, the number of GPS products more than doubled in 5 years worldwide – with predictions that about 40% of handsets would support GPS by the end of 2010 (Goth, 2007). The global satellite navigation products and services are said to be growing at an annual rate of 25% – with an estimated 3 billion satellite navigation receivers expected to be in service by 2020 (Flament, 2004). Around 90% of the equipment is believed to be sold in the civilian market (EVP Europe, 1999).

Receivers for personal applications incorporate a user-friendly interface and reduced power consumption in conjunction with advanced battery technology. Although the focus here lies in personal applications, we note that satellite navigation applications span conventional location-based service, road transport, maritime transport, aviation, rail, etc., to less traditional applications including agriculture, land survey, energy, scientific research, etc.

The global navigation systems are generically known as the Global Navigation Satellite System (GNSS). In addition to GPS and GLONASS, several global and regional systems are planned or are under development towards a multidimensional GNSS. This chapter introduces the underpinning navigation technology, including those of a GPS receiver; additionally, it introduces the related topics of distress and safety systems and location-based services, although the reader may view them merely as applications of navigation systems.

13.2 Categorization

Navigation systems are an essential aid in all environments of human interest – air, land, sea and space. Table 13.2 summarizes the salient attributes in each environment (adapted from Kayton, 2003). For obvious reasons, navigation systems that provide ubiquitous land coverage offer the greatest potential for personal applications, while maritime and airborne systems have a rather restricted appeal to the personal domain.

Table 13.1 Features of navigation systems and their suitability as personal navigation aids [Key: [1]VOR = Very-high-frequency Omni-directional radio Range; [2]DME = Distance Measuring Equipment; [3]Loran = Long Range Navigation]

System	Estimated accuracy (m)	Suitability	Radio system	Navigation principle/ orbit type	Comments
Differential GPS	~2 (95%), in practice	Yes	Satellite or Hybrid	Range triangulation with differential correction	Accuracy depends on distance between user and reference station
Instrument Landing System (ILS)	5–10	No	Terrestrial	Differential depth of modulation between two transmitters as seen at receiver for azimuth, elevation, and marker beacon for distance	Used for aircraft landing
Microwave Landing System	5–10	No	Terrestrial	Azimuth and elevation derived from scanning beams from a reference source; Distance measurements – as in DME	Developed to replace ILS
GPS (P-code)	~6 (95%), in practice	Limited to military personal systems	Satellite	Range triangulation; high range resolution	US military system
GLONASS (P-code)	≤20	Limited to military personal systems	Satellite	Range triangulation	Former Soviet Union's military system
VOR[1]	99.94% of the time <±0.35° of error	No	Terrestrial	Bearing derived by comparing reference phase transmitted by a static omni-directional antenna against the phase transmitted by a rotating directional antenna	
DME[2]	60–180	No	Terrestrial	Distance estimated by travel time of radio waves from aircraft to a land located transponder	Terrestrial system: used for aircraft navigation; DME may serve as a GPS back-up in future
GPS (C/A) code/ GLONASS (C/A) code	~7 (in practice, without Selective Availability)/100	Yes	Satellite	Range triangulation; high range resolution	Civilian version of GPS and GLONASS with code dithering switched on
Euteltracs/ Omnitrack	100	Yes	Satellite	Multi-lateration based on range measurements from two geostationary satellites	Includes a low-bit-rate mobile communication service
TRANSIT/ TSIKADA	450	Not applicable	Satellite	Derived from Doppler curve and satellite ephemeris	TRANSIT phased out (1996); TSIKADA replaced by GLONASS
Loran-C[3]	185–463	No	Terrestrial	Time difference in received signals from two transmitting sources	Terrestrial first-generation maritime and airborne system; modernized version continues to be in use (2009); a potential back-up to GPS

(*continued overleaf*)

Table 13.1 (*continued*)

System	Estimated accuracy (m)	Suitability	Radio system	Navigation principle/ orbit type	Comments
Argos	<200->1500	Yes	Satellite	Doppler principle	Targeted to study and protect the environment and protect human life or support programmes of declared government interest
Cospas-Sarsat	≤2000	Yes	Satellite	Doppler principle applied on satellite or ground	Provision to use GPS derived position estimate
Celestial observation	~3000	Yes	Visual	Sightings of heavenly bodies	Historically used by mariners
Omega	3600–7200	Not applicable	Terrestrial	Time difference between two stations observed on receiver	Decommissioned in 1997
Aircraft inertial navigation	300–20 km	No	Local	Derived from onboard sensors	Best accuracy near take-off and touchdown
Cellular location systems	Various possibilities	To be determined	Satellite or Terrestrial	Various techniques used: Distance triangulation, GPS, etc.	Various technologies (including GPS) under investigation

When comparing alternative navigation solutions, the following aspects require evaluation: receiver (complexity) and associated costs, fix accuracy, delay in obtaining a fix, coverage area and operational environment (vehicular, on-person, etc.).

Some applications require a position fix only (e.g. marking a boundary), whereas others incorporate guidance where the system uses the fix to steer a course. A navigation system fixes the location of an observer in an absolute sense, and the guidance system assists in steering the user to a predefined course or route.

A fix can be *absolute*, that is independent of the previous location (no memory), or *relative* with respect to a known, or *dead-reckoning* where the fix at time t is derived from a known trajectory given the initial fix at time t_0 and the users' velocity and acceleration. According to the theme of the book, we will confine our treatment to satellite-enabled absolute-fix navigation systems.

A satellite navigation system is said to be *active* when users transmit signals that are processed at a hub, or *passive* when the user processes the received signals only. Navigation systems may use a single or multiple satellites operating in geostationary or non-geostationary orbit. The estimate can be based on *Doppler* or *range* measurements. The Doppler method was used in the first-generation navigation system, such as TRANSIT, and continues to be in use in systems such as Argos; the range-enabled triangulation method is deployed in the GPS and GLONASS system.

13.3   Doppler-Assisted Navigation

In Chapter 7 we explained that a Doppler-enabled navigation system is based on the concept that it is possible to estimate a user's location when the Doppler frequency shift Δf at the location due to satellite motion and the ephemeris of a satellite are known.

Table 13.2 Main attributes of physical environments encountered in practice (adapted from Kayton, 2003)

Environment	Attribute
Land	Widely used by individuals (tourists/hikers/mountaineers, etc.)
	Receiver movement generally slow for personal systems
	Altitude determination not crucial
	Areas of major interest: city, highways and tourist spots
	Supplementary data readily available through maps, etc.
	Wide variations in receiver temperature range
	Vibrations/impacts may be severe in some applications
Sea	Vital for individuals and groups for activities such as exploration, sports, etc.
	Receiver movements slow
	Altitude is near zero
	Wide coverage system preferred owing to international nature of sea transport
	Severe environmental conditions and wide movements in pitch and roll
Air	Vital for individuals interested in aviation
	Necessitates altitude measurement
	Velocity component high
	Wide coverage system preferred owing to large travelling distances
Space	Not used for personal systems

Typically, the Doppler observed by a fixed observer on the ground follows an 'S' curve, with the highest positive Doppler experienced at the rise time and the highest negative Doppler at the set time. The Doppler shift is zero at the instant when it is closest to the user – beyond which it becomes negative as the satellite moves away from the observer.

We observed that, given a velocity and wavelength, the Doppler frequency shift depends solely on the offset angle α. The locus of the Doppler shift observed on the ground at a given instant therefore comprises all the points where the direction vector is offset by α (see Chapter 7). The Doppler circle intersects the Earth at two locations, one of which is the user location. Knowing the satellite position through orbit ephemeris, the location solution (latitude and longitude) is obtained by iteratively fitting Doppler estimates from various locations with the measured Doppler, as explained in Chapter 7 (see also, Guier and Weiffenbach, 1960; Argos User Manual, 2008). The solutions at each point of the Doppler curve are averaged to provide a typical accuracy of the order of half a mile.

The main elements of a Doppler-enabled navigation system are as the follows:

- A satellite constellation designed to achieve the desired availability (i.e. time interval between successive fixes); Orbital inclination and altitude of the constellation govern the service area. For example, a constellation of four satellites in a circular polar orbit can provide a fix every 1.5 h to most locations on the Earth. An equatorial orbit constellation, on the other hand, would cover an equatorial belt.
- A network of ground stations dispersed around the world for accurate estimation of the satellite ephemeris.
- Fixed station(s) to upload the satellite ephemeris (and auxiliary data) to each satellite regularly because the ephemeris begins to deviate with time. The satellite must provision memory to store the data and downlink them regularly at adequate power to the user community and transmit an unmodulated signal at an appropriate frequency for Doppler detection.

Other provisions include a highly accurate transmission frequency and a means to correct the receiver's oscillator drift; a time synchronization signal to obtain a time standard for receiver frequency and time correction; a mechanism to correct Doppler distortion caused by ionospheric refraction. The receiver can apply corrections induced by ionospheric distortions by algorithmically processing when the satellite transmits beacons at two harmonically related frequencies.

13.3.1 Errors and Countermeasures

Doppler Position Errors

Levanon and Ben-Zaken have shown that, for the intermittent type of Doppler positioning system, the random error is very dependent on the frequency stability of the transmitter. For example, in the Argos system, a transmitter with a short-term frequency stability of 2.5×10^{-9} yields a random positioning error of between 0.1 and 2 km (Levanon and Ben-Zaken, 1985). By contrast, in the defunct TRANSIT Doppler navigation system, which used (optimum) phase tracking, the position error was $\sim$90 m for a single-frequency receiver, owing to uncertainties in the path delay due to the ionosphere, and $\sim$30 m in a dual frequency receiver.

We noted that, because of the large numbers of navigation fixes extracted during each pass, it is feasible to minimize the statistical uncertainty significantly. Non-statistical biases cannot be eliminated in this way.

One of the main error sources occurs in frequency measurement. Theoretical investigations show that statistical errors do not contribute significantly to the accuracy of the results for relatively coarse accuracy requirement ($\sim$0.5 miles), except for the polar orbit systems at the two extreme edges of the range (i.e at the lowest elevation angles) and when passing overhead because of the difficulty in estimating user location in both situations (Guier and Weiffenbach, 1960). However these results are based on very low computational processing.

The non-statistical sources of error include:

- changes in refractive index of the ionosphere;
- frequency drifts;
- uncertainty due to user velocity;
- uncertainty due to user altitude;
- error caused by signal multipath.

Changes in refractive index of the ionosphere cause a deviation in the path length and hence in estimating the Doppler frequency. The accuracy of the estimate can be improved by transmission of beacons at two frequencies, which allows the error term to be removed algorithmically. The remaining higher-order terms do not contaminate the fix significantly in typical conditions; nevertheless, the technique is inefficient in anomalous ionosphere conditions such as a magnetic storm.

Transmitter and/or receiver oscillator drift also cause error. The bias becomes insignificant if, in addition to the latitude and longitude, satellite oscillator frequency is included as a variable during the fitting process. It has been shown that a drift rate of the order of 1 part in 10^8 per hour is adequate for a position fix accuracy of the order of 0.5 miles.

Error due to velocity of the user location is relatively minor for slow-moving vehicles but not for fast-moving vehicles such as aircraft.

Other potential sources of error include those introduced by multipath and unfactored altitude of the user. Multipath tends to occur at low elevation angle, and hence one possibility to minimize its impact is to block signals arriving from low elevation angle. A more elaborate method would be to use multipath rejecting antennas. Errors caused by altitude uncertainty are minor for fixes on ships or flat terrain.

13.4   Range-Assisted Navigation

As explained in Chapter 7, in this technique the user position is estimated by measuring the range of the location from two or more satellites and solving the mathematical equations representing the system. OmniTracks (USA) and EutelTRACS (Europe) systems use two satellites for obtaining a position fix. In a three-satellite system the user lies at the intersection of three or more spheres of radii r_1, r_2, etc., centred at each satellite and the Earth's surface. Owing to the additional range measurement from a third satellite, the system can also determine the altitude of the user. A fourth satellite resolves uncertainties in pseudorange estimation. The GPS and GLONASS systems use this principle, as will the Galileo system (when completed).

13.4.1   Reference Frames

13.4.1.1   System Time

Each satellite carries a number of atomic clocks to provide accurate transmitter timing. A number of atomic clocks are used in the navigation system's ground control segment to form navigation system time (reference time). The drift of the satellite clocks from this system time is monitored, and a correction is broadcast to the user. User receiver time is synchronized to this system time, which is usually a statistical average of a number of atomic clocks, steered (gradually adjusted) so as to track International Atomic Time (TAI). TAI is itself a statistical combination of the global ensemble of atomic clocks.

13.4.1.2   Geodetic Reference

All position determination will be relative to some reference coordinate system (of which there are several). A geodetic reference system comprises a model of the gravitational equipotential surface (because the Earth is not a perfect sphere) and a standard coordinate frame. It is common to use an oblate (flattened) reference spheroid, characterized by two parameters: semi-major axis (radius) and inverse flattening, as discussed in Chapter 2. The coordinate origin is the Earth's centre of mass. By far the most common geodetic system, and that used by the global positioning system, is the World Geodetic System 1984 (WGS84).

13.4.1.3   Navigation Data Message

In order to indicate the transmitter Pseudo Random Number (PRN) code epoch, a low-data-rate navigation message is also transmitted along with the PRN code. The navigation message includes a number of other data used to work out the user position and system time:

- *System time.* This is needed to work out the transmitted time.
- *Satellite ephemeris.* Satellite ephemeris data are used to determine where each satellite is at a given time.
- *Satellite clock errors.* Differences between the satellite clocks and system time.

As explained in Chapter 7, there are numerous sources of error and corresponding countermeasures. Figure 13.1 (Misra, Burke and Pratt, 1999) presents the accuracy available in practice from various types of GPS receiver (S/A switched off) – the standard GPS receiver and three variants of differential GPS (discussed next). At that time, the accuracy ranged from tens of metres down to millimetres, depending on receiver capabilities and availability of augmentation signals, measurement time (in real time to hours) and whether the P-code signals are available to the user.

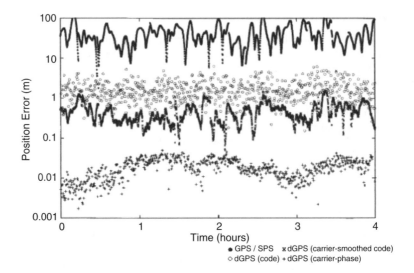

Figure 13.1 Accuracy of GPS-based position estimates in real time can range from tens of metres to centimeters; dGPS (carrier-smoothed code) curve is the third from the top; dGPS (carrier-smoothed code curve is the third from the top (Misra, Burke and Pratt, 1999). Reproduced by permission of © IEEE.

13.4.2   Error and Countermeasures

13.4.2.1   Pseudorange Error Contributions

We will consider the GPS system in formulating the error budget and to explain countermeasures (see Chapter 14 for a GPS description):

- *Satellite clock offset relative to GPS system time.* Highly accurate cesium clocks are used for maintaining GPS time at a Master Control Station (MCS). Although each satellite incorporates a highly stable atomic clock, the clock frequency drifts with time; the MCS measures clock offsets of each satellite in the constellation daily and uploads the corrections, which are subsequently transmitted to the receiver to utilize.
- *Ephemeris error.* Certain components of ephemeris errors cannot be isolated from a satellite's clock offset error and hence are left together with the satellite clock error.
- *User clock offset.* For the receivers to be cost effective, the receiver clocks are low in accuracy. The receiver clock uncertainty is eliminated by utilizing the pseudorange measurements from the fourth satellite.
- *Propagation delays.* Ionosphere causes a delay in RF propagation owing to refractive effects (ray bending and velocity reduction); the delay is inversely proportional to the square of the frequency. For authorised users, the correction can be estimated by comparing the propagation behaviour at two frequencies (for GPS at 1227.6 and 1575.42 MHz). For single frequency (civilian) users the approximate correction mode parameters are calculated at the MCS and downloaded to users with other navigation data. Tropospheric delays are frequency independent and modelled at the receiver as an elevation-angle-dependent correction.

- *Satellite processing delay.* A processing delay is incurred at the satellite. The delay is measured during the ground tests and included with other data sent to the users.
- *Multipath errors.* Multipath noise is caused by signals arriving from different directions at the receiver. Multipath can be caused in the ionosphere, troposphere and locally owing to scattering around the user receiver. Local multipath caused by scattering around the receiver can be mitigated to some extent by antennas that reject multipath that occurs at low elevation angles.
- *Scintillation.* Tropospheric scintillation is negligible at L band. However, ionospheric scintillation can be problematic; it is time-of-day, season and latitude dependent, with the worst effects in the equatorial region after sunset.

13.4.2.2   Satellite Ephemeris Errors

Estimation of pseudorange clearly relies on an accurate estimate of satellite position at any given time. Errors arise in this estimate for two reasons: there may be uncertainty in the measurement of satellite position (via ranging from ground stations) and subsequent modelling of orbital parameters. Secondly, the ephemeris data age as the time increases between the original satellite ranging measurements and their use in determining user position. The *age* of the ephemeris is usually broadcast as a part of the navigation message so that receivers know how old the orbital parameters are.

13.4.2.3   Satellite Clock Errors

Atomic Clocks
Central to the use of navigation satellites for pseudoranging is the deployment of highly accurate, space-qualified atomic clocks on-board the navigation satellites for accurate determination of satellite time (broadcast transmit time). These clocks are used to generate the carrier and PRN code, as well as to define the code epochs. Atomic clocks are microwave oscillators locked to an atomic transition of a gas molecule, the gas in question being housed inside the microwave cavity.

Typically, navigation satellites employ space-qualified cesium and/or rubidium atomic clocks, although the trend is to employ passive hydrogen masers (Microwave Amplification by Stimulated Emission of Radiation – the microwave equivalent of lasers). In general, rubidium clocks are more compact and cheaper to manufacture, and offer very good short-term stability, while cesium clocks provide better long-term stability; passive hydrogen masers have superior short-term stability to either cesium or rubidium. Two different types of atomic clock may be used together to provide both short- and long-term stability.

System Time
To reiterate, several additional atomic clocks are used to evolve the navigation reference *system time*; the drift of the satellite clocks from the system time is monitored, and a correction is broadcast to the user receivers. User receiver time is synchronized to this system time, which is usually a statistical average of a number of atomic clocks, steered (gradually adjusted) so as to track International Atomic Time (TAI).

The principal contribution to satellite clock error is the difference between the satellite clock and system time. The atomic clocks in navigation satellites are free running but are steered to align with system time – typically over a 24 h period. The difference between these clocks and system time must thus be estimated, and the offset broadcast (in the navigation message).

Relativity

Owing to the precision required of satellite navigation, it is also necessary to account for the effects of relativity (Kaplan and Hegarty, 2006; Ashby, 2002). A discussion on relativistic effects is beyond the scope of this book, but the major relativistic effects on navigation satellites are:

- *Special relativity.* Owing to satellite velocity, the theory of special relativity predicts that the on-board atomic clocks on satellites run slower than clocks on the ground. For example, GPS satellite speeds are about 4 km/s, which results in time dilation of the order of 1 part in 10^8, or 7.1 µs/day. This may be corrected by adjusting the speed of the satellite clocks.
- *General relativity.* The curvature of the gravity potential due to the Earth's mass in orbit is less than it is at the Earth's surface, and the clocks on satellites appear to be ticking faster than identical clocks on the ground. For GPS satellites, the theory of general relativity predicts that the clocks in each GPS satellite will get ahead of ground-based clocks by order of 5 parts in 10^8 or 45.7 µs/day. This may also be corrected by adjusting the speed of the satellite clock. A clock closer to a massive object will be slower than a clock further away.
- *Sagnac effect.* In special relativity, the speed of light is constant when measured in any inertial frame, but navigation systems tend to use an Earth-Centred Earth-Fixed (ECEF) reference frame, that is, one that rotates relative to the ECI frame (Ashby, 2004; Fliegel and DiEsposti, 1996). This difference in reference frames leads to the Sagnac effect, which may be thought of as accounting for the motion of the receiver during the propagation of the navigation signal from the transmitter to the receiver (as visualized in an ECI frame). A secondary Sagnac effect arises from the broadcast of satellite ephemerides in an ECEF frame (for convenience) rather than ECI. The maximum error due to the Sagnac effect amounts to 133 ns (Nelson, 2007). Correction for the Sagnac effect is applied at the receiver.

In the case of GPS, the net effect of special and general relativity is that, if not corrected, satellite clocks would tick faster than identical clocks on the Earth's surface by about 38 µs/day. To account for this difference in clock speed in orbit, the frequency standard on-board each GPS satellite is slowed from 10.23 to 10.22999999545 MHz prior to launch (Kaplan and Hegarty, 2006).

13.4.2.4 Atmospheric Delay

Atmospheric delay (comprising both ionospheric and tropospheric effects) is the dominant error in satellite navigation systems using a single-frequency receiver. It should be noted that the effect of atmospheric delay is different for PRN code pseudoranging and carrier-phase pseudoranging techniques. In the case of the carrier-phase pseudoranging, the relevant atmospheric delay is the phase delay, while in the case of PRN code pseudoranging the code is obtained from modulation (information) on the received signal and thus experiences the group delay.

Ionospheric Delay

One of the largest sources of uncorrected error in single-frequency pseudorange measurements is the effect of propagation through the ionosphere. Although the propagation and group delays may be estimated given a knowledge of the Total Electron Concentration (TEC), errors arise owing to the difficulty in predicting the local TEC at a given location and time.

For single-frequency receivers the TEC must be estimated, and this is a significant potential source of error. The GPS system uses a relatively simple semi-empirical ionospheric delay model for single-frequency measurements (GPS ICD, 1984) that is a variant of the semi-empirical model described by Klobuchar (1987). The Klobuchar model approximates the diurnal variation in ionospheric time delay as a half-wave rectified sine wave. The eight parameters needed to implement the model are broadcast in the navigation message (four to describe the variation in amplitude of the vertical ionospheric delay and four to describe its cyclical phase). This simple model is estimated to reduce the single-frequency RMS error due to ionospheric delay by at least 50%.

Dual-Frequency Receivers

It was established in Chapter 3 that the excess propagation delay due to the ionosphere varies as $\frac{1}{f^2}$. It is therefore apparent that, if two measurements can be made at two different frequencies, the dispersion may be estimated and the error due to ionospheric propagation delay may largely be removed.

After some manipulation, it can be shown that the ionospheric correction to the pseudorange for frequency i ($i = 1, 2$) is given by

$$c\,\Delta t_{\text{IONO},k}\left(f_{1,2}\right) = \frac{f_{2,1}^2}{f_2^2 - f_1^2}\left(\rho_k\left(f_1\right) - \rho_k\left(f_2\right)\right) \tag{13.1}$$

where $\rho\left(f_i\right)$ is the pseudorange measurement for frequency i.

Until recently, dual-frequency navigation receivers had largely been the preserve of military and specialist users, as the PRN code at the GPS L2 frequency is reserved for authorized users only. However, with the recent advent of additional civilian GPS signals and (soon) multiple civilian Galileo signals, one may expect dual-frequency receivers to become much more widespread, and consequently the position error due to the ionosphere will become much less significant.

Tropospheric Delay

Unlike the ionospheric delay, the tropospheric delay is not frequency dependent and so cannot be mitigated by using dual-frequency receivers. The correction of tropospheric errors is achieved solely in the receiver. Typically, no parameters are broadcast from the satellite to assist with this correction, as the correction required depends primarily on user location.

In principle, tropospheric delay depends mainly on receiver altitude and satellite elevation angle, which may be easily determined because these are part of the normal position estimation – as well as on meteorological parameters such as temperature pressure and water vapour content, which are not. The refractive index (and hence tropospheric delay) has both wet and dry gas contributions – the latter being both geographically and seasonally dependent via the local water vapour density. Fortunately, the dry component, which is more easily estimated, comprises approximately 90% of the observed tropospheric delay.

Residual tropospheric delay errors using such simple models are estimated to be a few nanoseconds – roughly equivalent to 0.3 m pseudorange error (Lewandowski, 1994).

Differential Pseudorange Systems

The GPS system originally included a Selective Availability (S/A) mode to prevent unwarranted military use of the GPS system. When switched on, it corrupts the data sufficiently so that accuracy is no better than ~100 m (as specified; ~60 m in practice). Figure 13.2(a) illustrates the 24 h scatter plot taken on 1 May 2000 with S/A switched on, and Figure 13.2(b) shows a plot taken on 2 May 2000 with S/A switched off (S/A set to zero). The plots were taken at Continuously Operating Reference Stations (CORS) operated by the NCAD Corp. at Erlanger, Kentucky (NOAA, 2009). The error introduced by S/A mode is clearly visible in Figure 13.2(a).

To circumvent the S/A 'problem', a technique known as differential GPS was deployed by the US Coastguard and others, which used corrections transmitted from fixed ground stations whose positions were accurately known. It was observed that the inaccuracy due to S/A mode were relatively insensitive to the distance.

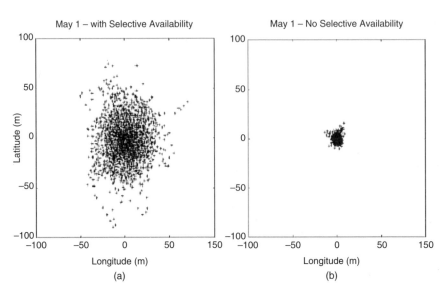

Figure 13.2 (a) Position estimates over 24 h with GPS S/A on – as observed in Kentucky on 1 May 2000 from a fixed location; (b) position estimates over 24 h with GPS S/A off – as observed in Kentucky on 2 May from a fixed location. Reproduced by permission of © NOAA, National Geodetic Survey.

The differential correction, when transmitted to other users located in areas sufficiently close to the reference station, can be utilized to correct their fix by calibrating out the error. However, since ephemeris and ionospheric errors degrade with distance, the accuracy of the differential correction decreases with distance relative to the reference station. Fortunately, the decorrelation distances are quite large; estimates in error growth vary, ranging from 0.67 m per 100 km to 0.22 m per 100 km.

In an operational system, a number of reference stations are used; these stations broadcast the corrections terrestrially or via satellite. The Satellite-Based Augmentation System (SBAS), explained in the next section, utilizes the same general principle to improve the accuracy of SBAS-enabled receivers.

13.5   Satellite Augmentation System

A committee set up by the International Civil Aviation Organization (ICAO) to develop and plan the Future Air Navigation System (FANS) recommended that ICAO move to a satellite-based Communications, Navigation and Surveillance/Air Traffic Management (CNS/ATM) concept. The introduction of new technologies such as a Global Navigation Satellite System (GNSS) for Air Traffic Management (ATM) would 'improve efficiency, maintain safety and reduce costs' (FANS(II)/4, 1993) through increased airspace and airport capacity, coupled with dynamic flight planning and reduced controller/pilot workload.

However, it was understood that the existing GNSS candidates at the time, namely GPS and GLONASS, fell short of the necessary performance in terms of accuracy, integrity, time to alert, continuity and availability (ICAO, 2000; Tiemeyer, 2002; CAA, 2004). The limitations could be overcome by introduction of aircraft-, ground- or satellite-based augmentation techniques. Receiver Autonomous Integrity Monitoring (RAIM) uses the redundancy of simultaneous measurements to more than four satellites to check consistency. The Aircraft Autonomous Integrity Monitoring

(AAIM) system improves integrity and availability by combining the measurements obtained from satellite receivers with those of on-board sensors. Ground-based Augmentation Systems (GBAS) improve accuracy and integrity for precision approach operations using local-area ground stations to monitor the satellite system status and calculate correction terms and uplink them to the approaching aircraft for accuracy enhancement. Another technique is to install beacons on aircraft approach paths to provide additional ranging information for improving positioning accuracy and increasing measurement redundancy.

Satellite-based Augmentation Systems (SBAS) enhance global navigation satellite GNSS system integrity and accuracy by broadcasting system status, corrections and positioning information from an independent satellite network. SBAS-enabled receivers utilize the information to augment the accuracy. The concept is illustrated in Figure 13.3. The main components of a SBAS system are:

- a network of ground stations equitably dispersed in the region of interest to monitor status of the GNSS constellation in real time (e.g. <6 s), collect data essential to estimate error differential and send them to a central facility;
- a central facility to process the data and estimate the differential errors and corrections which the users can apply for accuracy enhancements;
- an uplink station to broadcast the processed data multiplexed with GNSS-compatible ranging signals to augment system robustness.

Although the SBAS concept was primarily driven by the needs of civil aviation, its utilization now extends to other industries. Benefits to the civil aviation community are:

- precision guidance of aircraft during landing and in remote areas such as the oceanic routes;
- avoidance of collision with ground;
- greater route flexibility;
- reduced aircraft–aircraft separation, thereby allowing denser packing of the routes.

Examples of social and economic and environmental benefits of SBAS include (ESA/ EGNOS, 2009):

- in a maritime environment, tighter monitoring of ships such as collision avoidance, monitoring approaches in heavy traffic areas, identification and tracking of containers to improve efficiency, etc.;

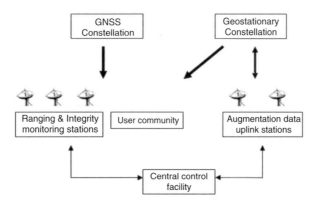

Figure 13.3   The SBAS concept.

- for railways, reduced spacing between trains;
- for road transport, more efficient and economic fleet management, tighter management of dangerous goods and more accurate personal navigation;
- availability of accurate timing to assist scientists, software developers and distributed software functioning.

Several regional SBASs have been implemented or are in development. There is a general agreement that the systems should be capable of interworking with one another because of the wide areas traversed by aeroplanes. The Wide Area Augmentation System (WAAS), operated by the Federal Aviation Administration (FAA), covers the American region. The European Geostationary Navigation Overlay Service (EGNOS), operated by the European Space Agency, covers the European region. The Multifunctional Satellite Augmentation System (MSAS), operated by the Japan Civil Aviation Bureau (JCAB), covers the Japanese airspace. A system called GPS and GEO Augmented Navigation (GAGAN), to cover the Indian region and fill the coverage gap between EGNOS and MSAS, will be implemented by India. Details of some of these systems are given in Chapter 14.

13.6   Navigation–Communication Hybrid Architecture

Satellite navigation receivers are an indispensible tool for wide-area location-based services. In a navigation-only application, the user is mainly interested in obtaining a *location fix*, which is then used for a variety of applications – typical examples are: 'SATNAV' car equipment and handheld units used by hikers, mountaineers, etc. In a *location-based service* the navigation fix is conveyed to a remote server via a satellite or a terrestrial radio link to obtain the necessary information. Examples of this class of service are: retrieval of local traffic conditions; tracking containers or cars of a fleet; stolen car recovery; assistance in search and rescue operations (see the Appendix).

A location-based service requires a radio communication link to convey the location and other details to the application server. Terrestrial mobile systems are commonly used in areas where coverage is well entrenched; however, satellite systems are attractive when the service area is wide with patchy terrestrial coverage.

A typical hybrid architecture is illustrated in Figure 13.4. A satellite navigation receiver embedded within a personal device feeds its location (latitude/longitude) to a communication transceiver for transmission to a host over a terrestrial or a satellite system, along with the service request for application processing at the destination. The host ('the service provider') processes the request and provides the desired information on a return communication link. The service could, for example, convey local information about traffic conditions, weather report, amenities or events. The service can also be triggered by a fixed user to retrieve information from a remote user, for instance the location of a container.

One of the visions of an integrated navigation–communication receiver is to incorporate the capability to support multiple communication standards for ubiquitous service ('go anywhere'). The user may utilize a terrestrial system in cities and a mobile satellite system when travelling to or resident in a remote area. Later in this chapter we discuss the concept of a hybrid receiver system that supports multi-standard communications for road applications. One of the most prolific applications of the hybrid concept is the fleet management system, where the movement of vehicles is managed and controlled by a central facility – this type of system has been applied for a number of applications ranging from the management of truck fleets to a take-away home delivery service. Numerous location services are on offer and are emerging rapidly. Novel applications in development include provision of details of a tourist attraction by clicking at the object of interest on a mobile device, a description of the celestial sky (stars, planets) on request, assistance to visually impaired individuals, etc.

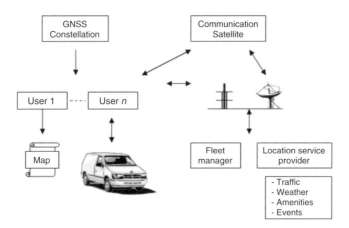

Figure 13.4 A conceptual diagram of a hybrid navigation–communication system. The GNSS constellation provides location fix. The user sends a request for service through a satellite or terrestrial radio link. The service provider responds back with the requested information.

13.7   Receiver Architecture

The user segment of the GPS system has grown phenomenally in recent years and remains a dominant satellite-enabled personal aid device at present. In this section we present the architecture of a GPS receiver (Braasch and Van Dierendonck, 1999). The GPS system is described in Chapter 14.

Consider first a future scenario. The GLONASS system is being replenished and enhanced; Galileo, a European initiative, is progressing towards the implementation phase; several regional systems are under consideration. In order to accommodate the proliferation of systems, receiver designers are considering architectures that are flexible to meet the demands of the next-generation GNSS alternatives. We shall present the issues and considerations essential in enhancing the receiver architecture to cater for the next-generation GNSS systems (Dempster, 2007).

A navigation receiver estimates the position, velocity and time by measuring range, range rate, Doppler, carrier phase and supplementary data. Receiver design alternatives and emphasis vary depending on the application (and hence the user). Table 13.3 summarizes a typical set of user requirements.

There will be additional requirements expected of multimode and multi-tasking receivers capable of supporting multiple navigation systems and/or hybrid system architectures.

Table 13.3 Receiver requirements and corresponding example users

Requirement	Example user
Long period between battery charges	Hiker
High location accuracy	Surveyor, pilot
Light weight	Recreational user
Embedded	Embedded cell phone

Receiver Architecture

GPS satellites transmit direct-sequence spread-spectrum signals for ranging, combined with low-bit-rate auxiliary navigation data that include start time of transmission, satellite clock offset with respect to GPS reference time, satellite ephemeris and supporting data such as constellation status. The pseudorandom noise sequence uses Gold codes, which are formed by the combination of two low cross-correlation maximum-length sequences. Each satellite is assigned a unique code, and all the satellites in the constellation can transmit at the same frequency without causing noticeable interference to each other.

Conceptual transmit-and-receive arrangements are depicted in Figure 13.5(a). A low-bit-rate 50 b/s navigation data BPSK modulates the carrier. The carrier is spread by two types of high-rate pseudorandom noise (PRN) sequence – a 1.023 MHz chip rate Clear Acquisition (C/A) code for civil applications and a 10.23 MHz Precise (P) code for military use. These signals are transmitted to each satellite in view. Each satellite stores the uplinked information and continues to transmit the same until the next upload.

The main constituents of a receiver are illustrated in Figure 13.5(b), and its features summarized in Table 13.4. The received navigation signal is amplified, down-converted and correlated with a replica of the transmitted pseudorandom noise code. The correlation process effectively amplifies the BPSK signal by approximately a factor determined by the degree of spread. The modulated carrier is demodulated in a conventional manner to extract the navigation data. The estimated range is given as the difference between the time of transmission, which is *time stamped*, and the time of receipt, given from tracking the correlation peak at the receiver.

For successful operation it is necessary for the locally generated carrier and the code to synchronize to those of the incoming signal. The drift in the receiver local oscillator is compensated for by solving the pseudorange equations – as explained in Chapter 7. The carrier frequency at the receiver differs from the original owing to Doppler effects and clock uncertainties. A frequency or phase lock loop is used to keep track of the receiver frequency after initial synchronization. To acquire the code, an initial search is made for all possible starting times until maximum correlation is achieved. The lock must be maintained to track fluctuations in the time of arrival. This is

Table 13.4 Main characteristics of a receiver corresponding to Figure 13.5

Component	Features
Antenna	Right-hand circularly polarized, hemispherical pattern, low profile: microstrip patch for personal systems, helix or quadrifillar arrangement for vehicles
Preamplifier	Gain: 25–40 dB; system noise temperature: 3–4 dB; minimum receiver power: −160 dBW
Down-converter	Single- or multistage – increasing emphasis on single; high out-of-band rejection filter at the front end (typically, a SAW filter is used)
Analogue-to-digital converter (A/D)	Sampling bandwidths vary from 2 MHz for low-cost receivers to 20 MHz for high-end receivers
In-phase–Quadrature phase (I/Q) conversion	Necessary for phase detection; can be performed before or after A/D conversion
Signal processing 1	First-stage signal processing provides raw data: pseudorange, delta range, carrier phase and associated navigation data
Navigation processing	Converts raw data to user data – location, velocity and time, as required

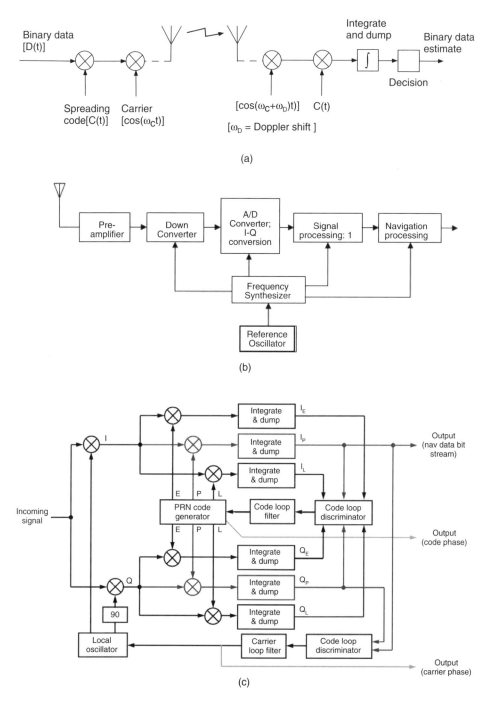

Figure 13.5 (a) Conceptual transmit-and-receive arrangements for a direct sequence CDMA navigation system. Adapted from (Braasch and Dierendonck, 1999). (b) The main components of a navigation receiver. Table 13.4 lists details of each block. (c) A typical digital pseudorange navigation receiver channel DSP functionality. Reproduced by permission of D. Plaušinaitis of the Danish GPS Centre.

achieved by a PRN code phase detector, which controls the phase of the local code generator. A typical phase detector correlates the received signal with multiple versions of the locally generated code to estimate the optimum phase.

Figure 13.5(c) illustrates a Digital Signal Processor (DSP) based implementation scheme (Plaušinaitis, 2010). The I and Q channels (explained later) are each split into three parallel signal paths, corresponding to three different code phase (bit transition) alignments: early, normal and late. The received baseband signal in each path is correlated by an estimate of the pseudorandom code – the PRN code transition phase being advanced for the early path and delayed for the late path by a fraction of a PRN chip. The output of the subsequent integrate-and-dump stage will be a maximum for the correct code alignment, and the code tracking error will be proportional to the difference between early and late samples. Hence, for the correct code alignment, their difference becomes zero. A code loop discriminator controls the precise adjustment of the PRN code generator phase in a Delay Locked Loop (DLL). Similarly, a carrier loop discriminator compares the main samples and controls the carrier phase in a Phase Locked Loop (PLL).

Antennas in personal navigation devices tend to use a single-patch microstrip. Both dual- and single-frequency converters are in use. Single-frequency designs support only the C/A signals at L1 (1575.42 MHz), whereas dual-frequency designs support L1 and L2 (1227.6 MHz). Analogue-to-Digital (A/D) converters may be single-bit (i.e. hard limiting), multi-bit and adaptive threshold types. The signal is split into in-phase (I) and quadrature (Q) phase to facilitate demodulation – this function can occur either before or after the A/D converter. Signals from each satellite are processed sequentially in single-channel design, leading to slow acquisition (1–20 min), whereas in a multichannel receiver several satellites can be processed in parallel. With large improvements in DSP processing power and lowering in cost, single-channel receivers have now been superseded by the multichannel architecture. Stage 1 of signal processing provides raw data; these data are further processed in stage 2 to extract user location, time and velocity.

Reacquisition performance is vital in situations where satellite signals are lost because of shadowing. For example, vehicular and urban users lose satellite signal regularly through shadowing. The reacquisition time depends on the combined effects of the local oscillator frequency drift and the Doppler introduced by platform movement.

The accuracy of a fix depends on the internal and external errors. The dominant sources of errors due to external factors are associated with ephemeris estimation uncertainty and refractive index fluctuations in the ionosphere, and those attributed to the receiver are caused by thermal noise, multipath and local interference. The high-end receivers mitigate these receiver errors by incorporating measures such as increased receiver sensitivity, a high-resolution A/D converter, an oven-controlled local oscillator, and high-performance carrier and code-tracking loops.

Contrary to the military receivers, where jamming can be intentional, interference in personal systems arises from a variety of unintentional interfering sources. For instance, harmonics from local transmitters are known to jam navigation receivers. Considerable numbers of interference events have been reported on the European continent. Amateur radio transmitters operating in 1240 MHz are known to cause complete blockage of GPS L2 frequency transmission (Butsch, 1997). The degradation in performance depends on the level and type of interfering source and the receiver. The interference rejection capability of spread spectrum becomes relatively ineffective for C/A code owing to the low level of received signals. Even with the expected rejection capability of about 40 dB, the received power level (of the order of −160 dBW) is susceptible to narrowband carriers of level −120 dBW, that is, 1 pW (Ward, 1996). To minimize the impact of out-of-band interference, the front end employs narrowband filters. Another countermeasure is to the use multi-bit A/D architecture, which provides a better rejection of narrowband interference.

In a multipath environment the received signals comprise a direct component and other components caused by reflections and scattering due to local environment (trees, vehicles, road surface, buildings, water bodies, etc.). The multipath noise induces errors in estimation of phase and range because of a degradation of carrier and code-tracking performance, and in carrier-phase estimation.

Multipath signals arriving from distances of $>20-30$ m can be cancelled in the receiver by echo cancellation and similar techniques, but those arising from shorter distances cannot be cancelled as they get embedded within the main signal. Theoretical assessment indicates that the range error due to severe multipath can approach $\sim$150 m; however, more likely errors are of the order of $\leq$10 m – provided the receiver is away from large obstacles such as skyscrapers where the pseudorange errors may approach 100 m (Van Nee, 1995). This order of error is unacceptable in various applications – for instance, personal or in-car navigation systems.

An alternative solution can be to use multipath-rejecting antennas. The antennas should be designed so that signals emanating below the local horizon (where the majority of local multipath arises) are mitigated. Another aspect is to ensure that antenna mounts and other objects do not perturb the antenna pattern, which could potentially lead to direction-dependent pseudorange error (Counselman, 1999). A choke ring ground plane is a common design where the antenna is surrounded by concentric rings which suppress tangential signals. Signals arriving from low elevation angles are cancelled by the rings, thereby rejecting multipath that emanates from low and negative elevation angles (Bailey, 1977). Conventional techniques may not be adequate for specialized application like that of a geodetic survey, where a tolerance of typically $\leq 2°$ in phase (1 mm of path length) is necessary. Improved designs have been proposed in the literature for specialized applications. An arrangement proposed by Counselman (1999) comprising a lightweight, compact, three-element vertical array antenna of 0.1 m diameter and 0.4 m high radome rejects multipath interference by 5 dB more than a conventional, 0.5 m diameter, ground plane antenna developed for precise differential GPS applications.

Receivers are implemented in various ways. Increasingly, DSP-based software solutions are replacing hardware components, leading to programmable receivers capable of in-field reconfiguration. This would allow receiver reconfiguration and/or offer the possibility of combining fixes from two or more systems. Moving the software functionality further up towards the front end leads to the concept of a software radio. Thus, in order to cater to various user profiles and system alternatives, a multitude of advanced architectures is likely to emerge as new GNSS systems are introduced and the existing ones enhanced.

GNSS Receivers

The conglomeration of global navigation systems is generically known as Global Navigation Satellite Services (GNSS). It is envisaged that, with the availability of a growing number of systems, the user will have the opportunity to benefit from a more robust navigation solution by using signals from more than one system. Thus, one of the goals of the receiver manufacturers is a generic design with the capability of multisystem operation. Design challenges in this emerging paradigm include management of different signal formats, new and diverse frequency bandwidth range, strategies for synergistic solutions, sensible cost-complexity trade-off, etc. In the following paragraphs, we highlight issues related to some of the more significant architectural aspects (Dempster, 2007).

Multifrequency Reception

Figure 13.6 illustrates the numerous frequency bands covered by the three prime GNSS candidates – GPS, GLONASS and Galileo. Some of the enhanced aspects include:

- second civilian transmissions from GPS satellites;
- second frequency transmissions for civilians from GLONASS satellites;
- galileo transmissions on four civil frequencies.

Thus, there arises the need to receive multiple transmissions from the same constellation. Taking GPS as an example, the extension of capability from L1 to L2C and possibly L5, which also offers

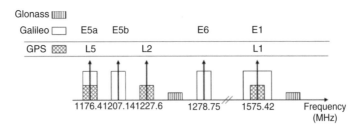

Figure 13.6 Frequency of transmission of GPS, Galileo and GPS navigation systems (after Demp-ster, 2007). Reproduced by permission of © IEEE.

a wider bandwidth, would clearly be advantageous in terms of pseudorange accuracy, redundancy and ionospheric error correction. Extending a receiver's reception to more than one system offers the potential to improve the accuracy and reliability, but they must be traded against the added complexity in receiver architecture and an increase in size, power consumption and cost. Existing designs for low-end receivers tend to use a microstrip patch antenna, which, being narrowband, typically cannot provide a multiband facility. A dual- or triple-band system would, for example, necessitate stacking of two or more patch antennas, thereby increasing receiver cost and size. However, alternative designs are evolving that may not be ideal at present, but one would expect them to be refined and optimized in the future – for instance, one proposed multiband design comprises a 4.7" square antenna (Rao *et al.*, 2002).

A multiband down-converter necessitates a number of bandpass filters with sufficiently high rejection to avoid interference from local oscillators. Correlation is one of the core functionalities of the receiver. Its performance depends on the signal-to-noise ratio and code characteristics. In a mixed signal environment, the code parameters differ, and hence performance optimization criteria require a different optimization strategy. For example, the GPS L2C signals are received at lower power but a larger bandwidth, and consequently, in a combined L1 and L2C receiver, the L1 signals would be acquired earlier. The newer signal, described in Chapter 14, include dataless parts in the signals to facilitate improved correlation. Modulated data tend to cause degradation during the correlation. The search space of longer code lengths is larger, which leads to a longer acquisition time. Similarly, tracking of multiple systems requires different solutions depending on the signal characteristics. Strategies likely to be of interest include ways to offer mutual assistance in acquisition, tracking and carrier-phase positioning.

Combined Navigation–Communication Receiver Architecture

Owing to a growing awareness of location-dependent applications, there is an interest in augmenting a navigation receiver system with communication capability. In fact, embedding GPS chips in consumer items such as cell phones is now common. In a more generic sense, it would be quite beneficial if a navigation receiver could integrate seamlessly with a variety of communication systems. In this section, we present an example of a hybrid receiver that integrates a multiple navigation receiver with a flexible multistandard communications transreceiver for road applications (Heinrichs and Windl, 2000).

The receiver consists of a combined GPS–GLONASS receiver interfaced to a multistandard communication system at a chip level. The terrestrial components of the communication systems included support to numerous standards: Tetra/Tetrapol (925–960 MHz), GSM 900 (925–960 MHz) and DCS 1800/DECT (1710–1900 MHz). The satellite communications systems support comprised: Iridium and Inmarsat (1525–1626.5 MHz), Globalstar (2483.5–2500 MHz), ICO (1980–2010 MHz) and Orbcom (137–138 MHz).

An analysis of user needs identified the application areas of fleet management and differential correction transmissions as the dominant applications. The market for position tracking and traffic control were found to be speculative. These dominant applications would need a maximum bit rate of 9.6 kb/s and would work with a partner GSM system. Fleet management was considered as the only application with a clear need for satellite communications.

For developing a dual-mode integrated receiver, it would be necessary to estimate the processing needs, and hence it was necessary to establish whether concurrent processing would be necessary or an alternate processing scheme – that is either navigation or communication at a given instant – would suffice.

It was concluded that satellite systems could be integrated to the navigation system with the least complexity of all. The highest integration would be feasible if navigation and communication functions were not conducted concurrently. The transmitter concepts were also studied but are not included here. Conclusions of the various alternatives are summarized as follows:

1. *Navigation processing functions continuously.* The architecture consisted of five existing (or planned) mobile satellite systems – Inmarsat, Iridium, Globalstar, Orbcom and (the then planned) ICO, combined with a GPS–GLONASS GNSS system. Assuming that only one satellite communication link would be used at a time, only those systems that operate in bands with close proximity could share the front end – the Inmarsat-C system and Iridium frequency bands were considered to be close enough. ICO, Globalstar and Orbcomm operate at widely separated bands and hence would require individual front ends. However, assuming that only one communication channel is used at a time, the hardware for units beyond the front end could be shared with addition of a correlator for the Globalstar system. Further, owing to the heavy processing needs of the navigation unit, it would be possible for navigation and communication systems to share only the baseband processor.
2. *Some correlators used for navigation signal processing can be shared with communication processing.* The front ends have a similar configuration to that in alternative 1. However, some digital processing functions are shared. After the initial acquisition of navigation signals, when processing is most intensive, the freed processing power of the correlation processors is utilized for communication processing. The correlation processors used for multi-tasking must have additional flexibility to support programmable digital filters with loadable coefficients, and correlators with reconfigurable code generators of adjustable code length.
3. *Navigation on the L2 band can be stopped for communication.* This option provides the highest level of hardware integration. Digital processors are shared; the front ends for Iridium, Inmarsat-C and GPS L2 band are combined.
4. *Navigation combined with satellite and terrestrial communications.* This alternative was considered as feasible. Architectures that use IF sampling were considered as a promising way forward. The referred paper (Heinrich and Windl, 2000) presents a possible concept of 'combining the RF sampling architecture for navigation with a low IF sampling structure for the reception of satellite-based and terrestrial standards'.

13.8   Distress, Safety and Location-Based Services

Owing to their vast coverage area, satellite systems are ideally suited to assist people in distress in the remotest places of the world where terrestrial systems are inaccessible. Hence, considerable international effort has been harnessed in the development and maintenance of the distress and safety system (also referred to as the search and rescue system). These systems – because of their international nature and the humanitarian perspective – are given the highest regulatory status internationally.

We observed that location-based services have received considerable attention in recent years in commercial circles owing to the massive proliferation of mobile communication services and a dramatic reduction in GPS processing chips. Advanced services and solutions have been the subject of considerable research activity in recent years on account of the huge potential they offer to the personal technology arena.

13.8.1   Distress and Safety Service

The Global Maritime Distress and Safety System (GMDSS), regulated by the UN (IMO/GMDSS, 2009) is at the forefront in provision of distress and safety wireless services globally. Regional and local systems that offer this type of service also exist. They are governed either through commercial arrangements or by local authorities responsible for safety.

Consider that an alert is triggered by a party in distress. The alert is communicated over a radio communication system to a search and rescue centre which takes the appropriate action in aid of the distressed party. The distressed party may be an individual, a pilot, a ship or a vehicle. A variety of radio communication systems are used, depending on the infrastructure of the area and the user preference.

Another type of service, called the Personal Locator Service (PLS) dealing with safety and distress applications specifically to individuals, is also of interest. A PLS service assists in locating people – for instance, a child or a mentally handicapped individual. In this section, we discuss the architectures of:

- the Global Maritime Distress Safety System (GMDSS);
- a satellite-enabled personal locator/safety system (ESA, 2005).

In Chapter 14 we address the implementation aspects.

Global Maritime Distress Safety System (GMDSS)

To ensure safety of life at sea, the International Maritime Organization (IMO) completed implementation of the Global Maritime Distress and Safety System (GMDSS) on 1 February 1999 (IMO/GMDSS, 2009). GMDSS regulations are defined in the International Convention for the Safety of Life at Sea (SOLAS). Its compliance is mandatory for ships above 300 gross tonnage. The system provides an alert to ships that lie in the vicinity of a distressed ship, and to rescue authorities on shore. Various types of communication are used – distress alerting, search and rescue coordination communications, signal location, transmission–reception of maritime safety communication and general radio communication. The communication equipment is recommended on the basis of the sea areas (categorized as A1 to A4 – see below). In addition there are various associated requirements to ensure that communication can be established reliably (e.g. availability of trained personnel, equipment or system redundancy, back-up power supply, etc.). Figure 13.7 illustrates the GMDSS concept.

Although primarily a maritime service for ships, the GMDSS system can support boat, aircraft and individual originated distress calls through the Inmarsat-C or Cospas-Sarsat satellite systems (Cospas-Sarsat, 2009) the two satellite systems constituting a part of the GMDSS.

The users should be equipped with appropriate equipment in the area of operation, designated as follows:

- Sea area A1 lies within coverage of at least one equipped VHF coast station; typically, this region extends up to 30 km out to sea.

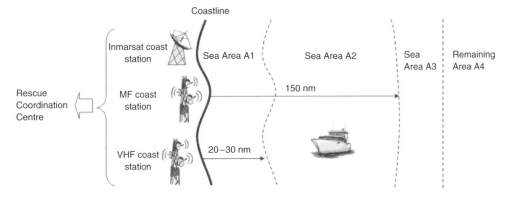

Sea Area A1: within range of shore-based VHF coast stations
Sea Area A2: within range of shore-based MF coast stations, excluding A1
Sea Area A3: within Inmarsat satellite system coverage
Sea Area A4: regions extending beyond A1–A3, mainly the polar region

Figure 13.7 The GMDSS concept.

- Sea area A2 excludes sea area A1 and lies within coverage of at least one equipped medium-frequency (MF) coast station. This area can extend up to 740 km in practice, but a conservative figure would be 190 km.
- Sea area A3 is an area excluding sea areas A1 and A2 within coverage of a geostationary satellite footprint that extends up to about ±76° latitude.
- Sea area A4 is the area outside A1, A2 and A3. For practical purposes, this covers the region beyond the coverage of a geostationary satellite system.

Areas A3 and A4 are serviced by the Inmarsat and Cospas-Sarsat satellite systems respectively. Chapter 14 introduces these two systems.

A Satellite-Enabled Personal Safety System

The concept of a personal safety system called the Personal Safety System Utilizing Satellite combined with Emerging Technologies (PERUSE) is being developed with the sponsorship of the European Space Agency's Advanced Research in Telecommunications Systems (ARTES) programme, elements 3 and 4 (ESA, 2005). The project aims to develop a wireless personal alarm (WPA) system that a user can activate in the case of a personal emergency. The alert includes the user's identity and location and supplementary information such as the user's health condition, local environmental parameters, etc. The WPA would be worn on the person – hence, it should be small and stylish with a long battery life. The implication was that the device would be unable to generate enough power for direct communication with a satellite.

Figure 13.8 illustrates the concept developed in the project's framework. The WPA message is transferred through a low-power radio link to a satellite phone and thence to the appropriate destination via the mobile satellite service (MSS) network. The destination may be an emergency assistance centre, a parent or a personal attendant. The prototype uses the Thuraya MSS system along with a modified Thuraya phone. The specific task undertaken was the development of the WPA and its radio interface to the Thuraya phone.

The concept can alternatively be implemented through a terrestrial cellular system where WPA would be replaced by a cellular phone. The location fix can be obtained by schemes other than

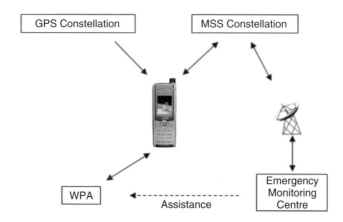

Figure 13.8 Block schematic of a personal safety system utilizing satellite combined with emerging technologies.

GPS, such as base-station triangulation (Koshima and Hoshen, 2000). These types of terrestrial system are widely used in Japan for child monitoring, location of handicapped relatives or even locating criminals in metropolitan areas.

13.8.2 Location Based Service

Wireless service operators offer a wide variety of value-added services to enhance the utility of their networks. An example of note is the phenomenally successful text and multimedia messaging service on cellular phones. The demand for real-time personalized services tailored to the user's location is growing steadily. Fleet management service is a prime example of this class of service; so also is the personal locator service described in the preceding section. These services may be classified as emergency (e.g. security alert), informational (e.g. news), tracking (e.g. fleet management) and entertainment/social (e.g. games) (Mohapatra and Suma, 2005).

Figure 13.9 shows the framework of a network to support a location-based service (adapted from Krevl and Ciglaric, 2006). User location is determined by one or more methods. The user's mobile unit connects with a gateway on a satellite or terrestrial radio link. The gateway interfaces with the Internet which supports a database server, a web server and an applications server. The database server contains the location map or a geographical database and associated information on offer (e.g. tourist spots, roadside service stations); the applications server extracts the information and formats it; the web server presents the information on a website in a user-friendly visual or text format.

Location-based services are generally transported over terrestrial cellular systems and may utilize a variety of location determination methods – satellite based (GPS or a GPS derivative such as Assisted GPS), network based (Cell Identification, estimated observed time difference between two cell sites, etc.) or local (e.g. Bluetooth or Wi-Fi assisted). Similar services are also conveyed by satellites, but their penetration is rather scarce at present.

Satellite positioning methods are accurate but typically cannot work indoors. Assisted GPS provides a more accurate estimate using a network of ground stations to augment the accuracy. Network positioning methods provide better accuracy when cell sizes are small, but the best accuracy would typically be of the order of 100 m. Local positioning (e.g. base station triangulation) methods are accurate but, owing to their limited coverage area, are best suited for indoor application. Hence, a hybrid location-dependent solution would appear to be an optimal approach.

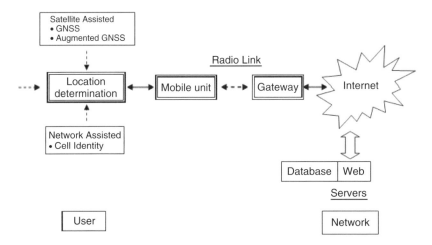

Figure 13.9 The architecture of a location-based service with an example of the location determination technique.

Location-based services may be user requested or automatic – subject to compliance with a condition (e.g. the cell phone roaming facility). In a user-requested application, the user's location is determined (this may be embedded in the request itself), the position is mapped onto a geographical grid, the requested information is extracted from a database and the desired information is sent back to the user. Now consider the functionality of a fleet management system in which vehicles are programmed to send their position automatically at regular intervals. A vehicle determines its position immediately prior to sending the information. The received information is sent over a transmission medium (e.g. the Internet) and presented to the fleet manager in an easily comprehensible display (such as a symbol on a map).

Considerable research and commercial activity is in progress to develop, consolidate and refine the system architecture (Krevl and Ciglaric, 2006; Mohapatra and Suma, 2005; Zeimpekis, Giaglis and Minis, 2005; Choi and Tekinay, 2003). Some issues under investigation include concerns regarding the security and privacy of users, protocols and interface definition for interaction between the functional entities, and their standardization to accrue the inherent advantage of open standards.

Revision Questions

1. The material in this chapter (together with Chapter 7) has outlined the principles and techniques of two well-known satellite navigation systems. (a) Outline the principle of a Doppler-assisted satellite navigation system. (b) Discuss the sources of errors in this scheme and suggest suitable countermeasures. (c) Suggest environments and applications suited to this type of navigation system, mentioning the reason for your choice.
2. Repeat 1(a) to 1(c) for a range-assisted satellite system.
3. (a) Discuss the principle of a differential GPS system. (b) Outline the principle of a satellite-based augmentation system to demonstrate the applicability of differential GPS principles.
4. A distress and safety communication system provides access to search and rescue coordination facilities, while a system that transports location-based communication services attaches the user to an application. Highlight the differences between the architecture of these classes of system on the basis of the examples given in the chapter.

References

Argos User Manual (2008) http://www.argos-system.org/html/userarea/manual_en.html [accessed June 2009].

Ashby, N. (2002) Relativity and the gloabl positioning system. *Physics Today*, May.

Ashby, N. (2004) The Sagnac effect in the global positioning system, in *Relativity in Rotating Frames*, ed. by Rizzi, G. and Ruggiero, M.L. Kluwer. Dordrecht, The Netherlands.

Bailey, M. (1977) A broad-beam circularly-polarized antenna, Dig. IEEE Antennas and Propagation Society (AP-S), Int. Symposium, Stanford University, Stanford, CA, June, pp. 238–241.

Braasch, M.S. and Van Dierendonck, A.J. (1999) GPS receiver architectures and measurements. *Proceedings of the IEEE*, **87**(1), 48–64.

Butsch, F. (1997) GPS interference problems in Germany, in Proceedings of Institute of Navigation Annual Meeting, Albuquerque, NM, June, pp. 59–67.

CAA (2004) *GPS Integrity and Potential Impact on Aviation Safety*, CAA Paper 2003/09, April. Available: http://www.caa.co.uk/docs/33/capap2003_09.pdf [accessed November 2009].

Choi, W.J. and Tekinay, S. (2003) Location based services for next generation wireless mobile networks. Vehicular Technology Conference, VTC 2003-Spring, The 57th IEEE Semiannual, 22–25 April, Vol. 3, pp. 1988–1992.

Cospas-Sarsat (2009) http://www.cospas-sarsat.org/Description/overview.htm [accessed June 2009].

Counselman, C.C., III, (1999) Multipath-rejecting GPS antennas. *Proceedings of IEEE*, **87**(1), 86–91.

Dempster, A.G. (2007) Satellite navigation: new signals, new challenges. ISCAS 2007, IEEE International Symposium on *Circuits and Systems*, 27–30 May, New Orleans, LA, pp. 1725–1728.

ESA (2005) (http://telecom.esa.int/telecom/www/object/index.cfm?fobjectid=14443) [accessed June 2009].

ESA/EGNOS (2009) http://www.egnos-pro.esa.int/br227_EGNOS_2004.pdf [accessed September 2009].

EVP Europe (1999) *A Beginner's Guide to GNSS in Europe*. Available: http://www.ifatca.org/docs/gnss.pdf [accessed August 2009].

FANS(II)/4 (1993) FANS Phase II Report; ICAO Document 9623, Montreal, Canada, September.

Flament, P. (2004) Galileo: a new dimension in international satellite navigation. Proceedings Elmar 2004, Electronics in Marine, 46th International Symposium, Zadar, Croatia, 16–18 June, pp. 6–14.

Fliegel, H.F. and DiEsposti, R.S. (1996) GPS and RElativity: An Engineering Overview. Available: http://Tycho.usno.navy.mil/ptti/1996/vol%2028_16.pdf [accessed February 2010].

Goebel, G. (2009) *International Navigation Satellite Systems*. Available: http://www.vectorsite.net/ttgps_2 .html#m1 [accessed June 2009].

Goth, G. (2007) You are here. *IEEE Distributed Systems Online*, **8**(5), art. no. 0705-o5007, 7 pp.

GPS ICD (1984) NAVSTAR global positioning system interface control document, ICD GPS-200, Rockwell International Corporation.

Guier, W.H. and Weiffenbach, G.C. (1960) A satellite Doppler navigation system. *Proceedings of the IRE*, **48**(4), 507–516.

Heinrichs, G. and Windl, J. (2000) SANSICOM – combination of GNSS satellite navigation and multi-standard communication for road applications. EUROCOMM 2000 *Information Systems for Enhanced Public Safety and Security*, IEEE/AFCEA, Munich, Germany, pp. 106–110.

ICAO (2000) Validated ICAO GNSS Standards and Recommended Practices (SARPS), November.

IMO/GMDSS (2009) http://www.imo.org/About/mainframe.asp?topic_id=389 [accessed June 2009].

Kaplan, E.D. and Hegarty, C.J. (eds) (2006) *Understanding GPS Principles and Applications*, 2nd edition. Artech House, Norwood, MA.

Kayton, M. (2003) *Navigation: Land, Sea, Air, Space*. Presented to Long Island Chapter, IEEE Aerospace and Electronic Systems Society, 20 November. Available: http://www.ieee.li/pdf/viewgraphs/navigation.pdf [accessed May 2009].

Klobuchar, J.A. (1987) Ionospheric time delay algorithm for single-frequency GPS users. *IEEE Trans, Aerospace and Electronic Systems*, **AES-23**(3).

Koshima, H. and Hoshen, J. (2000) *Personal Locator Services Emerge*, IEEE, February, pp. 41–48. Available: http://telecom.esa.int/telecom/www/object/index.cfm?fobjectid=14443 [accessed February 2010].

Krevl, A. and Ciglaric, M. (2006) A framework for developing distributed location based applications. Parallel and Distributed Processing Symposium, IPDPS 2006, 20th International, 25–29 April, 6 pp.

Levanon, N. and Ben-Zaken, M. (1985) Random error in ARGOS and SARSAT satellite positioning systems. IEEE Trans Aerospace and Electronic Systems, **AES-21**(6), 783–790.

Lewandowski, W., Klepczynski, W.J., Miranian, M., Griidler, P., Baunlont, F. and Inlae, M. (1994) Study of troposheric correction for intercontinental GPS common-view time transfer. Proceedings of the precise time and time interval (PTTI) Systems and Applications (formerly Applications and Planning) Meeting.

Misra, P. Burke, B.P. and Pratt, M.M. (1999) GPS performance in navigation. *Proceedings of the IEEE*, **87**(1), 65–85.

Mohapatra, D. and Suma, S.B. (2005) Survey of location based wireless services. IEEE International Conference on *Personal Wireless Communications*, 23–25 January, New Delhi, India, pp. 358–362.

Nelson, R.A. (2007) Relativistic time transfer in the solar system. IEEE International Frequency Control Symposium, 29 May–1 June, Geneva, Switzerland, pp. 1278–1283.

NOAA (2009) http://www.ngs.noaa.gov/FGCS/info/sans_SA/compare/ERLA.htm [accessed May 2009].

Plaušinaitis, D. Online, 'Code Tracking; Multipath', Course: GPS Signals and Receiver Technology MM13, Danish GPS Center, Aalborg University, Denmark. Available: http://kom.aau.dk/~dpl/courses/mm13_slides.pdf [accessed February 2010].

Rao, B.R., Smolinskii, M.A., Quach, C.C. and Rosario, E.N. (2002) *Triple Band GPS Trap Loaded Inverted L Antenna Array*. MITRE Corporation. Available: http://www.mitre.org/work/tech_papers/tech_papers_02/rao_triband/ [accessed June 2009].

Richharia, M. (2001) Mobile Satellite Communications. Addison-Wesley, New York, NY.

Schänzer, G. (1995) Satellite navigation for precision approach: technological and political benefits and risks. *Space Communications*, **13**, 97–108.

Tiemeyer, B. (2002), Performance evaluation of satellite navigation and safety case development. *PhD Thesis*. Available: http://ub.unibw-muenchen.de/dissertationen/ediss/tiemeyer-bernd/inhalt.pdf [accessed November 2009].

Van Nee, R. (1995) Multipath and multi-transmitter interference in spread-spectrum communication and navigation systems. *PhD. Dissertation*, Faculty of Electrical Engineering, Delft University of Technology, Delft, The Netherlands, 1995.

Ward, P. (1996) Effects of RF interference on GPS satellite signal receiver tracking, in *Understanding GPS: Principles and Applications*, ed. by Kaplan, E. Artech House, Boston, MA, Ch. 6, pp. 209–236.

Zeimpekis, V., Giaglis, G.M. and Minis, I. (2005) A dynamic real-time fleet management system for incident handling in city logistics. 61st IEEE Vehicular Technology Conference, 30 May–1 June, Stockholm, Sweden, pp. 2900–2904.

14

Navigation, Tracking and Safety Systems

14.1 Introduction

In this chapter we address the application of the concepts introduced in Chapter 13. We describe prominent global navigation systems, emphasizing the GPS (Global Positioning System) because of its wide acceptance and appeal in the consumer mainstream. GLONASS (English translation: Global Navigation Satellite System) is a rival satellite navigation system that, after a period of decline, is beginning a revival. Galileo is a European system in development, aiming to complement the GPS and provide an independent civilian solution.

The global navigation systems are generically known as the Global Navigation Satellite System (GNSS). It is anticipated that, in future, users will be able to utilize the harmonized benefits of the GNSS to provide increased reliability and accuracy.

The Argos system, while not as robust as GPS and other global systems, contributes to numerous international environment and personal safety initiatives throughout the world. The GNSS relies on the range-assisted navigation principle, whereas the Argos system is based on the Doppler-assisted principle.

Various regional navigation systems are in development. A regional system provides more local control and avoids dependence on other countries. It also offers technical and commercial opportunities to the local industry. They can be deployed relatively economically, as the service area is confined and the accuracy requirements can be tailored for civilian/commercial applications. We describe the Chinese system Beidou and Compass (planned), and GAGAN, a system planned by India.

Next we introduce two satellite-based augmentation systems (SBASs). The Wide-Area Augmentation System (WAAS) is an operational US system, and the European Geostationary Navigation Overlay Service (EGNOS) is a European system under construction. In Chapter 13 we observed that SBASs provide the extra accuracy and integrity necessary in certain applications, for example, flight safety, road transport and personal navigation such as a guidance aid for visibly impaired individuals.

Distress and safety communication was one of the earliest humanitarian contributions of satellite systems; we discuss two distinguished global systems responsible for saving thousands of lives around the world – the COSPAS-SARSAT system and the Inmarsat system.

The concluding section of the chapter presents a case study of a fleet management system – a well-established application of a hybrid navigation–communication system architecture. EutelTracs is a European system that provides a variety of location-based services.

Satellite Systems for Personal Applications: Concepts and Technology   Madhavendra Richharia and Leslie David Westbrook
© 2010 John Wiley & Sons, Ltd

14.2   Global Navigation Satellite Systems

The International Civil Aviation Organization (ICAO) has defined the Global Navigation Satellite System (GNSS) applied to civil aviation as 'a worldwide position and time determination system that includes one or more satellite constellations, aircraft receivers and system integrity monitoring, augmented as necessary to support the required navigation performance for the intended operation' (ICAO, 2007). This definition is general enough to be extended to other spheres where satellite navigation systems are applicable – including personal navigation systems.

According to the definition of the ICAO standards, GPS and GLONASS fall in this category, while Galileo, in development, is likely to be included in due course. Other systems in various stages of development may be GNSS candidates in future (Hegarty and Chatre, 2008).

14.2.1   Global Positioning System (GPS)

14.2.1.1   Overview

The GPS system is owned and operated by the US military. However the system is instrumental in promoting the uses of satellite navigation technology to the civilian market.

The first major test of the military effectiveness of GPS under combat conditions came with the first Gulf War in 1990–1991. Indeed, such were the benefits of GPS in desert warfare that the demand for GPS receivers far outstripped the available military inventory, and more than 10 000 additional civilian receivers were purchased for use by coalition forces (Rand Corporation, 2007). Today, such is the ubiquity and utility of GPS to the military, the loss of GPS would have an almost inconceivable impact on the conduct of military operations.

From 1980, GPS satellites have had a dual role, with the inclusion of a Nuclear Detonation (NUDET) sensor payload on the spacecraft, which is used to monitor the testing of nuclear weapons anywhere in the world, and thus help enforce nuclear test band treaties. All current GPS satellites incorporate a NUDET payload.

GPS is based on the range triangulation method described in Chapter 7. The system uses range estimates from four satellites to obtain latitude, longitude and altitude of a location on the Earth or in space (within visibility and power limits). Range obtained from three satellites is sufficient to obtain latitude, longitude and altitude. Range from the fourth satellite provides user clock correction. It is also feasible to obtain user velocity.

Users obtain a fix by determining their range to each satellite by a correlation technique aided with accompanying data. The process involves transmission of spread-spectrum pseudorandom noise (PRN) sequence from each satellite. The received signal is correlated with an identical locally generated code provided that the code characteristics are known at the receiver. This feature also provides the encryption essential in military systems (by using a secret PRN code); furthermore, it minimizes the interference potential between GPS satellites and provides resistance to RF multipath. The travel time is obtained by comparing the instant of code transmission with the received instant, that is, when the correlator peaks. Correlation is achieved by time sliding the local sequence against the received signal until the correlator peaks. The transmission instant is time stamped and embedded within the accompanying navigation data which are multiplexed with the PRN sequence.

Replenishment of the GPS satellite constellation took place with the launch of 12 Block IIR (Replenishment) satellites between 1997 and 2004. Block IIR satellites include an ability autonomously to generate their own navigation message data, should the link with the GPS ground segment be unavailable. Subsequently, six Block IIR-M (Replacement-Modernized) satellites, which incorporate additional navigation signals – one new military signal on L1 and L2 (M-code) and a new civil signal on L2 (L2C) – have been successfully launched.

In order to improve correlator accuracy, the new signals include a dataless portion to assist in tracking of the received signal, and in addition introduce a robust coding scheme to the navigation data to facilitate low-error data demodulation. The next series of satellites, called Block IIF, add a new carrier at 1176.45 MHz, known as Link 5 (L5), at 10.23 MHz chipping rate for civil users. The addition of the new signal will provide a more accurate estimate of ionospheric correction as well as improving radio link reliability and accuracy.

The next-generation satellites Block III, expected to be available in 2014, will have one more signal at L1, called L1C, at 1.023 MHz chip rate. Block II and Block IIA satellites are equipped with two rubidium and two cesium atomic clocks each (White, Rochat and Mallet, 2006). Block IIR and IIR-M are equipped with three improved rubidium atomic clocks (White, Rochat and Mallet, 2006), while Block IIF satellites are expected to be equipped with two rubidium clocks and one cesium clock.

Block IIF GPS satellites are due to be launched from 2010, and will add a third civilian GPS frequency specifically for use in aeronautical applications. Nevertheless, the US Air Force has also begun procurement of the next generation of Block III satellites, with the first Block III launch planned for 2014. The first Block III GPS satellites – Block IIIA – will provide new signals for improved interoperation with other GNSS, notably Galileo, as well as increased transmitter power (Goldstein, 2009). Subsequent Block IIIB satellites are expected to introduce satellite cross-linking for improved system integrity, while Block IIIC will provide a spot beam capability for increased power budget in order to overcome regional interference. Eight GPS IIIB and 16 GPS IIIC satellites are planned. Block III satellites are expected to use a combination of rubidium clocks and passive hydrogen masers.

14.2.1.2   GPS Space Segment

The architecture of GPS can be conveniently categorized into space, ground control and user segments. The space and control segments are both under the control of the US Air Force, Space Command.

The GPS satellite constellation comprises a minimum of 24 operational satellites, usually referred to as Space Vehicles (SVs). Individual satellites are identified by their SV number (SVN1, SVN2, etc.) and their radio signals by their PRN code index (PRN1, PRN2, etc.). The satellites are located in nominally circular inclined Medium Earth Orbit (MEO), in six equally spaced orbital planes, each having an inclination of approximately 55° relative to the equatorial plane and with nominally four satellites in each plane. This arrangement (which is denser than a Walker constellation) is illustrated in Figure 14.1. The GPS orbital planes are labelled 'A' to 'F', while the satellite slots in each plane are nominally numbered 1 to 4, although there are today typically more than four active satellites per plane.

GPS satellites orbit at an altitude of approximately 20 200 km, and each satellite makes two orbits in a sidereal day (the period in which the Earth completes one rotation about its axis with respect to the stars). The resulting orbital period in 'Earth' time is 11 h 58 min. The satellites are therefore located approximately at the same point at the same time each day. The GPS satellite antennas provide a nominal service coverage angle of 41.3°, corresponding to a maximum user altitude of 3 000 km. Additional parameters for the GPS constellation are given in Table 14.1

14.2.1.3   GPS Control Segment

The GPS ground control segment comprises a Master Control Station (MCS) located at Schriever Air Force Base in Colorado Springs, Colorado, USA, with a back-up facility located in Gaithersburg, Maryland, USA, together with a number of L-band tracking stations and S-band ground (uplink) antennas. Satellites are tracked by six unmanned L-band monitoring stations. Each L-band monitoring station is equipped to collect satellite status and pseudorange, together with local meteorological data. These stations are now augmented by US National Geospatial-Intelligence Agency (NGA)

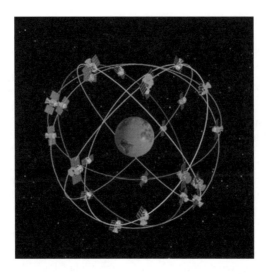

Figure 14.1 GPS constellation. Reproduced by permission of © Aerospace Corporation.

Table 14.1 GPS constellation summary

Parameter	Value
Number of satellites	24+
Orbit type	Circular Medium Earth Orbit (MEO)
Altitude	20 200 km
Constellation type	55°: 24/6
Orbital period	~12 h
Mass	1 816 kg (Block II/IIA), 2217 kg (IIR/R-M), 1 705 kg (IIF), 844 kg (III)
Predicted life	7.5 years (Block II/IIA/IIR/R-M), 12 years (IIF), 15 years (III)
DC power	800 W (Block II/IIA), 2450 W (IIF)
EIRP	26 dBW (L1 C/A code)
Transmit polarization	RHCP

GPS monitor stations (NGA GPS Division Web Page, 2010). Continuous tracking of all satellites is possible using these tracking stations. Four S-band ground uplink antenna stations are used to uplink GPS navigation messages and provide spacecraft Tracking, Telemetry and Command (TT&C).

The most recent GPS satellites incorporate precision optical retroreflectors, permitting them to be tracked using Satellite Laser Ranging (SLR). Use of retroreflectors for SLR was pioneered on GLONASS satellites to achieve satellite tracking with cm precision.

GPS System Time

GPS system time comprises a filtered, weighted mean of all the atomic clocks within the GPS system, comprising clocks on the GPS satellites and those in ground control stations (Moudrak *et al.*, 2005; Kaplan and Hegarty, 2006). GPS time is steered to (i.e. is gradually adjusted to track) Coordinated Universal Time (UTC) on a daily basis (UTC is, in turn, derived from International Atomic Time (TAI)).

GPS system time is referenced to midnight on 5/6 January 1980, UTC. However, unlike UTC, GPS system time does not include the addition of leap seconds. The GPS navigation message contains the GPS time (given as an integer GPS week number and the number of seconds since midnight on the previous Saturday/Sunday), together with the difference between GPS system time and UTC.

GPS Geodetic Reference
The GPS geodetic reference frame is that of World Geodetic System 1984 (WGS-84).

14.2.1.4 GPS User Segment

The user segment comprises the numerous GPS receivers (and associated antennas) used around the world by both civilian and military users. These range from highly accurate receivers used for surveying and geodesy to handheld consumer equipment.

Military GPS Mapping Terminal
The primary function of GPS is to support navigation and targeting for the military. A wide range of rugged military GPS receivers are available for use on aircraft, ships and land vehicles, and for use by dismounted troops. An example of a military handheld GPS receiver, the Defense Advanced GPS Receiver (DAGR), is illustrated in Figure 14.2. The DAGR is a 12-channel dual-frequency (L1/L2) GPS receiver, capable of direct acquisition of the encrypted GPS precision P(Y)-code. The DAGR 95% horizontal position accuracy using the military P(Y)-code is quoted as <10.5 m (DAGR Specifications). The unit can also use differential GPS signals, where available, and includes a mapping mode.

Civilian GPS Mapping Terminals and SATNAV
A single-frequency consumer handheld mapping GPS receiver that uses unencrypted C/A code is illustrated in Figure 14.3 (left). One of the most widely used consumer applications of GPS is for

Figure 14.2 Rockwell Collins DAGR handheld military GPS terminal. Reproduced by permission of © Rockwell Collins.

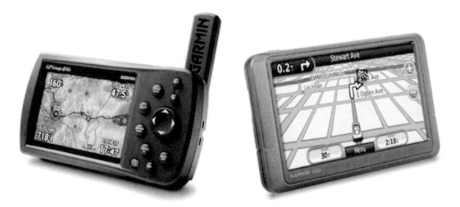

Figure 14.3 Consumer handheld GPS mapping terminal (left) and in-car SATNAV (right). Reproduced from © Garmin (Europe) Ltd.

in-car SATellite NAVigation (SATNAV), which combines a GPS map display with a route planner. An example GPS SATNAV unit is shown in Figure 14.3 (right). Consumer mapping GPS receivers are also available for use in small boats and private aircraft.

Miniature GPS Receivers
Basic GPS receivers are widely available for direct connection to laptop Personal Computers (PCs) and Personal Digital Assistants (PDAs), and work in conjunction with software loaded on the PC. These receivers are used to provide not only positioning information but also a highly accurate clock source. An example of such a receiver is illustrated in Figure 14.4 (left). The small size of the GPS receiver (and some modern GPS antennas) allows GPS receivers to be incorporated into body-worn items, such as the GPS-enabled sports watch illustrated in Figure 14.4 (right).

14.2.1.5 Navigation Signals

Block II GPS satellites broadcast, using CDMA, on two frequencies, known as L1 (1.57542 GHz) and L2 (1.2276 GHz). Each of these frequencies is obtained by multiplying up the 10.23 MHz

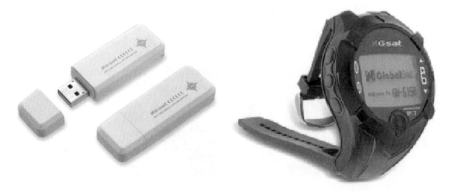

Figure 14.4 GPS receiver intended for use with a computer laptop (left) and a sports watch incorporating a GPS receiver (right). Reproduced from © GlobalSat.

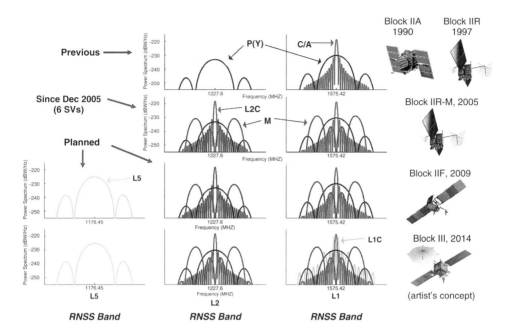

Figure 14.5 Current and future GPS signals (Goldstein, 2009). Top to bottom: Block IIA/IIR; Block IIR-M; Block IIF (2009+); Block III (2014+). Reproduced courtesy of U.S. Air Force.

frequency reference derived from the on-board atomic clocks (for example, $L1 = 154 \times 10.23$ MHz; $L2 = 130 \times 10.23$ MHz). These signals will, in future, be augmented by a further frequency L5 (1176.45 MHz), starting with Block IIF. The signals transmitted by current and planned GPS satellites are illustrated in Figure 14.5 (Goldstein, 2009).

GPS was originally intended to provide two levels of service:

- *Standard Positioning Service (SPS).* This service, available to all users, without charge, is currently provided by the so-called Coarse/Acquisition (C/A) waveform at the L1 frequency.
- *Precise Positioning Service (PPS).* This service is available to authorized (US military, US Federal and selected allies) users only, and is provided by the military P(Y)-code using both L1 and L2 frequencies. The P(Y)-code incorporates encryption for antispoofing.

GPS signal frequencies and corresponding wavelengths are indicated in Table 14.2.

On GPS satellites prior to Block IIR-M, the L1 signal comprises the Coarse/Acquisition (C/A) signal and navigation message, together with the military P(Y)-code signal, while the L2 signal

Table 14.2 GPS navigation signal frequencies and wavelengths (** indicates planned signal)

Signal	Frequency	Wavelength
L1	1.57542 GHz	0.190 m
L2	1.2276 GHz	0.244 m
L5**	1.17645 GHz	0.255 m

comprises the military P(Y) signal only. Block IIR-M satellites transmit two new codes, a second unencrypted civilian code at the L2 frequency (L2C) and a new military code (M-code) on both L1 and L2 frequencies. Block IIF satellites will transmit a new civilian signal, L5, intended for use in safety-critical applications, while, under the GPS modernization programme, Block III satellites will introduce a new civilian code at the L1 frequency.

The GPS navigation signals are:

- *Coarse Acquisition (C/A) signal.* The coarse/acquisition CDMA code transmitted on the L1 frequency using a relatively short repeating code is primarily intended to facilitate acquisition of the approximate code phase prior to acquiring the longer military P(Y)-code but has become the most widely used GPS signal. As the C/A signal is not encrypted, it provides the basis for the civilian standard positioning service. Each satellite uses a different, mutually orthogonal code, known as a Gold code for the C/A CDMA signal. Each code has a code length of 1023 chips and a period of 1 ms. The code bit rate is therefore 1.023 Mb/s. The modulation format is BPSK-R (BPSK with rectangular pulses).
- *Precision P(Y) signal.* The P-code navigation code provides the basis of the GPS military Precise Positioning Service (PPS). If required, the P-code may be encrypted to prevent spoofing (the transmission of a bogus GPS signal), and the encrypted P-code forms the so-called Y-code. The ability to decode the Y-code with the correct cryptographic key indicates the authenticity of the signal. The unencrypted P-code or encrypted Y-code is usually denoted as P(Y). The CDMA precision P-code signal is transmitted on both L1 and L2 frequencies, thereby allowing better mitigation of errors due to ionospheric delay. Each P(Y)-code is 6 187 100 000 000 chips long – that is much longer than the C/A code. However, all P(Y)-codes are different segments of the same code sequence. The P(Y)-code is transmitted at a chip rate of 10.2 Mb/s, thereby making code tracking more accurate and providing additional processing gain over the C/A code – yielding additional resilience to interference. The modulation format is BPSK-R. Because of its long length, it was anticipated that receivers would have difficulty synchronizing with the P(Y)-code and would thus need first to achieve coarse code synchronization using the L1 C/A code.
- *L2C signal.* From Block IIR-M, GPS satellites have had the ability to transmit a new unencrypted civilian GPS signal at the L2 frequency, known as L2C. L2C is currently planned to be available on 24 GPS satellites by 2016. These new signals are primarily intended to allow civilian users to benefit from the greater accuracy of dual-frequency GPS receivers. They also provide some redundancy in the event of local interference to the legacy L1 C/A signal. The new L2C code is longer than the existing C/A code. The code-tracking accuracy of civilian receivers using the C/A code, while generally very good, is ultimately limited by its relatively short code length. The L2C signal actually comprises not one but two PRN codes, the Civilian Moderate (CM) code and the Civilian Long (CL) code, with the two codes being time multiplexed together, on a chip-by-chip basis, to create a 1.023 Mb/s CDMA signal. The CM-code is 10 230 chips long, repeating every 20 ms. The CL code is longer still at 767 250 chips long, repeating every 1500 ms. Of these two codes, only the CM code is modulated by navigation data – thereby improving the PRN code autocorrelation accuracy. The modulation format for both CM and CL codes is BPSK-R.
- *L5 signal.* A significant limitation of the exploitation of GPS in safety-of-life application is the fact that the L2 GPS frequency is shared with other services in some parts of the world, and being unprotected the potential exists for interference from other legitimate spectrum users. For this reason, future Block IIF and Block III GPS satellites will provide a third navigation frequency, L5 (1.17645 GHz) in the (protected) Aeronautical Radio Navigation (ARN) band. L5 is currently planned to be available on 24 GPS satellites by 2018 (L5 = 115 × 10.23 MHz). The L5 signal will comprise two PRN codes of 10 230 chips modulated in phase quadrature. The I5 (In-phase L5) code will comprise a 10-bit PRN code known as a Neuman–Hofman code, while the Q5 (quadrature L5) code will comprise a 20-bit Neuman–Hofman PRN code. The L5 navigation

message, known as L5 CNAV, is modulo-2 added to the in-phase L5 signal only. As the L5 chip rate is 10 times that of the C/A code, the CDMA processing gain of the L5 codes is 10 times (10 dB) greater and the sharper autocorrelation function potentially offers greater accuracy. In addition, the L5 signals are twice as powerful as the C/A code. The L5 code should therefore provide additional receiver sensitivity and rejection of interference.

- *M-code signal.* From Block II-RM, GPS satellites provide a new military code at both L1 and L2 frequencies, the M-code, known as L1M and L2M. The M-code is designed to provide at least as good accuracy as the existing P(Y)-code while offering improved resilience to interference, next-generation cryptography and increased flexibility. The M-code modulation is Binary Offset Carrier (BOC) with a 10.23 MHz subcarrier frequency and 5.115 Mb/s spreading rate. Unlike the P(Y)-code, the M-code is designed for direct acquisition (Barker, 2000).

- *L1C signal.* Block III satellites will provide a new L1 signal, known as L1C. L1C has been designed to enhance interoperability with Galileo (see later in this chapter). L1C is currently planned to be available on 24 GPS satellites around 2021. The L1C modulation format is Multiplexed BOC (MBOC).

Navigation Messages

There will be two GPS navigation messages:

- *Navigation (NAV) message.* The C/A signal incorporates a navigation (NAV) message, transmitted at 50 b/s, modulo-2 added to the PRN signal. The navigation message comprises 25 data frames, of 1500 bits each, with each frame being made up of five subframes of 300 bits each. At a rate of 50 b/s, a subframe takes 6 s to transmit, and each frame takes 30 s. These frames are illustrated in Figure 14.6. Transmission of the complete navigation message therefore takes 12.5 min. Each NAV message comprises the following subframes:
  - Subframe 1 contains the GPS time and week, and satellite status and telemetry data.
  - Subframes 2 and 3 contain the satellite ephemeris data (orbit parameters) needed to calculate the satellite's position.
  - Subframes 4 and 5 contain parts of the almanac. The almanac provides information needed to correct for satellite clock drift and satellite ephemeris errors and contains ionospheric model parameters (for use in single-frequency GPS receivers) and information needed to convert between GPS time and Universal Coordinated Time (UTC). Subframes 4 and 5 taken from successive frames build up into an almanac comprising 15 000 bits (50 subframes).

- *CNAV navigation message.* The new civilian codes will include a new civilian navigation message, known as CNAV. Unlike the legacy NAV message, the CNAV signal will incorporate forward error correction, using half-rate convolution encoding (the actual CNAV data rate is therefore reduced to 25 b/s as a result of coding).

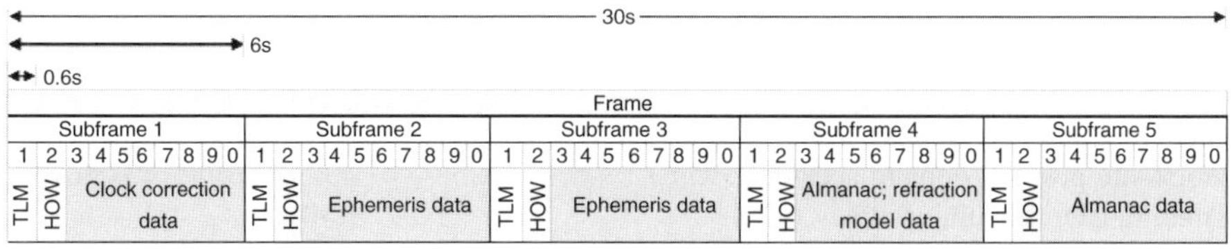

Figure 14.6 GPS NAVMSG navigation message structure. Reproduced by permission of © Michael Wößner.

14.2.1.6 GPS Position Accuracy

Selective Availability (SA)

Prior to May 2000, the accuracy of the C/A code was deliberately degraded through the introduction of pseudorandom errors to the transmitted data, using the Selective Availability (SA) feature. SA is designed to permit degradation of the horizontal positional accuracy of the C/A signal by up to 100 m, and the primary purpose of the SA was to prevent enemies of the United States exploiting the very accuracy of the GPS system against them. SA was accomplished by manipulating navigation message orbit data and dithering satellite clock frequencies.

With the widespread growth in the use of commercial GPS receivers, the continued use of SA proved highly unpopular and counterproductive, restricting the exploitation of GPS in certain key areas. Indeed, the use (or the potential use) of SA was one factor behind the decision of the European Union to develop its own navigation system, Galileo.

The US DoD temporarily turned off SA during the first Gulf War (1991) owing to the widespread use of civilian handheld GPS receivers by US and coalition forces, who were experiencing a shortage of military receivers (Pace *et al.*, 1993). Furthermore, agencies such as the US Coast Guard had begun broadcasting GPS corrections (differential GPS), which mitigated the deliberate errors caused by SA. In the end, the widespread use of differential GPS techniques has largely made SA irrelevant in regions such as the United States and Europe.

On 1 May 2000, the US DoD turned off SA, with the initial intention to review this decision annually. The US government has since stated that there are no plans to turn SA back on, and that new GPS satellites will no longer include this feature. The decision has substantially enhanced the usefulness of GPS to civilian users.

Accuracy

The position accuracy of GPS may be expressed as the product of the User Effective Range Error (UERE) times the geometric Dilution Of Precision (DOP) due to the available satellite positions at any time. GPS specifies a PDOP of ≤ 2.6 m at 95% availability and a PDOP of ≤ 6 m at 98% (GPS Performance, 2008). SPS accuracy is defined in terms of the 95% percentile Signal In Space (SIS) User Range Error (URE), and its time derivatives, plus the UTC Offset error (UTCO). SPS accuracy and availability standards (which apply to the use of L1 C/A signal only) are given in Table 14.3 (GPS Performance, 2008). These parameters are defined for signal in space and neglect ionospheric, atmospheric and multipath effects.

UERE is assumed to follow a normal (Gaussian) distribution with zero mean, and a GPS error budget is illustrated in Table 14.4 for the two extremes of the Age Of Data (AOD) – although it should be noted that the user segment errors shown are those for 1980s equipment and may not reflect the performance of modern receivers. The user range error (excluding error contributions originating in the user segment) is 4 m RMS (7.8 m at 95%) (GPS Performance, 2008). It is apparent

Table 14.3 GPS SPS performance standards (2008) (GPS Performance, 2008)

SPS standard	Value
Global av. position domain accuracy (95%)	<9 m horiz.; <15 m vert.
Worst site position domain accuracy (95%)	<17 m horiz.; <37 m vert.
Time transfer domain accuracy (95%)	<40 ns
Availability (17 m horiz.; 37 m vert.), av. position/24 h	>99%
Availability (17 m horiz.; 37 m vert.), worst site position/24 h	>90%

Table 14.4 GPS SPS (single-frequency) error budget (adapted from GPS Performance, 2008) for the extremes cases of AOD (* indicates user equipment performance based on 1980s equipment)

Segment	Parameter	Zero AOD (95%)	Max. AOD (95%)
Space	Clock stability	–	8.9 m
	Group delay stability	3.1 m	3.1 m
	Acc. uncertainty	–	2 m
	Other	1 m	1 m
Control	Clock/ephemeris estimation	2.8 m	2.8 m
	Clock/ephemeris prediction	–	6.7 m
	Ionospheric delay model	9.8–19.6 m	9.8–19.6 m
	Group delay time correction	4.5 m	4.5 m
	Other	1 m	1 m
User*	Ionospheric delay compensation	–	–
	Tropospheric delay compensation	3.9 m	3.9 m
	Receiver noise	2.9 m	2.9 m
	Multipath	2.4 m	2.4 m
	Other	1 m	1 m
TOTAL	95% SPS UERE	12.7–21.2 m	17.0–24.1 m

from the table that the dominant contribution to the SPS UERE budget arises from the uncertainty in predicting ionospheric delay.

14.2.2 GLONASS

The GLONASS system was conceived in the early 1970s, drawing on the experience of the TSIKADA system. It is operated by the Russian Federation.

After a period of decline, the programme was revived, with plans of a full deployment by 2010 (Space Daily, 2005). GLONASS satellites have a lifetime of 1–3 years, and, consequently, 81 GLONASS and 14 modified GLONASS (GLONASS-M) satellites were launched until mid-2007.

The 24 satellites of its operational constellation are evenly arranged in three orbital planes that are inclined at 64.8° to the equator and spaced 120° apart at an altitude of 19 100 km.

GLONASS, like GPS, is a pseudoranging system. However, there are differences. The orbital radius of the constellation is 1050 km less than that of GPS. GLONASS employs Frequency Division Multiple Access (FDMA) and hence does not need different codes to distinguish satellites, thereby allowing the same code to be reused. GLONASS coordinates are expressed in the geodetic system 'Parameters of the Earth 1990' (PZ90), whereas those for GPS are given in the 'World Geodetic System 1984' (WGS 1984); WGS84 has been promulgated by the International Civil Aviation Organization (ICAO) as the global coordinate system for aviation with effect from the 1 January 1998. GLONASS is referenced to the Universal Time Coordinated (SU) (UTC-SU).

The GLONASS satellites emit signals around two carrier frequencies: approximately 1602 MHz (L1) and approximately 1246 MHz (L2), with the exact frequency being a function of the GLONASS satellite's channel number.

The civilian C/A code uses a 511 maximum-length pseudorandom sequence at 0.511 MHz chipping rate. The higher accuracy P-code transmission operates at 5.11 MHz chipping rate but with restricted availability. Navigation data are modulated on the spread-spectrum signals at 50 bps without forward error correction. The data symbols are Manchester encoded. The C/A-Code is modulated on L1 only, and the P-code is modulated on both frequencies. The C/A code can be deliberately degraded by the military. Accuracies with degraded C/A code and P-code are respectively of the order of 100 and 10–20 m.

The modified GLONASS (GLONASS-M) satellites broadcast both signals on L2 and in addition have an enhanced life time of 7 years. GLONASS-K, the next-generation satellites (2009–2010), add transmissions in a new L3 band. The new sub-band, called L3, operates at 1202–1208 MHz. The L3 band employs a higher chipping rate of 4 MHz for civil use. Additional transmissions of CDMA signals are under consideration because the FDMA scheme increases the complexity of the receiver and biases can be added when bands are changed. The bands under consideration are at 1575 and 1176 MHz, using a signal structure similar to the GPS L1C and L5. Figure 14.7 demonstrates enhancements in the GLONASS satellite signal features (Hegarty and Chatre, 2008).

The GLONASS system architecture consists of a control segment with 10 monitor stations distributed throughout Russia, and facilities to command and control the satellites. GLONASS user equipment provides positions in the Earth parameter system, and hence a transformation was required to convert from PZ-90 to WGS-84 coordinates and adjust the PZ-90 system. The difference in the terrestrial coordinates between the two systems is now within 2 cm.

14.2.3 Galileo

In the early 1990s, the European Union (EU) identified a need for Europe to develop an independent GNSS, under the control of a civilian administration. In 2002, the EU and the European Space

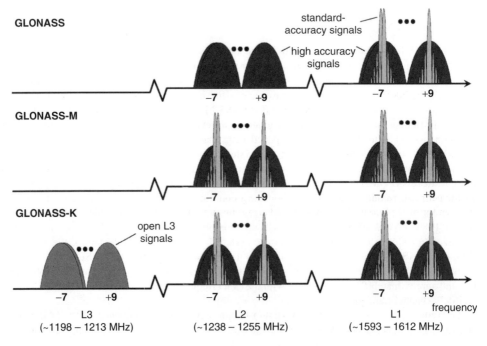

Figure 14.7 Evolution of the GLONASS frequency regime (Hegarty and Chatre, 2008). Reproduced by permission of © IEEE.

Agency (ESA) jointly agreed to fund a European GNSS to be called Galileo. Significantly, unlike GPS or GLONASS, Galileo would be under civilian control and would be guaranteed to operate at all times, except in the direst emergency. While the principal driver for the development of Galileo is to protect European economies from dependency on other states for access to a GNSS, other reasons cited for development of Galileo are:

- to allow user positions to be determined more accurately – even in high-rise cities – because the number of satellites available from which to take a position will be more than doubled through interoperability with GPS (and GLONASS);
- better coverage at high latitudes, particularly northern Europe, through the use of satellite orbits with greater inclination to the equatorial plane (than GPS);
- to secure an increased share for Europe in the equipment market and novel business opportunities for EU application providers and service operators.

There is a US–EU agreement on GPS–Galileo interoperability. Additionally, the Galileo programme has international backing from a number of non-EU nations, including China, Israel, India and South Korea. Although initially conceived as a joint public/private venture, in 2007 the EU took over direct control of the project owing to lack of progress. A budget of €3.4 billion was agreed, and procurement of the Galileo satellites and ground infrastructure began in 2008.

In 2005, an experimental Galileo satellite, Galileo In-Orbit Validation Element (GIOVE)-A, was launched, with the aim of validating and characterizing critical Galileo technologies and reserving the Galileo ITU-R frequency allocations. Subsequently, a second test satellite, GIOVE-B, was launched in 2008, which is transmitting representative Galileo navigation signals. The development plan calls for four In-Orbit Validation (IOV) satellites to be launched in order to validate (together with GIOVE-B) the space and ground control segments, prior to launch of the full Galileo constellation.

Galileo will be interoperable with the GPS system. It specifies four classes of service – the Open Service (OS), the Safety-of-Life Service (SoL), the Commercial Service (CS) and the Public Regulated Service (PRS). The OS provides position and timing services akin to the GPS system; the SoL additionally provides integrity information by including warnings when performance goals cannot be met; the CS allows user terminals to improve the position and timing accuracy by access to two additional signals; the PRS offers position and timing information with a higher continuity of service but with controlled access.

The OS and SoL are likely to be used by the civil aviation industry. Hence, their performances are also specified in terms of accuracy and integrity, that is, risk, alert limit and time to alert. The SoL is a certified service that primarily targets civil aviation. The accuracy available for OS using a single frequency is 15 m horizontal (95%), 35 m vertical (95%), which enhances to 4 m and 8 m respectively, identical to the SoL accuracy specification, when two frequencies are used.

The space segment of Galileo consists of 27 satellites in a medium Earth orbit constellation distributed in three, 56° inclined planes at an altitude of about 23 000 km with one active spare satellite per plane (Zandbergen *et al.*, 2004). The ground segment will consist of two control centres, five monitoring and control centres and five mission uplink stations to provide global coverage. The facilities are interlinked to facilitate data flow across the network.

The control centre provisions constellation management and navigation system control by receiving data from about 30 sensor stations that monitor the navigation signals of the constellation. The control centre is responsible for orbit and time synchronization and provision of integrity data for the SoL service uplinked through five ground stations.

Four frequency bands, designated as E5a, E5b, E6 and E1 lying within the Radio Navigation Satellite Services (RNSS) bands, are used for transmission. The bands either overlap or are close to GPS transmissions to ensure the possibility of synergistic operation to enhance robustness and accuracy. The signals use direct-sequence spread-spectrum modulation or its variant (known as binary offset carrier). The signals for the OS and SoL services are transmitted in the E1 band, with identical characteristics to GPS L1C transmissions (Galileo SIS ICD, 2010).

Galileo Services

When completed, the Galileo navigation system will offer five services:

- An Open Service (OS) for mass-market and recreational users.
- An encrypted Commercial Service (CS) for specialized applications, with guaranteed service and higher accuracy.
- An encrypted Safety-of-Life (SoL) Service. SoL will provide higher reliability and additional integrity data (to provide a warning to the user if the accuracy of the service is not sufficient for the intended application).
- An encrypted Public Regulated Service (PRS) for government-approved users (police, civil protection, emergency services, etc.).
- A Search and Rescue (SAR) Service. Galileo satellites will be equipped with a COSPAS-SARSAT-compatible transponder (see Section 14.5.1) in order to transfer distress signals from the user transmitters to a rescue coordination centre. Uniquely, the Galileo service will also provide a return signal to the user, via the appropriate service message data, informing him/her that help is on its way.

14.2.3.1   Space Segment

When completed, Galileo will comprise 30 satellites (27 plus three in-orbit spares) located in Medium Earth Orbit (MEO). The constellation type is of the Walker Delta 56°: 27/3/1 type, that is, having an inclination of 56° relative to the equatorial plane, with 27 satellites in three equally spaced orbital planes. The Galileo orbit inclination was selected to ensure good performance at polar latitudes and, in particular, northern European latitudes. Each satellite will take approximately 14 h to orbit the Earth. The Galileo satellite payload includes two pairs of redundant clocks, each pair comprising a rubidium atomic clock and a passive hydrogen maser.

Further parameters of the Galileo satellite constellation are given in Table 14.5.

14.2.3.2   Ground Control Segment

The Galileo ground control segment will comprise two Galileo Control Centres (GCCs), one near Munich (Germany) and the other in Fucino (Italy). The satellites will be monitored by a global

Table 14.5   Galileo constellation summary (** based on $-157\,$dBW received power with 0 dBi antenna)

Parameter	Value
Number of satellites	30 (27 plus three active spares)
Orbit type	Circular MEO
Altitude	23 222 km
Constellation type	Walker Delta 56°: 27/3/1
Orbital period	14 h 5 min
Mass	675 kg
Predicted life	>12 years
DC power	1500 W
EIRP	∼26 dBW **
Transmit polarization	RHCP

network of between 20 and 40 Galileo Sensor Stations (GSSs), linked to these Galileo control centres. In addition, 10 C-band mission uplink stations will facilitate uplinking navigation and integrity data, while five collocated S-band uplink stations will provide satellite TT&C facilities. All Galileo satellites will also be fitted with laser retroreflectors permitting highly accurate satellite laser ranging (likely to be used about once a year).

Galileo Time

Galileo will use a reference timescale known as Galileo System Time (GST), which will be steered to International Atomic Time (TAI). Galileo system time generation will be the responsibility of the ground-based Galileo precise time facility, which will employ an active hydrogen maser (with hot spares) together with an ensemble of cesium atomic clocks. Unlike GPS, the satellite clocks are not used to derive Galileo time.

Galileo system time will be steered to within 50 ns of TAI (with 95% probability), and this time offset will be known with an accuracy of 28 ns (95%). The offset between TAI representations derived from GPS and Galileo broadcast can be expected to be about 33 ns (95%) (Moudrak *et al.*, 2005). The time offset between Galileo time and GPS system time will be broadcast by both Galileo and enhanced GPS satellites.

Galileo Geodetic System

Unlike GPS, which uses the WGS-84 datum, Galileo will use the Galileo Terrestrial Reference Frame (GTRF) geodetic datum, itself based on the International Terrestrial Reference System (ITRS) datum. However, the difference between the ITRS and WGS-84 datums (as used by GPS) is only of the order of a few centimetres (Kaplan and Hegarty, 2006), and hence, for the vast majority of applications, the geodetic frames will be compatible, with no transformation between the reference frames being necessary.

14.2.3.3   User Segment

At the time of writing, Galileo-specific consumer devices had yet to be released. The general availability of Galileo-specific devices may be expected to coincide with the launch of the full constellation. Nevertheless, at the time of writing, a number of GNSS receiver chip manufacturers have released components for integration into Galileo consumer devices.

A number of GPS products that incorporate the ability to operate with Galileo have also been released. The expectation is that most Galileo receivers will also include a GPS capability, with more advanced products providing the ability to exploit both constellations for improved availability and accuracy.

14.2.3.4   Galileo Signals

Galileo satellites will broadcast, using CDMA, on four L-band frequencies:

- E1/L1 (1.57542 GHz). This is the same centre frequency as for GPS L1 (Galileo signals are broadcast using a different modulation). The navigation signals transmitted at this frequency are known as L1F, supporting Open Service (OS) and Safety-of-Life (SoL), and L1P, supporting Public Regulated Service (PRS).
- E5a (1.17645 GHz). This is the same centre frequency as for GPS L5 (again a different modulation is used). The navigation signal transmitted at this frequency is known as E5A, supporting Open Service (OS) and Safety-of-Life (SoL).

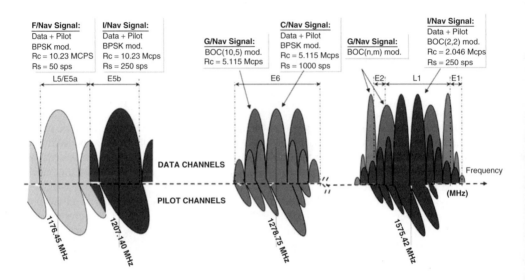

Figure 14.8 Galileo signals (Steciw, 2004). Reproduced by permission of © ESA.

- E5b (1.20714 GHz). The navigation signal transmitted at this frequency is known as E5B, supporting Open Service (OS) and Safety-of-Life (SoL).
- E6 (1.27875 GHz). The navigation signals transmitted at this frequency are known as E6C, supporting Commercial Service (CS) and Public Regulated Service (PRS).

The Galileo navigation signals are illustrated in Figure 14.8. E5a, E5b and L1 frequencies are contained within the bands reserved for Aeronautical Radio Navigation Services (ARNS), which are protected from interference from other spectrum users, thereby allowing their use for dedicated safety-critical applications.

The L1F, E5A, E5B and E6 signals are actually multiplexes of navigation signal channels, with 10 navigation channels in all (Galileo Joint Undertaking, 2005). Four channels are broadcast on the E5 frequency, three on the E6 frequency and three on the L1 frequency (Progeny, 2010):

- *L1P signal.* The L1P (also known as L1-A channel) navigation signal is a restricted access signal used by the Public Regulated Service (PRS), transmitted in the L1 band. Its ranging codes and navigation data are encrypted using a governmental encryption algorithm. The L1P modulation type is 'BOC(15, 2.5) cosine'.
- *L1F signals.* L1F signals are the open-access signals transmitted in the L1 band. The L1F signals consist of:
  - A data channel, L1-B, comprising a PRN ranging code and a navigation data message combined using modulo-2 addition. The L1-B code is 4092 chips long.
  - A pilot channel, L1-C, comprising a PRN ranging code only (for improved PRN code synchronization).The L1-C primary code is also 4092 chips long; however, this pilot signal has a secondary PRN code of 25 chips, resulting in a longer effective PRN code for enhanced accuracy.
- *E6P signals.* E6P is a restricted-access signal transmitted in the E6-A signal channel for use by the Public Regulated Service (PRS). Its ranging codes and navigation data are encrypted using a governmental encryption algorithm. The modulation type for the E6P signal is 'BOC(10, 5) cosine'. The E6-A channel navigation data stream is the G/NAV message type.

- *E6C signal.* E6C is a commercial access signal transmitted in E6 that includes data and pilot (dataless) channels (E6-B and E6-C respectively). Its ranging codes and navigation data are encrypted, but, unlike the PRS, they are encrypted using a commercial-grade algorithm. The modulation type for the E6C signal is BPSK-R. The E6-B channel navigation data stream is the C/NAV message type.

- *E5a signal.* E5a is an open-access signal transmitted in the E5 band that includes data and pilot channels. The data and pilot channels are modulated as in-phase (I) and quadrature (Q) components (E5a-I and E5a-Q respectively). The E5a signal has unencrypted ranging codes and navigation data, which are accessible by all users. The modulation type for E5a is Alt-BOC(15, 10). The E5a-I channel navigation data stream is the F/NAV message type.

- *E5b signal.* E5b is an open-access signal transmitted in E5 band that comprises data and pilot channels (E5b-I and E5b-Q signal components respectively). It has unencrypted ranging codes and navigation data accessible to all users. The E5b-I navigation data stream corresponds to an I/NAV message type and contains integrity messages as well as encrypted commercial data.

Under the US–EU agreement, the spectral shape of the Galileo L1F and GPS L1C signals will be the same (as defined by MBOC); however, the time-domain implementation of these signals differs between the two systems. Whereas the GPS L2C signal is time multiplexed BOC, the total Galileo L1-A, L1-B and L1-C signals are multiplexed together and modulated using a method known as Coherent Adaptive Subcarrier Modulation (CASM) – also known as tricode hexaphase modulation.

Galileo Navigation Messages

Galileo will transmit four different message types, according to the particular service and signal content:

- *Freely Accessible Navigation (F/NAV) message.* This contains navigation only. The F/NAV message type transmits the basic data to support navigation and timing functions.
- *Integrity Navigation (I/NAV) message.* This contains navigation, integrity and service management data and SAR return (L1-B/C channels).
- *Commercial Navigation (C/NAV) message.* This contains service management and supplemental data.
- *Governmental Access Navigation (G/NAV) message.* This contains navigation, integrity and service management data for the PRS.

The Galileo message format is designed to be flexible. The basic building block of the navigation message is a subframe that comprises a unique synchronization word, followed by data bits encoded using half-rate convolution encoding and block interleaving (Kaplan and Hegarty, 2006). A collection of subframes forms a frame, and a collection of frames forms a superframe. The hierarchical superframe/frame/subframe format facilitates flexibility in message content. For example, Safety-of-Life (SoL) services use integrity data carried in special messages designated for this purpose.

14.2.3.5   Galileo Accuracy

When considering the accuracy of civilian Galileo services, it should be noted that several Galileo signal combinations are possible, including:

- single-frequency services at L1, E5a or E5b (using an ionospheric error model);
- dual-frequency services using L1 and E5a signals (for best ionospheric error cancellation);
- triple-frequency services using all signals together (L1, E5a and E5b).

Table 14.6   Galileo open service performance (Steciw, 2004)

Parameter	Single frequency	Dual frequency
Global av. position domain accuracy	<15 m (hor. 95%); <35 m (ver. 95%)	<4 m (hor. 95%); <8 m (ver. 95%)
Time transfer domain accuracy	<30 ns (95%)	
Availability	>99.8%	

Table 14.7   Number of GPS and Galileo satellites visible for various mask angles (after Steciw, 2004)

Mask angle	Galileo	GPS	Total
5°	13	12	25
10°	11	10	21
15°	9	8	17

The anticipated performance of the Galileo open service is indicated in Table 14.6 for both single-frequency and dual-frequency receivers. Using a dual-frequency Galileo receiver, the initial 95% (i.e. two-sigma) horizontal positioning accuracy for the open service is expected to be 4 m – equivalent to an RMS (one-sigma) accuracy of approximately 2 m.

14.2.3.6   Galileo/GPS Satellite Visibility

The accuracy and availability of a GNSS depends on the number of visible satellites. By utilizing both GPS and Galileo constellations, the total number of satellites is increased. Table 14.7 illustrates the typical number of visible satellites for each constellation and for both constellations combined, with different terminal elevation angle masks (Steciw, 2004).

14.2.4   Argos

The Argos satellite navigation system (Argos, 2008) operates in an active mode to provide location services and remote data collection–processing facilities to study and protect our environment. The system comprises Argos user platforms which transmit signals to the Argos satellite in view; the satellite relays the signals to a fixed Earth station which processes the signals to derive the user location, or, in the case of data collection platforms, processes the sensor data; the Earth station results are transferred to the user by a variety of techniques:

- secure connection to the Argos website;
- ArgosDirect automatic distribution (e-mail, ftp, fax or data transmission network, or CD-ROM);
- connection to the ArgosServer transmission network (Telnet);
- custom types of dispatch (archiving data for up to 12 months, during which they can be sent on request).

The user platforms transmit periodic messages with a repetition rate of 90–200 s, depending on the use, and lasting less than 1 s at a stable transmission frequency of (401.650 MHz ±30 kHz), along with its identification number and collected data. The Argos system achieves an accuracy of ~250–1500 m, depending on Geometric Dilution of Precision (GDOP)[1], the number of messages received by the processing system and the accuracy of the user transmitter. As mentioned earlier GDOP is a measure of how well the user is placed geometrically for navigation – smaller values of GDOP yielding more accurate results.

The satellites are located in Sun-synchronous polar orbit at an altitude of 850 km (5000 km diameter footprint). Satellites in this type of orbit rise and set at a fixed time at any location on the Earth and exhibit an average visibility of about 10 min. The Argos payload is carried aboard Polar-Orbiting Environmental Satellites (POESs) belonging to the National Oceanic and Atmospheric Administration (NOAA) and MetOp satellites of the European Organization for the Exploitation of Meteorological Satellites (Eumetsat).

On receipt, the user messages are stored on the satellite and transmitted to one of the three main receiving stations located in the United States (Wallops Island, Virginia, and Fairbanks, Alaska) and in Norway (Svalbard) as the satellite transits. The message is also transmitted instantly for reception by any of the 50 regional reception centres.

There are two processing centres – one near Washington, DC (United States) and the other in Toulouse (France). The centres perform the following functions:

• a check of message quality, its reception level, time-tag (in UTC), transmitter identification number, sensor message length and received frequency (to compute the location);
• message classification by platform and by chronological order;
• navigation and sensor data processing.

The Argos system supports the following modes of location (In both cases, the coordinates are the latitude and longitude in the WGS 84 reference system):

• *Argos location.* Argos processing centres estimate a transmitter's location using the Doppler principle discussed earlier, and illustrated in Figure 14.9.
• *GPS positioning.* On the user's request, the GPS location sent by the user is extracted and formatted in the same way for user distribution as the Argos-based location.

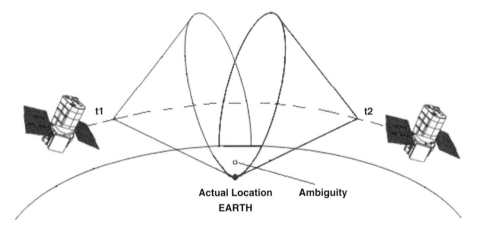

Figure 14.9   Argos system.

---

[1] In the case of Doppler navigation systems, GDOP may relate to the geometry of multiple data points for one satellite in a single pass.

14.3   Regional Navigation Systems

Several countries have initiated the development of regional navigation systems – some with plans of expanding into a global system. In this section we introduce the systems being developed in China and India as examples. We note here that a satellite-based GPS augmentation system called the Quasi-Zenith Satellite System (QZSS), which utilizes a three-satellite constellation of inclined geosynchronous orbit, is being planned in Japan. The orbital configuration provides a better visibility to the users in Japan than a geostationary orbit (Wu, Kubo and Yasuda, 2004).

14.3.1   Beidou and Compass

The Beidou system, owned and operated by China, is a regional geostationary active navigation–communication system, operational since 2003, to provide positioning, short-data communications and time service in eastern and southern Asia (longitude 70° E–140° E, latitude: 5° N-55° N). China is developing a second-generation satellite navigation system, called Compass, which will incorporate the capability of global coverage. The system will deploy a constellation of 35 satellites (Hongwei, Zhigang and Feng, 2007).

The Beidou system capacity supports 150 users per second; the user–satellite uplink frequency is 1610–1626.5 MHz, and the satellite–feeder downlink frequency is 2483.5–2500 MHz. The positioning accuracy is <100 m, improving to <20 m with differential positioning, with a system response time of 1–5 s and a success rate of 95%. The system can also support short data messages of up to 150 Chinese characters.

The distance ranging principle is applied by measuring the distance of a user from two satellites. Knowing the location of the satellite, the user lies at the intersection of the Earth and distance spheres around each satellite (see Chapter 13).

The system is said to be useful for a variety of applications and services, including asset tracking, transportation, meteorology, petroleum production, forest fire prevention, disaster forecast and public security and military.

The active nature of the system leads to user expenses, vulnerability of detection for the military users and incompatibility with other navigation systems. Hence, a second-generation system, known as Compass, was conceived. It comprises a constellation of 27 satellites in a medium Earth orbit (altitude 21 500 km, inclination 55°), three satellites in an inclined geosynchronous orbits (inclination 55°, equatorial crossing 118°) and five satellites in a geostationary orbit (location 58.75° E, 80° E, 110.5° E, 140° E and 160° E). Two types of service are planned – the open service, which offers accuracies of 10 m for location, 0.2 m/s for velocity and 50 ns for timing, is available to all, whereas the authorized service, offering a safer positioning, velocity and timing over a secure network, tagged with system integrity data, is available only to authorized users. The system will use CDMA with DSSS signals and operate on four carriers at frequencies of 1207.14 MHz (shared with Galileo E5b), 1268.52 MHz (shared with Galileo E6), 1561.098 MHz (E2) and 1589.742 MHz (E1). Its implementation is planned in phases, beginning with the Chinese region and later expanding the constellation gradually to achieve a global coverage.

14.3.2   Indian Regional Satellite System

The Indian Regional Navigation Satellite System (IRNSS) is planned by the Indian Space Research Organization for implementation by the year 2012 to serve the Indian region (Rao, 2007). The system will comprise a hybrid constellation of seven satellites – three in geostationary orbit (34° E, 83° E, and 132° E), and the remainder in a geosynchronous orbit at an inclination of 29° such as to centre their footprint at 55° E and 111° E longitudes. The service areas lie within longitude 40–140° E and latitude 40° S–40° N to offer a positional accuracy of ~20 m.

The standard positioning service will utilize Direct-Sequence Spread-Spectrum (DSSS) signals at 1.023 MHz chip rate, broadcast at 1191.795 and 2491.005 MHz; a precise positioning service will operate on the same carrier and also on 1191.795 MHz at a chip rate of 10.23 MHz.

14.4 Satellite-Based Augmentation Systems

The accuracy and integrity available from the global navigation systems is insufficient, particularly for safety critical aeronautical applications. As explained in Chapter 13, a satellite-based augmentation system (SBAS) can improve the receiver performance considerably. A majority of the existing SBAS systems monitor GPS signals from a network of ground stations located strategically within the region of interest and dispatch the raw data to a central facility. The data are processed to estimate the differential correction applicable to the region and extract the real-time GPS constellation status (integrity). These data are combined with a GPS PRN-like sequence and uplinked via one or more stations to one or more geostationary (or quasi-stationary) satellites for broadcast. The SBAS-enabled receivers utilize this information to improve the accuracy of their fix.

Various SBASs are in place, and others are under study or in development. The United States implements the Wide-Area Augmentation System (WAAS), Europe is in the process of introducing EGNOS and Japan operates the Multi-functional Satellite Augmentation System (MSAS), with plans to enhance the system with the QZSS system mentioned in Section 14.3. India is in the process of implementing GPS-Aided Geo Augmented Navigation (GAGAN). Besides, there are various commercial systems in operation. We shall introduce the salient aspects of WAAS, EGNOS and GAGAN.

14.4.1 Wide-Area Augmentation System

The Wide-Area Augmentation System (WAAS) was deployed by the US Federal Aviation Administration primarily to serve air traffic in the region during nearly all phases of flight, that is, oceanic routes, en route over domestic airspace, including crowded metropolitan airspaces, and during airport approach (Enge *et al.*, 1996). WAAS has been operational since the year 2000. The Safety of Life (SoL) feature was introduced in 2003. WAAS was conceived to comprise over 35 Wide-area Reference Stations (WRSs) that would be precisely referenced and dispersed throughout North America, Hawaii, Canada and Mexico to monitor the constellation and transfer the data to one of the two master facilities. The reader should visit an appropriate website for the latest configuration (e.g. FAA William J. Hughes Technical Centre online, 2010).

The processed information consists of differential correction and constellation status by region. These data are relayed via two GEO satellites located at 133° W and 107° W and relayed at GPS L1 and L5 frequencies to WAAS-capable receivers. The correction and integrity data stream are modulated onto a C/A code. The WASS-enabled receivers typically improve the position accuracy of single frequency receivers to less than 3 m. The constellation updates are delivered in less than 6 s. In precommissioning flight trials, the positional error was measured as ±2 m in the vertical direction near touchdown (Enge *et al.*, 1996). In practice, the system can be utilized for en route to approach guidance up to a height of around 200–250 feet (~61–76 m) – the so-called Localizer Performance with Vertical guidance (LPV) limit. Recent measures (fourth quarter, 2009) demonstrate LPV availability of 99% within the continental United States (FAA/William J. Hughes Technical Centre, 2010). With the modernized version of GPS, where two frequencies are available, the vertical errors can be reduced through improved error cancellation of ionospheric errors. The corrections available from the WAAS system can be utilized by the public for non-aviation

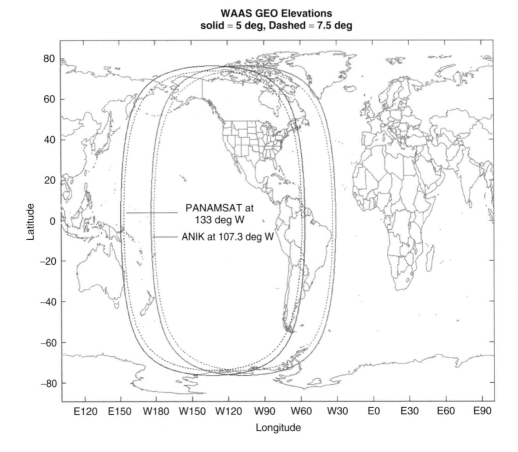

WAAS GEO Elevations
solid = 5 deg, Dashed = 7.5 deg

PANAMSAT at
133 deg W

ANIK at 107.3 deg W

Figure 14.10 The corrections available from the WAAS system can be utilized by the public within the footprint of the WAAS geostationary satellites-PANAMSAT (133 deg W) and ANIK (107.3 deg W) to be augmented in future with Inmarsat I4-F2 (98 deg W). Reproduced with the permission of FAA/William J. Hughes Technical Center.

use throughout the footprint of the WAAS system's geostationary satellites (Figure 14.10). It is perceived that, with the availability of Galileo, the reliability and accuracy will improve further owing to an increase in the number of ranging signals (Walter, Enge and Reddan, 2004).

14.4.2 European Geostationary Navigation Overlay Service

The European SBAS, known as the European Geostationary Navigation Overlay Service (EGNOS), will augment the accuracy and availability of the GPS and GLONASS navigation systems to enhance accuracies to 1–2 m, horizontal and 2–4 m vertical. Integrity and safety are enhanced by transmitting a GPS or EGNOS constellation malfunction within 6 s. Navigation system availability is augmented by transmitting the signals over three geostationary satellites – two Inmarsat-3 satellites (AOR-E and IOR-W) and ESA's Artemis payload (EGNOS, 2004).

The EGNOS SBAS development is coordinated by the European Commission, Eurocontrol and the European Agency, supported by wide participation from European research institutes and industry. EGNOS is the first step towards development of a European GNSS system; the second step is the deployment of the Galileo system described in Section 14.2.3.

The system consists of 34 Ranging and Integrity Monitoring Stations (RIMS) spread over 22 countries that collect data from the navigation satellites and transfer them every second to the four master control stations (MCSs); the MCSs, knowing the precise location of each RIMS, calculate the range corrections for each satellite, a confidence level of the data at the given instant and up to 150 s later (integrity data) and the differential correction. Three Navigation Land Earth Stations (NLESs), each backed up by another, transmit the supplementary navigation signals, integrity data and differential correction to the user segment. A terrestrial network interconnects the ground facilities. The system is managed centrally to ensure proper monitoring of the network.

14.4.3 GAGAN

To cover the gap between the EGNOS and MSAS systems and improve the reliability in the region, India has demonstrated the SBAS technology in a project known as the GAGAN (GPS-Aided Geo Augmented Navigation) system.

The goal of the GAGAN system is to provide an SBAS service in the Indian region in such a manner that the system can interwork with other SBASs in the region. The project was conceived in two phases – the proof-of-concept demonstration phase and the operational phase. Phase 1 of the project, called GAGAN-Technology Demonstration System (GAGAN-TDS), was completed successfully in 2007. An accuracy well within 7.6 m was reported for both horizontal and vertical directions in these trials (Nandulal *et al.*, 2008). The preparations for the operational phase began in 2009 (Kibe, 2003).

The GAGAN-TDS ground system consisted of a master control station, one uplink station and eight reference stations, called Indian Reference Stations (INRESs). The widely dispersed INRESs received and collected data from GPS satellites and transferred them to the Master Control Station (MCS), where the differential corrections and residual errors for each predetermined ionospheric grid point were estimated. The MCS additionally performed the functions of network management, integrity monitoring, ionospheric and tropospheric delay estimation, wide-area corrections, separation of errors, orbit determination; and command generation.

The correction data were transferred to the uplink station, called the Indian Land Uplink Station (INLUS), where they were transmitted via a geostationary satellite to the users in a GPS-compatible format at L1 and L5 frequencies. The INLUS additionally had the function of providing GEO ranging information and corrections to the GEO satellite clocks. Message formats and timing followed the international standards.

In the operational system, the navigation payload will transmit signals at L1 and possibly L2 frequencies in a global beam from a spacecraft located between 48° E and 100° E longitude. The spacecraft will also incorporate a C band – C band path for uplinking ranging signals from the INLUS to INRESs.

14.5   Distress and Safety

In Chapter 13 we introduced the concept of GMDSS, an international communication system to assist the maritime community. GMDSS elements include two complementary satellite communication systems –Cospas-Sarsat and Inmarsat. The Cospas-Sarsat system provides a true (but intermittent) global coverage (i.e. up to the poles) whereas the Inmarsat system operates in regions enclosed within about ±76° latitude.

14.5.1 Cospas-Sarsat

Cospas-Sarsat is a Search And Rescue (SAR) satellite communication service provider to assist distressed ships, aircraft and individuals by transporting the distress message and location of the distressed source to the appropriate points of contact. Its aim is to 'support all organizations in the world with responsibility for SAR operations, whether at sea, in the air or on land' (Cospas-Sarsat Online, 2009). The system constitutes a part of the GMDSS to serve remote areas not reachable by other communication services. It utilizes a hybrid orbit constellation comprising a mix of low Earth near-polar-orbiting satellites and geostationary satellites. The system, founded by a consortium of four countries (the United States, Canada, France and the former Soviet Union), was declared operational in 1985, although its use had begun in 1982. It is credited as having provided assistance in rescuing at least 24 798 persons between September 1982 and December 2007 across the world (Cospas-Sarsat, 2008).

Figure 14.11 presents the main components of the system: distress locator transmitters of various types, a space segment consisting of a hybrid orbit constellation (LEO-GEO) and a ground network consisting of a Local User Terminal (LUT), a Mission Control Centre (MCC) and a Rescue Coordination Centre (RCC) responsible for the provision of rescue facilities. An estimated 1 million LUTs were in use throughout the world in 2007, supported by 29 MCCs, 45 LEOLUTs (Low Earth Orbit LUTs) and 19 GEOLUTs (Geostationary orbit LUTs) (Cospas-Sarsat, 2008). The system operates in the 406–406.1 MHz frequency band reserved by the International Telecommunication Union exclusively for satellite distress and safety systems. The 121.5/243 MHz beacon transmissions were discontinued in February 2009.

The user triggers a message containing the country of origin, the originator's identification and, optionally, location data to the space segment. The user terminals are called Emergency Locator Transmitters (ELTs) for aviation users, Emergency Position Indicating Radio Beacons (EPIRBs) for

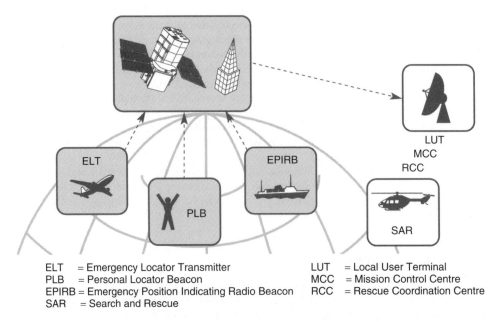

ELT = Emergency Locator Transmitter	LUT = Local User Terminal
PLB = Personal Locator Beacon	MCC = Mission Control Centre
EPIRB = Emergency Position Indicating Radio Beacon	RCC = Rescue Coordination Centre
SAR = Search and Rescue	

Figure 14.11 The main constituents of Cospas-Sarsat search and Rescue system. Reproduced by the permission of © Cospas-Sarsat programme.

maritime users and Personal Locator Beacons (PLBs) for individuals. ELT transmissions can be triggered manually or automatically in the case of an event involving an aircraft crash or a sinking ship. Each distress signal received by a satellite is relayed to a ground station called a Local User Terminal (LUT). The alert is transferred to the Mission Control Centre (MCC) which forwards it to the Rescue Coordination Centre (RCC) to arrange an appropriate rescue measure.

The space segment consists of a constellation of five low Earth orbiting (LEO) satellites and five geostationary (GEO) satellite overlays (2009 data). The LEO constellation comprises Cospas and Sarsat satellites. The LEO component of the system is known as the LEO Search and Rescue (LEOSAR) system, and the GEO component as the GEO Search and Rescue (GEOSAR) system. The constellation is illustrated in Figure 14.12.

The MCC has the function of system and distress data management, including dispatching data to its associated RCC, or the point of contact. All the MCCs are interconnected for the purpose. System data, useful in facilitating a healthy operation of the network, include satellite ephemeris, time calibration, status of space and ground segment and coordination messages. The GEOSAR component benefits from uninterrupted visibility within its footprint (between approximately ±76° latitude). The GEOSAR satellites operate at 406 MHz via a transparent transponder. The distress messages should contain the location of the sender, as the satellites cannot locate a sender. The LEOSAR coverage extends up to the pole, but the coverage is intermittent. The waiting time at mid-latitudes is less than an hour. A typical satellite pass time is of the order of 10–15 min.

The Russian Federation supplies Cospas (Russian acronym for: space system for the search of vessels in distress) and the United States supplies the Sarsat (search and rescue satellite-aided tracking) instrumentation on NOAA satellites. The instrumentation is supplied by Canada and France. The altitude of the near-polar orbit used by Cospas satellites is 1000 km, and the NOAA

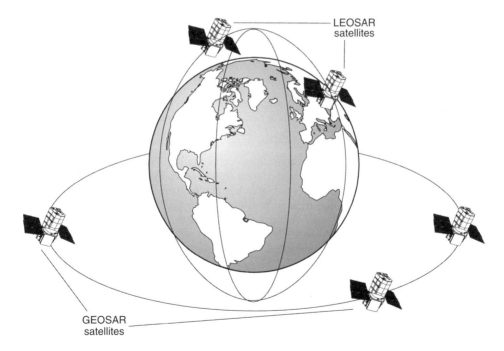

Figure 14.12 The Cospas-Sarsat constellation. Reproduced by the permission of © Cospas-Sarsat programme.

satellites operate in a Sun-synchronous near-polar orbit at an altitude of 850 km. The GEOSAR constellation comprises GOES satellites (United States), INSAT-series satellites (India) and Eumetsat Meteosat Second-Generation (MSG) series (Europe).

The LEOSAR satellites incorporate a Search And Rescue Processor (SARP) which receives the digitally modulated RF transmissions, recovers the data, measures the Doppler shift, time stamps, reformats and transmits the data stream to LEOLUTs. LEOSAR satellites thus allow estimation the position of a distress beacon by the Doppler processing method discussed in Chapters 7 and 13. The accuracy, of the order of 2 km, is generally adequate to narrow down a search. Cospas satellites include a transparent repeater called the Search and Rescue Repeater (SARR), where the signals are relayed without any processing. Hence, the location has to be extracted by the receiving LUT.

Distress beacon transmissions to LEO satellites are more resistant to blockage, as the movement of the satellites across the sky improves the probability of circumventing an obstacle. There are two coverage modes of the LEO system – global and local. In the global mode, each satellite stores and continues to transmit the distress message for several orbits, thus ensuring that the signal is received by all the LUTs tracking the satellite. In the local mode, the signal is transmitted only to a visible LUT.

14.5.2 Inmarsat Distress System

Inmarsat is a global mobile satellite system company mandated to provide a distress communication service to the maritime community. The Inmarsat system comprises four ocean regions, each served by a geostationary satellite. Various types of digital communication service are supported by the network. The ocean regions are designated as:

- Atlantic Ocean Region-East (AOR-E);
- Atlantic Ocean Region-West (AOR-W);
- Indian Ocean Region (IOR);
- Pacific Ocean Region (POR).

A Network Coordination Station (NCS) in each region manages the communications traffic flow of each service, while a network of Land Earth Stations (LESs) dispersed across the world carry the communications traffic. Inmarsat B and Fleet 77 share the NCS infrastructure, whereas Inmarsat C operates on a separate NCS.

The Inmarsat communication system constitutes one of the key elements of GMDSS coverage chart area A3, that is, the area beyond about 190 km from a coast, extending up to the edge of the geostationary satellite footprint (about ±76° latitude). The infrastructure that provisions GMDSS consists of a geostationary satellite fleet, each satellite being backed up by a spare satellite for contingency operation, and three types of communication system, designated Inmarsat B, Inmarsat C and Fleet 77 (F77) (Inmarsat Online, 2009). The network assigns the highest priority to distress messages to deal with the following situations:

- distress alerts from ship to shore;
- alerts to ships, initiated by RCCs;
- support to search and rescue operation;
- dissemination of maritime safety information;
- support of various distress-related events and in general to support the welfare of the maritime community.

Inmarsat B is a digital system that provides two-way direct-dial phone, telex, facsimile and data communications at rates of up to 64 kb/s. An Inmarsat-B terminal is a small self-contained satellite Earth station comprising above-deck and below-deck equipment and uses a stabilized parabolic antenna. Calls to and from an Inmarsat-B mobile Earth station are routed through one of the several Earth stations dispersed throughout the world. Each Earth station is bridged to the public network and thence to the rescue centres. Distress messages may be sent by telephone or telex.

The Inmarsat C system supports smaller, simpler, low-cost terminals, which may be hand-carried or installed on vessel, vehicle or aircraft. An omnidirectional antenna simplifies the terminal design. The GMDSS versions of user terminals are required to conform to the International Maritime Organization (IMO) performance standards. The system supports only data- or message-based communication at an information rate of 600 b/s. It can handle messages of up to 32 kB encapsulated in data packets.

The Inmarsat-C enhanced group call service permits broadcasts to predefined groups or a geographical area, and the SafetyNET service offers maritime safety information communications to vessels at sea, used by hydrographic, search and rescue, meteorological and coastguard coordination authorities. Messages can be addressed to mobiles within specific regions, including the sea area around a search and rescue incident. In the event of an emergency, the distress-alerting feature of the system automatically generates and sends a priority distress alert, incorporating position and other information.

Fleet 77 provides a voice and advanced data communication facility consisting of access to the Integrated Switched Data Network (ISDN) and mobile packet data service at data rates of up to 64 kb/s. A Fleet 77 terminal comprises an outdoor unit and an indoor unit that houses the communication equipment. The outdoor unit consists of a stabilized parabolic dish housed in a radome for protection against the hostile sea environment, attached to front-end electronics. The below-deck unit houses the communication equipment which includes interfaces to equipment such as telephone, computers, ISDN terminal adapters, etc.

14.6   Location-Based service

A fleet management system is one of the most successful and established location-based services provided by the satellite industry. We will outline features of a positioning and communication management system called EutelTRACS™, operated by Eutelsat Communications in the European region (EutelTRACS™ Online, 2009). The system is useful for vehicle tracking and for managing human resource and freight. Typical applications anticipated by the operators are: trailer management systems to monitor the status of trailers, performance reporting systems used for tasks such as monitoring on-board data for efficient running of vehicles, support of transport security conditions, management of drivers' needs including fuel requirements and regularly transmitting driver work data in line with EU regulations.

The service is available in the footprint of a geostationary satellite called Eutelsat W7 (Eutelsat Online, 2009) located at 36° E which covers western and eastern Europe, the Mediterranean basin, the Middle East and a large section of the Atlantic Ocean.

The positioning accuracy is of the order of 100 m, with the provision for improving accuracy through external, more accurate fixes than the EutelTracs internal system. The central switching facility located near Paris acts as a connecting node for the operator's network management centre and authorized mobile unit(s).

The system architecture (Figure 14.13) is a derivative of the OmniTracs system of the United States (Colcy and Steinhäuser, 1993). The customer terminal is connected through Eutelsat W-7 to the Service Network Management Centre (SNMC) which exchanges information between the fleet manager and each mobile. The SNMC interfaces with the Hub Network Management Centre

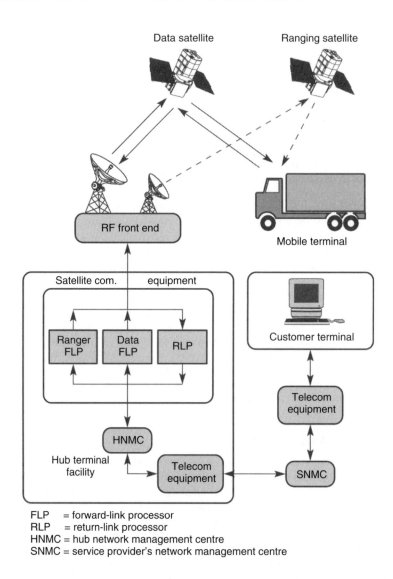

FLP = forward-link processor
RLP = return-link processor
HNMC = hub network management centre
SNMC = service provider's network management centre

Figure 14.13 EutelTracs system architecture. Adapted from Colcy and Steinhauser, 1993.

(HNMC) connected to forward- and return-link processor units and the satellite transmission equipment. The forward satellite link consists of a high-power TDM stream spread to 2 MHz with a chirp signal to minimize the effects of interference. The user data rate is 4.96 kbps or 14.88 kb/s. The return-link transmission occupies the full 36 MHz transponder to accommodate up to 45 000 users. The signals are half-rate FEC coded and interleaved to provide resistance against interference. The 32-ary FSK signal is combined with a 1 MHz rate direct spread spectrum. The spread signal is then randomly frequency hopped across the full band. The hop rate is synchronized at both ends to facilitate demodulation. A store-and-forward protocol with acknowledgement effects a robust data exchange mechanism.

Revision Questions

1. Outline the main features of GPS and GLONASS navigation systems.
2. Compare the main features of each, highlighting similarities and differences.
3. How are the positions of GPS satellites estimated? Explain why the GPS military P(Y)-code is potentially more accurate than the civilian C/A code.
4. What are the advantages of using navigation signals with and without navigation data?
5. Assuming that the error due to ionospheric delay can be completely eliminated by use of a dual-frequency receiver, using Table 14.4, estimate the total GPS two-sigma error for a twin-frequency receiver for both zero and maximum AOD.
6. Why do both GPS and Galileo satellites utilize phased-array antennas for their Earth coverage beams?
7. Why does Galileo provide several different navigation messages?
8. Explain the principle of the satellite-based augmentation system (SBAS). Describe how WAAS implements the SBAS principles in practice.
9. (a) Describe the architecture of the Cospas-Sarsat and Inmarsat systems as applied to the GMDSS framework. (b) What are the main differences in architecture of these systems? (c) Explain how the systems complement each other.
10. This suggestion is for those readers who wish to keep abreast of technology. Several programmes mentioned in the text are in a state of evolution. For example, the GPS and GLONASS systems are being replenished, EGNOS development is making rapid strides, WAAS is undergoing a steady evolution and various regional systems are progressing to implementation phase. Select at least two systems that interest you and research them in detail, concluding with remarks about their current state, problems experienced by each and a summary of technical achievements based on published papers, etc.

References

Argos (2008) *Argos User Manual*, 14 October.

Barker, B.C., Betz, J.W., Clark, J.E., Correia, J.T., Gillis, J.T. Lazar, S., REhborn, K.A. and Straton, J.R. (2000) Overview of the GPS M code signal. Proceedings of the ION 2000 National Technical Meeting, Institute of Navigation, January. Available: http://www.mitre.org/work/tech_papers/tech_papers_00/betz_overview/betz_overview.pdf [accessed February 2010].

Colcy, J.N. and Steinhäuser, R. (1993) Euteltracs: The European experience on mobile satellite service. IMSC 1993, 16–18 June, Pasadena, CA, JPL publication 93-009, pp. 261–266.

Cospas-Sarsat (2008) Cospas Sarsat System Data, No 34. Available: http://www.cospas-sarsat.org/DocumentsSystemDataDocument/SD34-DEC08.pdf [accessed February 2010].

Cospas-Sarsat Online (2009) http://www.cospas-sarsat.org/Description/overview.htm [accessed November 2009].

DAGR Specifications (2010) http://www.rockwellcollins.com/ecat/gs/DAGR.html [accessed February 2010].

EGNOS (2004) http://www.egnos-pro.esa.int/br227_EGNOS_2004.pdf [accessed February 2010].

Engle, U. (2008) A theoretical performance analysis of the modernized GPS signals. Position, Location and Navigation Symposium, IEEE/ION, 5–8 May, pp. 1067–1078.

Eutelsat Online (2009) http://www.eutelsat.com/eutelsat/eutelsat.html [accessed November 2009].

EutelTRACS™ Online (2009) http://www.eutelsat.com/products/mobile-fleet-management.html [accessed November 2009].

FAA William J. Hughes Technical Centre Online (2010) http://www.nstb.tc.faa.gov/ [accessed February 2010].

Galileo Joint Undertaking (2005) L1 band part of Galileo Signal in Space ICD (SIS ICD).

Galileo SIS ICD (2010) Galileo Open Service Signal in Space Interface Control Document (Galileo OS SIS ICD). Available: http://www.gsa.europa.eu/go/galileo/os-sis-icd [accessed February 2010].

Goldstein, D. (2009) *Modernisation and GPS III*. Available: www.ion.org/sections/southcalifornia/Goldstein_GPS_III.ppt [accessed November 2009].

GPS Performance (2008) Gloabl Positioning System Standart Positioning Service Performance Standard (GPS SPS PS), 4th edition, September. Available: http://www.navcen.uscg.gov/GPS/geninfo/2001SPSPerformanceStandardFinal.pdf [accessed February 2010].

Hegarty, C.J. and Chatre, E. (2008) Evolution of the Global Navigation Satellite System (GNSS). *Proceedings of the IEEE*, **96**(12), 1902–1917.

Hongwei, S., Zhigang, L. and Feng, P. (2007) Development of satellite navigation in China. Control Symposium 2007, Joint with the 21st European Frequency and Time Forum, IEEE International, Geneva, Switzerland, 29 May–1 June, pp. 297–300.

ICAO (2007) Annex 10 to the Convention of International Civil Aviation, Vol. I, Radio Navigation Aids, Amendment 82, Montreal, PQ, Canada, 17 July.

Inmarsat Online http://www.inmarsat.com/Maritimesafety/default.html [accessed November 2009].

Kaplan, E. and Hegarty, C, (eds) (2006) *Understanding GPS: Principles and Applications*. Artech House, Norwood, MA.

Kibe, S.V. (2003) Indian plan for satellite-based navigation systems for civil aviation. *The Current Science*, **84**(11), 1405–1411. Available: http://www.ias.ac.in/currsci/ jun102003/1405.pdf [accessed February 2010].

Moudrak, A., Konovaltsev, A., Furthner, J., Hammesfahr, J., Defraigne, P., Bauch, A., Bedrich, S. and Schroth, A. (2005) Interoperability on time. *GPS World*, March.

Nandulal, S., Babu Rao, C., Indi, C.L., Irulappan, M., Arulmozhi, S. and Soma, P. (2008) GAGAN: evaluation of real-time position accuracy and LNAV/VNAV service availability of GAGAN SBAS (Wide Area Differential GPS) over Indian region. Tyrrhenian International Workshop on *Digital Communications – Enhanced Surveillance of Aircraft and Vehicles*, 3–5 September, pp. 1–6.

NGA GPS Division Web Page (2010) http://earth-info.nga.mil/GandG/sathtml/ [accessed 14 February 2010].

Pace, S., Frost, G.P., Lachow, I., Frelinger, D.R., Fossum, D., Wassem, D. and Pinto, M.M. *GPS History, Chronology, and Budgets Rand Corporation 1993–2003*. Available: http://www.rand.org/pubs/monograph_reports/MR614.pdf [accessed February 2010].

Progency (2010) *Introduction to Galileo*. Project PROGENY (PROvison of Galileo Expertise, Networking and Support for International Initiatives). Available: http://www.nsl.eu.com/progeny/Introduction_to_Galileo.ppt [accessed February 2010].

Rand Corporation (2007) GPS History, Chronology, and Budgets.

Rao, K.N.S. (2007) Indian Regional Navigation Satellite System (IRNSS), Proceedings of 2nd Meeting of United Nations International Committee on Global Navigation Satellite Systems (ICG), Bangalore, India, 4–7 September.

Space Daily (2005) *India to Use Russian GLONASS Navigation System – Minister*. 18 November. Available: http://www.physorg.com/news8296.html [accessed February 2010].

Steciw, A. (2004) *Galileo Overview and Status*. Available: galileo.kosmos.gov.pl/en/images/stories/Workshop/steciw.ppt [accessed February 2010].

Walter, T., Enge, P. and Reddan, P. (2004) Modernizing WAAS. Proceedings of the ION GPS Meeting. Available: http://waas.stanford.edu/~wwu/walter/papers/Modernizing_WAAS.pdf [accessed February 2010].

White, J., Rochat, P. and Mallet, L.A. (2006) *History of Atomic Frequency Standards Used in Space Systems–10 Year Update*. Available: http://www.spectratime.com/product_downloads/master_ptti06.pdf [accessed February 2010].

Wu, F., Kubo, N. and Yasuda, A. (2004) Performance analysis of GPS augmentation using Japanese Quasi-Zenith Satellite System. *Earth Planets Space*, **56**, 25–37. Available: http://www.terrapub.co.jp/journals/EPS/pdf/2004/5601/56010025.pdf [accessed February 2010].

Zandbergen, R., Dinwiddy, S., Hahn, J., Breeuwer, E. and Blonski, D. (2004) Galileo orbit selection. Proceedings of Institute of Navigation ION GNSS 2004, Long Beach, CA, September.

15

Remote Sensing Techniques

15.1 Introduction

Remote sensing deals with 'acquiring information remotely' and interpreting the acquired data to accomplish an assigned task. For the purpose of illustration, we digress to introduce a simple daily-life scenario to introduce the constituents of a remote sensing system. Consider we are watching a sports event. The scene is illuminated by sunlight, sensed through our eyes and interpreted in our brain, which results in a reaction of some sort. The constituents of a remote sensing system in this scenario can be partitioned as follows:

- a target or scene (participants, the sports arena);
- a source to illuminate the scene (sunlight);
- a frequency band of interest (optical spectrum range);
- interaction between the target and the illumination (features and colours of the scene);
- characteristics of the propagation medium that transfers the reflected scattered energy (the transparent medium);
- a sensor that receives the energy (eyes);
- a process or means to collect and transfer sensed data to a processing centre (brain);
- data processing and interpretation (interpretation of the scene in our brain);
- application of interpreted data (reaction).

Having introduced the general idea with the aid of this simple example, Figure 15.1 shows the main constituents of a remote sensing system. The illumination source transmits energy towards the target scene at a frequency of interest. The intervening medium and the characteristics of the material affect the energy in various ways, depending on the frequency band. The intervening material may be transparent, partially transparent or opaque to the incident energy. We have explained earlier that only certain parts of the spectrum offer a transparent window for remote sensing. The target may absorb, pass, reflect or scatter the incident energy to various extents, depending on the characteristics of the target, and therefore the energy emanating from the target contains vital information regarding the objects. For instance, Figure 15.2 demonstrates the differences in the reflectance spectra of water and vegetation. The reflected or scattered signal received by a sensor is used for interpretation of the scene.

We may conceive of a remote sensing system in a wider perspective – remote sensing of planetary objects, remote sensing at a ground level (as in the scenario introduced above), remote sensing from an airborne platform (e.g. aerial photography), remote sensing through energy other than electromagnetic waves (e.g. ultrasonic). Our interest here lies in monitoring the Earth from satellites remotely by means of electromagnetic radiations. We have learnt that only certain parts of the electromagnetic spectrum are useful for remote sensing. To keep these discussions confined within the scope of the book, we will further narrow down the scope to those applications and systems

Satellite Systems for Personal Applications: Concepts and Technology Madhavendra Richharia and Leslie David Westbrook
© 2010 John Wiley & Sons, Ltd

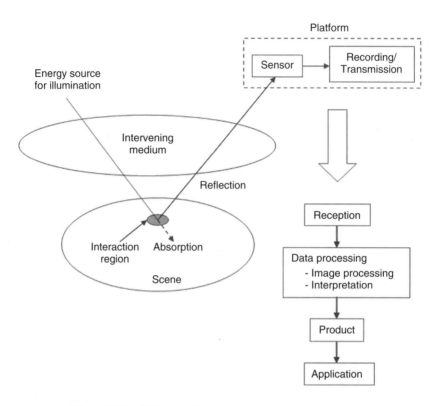

Figure 15.1   Main constituents of a remote sensing system.

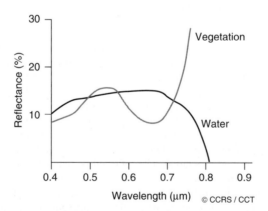

Figure 15.2   Reflectance of water and vegetation. Reproduced with the permission of the Minister of Public Works and Government Services, 2010 (CCRS, 2009).

that can potentially offer a service to individuals rather than serving the interests of organizations and governments. These 'personalized' services include high-resolution satellite imagery allowing users to see images in the areas of interest and meteorological services that provide accurate weather forecasts with visuals such as observed on television forecasts. Strictly, these services do not involve a direct interaction between the user and a satellite; however, we feel that these niche remote sensing applications are of direct relevance to individuals, and the book would miss an important contribution of satellite systems if they were ignored.

Satellites offer a number of advantages in remote sensing applications. A satellite has a large view of the Earth, potentially extending thousands of kilometres, and thereby allows a wide span to be sensed simultaneously. Following a launch, they continue to provide regular coverage of the regions of interest, often the whole world, for years.

The TIROS-1 (Television and Infrared Observation Satellite-1) satellite, launched by United States in 1960, heralded the era of satellite remote sensing. The satellite demonstrated the application of satellites to weather monitoring and forecasting. In 1966, US ATS-1 satellite demonstrated the utility of geostationary orbits in tracking weather fronts. The technology has since evolved such that today there are dozens of satellites in space owned and operated by numerous countries and organisation providing crisp weather images on our computer and television screens daily, and remote sensing has evolved to become a commercial service. More recently, satellite photography has caught the imagination of the public through the availability of photographic views of the Earth with very high resolutions allowing buildings and landscape to be recognizable, and coupled with aerial photography it has added a new dimension to industries such as real estate and tourism.

Table 15.1 (Canada Centre for Remote Sensing, 2009) illustrates examples of remote sensing applications to illustrate its wide scope. The literature is replete with excellent books and web tutorials on the subject. The references to this chapter list a few.

15.2   Remote Sensing Data

In the remote sensing literature, 'image' is a generic term that represents the data recorded at the sensor, irrespective of the wavelength, whereas 'photograph' refers only to those images that have been recorded on a photographic paper.

Remote sensing data are collected digitally or in analogue form. Satellite images are usually digital in nature, comprising a number of rows of picture units known as *pixels* that correspond to a specific area on the Earth typically denoted by latitude and longitude. The location of each pixel of an image is identified by a row and a column in an array. A brightness (intensity) level represents the measured radiation of the pixel. The resolution of the image depends on the pixel size

Table 15.1   Scope of remote sensing applications

Discipline/area	Example application
Agriculture	Crop growth monitoring, cataloguing crop type
Forestry	Mapping forest clearance, growth, monitoring forest fire
Geology	Structural mapping
Hydrology	Monitoring flood, soil moisture
Sea ice	Type and motion
Land cover	Rural or urban change
Mapping	Digital elevation models
Ocean and coastal	Oil spill detection
Personal application	Tourism, real estate, virtual tours, weather reports, hands-on school projects

when mapped on to the Earth, but it may degrade by signal processing. The intensity of the picture received by the sensor is digitized for signal processing. The resolution of the analogue-to-digital converter gives the 'radiometric' resolution of the data. Most of the remote imaging satellites carry a number of sensors to obtain a wider scope for data interpretation. A multilayer image is formed by stacking the radiation intensity from various sensors for each pixel.

The thematic information derived from the remote sensing images are often combined with auxiliary data to form the basis for a Geographic Information System (GIS) database, where each layer contains information about a specific aspect of the same area.

15.3   Sensors

15.3.1   Overview

Sensor systems are either fixed rigidly or incorporate a scanning mechanism. Some recent commercial imaging satellite systems include a capability to move the on-board camera rapidly to any desired location on the Earth to increase data capture in one pass or to reorient in response to specific needs.

The *swath width* of a sensor is the width of the area that a sensor views on either side of the orbital path. The point directly below the satellite is known as the nadir point (see Chapter 2). The Field of view (FOV) is the angle subtended by a sensor at a given instant.

A *scanning sensor system* comprises an electronic sensor that has a narrow field of view and builds up the image by scanning the scene rapidly along or at right angles to the direction of movement up to the swath. When the sensing is done in multiple bands, it is referred to as a MultiSpectral Scanner (MSS). In a scanning system, the sensor is moved (physically) to scan the scene point by point in a systematic way. The size of each discrete point (pixel) determines the spatial resolution of the system.

Scanning systems can be categorized by their scanning mechanism (Harrison and Jupp, 2009). In a *cross-track* electromechanical scanning system, the sensor oscillates from side to side capturing data at each point until the scene is fully captured. Such system benefit from a wide coverage.

In an electromechanical sensor, an oscillating or spinning mirror reflects the received radiation to a set of detectors which output proportional electrical signals. The signals are sampled, converted into a digital format and transmitted to the ground. The small dwell time available from the sensor implies that the signal-to-noise ratio of the signal is low – this implies that effective sensing is feasible only over relatively large spectral bands. The scanning mechanism introduces geometrical errors to the data that must be corrected for interpretation. The mechanism is used in Landsat's multispectral Thematic Mapper (TM) sensor in conjunction with the satellite's forward velocity. The TM is an Earth resource sensor capable of acquiring data in seven spectral bands – six bands for Earth reflectance (30 m spatial resolution) and one band for Earth temperature (120 m Earth resolution). The scanning mirror of the TM sensor collects data in the forward and return sweeps moving west to east as the satellite moves north to south. There are thus six reflectance lines on each scan.

Charge coupled devices (CCDs) are often used as the elements of the linear sensor array to form an image line in an *alongside* ('push broom') scanner. In a linear array, the detectors capture a line of a scene simultaneously with each element of the array representing a pixel. As the direction of the image line is that of the spacecraft motion, the scheme does not require a moving part, thus eliminating the error caused by timing that occurs in a scanning mirror. The arrangement allows a larger dwell time at each point and hence is amenable to supporting a narrower spectral line. Furthermore, sensor size and weight are small and the device power requirement is low. It also provides higher spatial and radiometric accuracy, although the arrangement requires very accurate calibration of a large number of detectors to avoid vertical striping in the image. Examples of this type of system are SPOT HRV (High-Resolution Visible) and Modular Optoelectronic Multispectral Scanner (MOMS) on the Space Shuttle.

In a *central perspective* scanning operation, either electromechanical or linear array technology is feasible. Here, the entire scene is captured simultaneously, such as in a photographic camera. Each line is imaged with respect to the centre of the image as depicted.

We note that sensors are used for recording the emitted data from a given scene in a chosen spectrum range and that the information is contained in the response of the target to the illumination. Hence, a sensor used in a land resource satellite is likely to differ from that of a meteorological satellite because the parameters of interest are widely different.

Sensors must be mounted on a steady platform to obtain an accurate representation of the target behaviour. The platform can be located on the ground or in air or space, however, remote sensing satellites have the advantage that they can view large parts of the Earth, or indeed the entire Earth (over multiple orbits), for several years. Even if they are an expensive alternative, they are ideal for certain types of mission. Urban growth pattern, changes to forest cover and progress of crops are some examples where the scene has to be revisited over months, years or even decades.

In general, three types of information are gathered by sensors – spectral, spatial and intensity (Short, 2009). We have already noted a distinction regarding the manner in which sensors collect data – to reiterate, active sensors transmit signals to illuminate the target and receive the reflected signals in a radar-like configuration, while passive sensors rely on a natural source of illumination.

The passive sensors may use a scanning mechanism or may be fixed. Sensors may be imaging, where an image of the scene is created from sensed data (e.g. on a photographic film or a monitor), or non-imaging, where the radiation from the scene is taken in entirety (integrated).

A *radiometer* is a sensor system that collects energy in a relatively large spectrum interval. A *spectrometer* collects energy at discrete wavelengths. A spectroradiometer collects energy in a band.

Sensors that capture the entire scene simultaneously (e.g. a camera) are known as *framing systems*, and those that scan the scene in discrete steps are known as *scanning systems*.

The *spatial resolution* of sensors provides the degree of detail discernible by the sensor. Objects that are smaller than the resolution are difficult to discern. The geometrical spatial resolution on the Earth is given as the product of the instantaneous field of view (IFOV) of the sensor (angular cone of visibility at a given instant) and the altitude of the sensor. The ground area for the IFOV is called the resolution cell. Note the size of a pixel may not be equal to the resolution cell. The spatial resolution of remote sensing satellites has improved considerably – down to a few metres or less.

The *spectral resolution* of a sensor is its ability to distinguish granularity of an object in the frequency domain. A finer resolution distinguishes features of a scene in a narrower spectral bin. Multispectral sensors are used for sensing in multiple bands – with more advanced versions permitting resolution into hundreds of bands.

The *radiometric resolution* is a measure of the intensity information content in an image, and hence a finer radiometric resolution shows a better ability to discriminate changes. A sensor using an 8-bit digitization code for recording brightness level has 2^8 (or 256) levels of brightness.

The *temporal resolution* deals with the repetition period of sensing an area. A satellite scans the area on the next pass, while parts of the area may be scanned from an adjacent pass if the swath is large. In a polar orbiting satellite, overlap between swaths increases at higher latitudes owing to the geometry of the orbit, while some types of sensor can be repositioned to view the same area from an adjacent pass. The temporal resolution allows variations in the landscape to be observed, such as the progress of a growth in a crop for agriculture monitoring.

Consider the spectral capabilities of Landsat, as an example – the satellites carry an MSS that senses four bands of spectrum (Harrison and Jupp, 2009):

- green 0.5–0.6 μm: band 4;
- red 0.6–0.7 μm: band 5;

- near infrared 0.7–0.8 µm: band 6;
- near infrared 0.8–1.1 µm: band 7.

Bands 1 to 3 were associated with another instrument carried on the first three satellites.

A scanning instrument called the Thematic Mapper (TM) and consisting of the following seven spectral bands operates on Landsat 4 and 5:

- blue/green: 0.45–0.52 µm;
- green: 0.52–0.60 µm;
- red: 0.63–0.69 µm;
- near infrared 0.76–0.90 µm;
- near middle infrared 1.55–1.75 µm;
- middle infrared 2.08–2.35 µm;
- thermal infrared 10.40–12.50 µm.

15.3.2 Optical Sensors: Cameras

Visual remote sensing (i.e. photography) is useful in situations where spatial resolution is more important than spectral resolution. Aerial and satellite photography are applied to an array of personal applications. Estate agents use satellite and aerial photographs to demonstrate areas of interest to clients; the tourist industry benefits through virtual tours of far-off locations; and applications like Google Earth, popular with web surfers, provide a convenient tool for visualization of any location on the Earth. Moreover, even for professional applications, visual images make interpretation easier for humans.

Conventional and digital cameras are in common use. Interestingly, Space Shuttle crews have on occasion photographed the Earth with handheld and high-quality zoom lenses. In a remote sensing satellite the cameras are mounted on a stabilized platform to reduce motion-induced distortion.

From a historical perspective, *conventional* cameras use photographic films, which, in addition to the visible band, may cover parts of the infrared and ultraviolet bands. Factors influencing the resolution of photographs are the altitude of the platform lens and film characteristics. The focal length of the lens determines the angular field of view, that is, the area photographed on the ground. Higher altitudes increase the swath area but reduce the resolution. Photographs are typically taken rapidly with the camera generally pointed vertically down, as this position exhibits the least distortion.

Digital cameras typically use a vast array of charge coupled device to capture a scene. Each device is charged in proportion to the brightness, and the resultant charge is encoded to provide the image in a digital format.

A *panchromatic sensor system* captures the full band and hence serves only as a black and white (intensity) sensing imaging system because the colour information is lost. Multiband photography uses filters to allow colours to be distinguished. It is used in situations where features must be identified more accurately. IKONOS PAN and SPOT HRV-PAN are examples of panchromatic sensor systems. LANDSAT MSS, LANDSAT TM and SPOT HRV-XS are examples of multispectral systems.

15.3.3 Non-Optical Sensors

We noted earlier that the absorptive and reflectance properties of targets differ. For instance, buildings and playing fields are easily distinguished because of differences in the reflectance and absorption properties of each. Sensors are optimized differently, depending on the absorption and reflectance properties and size of the targets. Thus, sensors for meteorology, land and marine observation have different spectral characteristics.

Land observation and meteorological sensing are broadly in line with the theme of the book. Hence, we will highlight the specific characteristics of sensors that apply to these areas. Reception and data processing of raw data are addressed later in the chapter. We present examples of a geostationary meteorological system and a polar-orbit constellation to illustrate the characteristics of remote sensing systems in Chapter 16.

However, a number of land and all of the maritime observations deal with multispectral sensing of land mass for applications that are primarily of interest at an institutional level. Land observation applications in this category relate to agriculture, forestry, urban planning, mapping, topographic information, vegetation discrimination, land cover mapping, natural resource planning, etc. Similarly, maritime applications relate to applications such as observing ocean colour, pollutants at the upper layer of ocean, nature of materials suspended in water columns, ocean influences on climate processes, etc.

Weather monitoring sensors have a relatively low spatial resolution but a high temporal resolution to capture the rapidly varying weather parameters such as atmospheric moisture, temperature, wind, cloud movement, movement of cyclones, tornadoes, etc., for forecasting. Depending on the mission, the spatial resolution varies from a few metres to several kilometres, while the temporal resolution varies from tens of minutes to several hours. Weather satellites typically use multispectral sensors to provide a variety of parameters needed for weather forecasting.

The visible portion is useful for applications such as detection of cloud and pollution, while infrared sensing provides for applications such as fog monitoring, fire detection, volcanic eruption, identification of cloud-drift winds, severe storms and heavy rainfall, identification of low-level moisture, determination of sea surface temperature, detection of airborne dust, etc. The spatial resolution for these types of monitoring is typically 1–4 km.

The Geostationary Operational Environmental Satellite (GOES) belongs to the National Oceanic and Atmospheric Administration (NOAA) and provides the United States National Weather Service with frequent, small-scale imaging of the Earth's surface and cloud cover for weather monitoring and forecasting in the United States. The GOES second-generation satellite imager has five channels to sense visible and infrared reflected and emitted radiation from the Earth. The infrared sensor is used for day and night monitoring. The sensor has the capability of scanning or pointing in any direction. The data have a 10-bit radiometric resolution. Table 15.2 lists the individual bands, spatial resolution and applications. Figure 15.3 illustrates an image of Hurricane Linda snapped on the afternoon of 9 September 2009 by the GOES-11 visible band sensor. (NOAA-NASA GOES Project, 2010).

Table 15.2   GOES wavelength, spatial resolution and application (CCRS, 2009)

Band	Wavelength (μm)	Spatial resolution (km)	Example Applications
1	0.52–0.72 (visible)	1	Cloud, pollution, haze, severe storm detection
2	3.78–4.03 (shortwave IR)	4	Night fog identification, water clouds, snow or ice cloud discrimination during day, fire/volcano detection
3	6.47–7.02 (upper level water vapour)	4	Mid-level atmospheric tracking
5	11.5–12.5 (IR window for water vapour)	4	Detection of sea surface temperature, airborne dust and volcanic dust.

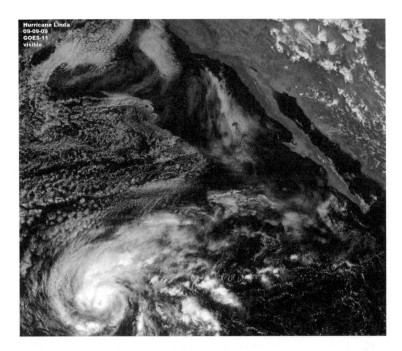

Figure 15.3 An image of Hurricane Linda taken on the afternoon of 9 September 2009 by the GOES-11 visible band sensor. Reproduced with permission of the NOAA-NASA GOES Project.

15.4 Image Processing

The data acquired by the Earth imaging sensors represent the image on the curved surface of the Earth. The accuracy of the data is compromised by the orbit, the Earth's rotation, the stability of the platform, the geometrical relationship between the sensor and the target and the motion of the scanning mechanism. These distortions should be corrected or minimized before the data can be interpreted accurately.

Relief distortion is caused in a camera snapshot by the difference in angle between the object and target. The distortion in the satellite image increases with offset between the object and the nadir – essentially, objects begin to lean in one direction. *Tangential distortion* is caused in a scanning system because of the larger area covered at the edges relative to the nadir, causing an elongation of the objects as the scan moves away from the nadir. *Skew distortion* is caused in a scanning system by the west–east rotation of the Earth, as the Earth has moved by a small amount to the east at the start of each new scan, causing a skew.

Digital processing involves preprocessing of data so as to effect image enhancements, image transformation, classification, and analysis.

Preprocessing prepares the data for image enhancement by applying radiometric and geometric corrections. *Radiometric corrections* apply to sensor or atmospheric noise, particularly when conditions vary while recording a scene, for example, when viewing a target on a different day or from a different view angle. Radiometric corrections attempt to minimize the impact of variations caused by: illumination levels; the geometry of the sensor with respect to the target and illuminating source; atmospheric conditions; sensor noise and sensor sensitivity. The corrections are essential in applications where a scene is to be assessed under a uniform illumination condition across various images, for example, when comparing images recorded on different sensors or the same sensor on different days. Techniques used for radiometric correction include:

- modelling of the geometry between the target, illuminating source and sensor;
- application of atmospheric models to correct atmospheric effects during data acquisition;
- calibration based on data contained within an image (e.g. a patch of known brightness level).

Noise in the image is caused by imperfect or erroneous sensor response, errors in the recording system or errors caused during transmission. Systematic errors such as 'banding' (striped images) can be corrected by applying an appropriate correction to the data. Random errors such as 'dropped lines' are corrected by replacing them with their best estimates.

Geometric corrections involve removal of various types of distortion mentioned earlier in the section, namely relief, tangential and skew distortions caused by sensor–Earth geometry and perspective, the scanning system, spacecraft motion, Earth motion and curvature. After elimination of the error, the targets can be placed at the corrected positions on the Earth. The predictable distortions can be modelled and cancelled. A technique known as the *geometric registration* process can be applied to determine and correct random errors. A few locations on the image are marked and compared against their actual location. The distortion is then modelled mathematically, and the transformation is applied to the acquired data. Instead of calibrating against real ground locations, an area in the image itself may be used to obtain a delta correction value. Correction of the original image necessitates that new pixel values be derived for assignment to the new coordinates. This exercise is known as *resampling*. Common algorithms used for this purpose are:

- *Nearest neighbour.* The pixel of the original image nearest to the new position is chosen. Although this is a simple method that retains the original pixel value, it leads to a 'disjointed or blocky' appearance.
- *Bilinear interpolation.* The weighted average of the four pixels nearest to the new position is taken, resulting in loss of the original pixel value but a clearer image than with the nearest-neighbour method. Loss of pixels could affect subsequent results of processing.
- *Cubic convolution.* Here, the weighted average of 16 pixels of the original image around the new position is taken resulting in loss of the original pixel value but a clearer image than with either of the two previous methods. Loss of pixels could affect subsequent results of processing.

Image enhancement improves the presentation to suppress unwanted features of images in order to assist visual interpretation by digitally processing specific characteristics of an image. The pre-processed data must be further enhanced to facilitate interpretation particularly of satellite images as they capture a massive area containing a vast variety of objects. An image interpreter is likely to be interested in only a certain aspect of the image. For example, a tour operator may wish to highlight the facilities in and around a resort. The image is modified to enhance the features of interest.

In *contrast enhancement* the range of brightness values of pixels is expanded to utilize the unused brightness levels. The resolution of the brightness level depends on the encoding bits used in the analogue-to-digital conversion. An 8-bit digitization, for example, supports 256 levels. An image is unlikely to utilize all 256 levels, and hence it is possible to expand the brightness range to the full capability of the system to give an increased contrast between image objects. In addition, the subject matter of interest may only span a limited range of brightness values within that of the whole scene. There are various techniques to achieve this transformation. In a *linear contrast stretch*, the given range is expanded uniformly to fill the available resolution. In a histogram-equalized stretch, the levels are expanded on the basis of their occurrence, with those occurring more often being expanded more. It would also be feasible to expand a specific range of brightness values. This allows specific areas of interest to be highlighted.

Spatial filtering deals with filtering of the image texture. It would, for example, involve moving a window across an image and filtering the pixels within the window using an appropriate transformation such as a low- or high-pass filter to enhance a specific feature of the texture. A low-pass filter filters out the finer details to emphasize homogeneous areas, thereby giving the image a smoother look. A high-pass filter, by contrast, emphasizes the finer details (edges). A directional filter emphasizes linear features of the image such as a row of houses or a road.

Image classification and analysis involves identifying the characteristics of each pixel in an image and assigning the pixel a class based on its statistical properties. The classification can either be supervised by an expert or be unsupervised through algorithmic processing.

Image transformation produces images by processing of data from multiple images – obtained from a multispectral sensor or from the same sensor taken at different times – to highlight features that are missed when analyzing the images separately. For instance, image subtraction can be applied to study the growth of crops or to assess urban development – the difference of the 'then' and 'now' images highlight the change. Another commonly used transformation is called ratioing, where the ratio of images of two bands is taken to capture variations across bands which would not be feasible otherwise. The resulting image highlights those targets where changes are significant. Other methods of ratioing are available.

The *principal components analysis* transform removes redundancy between several sensor outputs and compresses as much information as possible into fewer new bands to facilitate interpretation.

15.5 Image Interpretation

In many situations, remotely sensed images require human involvement in their interpretation. Although it is feasible to interpret the images algorithmically, human involvement is sometimes preferred. On the other hand, human involvement is time consuming (and costly) and susceptible to different interpretations and depends on the experience of the interpreter. While human interpretation requires skill and knowledge, computer interpretation requires complex and often expensive equipment and software. Computer assistance provides a powerful tool to manage complex sets of images, for example, involving several simultaneous channels.

Human interpretation requires special skills because the images are (generally) depicted in two dimensions from an unfamiliar perspective and scale, often as *false colour* or sensed at a wavelength outside the visible band. False colour represents a wavelength (colour) at a different wavelength – for instance, infrared images may be presented in the visible band. Generally, stereoscopic projections obtained by combining images from two different angles are better suited for human interpretation. Recognition of targets involves visual elements of tone, shape, size, pattern, texture, shadow and association. Table 15.3 summarizes the characteristics of these attributes, with examples of visual analysis and interpretation.

15.6 System Characteristics

Figure 15.4 shows the main components of a remote sensing satellite system. Low Earth orbit (LEO) and geostationary orbit (GEO) are in common use for the remote sensing space segment. In this context an orbit relates to the coverage area, geometric resolution and temporal repetition. As remote sensing satellites cover a variety of applications, the spectral band and resolution of the sensors vary. For instance, the spatial resolution may vary from about 1 km for weather monitoring to 5 m or less for photography. Spectral resolution ranges from mono-spectral to hyper-spectral systems segmented into hundreds of spectral bands.

Polar or near-polar LEO systems can provide higher resolution compared to GEO system owing to their closer vicinity to the Earth, together with full global coverage repeated every few days. Global

Table 15.3 Visual attributes and their characteristics

Visual attribute	Characteristics	Interpretation example
Tone	Relative brightness and color; fundamental property in determining an object	Compare shades of a playing field and a building in a photograph
Shape	Assists in object identification	Compare man-made features (e.g. rows of buildings) and natural features (e.g. a coastline)
Size	Object size	A playing field can scale other targets
Pattern	Object arrangement	Rows of houses
Texture	Variation in tone in an image	Forested areas have a rougher texture than a desert
Shadow	Relates to angle of illumination and object height	Can assist in estimating the height of an object
Association	Relates to surroundings with respect to a known object	A distinguished landmark (e.g. an airport) assists in identifying objects in the vicinity

coverage is ideal for applications that require regular worldwide or large-scale monitoring over long periods. Sun-synchronous orbit–due to constant relationship with the sun – has the advantage of appearing at the same time over each location, thereby collecting data repeatedly in similar lighting conditions to facilitate interpretation.

The GEO system provides a wide and continuous coverage of the Earth, covering nearly one-third of the globe. It therefore permits meteorologists to view wide-area movement of the weather

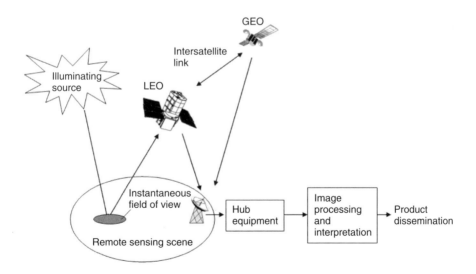

Figure 15.4 The main components of a remote sensing satellite system.

conveniently and repeatedly – for instance, monsoon progress, cloud and wind movements, hurricanes or tornado movement (Figure 15.3). The weather forecast images presented on television screens are typically derived from geostationary meteorological satellites.

A LEO imaging system typically consists of a constellation of 850–1000 km altitude low Earth orbiting satellites incorporating a high-resolution camera system. The pictures are transmitted back to the Earth where they are processed and made available to the users. In the past decade, satellite remote sensing has become increasingly commercialized, delivering images for various types of personal and other application. Chapter 16 presents samples of images taken from GeoEye-1, Quickbird and Landsat LEO systems.

The data captured by on-board sensors are transferred via traditional satellite communications links to the ground station visible from the satellite. In a geostationary system the data can be transferred in real time because the satellites remain stationary with respect to ground. In a LEO system the satellite does not have a continuous visibility with the feeder Earth station; hence, the satellite stores data until a ground station becomes visible. Another alternative (Figure 15.4) is to transfer the data to the Earth station in real time through intersatellite links, where the data are passed from satellite to satellite until a suitable ground station becomes visible.

The maximum data transfer rate depends on the telecommunication channel throughput and sensor capability. The volume of data is large for multispectral sensors and increases directly with an increase in sensor resolution, requiring a trade-off with radio link capacity. Consider the SPOT multispectral scanner. It covers a swath of $60\,\text{m}^2$ at a resolution of 20 m; hence, with an 8-bit A–D converter, the data volume would be 27 MB per image, equating to 81 MB for three bands. A panchromatic (broadband) sensor, on the other hand, requires only a single image and hence can provide a finer resolution for the same data throughput. The raw received data are processed as described in the preceding sections. Some satellite systems support quick turnaround of imagery for those who need access to the imagery rapidly, that is, in hours. The imagery in such systems tends to have a lower resolution.

Revision Questions

1. Explain the remote sensing concept with the help of a diagram.
2. What are the factors that govern the suitability of a band for remote sensing?
3. (a) What are the different types of sensor system? (b) Mention the main properties of these sensor systems. (c) Which properties of the sensor systems are used for remote sensing of weather fronts?
4. Describe the steps involved in image processing.
5. With the help of a diagram, discuss features of a remote sensing system, including the role of satellite orbits.

References

Canada Centre for Remote Sensing, (2009) http://ccrs.nrcan.gc.ca/resource/tutor/fundam/chapter5/01_e.php [accessed June 2009].

CCRS (Canada Centre for Remote Sensing) (2009) *Fundamentals of Remote Sensing*. Available: http://ccrs .nrcan.gc.ca/resource/tutor/fundam/pdf/fundamentals_e.pdf [accessed June 2009].

Harrison, B. and Jupp, D.L.B. (2009) *Introduction to Remotely Sensed Data – Part One of microBRIAN Resource Manual*. Commonwealth Scientific and Industrial Research Organization (CSIRO), Australia. Available: http://ceos.cnes.fr:8100/cdrom/ceos1/irsd/pages/datacq4.htm#anchor1329343 [accessed July 2009].

NOAA-NASA GOES Project (2010) http://goes.gsfc.nasa.gov/ [accessed February 2010].

Short, N. (2009) *Remote Sensing Tutorial. Version: February 26, 2009*. NASA. Available: http://rst.gsfc.nasa .gov/ [accessed July 2009].

16

Remote Sensing Systems

16.1   Introduction

A large number and variety of remote sensing satellite systems belonging to various countries and organizations are in operation and planned. Their characteristics – spatial resolution, spectral band, orbit, coverage area and repetition rate – depend on their mission objectives. A majority of the weather forecast imagery appearing on television and the Internet is obtained from geostationary satellite systems, whereas images of more personal interest – tourist locations, cities, interesting events around the globe – are generally collected by low Earth orbiting satellites. Major satellite remote sensing programmes have continued to upgrade satellites to benefit from technical advancements, or modify instruments on the basis of acquired knowledge, while continuing the use of the previous generation systems. In doing so, previous generations continue to provide their services. Table 16.1 summarizes features of representative systems. The reader is referred to the vast remote sensing system literature for a detailed exposition (e.g. Liew, 2009; Short, 2009).

The World Meteorological Organization (WMO – discussed later) considers the following operational meteorological satellites as candidates of a global observation system (Bizzari, 2009):

Geostationary series:

- Meteosat (Europe);
- GOES (United States);
- MTSAT-replacing GMS (Japan);
- GOMS-Elektro (Russia);
- FY-2 followed by FY-4 (China);
- INSAT and Kalpana (India);
- COMS under development (Korea).

Sun-synchronous series:

- POES, supported by DMSP, to converge into NPOESS (United States);
- MetOp (Europe);
- Meteor (Russia);
- FY-1 and FY-3 (China).

The WMO long-term concept of a global observation meteorological satellite system consists of satellites in both geostationary and Sun-synchronous orbits arranged as follows:

- six satellites regularly spaced in the geostationary orbit;
- four satellites optimally spaced in Sun-synchronous orbits (three, if feasible);
- a comparable imaging quality across systems.

Satellite Systems for Personal Applications: Concepts and Technology   Madhavendra Richharia and Leslie David Westbrook
© 2010 John Wiley & Sons, Ltd

Table 16.1 Example of remote sensing satellite systems

Satellite system	Country/operator	Objectives	Orbit	Coverage	Characteristics			Comments
					Bands	Resolution		
Geostationary Operational Environmental Satellite (GOES)	USA/NASA and National Oceanic and Atmospheric Administration (NOAA)	Meteorology: national weather service operation	Geostationary (GOES-12 at 75° W and GOES-10 at 135° W)	American Continent	Optical, infrared	Low		GOES-NO/P is next series of satellites; supports SARSAT, contributes to worldwide environmental warning services and provides useful data to extend knowledge of atmosphere (GOES, 2009)
Fengyun series	China	Meteorology	Geostationary (105° E)	Asia-Pacific	Visible infrared; water vapour (WV)	Visible: 1.25 km. Infrared and WV: 5 km		Full-view image every 30 min; refers to flight FY-2B; improved performance expected in later launches
Geostationary Meteorological Satellite (GMS series)	Japan/National Space Development Agency (NASDA).	Meteorology	Geostationary (140° E)	Asia-Pacific	visible; infrared	Visible: 1.25 km, Infrared: 5 km		Full-view image every 30 min
INSAT series	India/joint venture of several government organizations	Meteorology	Geostationary	Indian region	Visible; infrared; water vapour	Radiometer: Visible: 2×2 km; infrared: 8×8 km; water vapour: 8×8 km CCD Camera: Visible : 1×1 km Infrared: 1×1 km		Only some INSAT satellites support meteorology

NOAA-Polar Orbiting Operational Satellite (POES)	USA/NOAA	Meteorological observation, atmospheric profile sounding and energy budget	Sun-synchronous, near polar, ~102.1 min period; mean altitude = 851 km	Visible; infrared	1.1 km	
Satellite Pour Observation de la Terre (SPOT)	France/Centre National d'Etudes Spatiales (CNES)	Earth imaging: land use, agriculture, forestry, geology, cartography, regional planning, water resources, GIS applications	Sun-synchronous, near polar, mean altitude = 832 km, ~101 min period; 26 days repetition	Visible; infrared multispectral band	Panchromatic: 5 m; Multispectral band: 20 m	Commercial imaging system
Indian Remote Sensing (IRS)	India/ISRO and National Remote Sensing Agency	Earth resource	Sun-synchronous, 817 km mean altitude; 101 min period; 24 day repetition	Optical; infrared	Panchromatic: 10 m Multispetral: 23.5 m Short Wave IR: 70 m Widefield sensor: Red 189 km; Near IR 189 km	

We have chosen to expand the features of the following systems:

- two commercial imaging systems. namely GeoEye and DigitalGlobe;
- a meteorological system called Eumetsat, which provides meteorological and environmental measurements over the European region while also contributing to a global meteorological system;
- a well-known land resource monitoring system called LANDSAT.

16.2   Commercial Imaging

Remote sensing systems are generally owned and operated by large organizations supported by governments. With the maturing of remote sensing technology, the associated system infrastructure, and their proven economic benefits, the technology has entered the commercial domain. Various commercial imaging systems capable of delivering high quality images tailored to specific requirements of customers are in place. USA based DigitalGlobe and GeoEye commercial ventures provide a variety of imaging solutions that are used for numerous personal rechnology solutions.

16.2.1   DigitalGlobe

DigitalGlobe offers high-resolution commercial imagery with submetre resolution in panchromatic, colour and infrared for a wide range of applications (DigitalGlobe, 2009). Its products are supplied in various formats and delivery methods. The satellite constellation consists of WorldView-1 and QuickBird satellites, which have a panchromatic resolution of 50 cm and 61 cm respectively. QuickBird offers a multispectral resolution of 2.44 m.

DigitalGlobe products are used in several industries, including defence and intelligence, government, humanitarian, entertainment, insurance, location-based service/navigation, mining, natural resources, oil and gas, online mapping, real estate and telecommunication. Applications of interest to us include, but are not limited to, the humanitarian location-based information and entertainment sectors. They include emergency planning, damage assessment, relief coordination, recovery management, etc. The entertainment industry benefits from imagery solutions in a variety of applications such as 'simulated Earth' experience, video games to create 'fly-through' videos, flight simulators, as well as other simulation experiences for television and films. The news media use the imagery to place breaking stories into a visual context. Movies and gaming utilize the imagery to position the actors in a real-world perspective.

The salient features of the DigitalGlobe products are summarized below:

- Provides relevant spacecraft telemetry and camera models with satellite imagery.
- Large image swath collection size:
  - 17.6 km width at nadir: WorldView-1;
  - 16.5 km width at nadir: QuickBird.
- High radiometric response:
  - 11-bit digitization (up to 2 048 levels of grey scale);
  - discrete non-overlapping bands.
- Open systems:
  - spacecraft telemetry and camera model information supplied;
  - compatible with leading commercial software providers;
  - common image file formats;
  - products available in raw form allowing users to perform their own processing.

Table 16.2   QuickBird characteristics

Launch date	18 October 2001
Orbit	Altitude = 440 km, inclination = 97.2°, Sun-synchronous, descending node = 10.30 a.m., period = 93.5 min
Speed	7.1 km/s
Revisit time	1–3.5 days, depending on latitude
Swath width	16.5 km at nadir
Metric accuracy	23 m horizontal
Digitization	11 bits
Resolution (panchromatic)	60 cm (nadir) to 72 cm (25° off-nadir)
Resolution (multispectral)	2.44 m (nadir) to 2.88 m (25° off-nadir)
Image bands:	
Pan	725 nm
Blue	479.5 nm
Green	546.5 nm
Red	654 nm
Near IR	814.5 nm

Support Data files supplied for photogrammetric processing include attitude and ephemeris data, geometric calibration, camera model, image metadata, radiometric data and rational functions.

Tables 16.2 and Table 16.3 respectively list the major characteristics of the QuickBird and WorldView-1 satellites.

Figure 16.1 shows a grey-scale, 60 cm resolution QuickBird satellite image featuring Kauai, the northernmost of the main Hawaiian Islands taken on 14 October 2004. Note the clarity of the buildings, roads and other features.

Table 16.3   WorldView-1 characteristics

Launch date	18 September 2007
Orbit	Altitude = 496 km; Sun-synchronous, descending node = 10.30 a.m., period = 94.6 min
Mission life	7.25 years
Sensor bands	Panchromatic
Sensor resolution	0.5 m at nadir; 0.59 m at 25° off-nadir
Dynamic range	11 bits per pixel
Swath width	17.6 km at nadir
Attitude	Three-axis stabilized
Pointing accuracy	500 m at start and stop
Retargeting agility	acceleration = 2.5 deg/s; rate = 4.5 deg/s; time to slew 300 km = 10.5 s
Communication	800 Mb/s at X band
Maximum viewing angle	±45°
Accessible ground swath	1036 km
Revisit frequency	1.7 days at 1 m GSD or less; 4.6 days at 25° off-nadir or less [0.59 GSD (Ground Sample Distance – a measure of resolution per pixel]
Geolocation accuracy	6.5 m at nadir

Figure 16.1 A 60 cm resolution QuickBird satellite image featuring Kauai, the northernmost of the main Hawaiian Islands. Reproduced by permission of © DigitalGlobe.

16.2.2   GeoEye

GeoEye is a US-based company that provides geospatial data, information and value-added products to a variety of customers using an infrastructure consisting of a satellite constellation, aircraft, an international network of ground stations, an advanced image processing facility and an image archive (GeoEye, 2009). GeoEye products support mapping, environmental monitoring, urban planning, resource management, homeland defence, national security, emergency preparedness, etc. The satellite constellation consists of GeoEye-1, Ikonos and OrbView-2.

OrbView-2, launched in 1997, collects colour imagery of land and ocean surfaces, and, using this, GeoEye built the SeaStar Fisheries Information Service for commercial fishing vessels. OrbView-2 detects changing oceanographic conditions used to create fishing maps that are delivered directly to commercial fishing captains at sea. The satellite also provides broad-area coverage in 2 800 km wide swaths, which are routinely used in naval operations, environmental monitoring and global crop assessment applications (GeoEye, 2009).

The Ikonos satellite launched in 1999 provides panchromatic images with 1 m resolution and multispectral imagery with 4 m resolution, which can be merged to create 1 m pan-sharpened colour imagery. Ikonos imagery is used for national security, military mapping and air and marine transportation by regional and local governments. Ikonos has a revisit time of 3 days, and downlinks directly to more than a dozen ground stations spread around the globe.

The GeoEye-1 satellite (see Figure 16.4), launched in 2008, captures panchromatic images of 0.41 m and multispectral images of 1.65 m. Figures 16.2 and 16.3 respectively illustrate images of central London and the Taj Mahal captured by GeoEye-1. The satellite has a revisit time of 3 days. Table 16.4 compares the salient features of Geoeye-1 and Ikonos.

16.3   Meteorology

The World Meteorological Organization (WMO), established in 1951, is a United Nations agency that promotes the safety and welfare of humanity, the understanding of the state and behaviour of the Earth's atmosphere and its interaction with the oceans, the climate and the distribution of water resources. The global span of the atmosphere and the associated disciplines necessitates an international framework. WMO was conceived to be a platform for such an endeavour. The organization 'promotes the establishment of networks for making meteorological, climatological, hydrological and geophysical observations, as well as the exchange, processing and standardization of

Figure 16.2 City of London – Buckingham Palace, Hyde Park, the River Thames and other landmarks are clearly discernible in the foreground. Reproduced by permission of © GeoEye Inc.

Figure 16.3 Taj Mahal, Agra, India, photographed from a GeoEye satellite. Reproduced by permission of © GeoEye Inc.

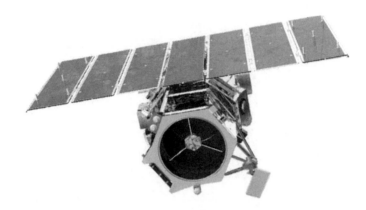

Figure 16.4   GeoEye-1 spacecraft. Reproduced by permission of © GeoEye Inc.

Table 16.4   A comparison of salient features of Geoeye-1 and Ikonos

Satellite feature	Geoeye-1	Ikonos
Spectral range (pan)	450-800 nm	526-929 nm
Blue	450-510 nm	445-516 nm
Green	510-580 nm	505-595 nm
Red	655-690 nm	632-698 nm
Near IR	780-920 nm	757-853 nm
Pan Resolution at nadir	.41 meters	.82 meters
Pan Resolution at 60° elevation	.50-meters	1.0 meter
Multi-spectral Resolution at nadir	1.64 meters	3.28 meters
Swath width at nadir	15.2 km	11.3 km
Launch date	06-Sep-08	24-Sep-99
Life Cycle	7 years	Over 8.5 years
Revisit Time	3 days at 40° latitude with elevation >60°	3 days at 40° latitude with elevation >60°
Orbital Altitude	681 km	681 km
Nodal Crossing	10:30 a.m.	10:30 a.m.

related data' (WMO, 2009). Space technology offers a powerful observation medium to facilitate observation of the Earth's surface and the atmosphere. WMO's space programme is therefore developing a space-based Global Observing System (GOS) and promoting the use of satellite data for weather, water, climate and related applications. It has developed a 'Vision for the GOS in 2025'. The satellite segment of such a system would comprise geostationary satellites spaced around the world to offer high elevation angle view and contingency back-up satellites in conjunction with a constellation of low Earth orbiting satellites to provide supplementary data and cover the polar region.

Table 16.5 and Figure 16.5 illustrate the present status of the geostationary component of an Earth observation system as agreed within WMO's Coordination Group for Meteorological Satellites (CGMS). The constellation includes six spacecraft to ensure global coverage from 50° S to 50° N with a zenith angle lower than 70° (WMO, 2009).

Table 16.5 The current global planning for operational geostationary satellites developed among WMO and satellite operators within the CGMS (Bizzari, 2009)

Region	Nominal operator	Nominal operational locations
North, Central and South America (GOES-West)	USA (NOAA)	135° W
East Pacific (GOES-East)	USA (NOAA)	75° W
Europe and Africa	Eumetsat	0°
Indian Ocean	Russian Federation (Roshydromet)	76° E
Asia	China (CMA)	105° E
West Pacific	Japan (JMA)	140° E

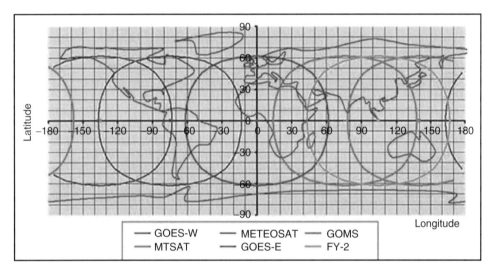

Figure 16.5 Footprints of a global operational geostationary meteorological satellite system as developed within WMO (Bizzari, 2009) Location: GOES-W: 135°W, MTSAT: 140°E, METEOSAT: 0°, GOES-E: 75°W, GOMS: 76.5°E, FY-2: 105°E. Reproduced by permission of © World Meteorological Organization.

In the following section, we describe Eumetsat's Meteosat system.

16.3.1 Meteosat

The European Organization for the Exploitation of Meteorological Satellites (Eumetsat), founded in 1986, delivers weather and climate-related data, images and products to the national authorities of the European member states and other countries. It operates a fleet of two generations of geostationary satellites known as Meteosat, complemented with polar non-geostationary satellites known as Metop and Jason-2 with associated ground segment. The data are used primarily for weather and climate monitoring (Eumetsat, 2009). The first Meteosat was launched in 1977 as part of an ESA (then called, ESRO) research programme. The success of the research initiative led ESA to convert the programme into an operational programme, resulting in the formation of the Meteosat Operational Program (MOP) and subsequently Eumetsat.

The first operational satellite Meteosat-4 was launched in March 1989, followed by launches of the Meteosat-5, Meteosat-6 and Meteosat-7 in 1997 to conclude the first-generation spacecraft series. The Meteosat Second-generation (MSG) satellites, which provide improved accuracy and better temporal and spectral resolution began to be introduced in 2002. In 2009, Meteosat-9 (MSG-2, the second in the series) provided the service in Europe and elsewhere. MSG-3 and MSG-4 are scheduled for launch respectively in 2011 and 2013. Meteosat-6 and Meteosat-7 serve the Indian Ocean as a part of the World Meteorological Organization's global observing system. Eumetsat and ESA are currently preparing for Meteosat Third-Generation (MTG) satellites.

Satellites

The Meteosat satellites are being improved in each generation in accordance with user expectations and improvements in technology.

The first generation, 800 kg spin-stabilized satellites, have a three-channel Meteosat Visible and Infrared Imager (MVIRI) with a repeat cycle of 30 min. The three channels operate in visible, infrared and water vapour absorption regions (5.7–7.1 μm) of the electromagnetic spectrum. The visible band, which corresponds to peak solar radiance, is used for imaging during the day; the water vapour band measures the water vapour content in the upper atmosphere; the thermal infrared band is used for 24 h imaging, including determination of the temperature of cloud tops and ocean surface. The radiation is gathered by a reflecting telescope that has a primary diameter of 400 mm.

As mentioned, the satellites are a part of the World Meteorological Organization's global observing system. A Meteosat first-generation satellite performs the Indian Ocean Data Coverage (IODC) service, providing operational data and images from over the Indian Ocean, backed up by another first-generation satellite.

Meteosat second-generation satellites – four in total – serve the European region and are planned to be in operation until 2018. The 2 tonne second-generation satellites are spin stabilized (100 rpm), comprising a 12-spectral-band radiometer known as the Spinning Enhanced Visible and InfraRed Imager (SEVIRI), which delivers 'daylight images of the weather patterns with a resolution of 3 km, plus atmospheric pseudosounding and thermal information'. Eight of the channels are in the infrared. One of the channels is the High-Resolution Visible (HRV) channel, which has a resolution of 1 km. The sampling rate of the Earth is 15 min to provide monitoring of rapidly evolving events such as thunderstorms, heavy rain, fog and intense depressions. There is also a rapid scanning mode, which provides a much shorter sampling period. In addition, the satellites carry the Geostationary Earth Radiation Budget (GERB) visible–infrared radiometer for climatology studies. It provides data on reflected solar and thermal radiation emitted by the Earth and the atmosphere. The satellites also carry a communications payload for conducting satellite operation, data communication and user data dissemination, along with a search and rescue transponder that can pass distress messages from a mobile terminal. At present, the second satellite of the series, Metosat-9, located at 0°, provides the prime operation with Meteosat-8 as an in-orbit spare.

The Meteosat Third-Generation (MTG) satellites, weighing 3 tonne, will be three-axis stabilized, each with a different observation mission. The sensors being considered include a 16-channel combined imager capable of full-disc high-spectral resolution imagery and fast imagery for the first satellite.

Ground Segment

The Eumetsat multi-mission ground segment of the second generation is illustrated in Figure 16.6.

The control and processing centre, located in Darmstadt, Germany, controls the Eumetsat satellites through primary ground stations located in Usingen, Germany, and pre-processes all acquired data. Raw data from the satellite are calibrated and geometrically and radiometrically corrected. The primary station is backed up by stations located in Spain and in Romania.

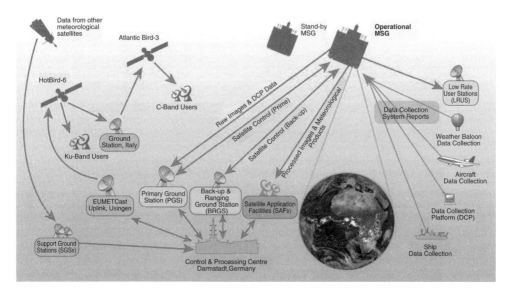

Figure 16.6   Eumetsat second-generation ground segment. Reproduced by permission of © 2010 Eumetsat.

The ground segment also supports the Global Earth Radiation Budget (GERB) mission, handling all communications with the instrument, and the reception of raw GERB data, which are then sent to the central GERB ground segment at the Rutherford Appleton Laboratory (RAL) in the United Kingdom for processing and forwarding to other European institutions. The primary ground station also supports all ranging functions and communications.

A products extraction facility analyses Meteosat images to provide a wide range of meteorological products. A Data Acquisition and Dissemination Facility (DADF) performs data acquisition from Meteosat satellites, formats and encrypts preprocessed image data in dissemination formats and monitors the MSG dissemination performance. A Unified Meteorological Archive and Retrieval Facility (UMARF) archives images and meteorological products from all Eumetsat satellite programmes and provides user access to these data.

The Satellite Application Facility (SAF) is a geographically distributed network. The SAF further processes satellite data into meteorological and geophysical products.

Products

The main service provided by the Meteosat system is the generation of images of the Earth, showing its cloud systems both by day and night, and the transmission of these images to the users in the shortest practical time. There are several other important supporting services summarized below.

Eumetsat services may be categorized as real-time and off-line services. Eumetcast disseminates MSG High-Resolution (HR) and Low-Resolution (LR) SEVIRI data. Low-rate SEVIRI image data are disseminated from Meteosat-9 satellites in Low-Rate Information Transmission (LRIT) mode directly to users possessing a Low-Rate User Station (LRUS). The real-time services include: SEVIRI image rectification and dissemination, meteorological product extraction and distribution, the data collection and retransmission service, Meteorological Data Dissemination (MDD) and search and rescue.

Offline data services include archive and retrieval services and an internet image service.

Figure 16.7 On 23–25 January 2009, Western Europe was hit by one of the worst storms of the decade which caused extensive damage. The progress of the storm was tracked by Meteosat-9, as illustrated by the snapshot above. Reproduced by permission of © 2010 Eumetsat.

Figure 16.7 demonstrates a typical image available from the Meteosat system. The image shows the progress of one of Western Europe's worst storms of the decade (Wettre and Roesli, 2009). The progress of the storm was tracked by Meteosat-9 for the period 23–25 January 2009, as illustrated above by a snapshot.

16.4 Land Observation

Land observation satellites are used for land resource survey, and hence their sensors are optimized to monitor land features like vegetation, rocks, urban areas, etc. The data are used by governments and organizations interested in gathering and utilizing knowledge related to the Earth land masses and coastal regions. Those benefiting from the land observation data include agribusiness, global change researchers, academia, state and local governments, commercial users, military and the international community. Here, we confine our discussion to the Landsat programme. Figure 16.8 shows Landsat 5 images of Las Vegas's growth over 25 years (1984–2009). The image was produced in two false colours – dark purple (city street grid) and green (irrigated vegetation). Although the effect is masked in the grey-scale reproduction, the expansion in boundary of the city is clearly discernible.

The programme, initiated in 1972, is managed by NASA and the US Geological Survey (USGS) (Landsat, 2009). NASA develops and launches the spacecraft and the USGS manages the operations, maintenance, ground data reception, processing, archiving, product generation and distribution.

The programme is an integral part of the US Global Change Research Programme and part of a long-term global research programme called NASA's Earth Science Enterprise, promoting 'a better understanding of natural and man-made environmental changes'.

Las Vegas   1984   2009

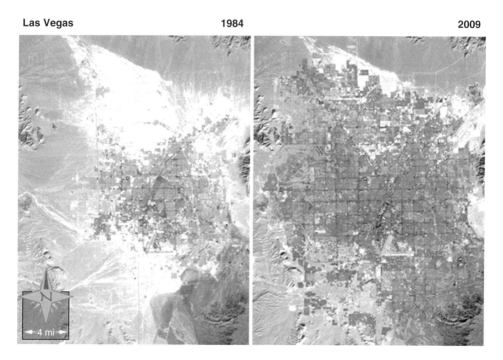

Figure 16.8 Landsat 5 images of Las Vegas's growth over 25 years (1984–2009). The expansion in boundary of the city is clearly discernible. Courtesy NASA.

16.4.1 Landsat

The programme has promoted the creation of a rich data archive that facilitates a wide range of studies by scientists and planners (NASA, 2009). The data have assisted them worldwide in identifying, monitoring and understanding forest fires, storm damage, agriculture trends, natural resources management, disaster preparedness, disease epidemic prevention, urban growth and urban planning. Landsat data are used extensively in education in disciplines such as geography, biology, ecology and Earth's evolution, where 'spatial thinking' adds value – understanding the water and carbon cycle, urbanization trends, deforestation, biological diversity, etc.

A series of satellites have flown over the years shown chronologically as follows:

- Landsat 1: 1972–1978;
- Landsat 2: 1975–1982;
- Landsat 3: 1978–1983;
- Landsat 4: 1982–2001 (data downlink capability failed in 1993);
- Landsat 5: 1984 – in operation (mid-2009);
- Landsat 6: failed launch, 1993;
- Landsat 7: 1999 – in operation (mid-2009);
- Land Data Continuity Mission (LDCM): a programme to ensure continuity of Landsat.

The Landsat satellites move in a north to south direction, crossing the equator in the morning hours. They cover the sunlit side of the Earth, rising each pass at a fixed time and revisiting each spot

Table 16.6 Characteristics of the Landsat system (MSS = MultiSpectral Scanner; RBV = Return Beam Vidicon – a television camera; Pan = panchromatic; TM = Thematic; ETM = Enhanced Thematic Mapper; ETM+ = Enhanced Thematic Mapper plus)

System	Instruments	Resolution (m)	Communication	Altitude (km)	Revisit interval (days)	Data rate (Mb/s)
Landsat 1	RBV	80	Direct downlink with	917	18	15
	MSS	80	recorders			
Landsat 2	RBV	80	Direct downlink with	917	18	15
	MSS	80	recorders			
Landsat 3	RBV	40	Direct downlink with	917	18	15
	MSS	80	recorders			
Landsat-4	MSS	80	Direct downlink and	705	16	85
	TM	30	TDRSS			
Landsat-5	MSS	80	Direct link (TDRSS	705	16	85
	TM	30	facility exists); no recording capability			
Landsat-6	MSS	15 (pan)	Direct downlink with	705	16	85
	TM	30 (ms)	recorders			
Landsat-7	ETM+	15 (pan)	Direct downlink with	705	16	150
		30 (ms)	solid-state recorders			

every few weeks. The platforms carry multiple remote sensor systems, improving over the years as new generations of satellites are introduced. Data relay systems on-board each satellite send acquired data to ground stations. Table 16.6 shows the main features of each satellite.

The Landsat ground segment, depicted in Figure 16.9, manages the spacecraft and the network, receives the image data, applies corrections and develops image products for users and archiving.

The ground segment consists of Landsat 7 unique components as well as institutional services. The unique components include the Mission Operations Centre (MOC), the Landsat Ground Station (LGS), the Landsat Processing System (LPS), the Image Assessment System (IAS), the Level-1 Product Generation System (LPGS), the EROS Data Centre Distributed Active Archive Centre (LP-DAAC) and the International Ground Stations (IGS).

The institutional support systems consist of the Landsat Ground Network (LGN), the Space Network (SN), the National Centres for Environmental Prediction (NCEP), the Flight Dynamics Facility (FDF) and the NASA Integrated Support Network (NISN).

The MOC, located at Goddard Space Flight Centre (GSFC) in Greenbelt, MD, is responsible for space vehicle operations. The Tracking Data and Relay Satellites (TDRS) control site, operated by NASA's Space Network, is also utilized in conjunction with TDRS. These sites operate to facilitate downlink real-time data, and manage stored housekeeping data and command the spacecraft.

The Land Ground Station (LGS) receives wideband X-band downlinks of data on 2 × 150 Mb/s links from the space vehicles. The data received at secondary ground stations are captured on tape and shipped to the LGS which serves as a front-end processor. The LGS supports S-band command telemetry operations, as well as tracking. The data are sent to the Land Processing System (LPS) where they are recorded at real-time rates. Each channel of raw wideband data is processed and binned into separate accumulations of Earth image data, calibration data, mirror scan correction data and Payload Correction Data (PCD). The LPS spatially reformats Earth imagery and calibration data into level-0R data.

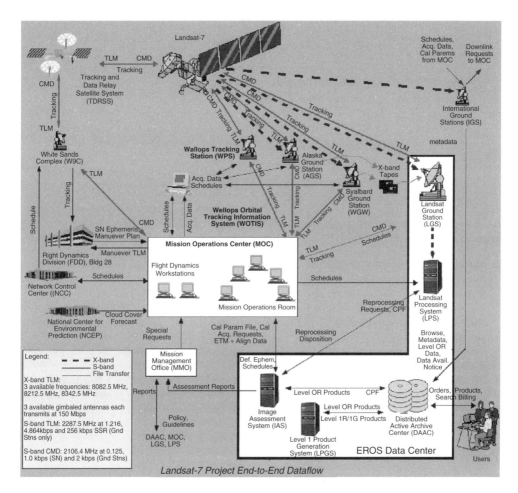

Figure 16.9 Main components of the Landsat-7 ground segment. Courtesy NASA.

The Level-1 Product Generation System (LPGS), located at the Earth Resources Observation and Science (EROS) Data Centre (EDC), generates level-1 products in response to user requests. Radiometric and geometric processing are performed by LPGS on level-0R data to create level-1 products.

The Image Assessment System (IAS) performs offline assessment of image quality to ensure compliance with the radiometric and geometric requirements of the spacecraft and Enhanced Thematic Mapper plus (ETM+).

The International Ground Stations (IGSs) are satellite data receiving stations located around the world that process and distribute services to their user community. The acquisition circles for the IGSs depict the Earth's land areas that are regularly imaged within the target region. The IGSs receive schedule and orbital element data from the MOC on request. The Level-1 Product Generation System (LPGS) and EROS Data Centre Distributed Active Archive Centre (LP-DAAC) are part of the Earth Observing System (EOS) Data and Information System (EOSDIS). They provide information management, user interface and data archival and distribution functions for a variety of data types including Landsat 7.

The Landsat Ground Network (LGN) consists of multiple communications sites that provide S-band and X-band communication support to the Landsat 7 mission. The Space Network (SN) comprises Tracking and Data Relay Satellites (TDRSs) for space-to-space and ground terminals for space-to-ground data relay services for Landsat 7 real-time command and telemetry monitoring.

The National Centres for Environmental Prediction (NCEP) provide worldwide forecast guidance products and generate weather-related products. For Landsat 7, the NCEP supply cloud-cover prediction data to the MOC for image scheduling. The Flight Dynamics Facility (FDF) provides workstations in the MOC operations, including orbit determination, attitude determination, ephemeris data generation, manoeuvre planning support and generation of planning and scheduling aids.

The NASA Integrated Support Network (NISN) is a global system of communications transmission switching and terminal facilities for NASA with long-haul communications services meant to replace the independent special-purpose networks.

Revision Questions

1. The chapter has introduced three varieties of remote sensing system – optical imaging, meteorological and land resources. Identify the similarities and the differences between them (Hint: Compare orbit, ground segment, sensors, applications, user base).
2. You must by now appreciate that personal applications are only a subset of remote sensing systems. Develop a comprehensive list of personal and mainstream applications of remote sensing systems.
3. Table 16.1 presents examples of a few remote sensing systems. Expand the list using the references cited in the text.
4. The World Wide Web is a rich source of remote sensing literature that offers a fascinating range of material and images. This chapter cites a few websites to begin the reader's voyage. Use them to learn and explore the remote sensing world, jotting down notes as you progress.

References

Bizzari, B. (2009). *The Space-Based Global Observing System in 2009 (GOS-2009)*. WMO. Available: http://www.wmo.int/pages/prog/sat/GOSplanning.html [accessed February 2010].

DigitalGlobe (2009) http://www.digitalglobe.com/index.php/5/Welcome+to+DigitalGlobe [accessed June 2009].

Eumetsat (2009) http://www.eumetsat.int/Home/Main/What_We_Do/index.htm?l [accessed June 2009].

GeoEye (2009) http://www.wmo.int/pages/prog/sat/GOSplanning.html [accessed November 2009].

GOES (2009) http://goespoes.gsfc.nasa.gov/goes/project/index.html [accessed November 2009].

Landsat (2009) http://landsat.gsfc.nasa.gov/ [accessed July 2009].

Liew, S.C. (2009) *Principles of Remote Sensing*. Centre for Remote Imaging, Sensing and Processing, National University of Singapore. Available: http://www.crisp.nus.edu.sg/~research/tutorial/rsmain.htm [accessed July 2009].

NASA (2009) *Landsat: A Global Land Imaging Project*. Available: http://www.usgs.gov/contracts/acq_opp/EROS_tech_library/LDCC_factsheets/Landsa_Global_Land_Imaging_Project.pdf [accessed July 2009].

Short, N. M (2009) http://rst.gsfc.nasa.gov/ [accessed July 2009].

Wettre, C. and Roesli, H. (2009) http://oiswww.eumetsat.org/WEBOPS/iotm/iotm/20090123_storm/20090123_storm.html [accessed July 2009].

WMO (2009) http://www.wmo.int/pages/about/index_en.html [accessed July 2009].

17

The Future

17.1 Introduction

In this chapter we conclude by exploring recent trends in the evolution of satellite systems within the confines of the theme and the length of the book. This is a vast and rapidly changing subject. The interested reader can find numerous authoritative and detailed surveys on the subject (Mitsis, 2009; NSR Online, 2009 etc.). We observe that the developments in the satellite personal applications sector are markedly influenced by terrestrial technology.

During the early phase in the introduction of space technology, the programmes were sponsored by governments because of the associated costs, complexity and political implications. The applications were confined to large-scale telecommunications, military and government projects. The technology has matured over the four decades since its inception, resulting in reduction in the cost of the user equipment, approaching that of mainstream consumer electronic products. A large number and variety of satellite systems operate commercially, although governments in many countries continue to own and operate satellite networks. Invariably all military satellite systems lie within the jurisdiction of the governments, for obvious reasons, but with some dependence on civil space systems and technology.

Satellite solutions for individuals and small groups are viable in broadcast, mobile and remote area communications and navigation applications. Commercial remote sensing is an emerging discipline with a growing awareness of the potential of remote sensing solutions in education, personal applications and non-scientific applications such as media reporting (Meisner, Bittner and Dech, 1998).

The commercial satellite industry is broadly segmented by its provision sector – launch, spacecraft manufacturing, ground equipment manufacturing, space segment provision, service provision and equipment retailers. Our interest here is broadly confined to the user equipment and service provision.

In this commercial environment, while some operators have established themselves, others have either failed or remain in financial difficulties. There has been a spate of privatization of government-owned companies, mergers, takeovers and bankruptcies. A majority of failures occur early in the growth cycle. Figure 17.1 (Richharia, 2001) illustrates a hypothetical model to illustrate the commercial viability of an operator during the introductory phase of a venture. The initial phase, lasting several years, requires massive investment. After implementation, during a period lasting several years a company can incur heavy losses. If the operator can sustain the operations financially and the product complies with user expectations, then the system becomes commercially viable (scenarios 'a' and 'c' in the figure); otherwise it incurs irreversible financial loss (case b) such as occurred with Iridium at the outset. However, that company was retrieved financially and is now a

Satellite Systems for Personal Applications: Concepts and Technology Madhavendra Richharia and Leslie David Westbrook
© 2010 John Wiley & Sons, Ltd

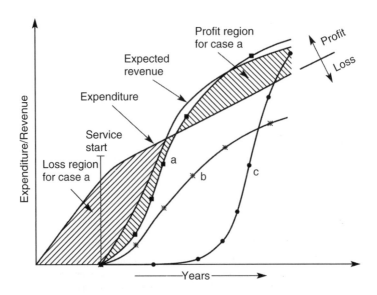

Figure 17.1 Commercial effectiveness of a business, illustrating its viability for various revenue levels and the rate of revenue generation. Reproduced by permission of © Pearson Education. All rights reserved.

successful commercial venture. Experience illustrates that well-established operators can introduce new products and satellites effectively owing to the strength of their finances, infrastructure and market hold. Satellite operators recognize the benefits of entering the consumer mainstream; but some analysts are of the opinion that the remarkable resistance of the satellite systems in the face of the recent financial turmoil was due to their insulation from the consumers. Some segments of the consumer market tend to be more resistant to economic downturn; for example, the broadband sector is more resistant than the entertainment sector because of people's dependence on broadband for revenue generation and communication (Engel, 2009).

In the remaining part of this chapter we discuss the influences guiding the growth of satellite consumer products, their impact on the evolution of the technology and the trend.

17.2   Influences

It is well recognized that satellite systems cannot grow in isolation. The factors likely to influence the growth and evolution of the industry include:

- competition between (a) satellite systems, and (b) satellite–terrestrial technologies;
- synergistic solutions (intersatellite segments and satellite–terrestrial);
- Internet pervasiveness;
- regulatory developments as enablers or barriers;
- commercialization;
- innovation;
- new concepts, solutions and opportunities in the wake of competition;
- technology progression;
- emerging societal needs.

Competition and Synergies

Satellite and terrestrial technologies compete with one another in numerous applications; and satellite operators and service providers compete with each other within the vertical satellite segment. Application developers attempt to obtain optimum solutions by combining the strengths of each technology and hence facilitate interworking and synergies to users' best advantage while fostering growth of the respective industries.

Consider some examples. A competitive threat to the satellite DTH industry comes from the growing penetration of IPTV. The increasing penetration of WiMax can erode the satellite broadband segment in developed regions of the world, while WiMax backhaul for interconnecting remote hot spots may provide a new market to satellites services.

Satellites are naturally suited to wide-area broadcasting because they cover vast expanses spanning thousands of kilometres. A prime competitive segment is that of radio and television broadcasting in regions where cable operators and DTH service providers have an equal reach. Satellite radio services experience considerable competition from personal music systems, the wireless-enabled Internet and conventional radio broadcasts. The choice of technology is personal, but nevertheless people are attracted by a combination of content, high audio quality and real-time news and chat, and hence satellite radio strives to excel in this respect.

Satellite communications combined with a navigation system and a terrestrial infrastructure constitute the core of location-based applications. Similarly, a combination of a satellite Earth observation system, a satellite communication system and a partner terrestrial transmission technology constitute vital parts of a disaster warning and management system.

Enablers

IP transport has been a key multiplatform enabler in recent years owing to its pervasiveness. The technology provides a seamless integration between terrestrial and satellite systems aided by evolving standards such as IP over Satellite (IPoS). Satellite operators have benefited considerably by supporting the IP transport protocol to the extent that some operators have staked their future growth primarily on this technology.

Supportive regulation offers a strong incentive in promoting the industry. For instance, new navigation bands have increased the scope of civil navigation systems in terms of availability and reliability and through introduction of an array of emerging GNSS systems and augmentation of new systems. The European Commission has mandated harmonized use of mobile services including video to foster unfragmented and harmonized growth of the technology.

Government-sponsored research and development initiatives continue to foster innovative solutions. The growth of terrestrial broadband systems – mainly owing to Internet pervasiveness – is spurring unprecedented growth in satellite fixed and mobile personal broadband sectors.

Barriers

In many instances, regulation and politics impose a barrier during and following implementation. Transborder use of VSATs and satellite phones are forbidden in several countries because of political sensitivities. Promising VSAT markets in various Asian countries are plagued by tight government regulation. Broadcasting of 'objectionable material' is a contentious issue in Europe and elsewhere, while direct broadcast services generally benefit through helpful regulations.

The high cost of satellite systems and long gestation periods have led to the downfall of numerous commercial enterprises, leading to bankruptcies, mergers and acquisitions.

Commercialization

Several financial institutions now own satellite communications companies; previously government-owned or sponsored organizations, such as Intelsat and Inmarsat, are now public companies. The advent of private equity has created a pressure for profit, and short-term objectives can often take precedence, to the detriment of long-term research.

Several companies have commercialized remote sensing and imaging, which were traditionally government owned, by offering data products suitable to the wants of small organizations, groups and individuals.

Innovation

Competition from terrestrial systems compels satellite operators and service providers to innovate. DTH operators are introducing high-definition interactive television aggressively. The extraordinary stride in GPS navigation receiver technology has enabled GPS chips to be embedded in myriads of personal electronic products and applications. Sirius XM Satellite Radio Inc. has developed a low-cost dock that turns an iPhone into a satellite radio to tap into the success of one of the fastest-growing consumer products.

Recent innovation combines mobile communication, satellite navigation, object recognition and a geographical database retrieval facility into a mobile tourist guide, The user photographs the feature of interest with an embedded camera, clicks at the object of interest (e.g. a building) and transmits a query about the object on a mobile communication channel to the application provider who sends back the desired information based on results of a database query. The concept was declared as a regional winner in the European Satellite Navigation Competition, 2008, sponsored by ESA's Technology Transfer Programme. Applications of the technology include tourism, education, remote healthcare, security, science, etc.

The European DTH service provider, BSkyB, has announced the launch of a 3D TV channel in 2010 across its existing HD infrastructure and the current generation of SkyPlus HD set-top boxes bridged to a 3D-ready TV. Those who have seen demonstrations of the 3D system have given favourable responses – 'seeing is believing'. However, 3D content and 3D-ready television sets are scarce at present (Bates, 2009).

Remote sensing products are available for seamless integration into mapping, enterprise solutions, hosting services, web services and mobile applications.

Another recent invention involves ultrasonic sensors, mobile communication and satellite navigation to estimate unused allocation in a truck, thereby optimizing the loading to save fuel and increase efficiency by accepting short-notice orders. Sensors mounted on each truck estimate the spare loading capacity, and the data, with precise GPS-derived location, are sent over a cellular system to headquarters. The invention was exhibited at the 2008 European Satellite Navigation Competition, backed by ESA's Technology Transfer Programme.

Efforts are in progress to develop personal navigation systems using a combination of techniques. Cavallo, Sabatini and Genovese (2005) have developed a GPS-based building block of a personal navigation system that embodies a Global Positioning System (GPS) receiver and an Inertial Measurement Unit (IMU). The disadvantages of inadequate signal strength indoors and in dense city areas that is experienced with GPS receivers suggests that the combined use of a dead-reckoning system based on inertial sensing with GPS-derived location would be a better answer to personal navigation needs.

17.3   Trend

Detailed analysis of the commercial satellite industry trends is a mammoth task requiring several volumes of work, as will be evident to the reader on surveying up-to-date literature of authoritative marketing and research consultation companies. There are major differences between regions in service demands, market drivers/enablers, affordability, regulations and infrastructure. For instance, direct-to-home television systems serve over a 100 million subscribers on over a 100 unique platforms spanning scores of countries. We will attempt to highlight the main features, focusing on developments at the leading edge and keeping within the scope and theme of the book. We encourage the interested reader to explore the literature to get to the depths of the research and development effort around the world.

Commercialization

The successful performance of numerous private companies points to the economic viability of satellite systems. As expected, the business continues to evolve, leading to interesting acquisitions and mergers for consolidation and mutual benefits. SES, which has strengths in broadcasting, has acquired New Skies to extend its reach in the FSS market. Intelsat has merged with its competitor Panamsat to consolidate its position in the international FSS market.

The private equity regime leads to pressures for short-term profitability while managing the long-term growth. More often, the emphasis on quick returns results in long-term goals being relegated.

To enhance profitability and ward off competition, many operators have moved down the value chain by taking a proactive role in content and content delivery. Triple-play offerings – that is the combined delivery of voice, data and video – serve as one example.

Space Segment

Communication satellite technology has evolved considerably, leading to a dramatic reduction in cost and size of user equipment. Satellites incorporate significantly more processing and transmit power supported by large and highly capable antenna systems. Terrestar-1 represents the present state of the art in satellites targeting individuals. It generates high transmitted power and up to 500 spot beams to provide a variety of multimedia, navigation and emergency services in the United States. Future satellites are expected to continue the trend. The development of 20 m satellite antennas to enhance the MSS services is under way (Stirland and Brain, 2006). Low Earth constellations are now well entrenched, with some of the operators already beginning to launch advanced second-generation LEO spacecraft.

There is a growing recognition of the advantage offered by flexible payloads to allow in-orbit change or to include multiple functions. There is also increasing interest within the government and military sectors in utilizing small satellites (<500 kg). Such satellites are placed in low orbits and are cost-effective for specific tasks, for example weather, navigation, Earth observation, imaging and remote sensing (and even potentially as a weapon). The SPOT remote sensing satellite is an example of a multiple-function payload. It incorporates a disaster warning facility. A ruggedized waterproof device known as the SPOT satellite messenger, weighing just over 200 g, provides consumers with the ability to send communications for emergency assistance and allows points of contact for visual tracking of the messenger's location and progress on a computer using Google Maps™ and the SPOT website. The Iridium next-generation system is likely to deploy multitasking satellites which could provide navigation, communication and imaging facilities.

High-Altitude Platforms

Several issues need to be resolved before HAP systems become a practical alternative for communication, surveillance and remote sensing applications. There is currently a lack of commercial push owing to the associated technical risk, uncertainty with regards to regulatory matters (both radio and aviation) and lack of a clear business case. Nonetheless, the emergence of such systems seems likely as commercial implementation of HAP-based broadband fixed access, security and surveillance services and WiMax augmentation come closer to realization. The technological issues in need of refinement are service specific and include, but are not be limited to, the following:

- optimization of platform technology, the antenna system, payload capacity/size/power trade-off and station-keeping;
- payload definition and development – communications only, multifunctional, transparent versus regenerative transponders and inter-HAP/HAP-satellite links;
- network issues – frequency planning, spot beam layout, roaming and hand-off, topology and architecture, inter-platform links, resource allocation, network protocols and interfacing and integration with the terrestrial node;
- characterization of the propagation environment in higher-frequency bands (27–28 and 47–48 GHz), optimization of modulation-coding schemes for optimum matching of the channel and service, radio link budget analysis and capacity estimates;
- user equipment design and development;
- backhaul link and transceiver design and development;
- adaptation of sensors and cameras for remote sensing and surveillance;
- communication traffic sizing.

Systems and Applications

Mobile television is believed to be among one of the largest potential growth areas. Asia is particularly active, with Korea already having operational system, and China intending to augment its terrestrial service. A satellite mobile multimedia interactive service operated by Terrestar is undergoing trials in the United States over Terrstar-1. Italy is leading the way in Europe through a terrestrial service, while Eutelsat and SES-ASTRA have combined in a joint venture to provide the service in Europe. Many systems are based on DVB-SH, and others on the DVB-MB standard. However, commercial success remains elusive – a Japanese system was shut down and the Korean system is plagued by financial uncertainties (Holmes and Bates, 2009). In a recent expert forum (Satellite, 2007), the experts considered mobile television to have a growth potential in the long term, with some citing it as a viable next-generation service. Nevertheless, some (optimistic) predictions place the number of mobile TV subscribers worldwide to reach 462 million by 2012, up from a base of 23.8 million in 2009. An indirect application of satellite systems in this arena relates to a backhaul role where satellites provide connectivity to distant regions. The distribution and transmission over a satellite–terrestrial hybrid network is another alternative.

As HD technology begins to take hold, the cinema and video industry is beginning to introduce *3D technology*. Although in its infancy for television, 3D has already demonstrated its potential in cinemas. The quality of 3D TV pictures has been rated highly by those who have viewed early demonstrations. According to the Society of Motion Pictures and Television Engineers, 3D movies generate 2–3 times the revenue of the conventional 2D showings (Bates, 2009). BSkyB of Europe introduced a 3D system in the United Kingdom in April 2010. However, at this stage, content and availability of 3D-enabled television sets remain limited. Transporting two HD video streams (left eye–right eye) over satellite would traditionally involve twice the bandwidth and hence is prohibitively expensive. Proprietary technologies attempt to minimize the bandwidth by compressing the streams to fit within a single HD channel. One approach is to remove the redundancy inherent in the two images to compress the streams into a single stream for transmission over a traditional HD channel. The satellite delivery is expected to follow on the heels of the cinematic adaptation of the technology, and the key ingredient in this respect will be to maximize reuse of the existing infrastructure.

Satellites are playing an increasingly important role in the advent of *digital cinema*. The application refers to the recent trend in the distribution of high-quality movies digitally in wide and often remote areas and hence is an indirect application that improves the cinematic viewing experience of a vast population. This mode of distribution offers a sales boost to the movie industry and provides access to widely distributed people simultaneously. In places like India, traditional methods of distributing films take months before a popular movie reaches small towns by which time it would have lost its appeal. Using satellite distribution, the end-users enjoy a timely high-quality cinematic experience. Technologies under refinement include higher-throughput and cost-effective VSAT software and middleware upgrades.

Cable companies and traditional telephone companies now offer a *video-on-demand* facility that allows the audience to view a programme at convenience using IP television (IPTV) technology. Similarly, iPOD downloads and distribution from websites are disruptive technologies for DTH and satellite radio. The satellite industry is investigating competitive solutions. For instance, in a 'push' technique, the popular contents are pushed during the off-peak hours to individual receivers for local storage such that the users get the feeling of an instantaneous video-on-demand service.

There is a growing demand for greater capacity, lower service charge and higher throughput in the maritime and aeronautical industry beyond L band, and hence Ku-band adaptation continues to grow.

The satellite *broadband sector* growth in the past year was led by US companies (mainly Wild-Blue and Hughes) and Thaicom's IPStar system in Asia. Governments in the United States, Europe and Australia are promoting the use of consumer broadband to bridge the digital divide with remote communities. In general, the cost of satellite broadband services is approaching that of identical terrestrial offerings.

Digital signage refers to the technique of broadcasting advertisements simultaneously to a large number of outlets – for example, promotional material transmissions to a chain of stores. The industry is investigating ways to provide a cost-effective and improved solution to existing terrestrial alternatives – through time-sensitive HD content that may hold the viewer's attention better.

Some US industry experts suggest that the most promising technologies for satellite communications in the next few years ($\sim$2012) are: video distribution, HDTV, IPTV, IP-enabled services in general, mobile TV, disaster recovery and innovative mobile applications. Video distribution services are the traditional services, and they are expected to continue to grow with better efficiencies as MPEG-4 penetration deepens.

Ancillary Terrestrial Component (ATC) technology has received attention in the regulatory bodies as MSS operators seek permission to reuse the MSS satellite band terrestrially. A major limitation of the traditional MSS is its inability to address subscribers resident in urban areas and indoors owing to intense shadowing of the already weak satellite signals. By retransmitting the signals through terrestrial transmitters, this limitation is eliminated. High optimism is exuded by the leading companies in the United States regarding the success of ATC technology. However, there are financial risks, as exemplified by the debacle caused to some MSS operators earlier in the decade when the terrestrial technology leap reduced the satellite market model dramatically. The earliest systems for communications were planned for deployment in 2008 but have yet to be functional. The technology is already applied to satellite radio systems in the United States and Canada.

At present, GPS is a fully populated *GNSS system*, GLONASS is striving towards full functionality, while Galileo is approaching implementation. It is anticipated that, in the next decade, the existing systems will be enhanced to offer a more diverse range of signals. Following the introduction of new GNSS systems, users will have the choice of selecting from several alternative systems. Utilization of several more ranging signals will improve the accuracy and reliability of navigation estimates. Additionally, an increase in the numbers of SBAS and GBAS ground facilities will enhance the capability and efficiency of civil aviation.

GPS receiver technology has made rapid and very effective strides in the past decades. GPS chips are now embedded in a vast number of mobile phones and personal devices. The manufacturers are already well on their way to developing architectures to accommodate the processing of a multitude of GNSS signals.

Location-based services are receiving considerable attention in the terrestrial arena. It is anticipated that these services will eventually extend to a wider area necessitating satellite services. Fleet management remains one of the most successful location-based solutions for the satellite communications industry. The majority of terrestrial systems favour satellite navigation solutions or their augmented versions, such as DGPS or terrestrially assisted GPS, for location determination owing to the ubiquity of, and the accuracy available from, the GPS system. We have cited an example of an innovation in this area (the 'tourist guide'). Other examples include reports of local weather or traffic conditions, events and services for patients and children. Equipping young children with personal locators can offer parents greater peace of mind, while sufficiently small locators could track pets. Personal locators could assist medical patients, where the locator would be combined with a detector that monitors the patient's health. Enabling technologies under scrutiny include – assisted GPS, locator miniaturization, increased battery life, multipath countermeasures, the ability to penetrate buildings, bundling of location services with two-way paging systems and the economical use of spectrum.

Remote sensing systems are traditionally of interest to the exploration and scientific communities and government bodies that process, present and interpret the images for agencies interested in formulating and understanding issues at large. Judging by the interest of educationists and the print and electronic media, it is clear that the data can be applied effectively for applications relevant to the public as the diverse capability of remote sensing is recognized (Meisner, Bittner and Dech, 1998). More recently, Internet applications such as GoogleEye, traffic route display and virtual tours organized by tourist agencies have caught the imagination of application developers and the public. In Chapter 16 we introduced two commercial companies providing visual satellite imagery. These and other operators are progressively attempting to improve the product range and image quality at an increasingly affordable price. However, a report has revealed that the main customers of imagery service providers are the US government (Wong, 2005); the growth in the commercial market is moderate, leading to concern about the commercial risk of relying on a single customer (the government). It was perceived that the growth in capacity of these systems would outstrip demand. Thus, the long-term evolution of this particular segment of the remote sensing industry remains uncertain for the moment – however, these observations should be viewed as pointers for corrective actions. The use of remote sensing data and imagery is now commonplace in education, facilitated by the vast support such as that provided by the Landsat mission (see Chapter 16) and the Canada Centre for Remote Sensing website (Langham, Botman and Alfoldi, 1996). There is a growing awareness of integrating the role of remote sensing with secondary school education in technically advanced societies like Germany to introduce pupils to the strengths of remote sensing technology. The exposure would facilitate inter-disciplinary fusion with ecology, disaster management or traffic control sectors of society (Voss *et al.*, 2007).

Integration

A recent announcement by a satellite DTH service provider (BSkyB) with a branded video games console provider (MicroSoft) to deliver satellite TV content on the console through the Internet is a concept where a satellite operator intends to tap into an enormous video games subscriber base. In addition to the new revenue source to the DTH provider, there may be indirect and long-term gains to the satellite industry through enhanced Internet traffic and subscribers continuing to use the DTH service later.

The permeation of IP in telecommunication networks has facilitated seamless interworking between satellite and terrestrial transmission modes – thus, the service providers are able to conceive and provide end–end IP-enabled solutions involving satellite, cable and terrestrial wireless technologies.

Specialist software companies benefit from sectors such as satellite imagery, broadband and location-based services, and therefore there is evidence of a more proactive involvement of these companies in promoting satellite technology. An example would be Google's foray into the satellite sector by investing in the development of a Ka-band global satellite constellation (called the Other 3 Billion or O3B) to provide broadband coverage throughout the world (to reach the 'other 3 billion' potential subscribers dispersed around the globe). The same company also has a commercial arrangement for images with GeoEye, and, by the same token, MicroSoft has a partnership with DigitalGlobe. Cisco initiatives created a non-profit organization called 'Planetary Skin' backed by NASA to develop an infrastructure for provision of information to study global adaptation to global climate change. The project, spanning 2010–2020, is supported by a number of companies of the satellite imagery sector (Hill, 2009).

The cost and size of MSS products now rival their high-end terrestrial counterparts. The trend is expected to evolve further with the introduction of ATC technology to provide seamless coverage, including difficult-to-access locations such as tunnels, cities and indoors, in conjunction with standardized technology like DVB-SH. Recent initiatives, such as that of Qualcom in the United States to integrate satellite and terrestrial components in a single chip for mass production, aim to bring MSS technology to the consumer mainstream. The Qualcom chip can support both L- and S-band satellite systems to allow a truly configurable multimode handset.

Handsets

The modern cell phone is already a multifunction device, which, in addition to voice, includes all or several of the following features: the Internet, camera, video and sound recorder, games, live television, GPS, music player, Bluetooth applications, address and notebook, etc. The S-band TerreStar-1 satellite (Figure 17.2 right) placed over the United States will deliver the voice, data and video services over TerreStar's all-IP mobile broadband core network and the 'next-generation' satellite phone technology. In addition to GMR-3 satellite standards, the worldwide 'smartphone', shown in Figure 17.2 (left), can handle various terrestrial wireless specifications and supports an array of rich personal features. It has an 'always on' connectivity. It does not require an external antenna and includes features such as a single phone number for satellite and cellular service, a touchscreen, a 3.0 megapixel camera, a MicroSD slot, WiFi, Bluetooth, GPS and a full QWERTY keypad. It supports satellite and terrestrial frequencies of 2000–2010 MHz and 2190–2200 MHz

 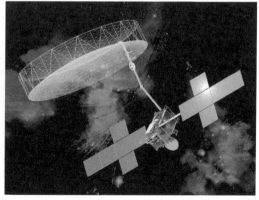

Figure 17.2   Left: Terrestar multimedia phone. Right: Terrestar-1 satellite with an 18 m reflector antenna. Reproduced by permission of © Terrestar Networks Inc.

using GMR-3G protocol operating over an ATC architecture. It also supports terrestrial systems at 850/900/1800/1900 MHz that operate over GSM/EDGE, WCDMA or HSDPA protocols (Terrestar Online, 2009).

Satellite Industry Recent Performance

According to a recent survey sponsored by the Satellite Industry Association, the global satellite industry reported an average growth of more than 14% per year during the period 2003–2008. In the satellite television sector, the growth was up 30% in 2007–2008 with some 130 million subscribers reported worldwide. The ground equipment sector, led by satellite TV, broadband, MSS and GPS products, reported a growth of over 25% during the year. Various broadband initiatives are in the offing throughout the world. Eutelsat and Viasat are planning connectivity in Ka band in Europe and the United States, which will result in a manifold increase in broadband capacity. HD TV grew by 170% between the end of 2006 and May 2009. One forecast expects the total number of HDTV subscribers to reach 255 million by the year 2013, out of which 38% are expected to be attributed to DTH (Holmes and Bates, 2009). The DTH sector, satellite radio and satellite navigation receivers have been the most rapidly evolving sectors according to a report released by the Space Foundation. FSS has shown a robust performance fuelled by the rapid uptake of the Internet (which is said to double every 2 years), growing penetration of HDTV and new applications such as communication on the move.

17.4   The Long Term

In spite of the economic downturns of the recent past, all the forecasts point to an optimistic outlook for the satellite industry in the next 10 years. Nevertheless, it is also said that this will require innovative research and development in a competitive environment, reinforced with sound commercial practices.

Targeting the period 2008–2018, the NSR forecasts that global revenue for commercial C- and Ku-band 36 MHz transponders will increase at an average annual rate of 4.2% (with regional variations), attributed to the combined effects of growth and aggressive pricing. The NSR report anticipates that the majority of this growth will be generated by video services, expecting that '83% of the newly leased transponders will be for video distribution, direct-to-home (DTH) and video contribution and occasional-use television services' (NSR Online, 2009). The current downturn will introduce a marginal short-term effect as the demand picks up beyond 2010. The report states that direct-to-home service is 'by far the single most important market for the satellite industry', attributed to a demand for new DTH platforms and the need to contain competition from terrestrial offerings by introducing HD channels to improve and expand the content of the SD channel bouquet.

WiMAX, a rapidly evolving terrestrial technology, is anticipated to grow considerably, with enhancements that will introduce better portability and mobility. The technology provides wideband access to people in traffic hot spots like an airport. WiMAX may pose a threat to rural satellite broadband in those developed countries where WiMax can be viable in these communities. However, in developing regions where the user affordability is in question, rural WiMax uptake is less likely. In a complementary role, satellites can provide the backhaul transport to isolated WiMax hot spots. There is also the possibility of a synergistic combination with the ATC component of satellite systems. An ongoing concern to the satellite community is the potential threat of C-band WiMax interference with satellite systems, in spite of the recent favourable regulatory outcome at the World Administrative Radio Conference (WARC).

A gradual move towards limited commercialization is clearly evident in the remote sensing arena, although it is imperative that governments and institutions will inevitably remain the largest sponsors and customers of data and services owing to their commitments to the public and the nature

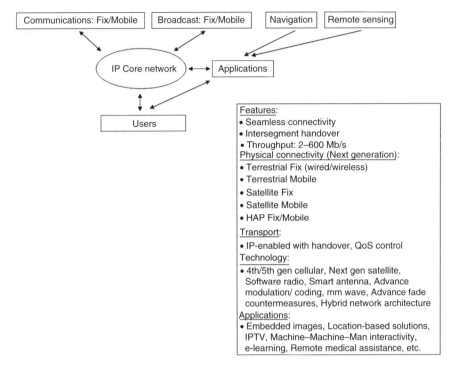

Figure 17.3   Concept of a network of systems for personal applications.

of collected data. The advent of high-resolution imagery into the public domain has encouraged the entry of a number of commercial companies in the value chains for image distribution and value-added services. Despite the global recession both GeoEye and DigitalGlobe (chapter 16) reported a healthy revenue growth in 2009. A numbers of Earth observation satellites are being planned in various countries, and that may boost the industry further.

The next-generation communication systems could provide user handsets with services ranging from voice to high-definition videos transported on a number of broad- or narrowband wireless technologies, depending on the user situation. Wireless technologies will range from broadband fixed networks to fixed, nomadic and mobile WiMax or its variant – as a component of fourth- and fifth-generation mobile systems. The platforms, ranging from narrow- and wideband LEO and GEO satellite systems to stratospheric high-altitude platform systems, may constitute a network topology that allows seamless connectivity (Ohmori, Yamao and Nakajima, 2000). Figure 17.3 shows elements of a wide-area network of the future, incorporating a multitude of technologies.

Satellite technologies promoting the evolution include: network convergence, next-generation spacecraft and handset electronics (software-enabled radio), next-generation space segment, next-generation Internet, advanced core technologies such as adaptive modulation/coding, robust and ubiquitous network protocols, next-generation digital signal processors and intelligent software agents.

17.5   Satellites and the Environment

People are concerned about the risks to the Earth's biosystem and our habitat because of long-term climate change to our planet. As the world's fuel requirements rise by industrial development and under the burden of population, it is perceived that the emission of harmful gases – particularly carbon dioxide – may cause an irreversible change to the Earth's environment. The satellite community is playing a role in contributing towards a better understanding of changes to the world's geographical landscape, the emission–absorption mechanism of greenhouse gases, the macroimpact of climatic change and monitoring of the ozone content. Satellites remain the sole means of collecting data on such a grand scale. Various remote sensing systems monitor the Earth's environment to provide Earth observation data to the international scientific communities for analysis and dissemination. Japan's greenhouse gases observing satellite (Gosat) measures sources of carbon dioxide and records greenhouse gas emissions around the globe. The data are made available by Japan to the worldwide scientific community. The National Oceanic Atmospheric Administration (NOAA) of the United States has been monitoring the Earth through the Polar Earth Orbiting Satellite (POES) system for over two decades. The last in the series of spacecraft, NOAA-19, provides a variety of data including reflected solar and thermal energy, global sea temperature, ozone concentration and soil moisture. The POES system will be replaced by another programme known as National POES (NPOSS). To provide weather forecast and climatic change data, an effort known as Global Earth Observation System of System (GEOSS Online, 2010) intends to pull together Earth observation data, tools and expertise of over 125 countries to build a knowledge base for understanding global environmental issues. Combining NPOSS and Eumetsat's MetOP polar satellite system will provide global data every 4–6 h repeatedly. Eumetsat will also contribute to environmental studies through its second- and third-generation geostationary satellite system under the auspices of ESA's Global Monitoring Environment and Security (GMES) programme. The recently launched Jason-2 satellite (a Europe–US joint mission) will constitute a part of the Ocean Surface Topography Mission, one of the aims of which is to monitor climate change. Alliances between Eumetsat, China and India are also being formed to benefit from each other's space monitoring and processing facilities (Kusiolek, 2009; Oberset, 2009).

There is criticism in some quarters regarding the adverse environmental impact of polluting emissions exhaled during a satellite launch. Considering Ariane as a test case, it is argued that heat and gas – generated for under 120 s during launch – is adequately compensated for by a pollution-free operation of the satellite lasting 10–15 years. Furthermore, experts reiterate that Ariane can launch two satellites simultaneously to double the advantage. It is claimed that a direct-broadcast satellite could replace up to 1100 UHF terrestrial transmitters in the United Kingdom alone. This step would reduce power consumption by over 50 MW considering only 50 of the most powerful transmitters. However, a recent regulatory study realized that satellite receivers are less efficient than their terrestrial counterparts, thereby largely offsetting the said advantage. Nonetheless, the launch industry is researching techniques to reduce gas emissions. NASA intends to develop an environmentally friendly rocket propulsion technology. Similarly, equipment manufacturers and operators such as BSkyB are developing technology to reduce the power consumption of set-top boxes and DTH front ends. VSAT manufacturers such as Gilat are improving their equipment technology in this respect by reducing or eliminating polluting material in the equipment.

17.6 Conclusion

We have attempted to present the satellite technology and applications in a different perspective – a view of it as it appears to the users. We hope you have enjoyed reading this book and share our belief that satellite technology is now an integral part of our society. Judging by the trends, it is evident that the technology will enrich our daily lives increasingly with time. We intend to keep you abreast of the evolving landscape through future editions of the book.

Revision Questions

1. What are the major external and internal influences likely to affect the evolution of commercial satellite systems? Justify your answer with examples.
2. Describe the trends in the evolution of satellite systems technology.
3. Elaborate upon the statement: 'Satellite technology contributes towards a better understanding of greenhouse effects'. You may conduct your own research to elaborate upon the content of the chapter.

References

Bates, J. (2009) 3D HD: the next big entertainer, money maker. *Via Satellite*, September, 20–26.

Cavallo, F., Sabatini, A.M. and Genovese, V. (2005) A step toward GPS/INS personal navigation systems: real-time assessment of gait by foot inertial sensing. International Conference on *Intelligent Robots and Systems*, August, Edmonton, Canada, IEEE/RSJ 2-6, pp. 1187–1191.

Engel, M. (2009) The satellite industry in bad times: broadband. *Via Satellite*, June, 16.

GEOSS Online (2010) http://www.earthobservations.org/geoss.shtml [accessed February 2010].

Hill, J. (2009) Major software companies: friend and foe of satellite? *Via Satellite*, May, 30–34.

Holmes, M. and Bates, J. (2009) Services drive satellite performance in 2008. *Via Satellite*, July, 26–29.

Kusiolek, R.T. (2009) Satellites going green. *Via Satellite*, July, 20–25.

Langham, C.W., Botman, A.M. and Alfoldi, T. (1996) From surfer to scientist: designing a Canadian remote sensing service for the Internet audience. Geoscience and Remote Sensing Symposium, IGARSS '96, *Remote Sensing for a Sustainable Future*, IEEE International, 27–31 May, Vol. 1, 324–326.

Meisner, R.E., Bittner, M. and Dech, S.W. (1998) Remote sensing based information for TV-, print- and multi-media applications. Proceedings of Geoscience and Remote Sensing Symposium, IGARSS '98, IEEE International, 6–10 July, Vol. 1, pp. 543–545.

Mitsis, N. (2009) Measuring the state of the Industry. *Via Satellite*, July, 20–24.

NSR Online (2009) *Global Assessment of Satellite Demand*, 5th edition, Available: http://www.nsr.com/Reports/SatelliteReports/SatelliteReports.html [accessed November 2009].

Oberset, G. (2009) The environmental impact of satellite communications. *Via Satellite*, September, 16.

Ohmori, S., Yamao, Y. and Nakajima, N. (2000) The future generations of mobile communications based on broadband access technologies. *IEEE Communications Magazine*, December, 134–142.

Richharia, M. (2001) *Mobile Satellite Communications*. Pearson Education Ltd, London, UK.

Satellite (2007) Satellite 2007 Conference and Exhibition, 20–22 February, Washington Convention Centre, Washington, DC.

Stirland, S.J. and Brain, J. R (2006) Mobile antenna developments in EADS. First European Conference on *Antennas and Propagation*, 6–10 November, Nice, France, pp. 1–5.

Terrestar Online (2009) http://www.terrestar.com [accessed November 2009].

Voss, K., Goetzke, R., Thierfeldt, F. and Menz, G. (2007) Integrating applied remote sensing methodology in secondary education. Geoscience and Remote Sensing Symposium, IGARSS 2007, IEEE International, 23–28 July, pp. 2167–2169.

Wong, F.C. (2005) Understanding the revenue potential of satellite-based remote sensing imagery providers. IEEE Aerospace Conference, 5–12 March, Big Sky, MT, pp. 4433–4440.

Appendix

A List of Personal Applications

Description	Example
Mobile and thin route telecommunications in remote areas	Communication with terminals on the move – sea vessels, aircraft, portable units used by journalists operating in remote parts of the world; messaging services with a remote on-the-move user community such as truck drivers; management of fuel and goods delivery; transmission of driver work load to a central facility
Broadband connectivity to individuals and small offices	Broadband initiatives in the United States and Europe; office extension; business continuity in case of emergency, etc.
Direct-to-home and individual broadcasts	Satellite radio systems, Direct to home SDTV and HDTV, 3D TV
Connectivity for closed user groups	Private VSAT networks: lottery selling points in the United Kingdom, military operations
Military broadcasts to dispersed forces	US Global Broadcast Service
Location-based solutions	Emergency, information and entertainment services; fleet management systems such as EutelTRACS™; real-time traffic information; parking space availability; local events; nearest gas station; shopping information; hotel guide; mobile yellow pages; security alerts; news; sports; weather; stocks; routing assistance; locate a friend; dating; games, etc. (see also, the novel GPS-assisted applications category)
Virtual private network	Remote office; office extension to mobile platforms
First response teams	Assistance in rescue operation such as tsunami, major earthquakes, shipwreck, etc.

Satellite Systems for Personal Applications: Concepts and Technology   Madhavendra Richharia and Leslie David Westbrook
© 2010 John Wiley & Sons, Ltd

Point-to-multipoint connectivity, data broadcast, IP multicast	Weather broadcast or storm warnings to a shipping community, such as the data alert facility of 1worldspace
Distance learning facility	Edusat initiative in India; 1worldspace classroom facility and virtual classrooms in remote locations
Wide area network webcasts	Entertainment, educational or training information – businesses, schools, hospitals and community centres (applies to a targeted audience at an arranged time)
Push and store	Daily reception of content directly into a receiver without the need for Internet access–for example 1worldspace service
Disaster warning	Group announcement on devices like a satellite radio or satellite phone; distress alerts from individuals and vehicles, vessels and aircraft
GPS (general)	Shipping, route guidance, law enforcement, farming, hitchhiking and outdoor activities
Novel GPS-assisted applications	Stolen-vehicle recovery systems; in agriculture to fertilize precise locations; GPS-auto steers on tractors to assist in planting, etc.; assistance in food drops to people in distress; tracking a runner or a boat during a race; electronic tour guides; restaurant programme beaming advertisements only to people in the neighbourhood; tracking criminals on parole; pinpointing people in distress; fire engine assistance in low-visibility scenarios; social networking and games; assistance in golf to obtain distances; locating best fishing locations
Imagery	3D Imagery for navigation on mobile sets (e.g. the 3DVU product); products that directly embed into applications such as route maps, video games, etc.
Interactive advice from experts	ISRO initiative–provides advice on agriculture, fisheries, land and water resources management and livestock management; interactive vocational training, telemedicine; etc.
Machine–Machine	Remote meter reading

Index